Ecological Studies in the Antarctic Sea Ice Zone

Wolf E. Arntz · Andrew Clarke (Eds.)

Springer-Verlag Berlin Heidelberg GmbH

Wolf E. Arntz · Andrew Clarke (Eds.)

Ecological Studies in the Antarctic Sea Ice Zone

Results of EASIZ Midterm Symposium

Springer

Professor WOLF E. ARNTZ
Alfred Wegener Institute for Polar and Marine Research
P. O. Box 120161
275515 Bremerhaven
Germany

Professor ANDREW CLARKE
British Antarctic Survey
High Cross, Madingley road
Cambridge CB3 0ET
UK

Library of Congress Cataloging-in-Publication Data

Eclogical studies in the Antarctic Sea ice zone: results of EASIZ midterm symposium/
Wolf E. Arntz, Andrew Clarke (eds.).
p. cm.
Includes bibliographic references (p.).
ISBN 978-3-642-63973-9 ISBN 978-3-642-59419-9 (eBook)
DOI 10.1007/978-3-642-59419-9
1. Marine ecology–Antarctica–Congresses. 2. Benthos–Antarctica–Congresses. I.
Arntz, Wolf. II. Clarke, A., 1949-

QH84.2. E25 2002
577.7′7–dc21

http.//www.springer.de

Softcover reprint of the hardcover 1st edition 2002

Cover design: Design & Production GmbH, Heidelberg
SPIN 10842137 31/3130-5 4 3 2 1 0 - Printed on acid-free paper

Foreword

Ten years ago *Polar Biology* published the book, *Weddell Sea Ecology*, containing the European "Polarstern" study EPOS in the Weddell Sea and Peninsula waters 1988/89. In certain respects, the present collection of papers, first published in *Polar Biology* in 2001, is a follow-up as it combines papers partly based on three "Polarstern" expeditions to the same region. Further articles relate to both land-based and shipborne studies, again primarily in the Atlantic sector and around the Antarctic Peninsula. The SCAR programme, "Ecology of the Antarctic Sea Ice Zone" (EASIZ), served as an umbrella for a truly international cooperation. Although funding came exclusively from national sources, 40% of the scientists on board "Polarstern" were foreigners. Out of the 35 papers of the present volume not less than 14 papers have multinational authorship. The scope of EASIZ is wider than the Southern Ocean Studies in JGOFS and GLOBEC. The Contents reflect emphasis on the study of benthos, which hitherto had not received the necessary attention in the attempt to understand key questions of evolution and zoogeography of fauna from the Southern Hemisphere. The information collected under EASIZ enhanced greatly our recognition of the rather high biodiversity of the Antarctic shelf benthos. In order to extend these studies to the deeper continental slopes and the deep sea, "Polarstern" is presently on her way for the first international survey of deep-sea benthos in the Atlantic sector of the Southern Ocean.

The volume also contains a wealth of ecological information on the coupling of the sea-ice, pelagic, and benthic subsystems. Furthermore, it deals with aspects of ecophysiology and survival strategies in many taxa and communities, ranging from bacteria to mammals. Studies in iceberg scourings have relevance to broad aspects of stability and resilience of benthic systems. The epifaunal suspension feeders are the most conspicuous element of the Antarctic benthos, but they have not been investigated exhaustively until EASIZ. Modern technology was extensively employed, including video ROVs, underwater cameras, and telemetry.

It has been a long way from the oral presentations at the First EASIZ Symposium 1999 in Bremerhaven to the publication of the papers in *Polar Biology* and, subsequently, the completion of this volume. Most contributions have been largely rewritten to include current results. The conveners of the Symposium, Wolf Arntz and Andrew Clarke, became guest editors. The usual game of "push and patience" resulted in a fine collection of manuscripts which then underwent final scrutiny by the editorial office of *Polar Biology*. A great number of referees were involved in the process. Their efforts and the patience and willingness of the authors to revise their manuscripts, often more than once, are gratefully acknowledged.

The Scientific Committee on Antarctic Research (SCAR) is to be congratulated on the outcome of the first part of EASIZ as documented in the present volume. *Polar Biology* will be happy to receive further first-class manuscripts produced under EASIZ and its international projects.

Bremen, April 2002 GOTTHILF HEMPEL

EASIZ so far

Some comments on the mid-term EASIZ Symposium

The midway point of the SCAR EASIZ (Ecology of the Antarctic Sea-Ice Zone) programme was marked by the first EASIZ Symposium, held at the Alfred-Wegener-Institut, Bremerhaven, Germany, between 22 and 25 June 1999. The symposium was attended by 130 scientists from 17 nations and, although the core of the meeting was concerned with work undertaken during the second EASIZ cruise to the Antarctic on RV *Polarstern*, the subjects presented covered the whole range of EASIZ science undertaken from both ships and shore stations in the past 4 years. A total of 140 talks and posters was presented. The subject presented most extensively at the symposium was biodiversity, but there were also a significant number of presentations in three other areas of particular interest to EASIZ, namely the biological effects of ice, cryopelagic and pelago-benthic coupling, and the ecophysiology of polar marine organisms.

This mid-term symposium showed clearly that the EASIZ programme has developed strongly and has already provided new insights into a number of the key ecological processes operating in the coastal and shelf ecosystem of Antarctica. In some cases, recent work within the EASIZ programme has changed fundamentally our view of the system, sometimes to an extent that will require the rewriting of textbook paradigms that have been in existence for decades.

A major change in our view of polar marine systems has come from the recognition that many (though not all) groups of marine organisms in Antarctica are taxonomically diverse. The previous paradigm of the bell-shaped curve determined by high diversity in the tropics and depauperate faunas towards the poles will have to be abandoned in favour of an asymmetric distribution of taxonomic richness in the two hemispheres. The greater species richness of the Southern Ocean compared with the Arctic Ocean may be the result of its greater age, the larger area and a higher level of structural heterogeneity formed by living organisms. Topics for future studies will include the role of spatial scale, e.g. local versus regional species richness, and the factors determining latitudinal patterns where these exist (J.S. Gray, keynote address).

The pycnogonids are a particularly rich and conspicuous group, with 31 genera and 251 species in the Southern Ocean. T. Munilla summarizes results from over 40 expeditions and suggests that Antarctic waters have been a long-term centre of pycnogonid radiation and dispersion. For many other taxa, however, the species inventory of the Southern Ocean is still far from complete, and recent finds are presented for bathypelagic hydromedusae (J. Bouillon et al.), octopodid molluscs and sabellid polychaetes. Both the number of octopodid species and their abundance increase with latitude, and the endemic genus *Pareledone* seems to have undergone extensive radiation on the Antarctic seafloor (L. Allcock et al.). The polychaete families Polynoidae and Sabellidae revealed a higher variability in reproductive features than hitherto assumed, with, for example *Harmothoe* species brooding their eggs under the elytra and *Myxicola* cf. *sulcata* incubating its embryos within the tentacle crown (M.C. Gambi et al.). The first quantitative data on Weddell Sea deep-sea polychaetes are provided by B. Hilbig. Species richness was found to be high compared to temperate deep-sea areas in the Atlantic and Pacific Oceans, whereas population densities were considerably lower. A striking difference in population density between Antarctic and Arctic waters was also detected in peracarid crustaceans from epibenthic sledge samples, with densities one order of magnitude higher in the Arctic (A. Brandt). This, combined with a preliminary assessment that species richness is high, may reflect the greater age and maturity of the Southern Ocean assemblages. The rate of speciation is an important component of biodiversity, but no evidence for a slowing of molecular substitution rates in serolid isopods under very low water temperatures could be detected by C. Held; temperature does not seem to be the key determinant of nucleotide substitution rate. A cytogenetic study of the bathydraconid fish *Gymnodraco acuticeps* (E. Pisano et al.) provides information on chromosomal diversification during the cladogenesis of this notothenioid family whose intra-familiar relationships are still very uncertain.

Using digestive tract analyses and relative species abundance data from three summer cruises, P. Dauby et al. have assessed the impact of amphipod populations on their prey in the Weddell Sea. Crustaceans, plankton particles and fish carcasses contributed the bulk of the diet. Another facet of diversity, that of epibenthic habitats occupied by gammaridean amphipods, was studied by C. De Broyer et al. The authors distinguish three epibenthic strata and a variety of symbiotic microhabitats associated with benthic suspension feeders. Studies such as these will be important in improving trophic flow modelling in Antarctic waters. Based on a scanning electron microscopy study of the mandibles and stomachs of three common Antarctic caridean shrimps whose diet was also studied using the experience of the various experts on board, V. Storch et al. conclude that stomach contents do not correlate well with mouth and stomach morphology, nor was there a clear relation with their observed life style or habitat preferences.

The two final biodiversity contributions in this volume deal with benthic assemblages. K. Jazdzewski et al. sampled a stony beach near Arctowski station during a complete annual cycle and found an unexpectedly high abundance of amphipods at the beginning of winter, possibly related to high autumn biomass of decaying algae on the beach. Assemblages of necrophagous animals were studied using baited traps on the continental shelf and in the deep sea of Enderby Land (Takeuchi et al.). Endemic Antarctic amphipods and isopods prevailed at the shelf sites, which revealed a higher diversity as compared with the deep-sea site, where the large cosmopolitan amphipod *Eurythenes gryllus* was dominant. The biodiversity discussions of the symposium benefitted considerably from the presence of biologists with first-hand experience of both northern and southern polar regions; the field of bipolar comparisons is clearly one which is potentially rich in insights and should be encouraged for the future.

The second major subject dealt with during the EASIZ Symposium were the interactions between the pelagial and the benthic ecosystem, and particularly benthic suspension feeders. In a keynote address, J.-M. Gili et al. summarise recent results from shallow and deep waters around Antarctica, emphasising that the traditional view that the Antarctic benthos depends entirely on the short seasonal plankton blooms with associated long periods of starvation during the major part of the year is obviously wrong. As in other oceans, many benthic suspension feeders in the Southern Ocean continue feeding in winter and can make use of the food web based around bacteria, nano- and picoplankton, unicellular organisms and organic detritus contained in the seston. In addition to this wide trophic range, they take advantage of the processes of advection and resuspension. Their feeding strategy thus does not differ much from that found in benthic suspension feeders elsewhere, and exchange processes between the pelagic and benthic realms may be of a similar order of magnitude as in other oceans.

Quite a different look – from above – is cast on the shallow-water ecosystem of Potter Cove (King George Island) by I. Schloss and G. Ferreyra. These authors measured phytoplankton photosynthesis and wind-induced turbulent mixing as well as light penetration in the inner and outer cove. They conclude that phytoplankton development in the area is restricted by low irradiance caused by suspended glacial flour and by wind-induced mixing, resulting in a low input to the benthos. Katabatic winds were also a key forcing factor for pelagobenthic coupling in Adelie Cove (Terra Nova Bay, Ross Sea). Over short time scales, they altered water column stability, vertical and lateral transport, nutrient evolution, primary production and degradation, and enhanced carbon flux to the shallow sediments, which had a three to four times higher organic matter content than sediments at the same depth offshore. This favoured the development of benthic detritus and large suspension feeders such as *Laternula elliptica* (P. Povero et al.). The flow of matter at even shallower depths was observed by Isla et al. at Johnson's Dock (Livingston Island). Erosion and transport of ice from the glacier situated inside the dock influenced fluxes to the sediment significantly, contributing above all to a high lithogenic content. Organic input was similar to that in deeper waters of Bransfield Strait, although in Johnson's Dock the residence time of particles in the column was too short to allow for the biological transformation found elsewhere.

Interesting evidence for opportunistic feeding was found by Orejas et al. in a study of the diet of four Antarctic benthic cnidarians. The food differed strongly among species and ranged from large zooplankters to the fine fraction of seston and resuspended material. Finally, Plötz et al. in their investigation of Weddell seal foraging showed that these seals feed predominantly on the abundant fish *Pleuragramma antarcticum*, and this determined much of their diel diving behaviour. Both pelagic and benthic diving occurred during daylight, but at night the seals foraged almost exclusively in the upper pelagic zone. This study was the first to combine time-depth recorder data and trawling, which turns out to be a promising approach for the future.

Sea ice is a special environment in the Antarctic, having many organisms associated with it and exerting a distinct influence (cryopelagic coupling) on the water column. Furthermore, the impact of icebergs and anchor ice has a major influence on the composition, stability and dynamics of the littoral and shelf marine ecosystems. Ice biota and ice impacts were therefore the third major subject treated during the EASIZ Symposium. In a keynote address, J. Gutt reviewed the effects of ice disturbance on Arctic and Antarctic benthos assemblages. In this comparative study, he found sea ice to be of greater relevance in the north and iceberg impact to be more common and evenly distributed in Antarctic waters. Iceberg scouring enhances diversity because the temporal pattern of disturbance allows for the co-existence of different successional stages. By contrast, glaciers cause faunal impoverishment, though primarily only in their close vicinity. Ice impact has also been important in the development of the present-day benthic fauna.

Fluxes of particulate organic matter from the platelet ice layer and through the water column were measured by D.N. Thomas et al. under the sea ice of the Drescher Inlet during austral summer. As there was little evidence of algal activity within the thick multi-year sea ice, algal growth must have taken place in the underlying platelet ice layer. The magnitude of the diel POC flux was not very high, with diatoms, mainly frustules of ice algae, sinking largely intact as faecal pellets to the benthos at about 400-m depth. This flux had important consequences for cryobenthic coupling. Fast-ice bottom algae production was measured in McMurdo Sound in the 1999 summer at a 30-m-deep site without a platelet layer, and oxygen microelectrode measurements were taken in situ by L.J. Trenerry et al. Their results complement former sea-ice production measurements in the area from the surface, the platelet layer and the bottom. The role of Antarctic sea ice for meiofaunal organisms is demonstrated in a study by S. Schnack-Schiel et al. Foraminifers dominated the sea-ice assemblages by number and turbellarians by biomass. The greatest concentrations of meiofauna were found in the lowest parts of sea ice in autumn and winter, but there may be high densities of these organisms in upper and intermediate layers of porous sea ice in summer as well as on the underside of flows and in the adjacent water.

Two further meiofauna studies by H.J. Lee et al. refer to the impact of iceberg scour on these organisms at the seafloor. The first describes the response of a shallow (8–9 m) meiofauna community off Signy Island to catastrophic iceberg scour. This reduced meiofaunal abundance by over 95% and diversity was also reduced. However, recovery was accomplished within 30 days, and over the longer term nematode community structure was affected very little, pointing to a high degree of resilience of this community to disturbance by ice. In a second study, three stations each at a different stage of recolonisation following iceberg scour were studied between 255 and 298 m depth off Kapp Norvegia (Weddell Sea); these comprised a fresh scour, an older scour and an undisturbed control site. Overall meiofaunal abundance and diversity were significantly lower in the fresh scour and highest in the older scour. Nematode abundance and diversity were also lowest in the fresh scour, however, whereas abundance had recovered to the undisturbed level in the old scour, generic diversity remained low, indicating that the continental shelf nematode community is sensitive to iceberg disturbance.

The final two contributions in this section deal with iceberg impact on the mega-epibenthos and on demersal fish. Gutt and Starmans used video transects on the eastern Weddell Sea shelf to classify the epibenthic megafauna into stages after iceberg impact and unaffected fauna, taking into account bottom topography and concentrations of grounded icebergs. Calculating impact factors from recently disturbed areas and recovery time suggested that iceberg scouring is among the most significant disturbances experienced by any large ecosystem on earth. Brenner et al. compared areas affected by icebergs and undisturbed bottoms in the same area and conclude that small-scale horizontal patterns caused by iceberg scour play a distinct role in niche separation of the fish genus *Trematomus.* Different species preferred either disturbed or undisturbed areas, as is indicated by total stomach contents and number of prey taxa per fish.

Although the overwhelming part of the contributions offered at this symposium was concerned with ecological questions and processes, physiology is also an important facet of the programme, often providing causal explanations for the mechanisms that underlie empirical ecological findings. In his keynote address, L.S. Peck reviews recent advances in the understanding of adaptation to low temperatures in marine organisms. He stresses the central role of energetics in this context. Polar ectotherms have low resting metabolic rates, which may be related to mitochondrial function at low temperature, slow growth and development rates, and low aerobic scopes. Of particular interest is the suggestion that oxygen supply to the tissues may determine thermal tolerance and also limit maximum size.

The question of how Antarctic macroalgae acclimate to seasonal changes in light was investigated by U.H. Lüder et al. using the photosynthetic apparatus of the alga *Palmaria decipiens* in culture experiments. Maximal quantum yield remained high during autumn, winter and spring, whereas maximal photosynthetic activity and pigment contents increased continuously in mid-autumn and winter, and were highest in spring. In summer, all these variables decreased to their lowest values and *P. decipiens* proved to be efficiently adapted to the short period of favourable light conditions in Antarctica. The identification of repair and protective mechanisms that allow Antarctic species to survive and reproduce under increased UVB flux is dealt with in the contribution by M. Hernando et al., who studied the role of mycosporine-like amino acids (MAAs) in the photoacclimation of *Thalassiosira* sp. using different experimental irradiance treatments. This marine diatom accumulated high concentrations of MAAs, which are important in protection against ultraviolet radiation, on a time scale of hours to days. Whereas Antarctic algae seem to have been quite successful in developing adaptive mechanisms to a changing environment, some faunal taxa appear to have been less successful. Frederich et al. link the striking absence of almost all reptant decapods from the waters south of the Polar Front to high magnesium levels in the haemolymph and a consequent failure of cardiac and ventilatory performance in very cold water, with Mg^{2+} acting as an anaesthetic. These authors investigated temperature-induced changes in oxygen consumption of two brachyuran crabs from southern Chile and present a model to describe how Mg^{2+} reduction in the haemolymph causes a change in the threshold temperature which characterises the onset of cold-induced failure in oxygen supply to tissues. Bluhm et al. address an old methodological problem related to the difficulty of age determination in crustaceans by measuring an autofluorescent pigment, lipofuscin, in the brain of ten polar species from various crustacean taxa. Lipofuscin granules occurred in easily detectable amounts in five species. Pigment concentrations were correlated with individual size within the species but did not indicate absolute body size when comparing different species. For those species that accumulate sufficiently high quantities of lipofuscin, the pigment is a potential indicator of age.

Ascidians are another polar taxon where growth and age are not easily determined, although some Antarctic species are believed to be relatively fast-growing. Combining results from colonisation panels and respiration measurements, Kowalke et al. show ascidians from Potter Cove to have high growth rates, which were favoured by a low basal metabolism and an extended longevity. These properties, together with their capability of surviving under intense sedimentation, made them effective competitors among benthic suspension feeders. Finally, Bock et al. used NMR techniques to compare temperature-dependent pH regulation in a stenothermal Antarctic and a eurythermal temperate eelpout (Zoarcidae). They conclude that, in contrast to previous reports in the literature, internal pH regulation in at least some Antarctic fish occurred as predicted by the alphastat hypothesis of acid-base regulation.

It is clear from the results presented at the mid-term EASIZ Symposium (not all of which could be included in this volume) that the EASIZ programme has been a success. In common with all biology programmes sponsored by SCAR (the Scientific Committee for Antarctic Research), support has been provided for developing the programme and holding symposia, but not for fieldwork or other scientific costs. The EASIZ programme has therefore had to rely on the National Programmes of SCAR nations for the scientific aims to be realised. Half-way through the programme it is clear that EASIZ work has been promoted particularly by work from a small number of ship platforms, most notably RV *Polarstern* and RV *Hespérides*, and a range of shore stations including (in no particular order) Arctowsky, Jubany/Dallmann, Signy, Rothera, Terra Nova and King Sejong Stations. One particular highlight of the EASIZ work has been the development of taxonomic expertise in key taxa, notably in Russia, Spain, Belgium and Germany. Whilst the wider programme of research proposed at the start of the EASIZ programme has been taken up enthusiastically by many national programmes, the core programme has been less successful. The core programme proposal was that a number of shore stations, preferably from all parts of Antarctica, should undertake a series of regular nearshore oceanographic and ecological measurements with the aim of documenting patterns of geographical and interannual variability in key ecological processes in the seasonal ice zone. Whilst some work has started, there is still room for better coverage and it is hoped that this will improve in the second half of the programme.

The mid-term symposium was followed by a series of concurrent workshops, the aims of which were to evaluate progress to date, and to determine those areas of EASIZ science that should receive particular attention during the final stages of the programme. Five workshops were run, and attendance was very encouraging, revealing a strong interest in the future progress of the EASIZ programme. The conclusion from these review workshops was that certain fields had been covered fairly satisfactorily (although not always exhaustively) to date. These included topics such as species composition and distribution in most macroinvertebrate, fish and macroalgal groups on the Antarctic shelf, characterization of typical Antarctic species assemblages in the benthos and plankton of some regions (Weddell Sea, Ross Sea, Antarctic Peninsula), and the structure of the different forms of sea ice and ice biota. Good progress had also been made in fields such as the impact of ice on benthic meiofauna and macrofauna (including fish), life-history strategies related to sea ice and in the pelagic and benthic subsystems, and the diving, foraging and migratory behaviour of pack ice seals. The latter topic is an area where the EASIZ programme has a strong overlap with the SCAR programme on Antarctic pack-ice seals (APIS). The wider programme proposed for EASIZ had laid specific emphasis on the need to understand ecophysiological aspects of organisms in the sea-ice zone, and the workshops recognised that significant progress had been made in this area. Examples of such progress included studies of the range of physiological tolerance related to mechanisms of molecular and physiological adaptation, photoinhibition and UV effects on marine algae, the role of oxygen consumption for behaviour and activity and, lastly, secondary production, trophic relations and energy flow in the benthos.

The workshops also identified some important gaps in our knowledge or practice, and a particular need identified was to co-ordinate the various existing taxonomic databases for Antarctic species. Areas for development included the application of molecular techniques to a variety of species in the context of biogeographical ranges, endemicity and adaptations to low temperature, and comparative assessments of biodiversity. Both of these topics relate well to the developing SCAR programme on the Evolutionary Biology of Antarctic Marine Organisms. On the technical side, the workshops identified the need for a wider use of modern, automated sampling gear such as cameras, remotely operated vehicles (ROVs), sediment traps and benthic landers.

We hope that these presentations give a taste of the science performed to date within the EASIZ programme, and a pointer to what can be achieved in the second half. By the time we come to our final symposium, in 2004, we are sure that those biologists who have worked within the EASIZ programme will have made a major contribution to our ever-improving understanding of the coastal and shelf ecosystem of Antarctica.

The guest editors would like to thank all those persons and institutions that contributed to these proceedings and the organization of the EASIZ Mid-Term Symposium. In particular, we would like to acknowledge the time and effort of the international referees who used their expert knowledge to improve the manuscripts selected for this volume. Gotthilf Hempel, chief editor of *Polar Biology*, kindly subjected them to another careful examination before they went to press. The Deutsche Forschungsgemeinschaft, the Senator für Wirtschaft of the State of Bremen, the Scientific Council on Antarctic Research, and the Alfred Wegener Institute for Polar and Marine Research provided funding to support participation of colleagues in the Symposium, its organization and the publication of the results. Lucie Marquardt, Sabine Ilsemann, Kerstin Beyer, Renate Kuchta, and Andrea Bleyer were very helpful in the preparation and organization of the Symposium and during the various stages of the finalization of these proceedings. Last, but not least, we wish to thank Springer-Verlag for kindly considering all our requests regarding the layout of this volume.

WOLF ARNTZ
ANDREW CLARKE
GUEST EDITORS

Contents

I. Biodiversity

II. Pelagobenthic coupling

III. Ice biota and ice impact on benthic communities

IV. Ecophysiology

I.
Biodiversity

Polar Biol (2001) 24: 633–641
DOI 10.1007/s003000100244

John S. Gray

Antarctic marine benthic biodiversity in a world-wide latitudinal context

Abstract In the 1950s and 1960s, the first data sets were assembled to examine whether or not there was a latitudinal gradient of species richness in the sea. These data comprised very few species and were from very small areas. However, recent data from large species lists covering broad geographical ranges suggest strongly that there is a gradient of increasing species richness from the Arctic to the tropics. However, the Southern Ocean has high species richness and in the southern hemisphere there is no clear evidence of a cline of increasing richness from pole to tropic. The great richness of the Southern Ocean compared with the Arctic is probably due to its great age, the fact that it covers a much larger area and that it has higher structural heterogeneity formed by living organisms. The importance of area as a determinant of species richness needs to be studied in more detail since most studies have been confined to small areas. A number of hypotheses have been proposed to explain the species:area relationship and these are discussed. An alternative explanation for the latitudinal cline in the northern hemisphere is the energy-input hypothesis, but again this has not been adequately tested. Two studies on the relationship between local and regional species richness show a significant positive correlation. These findings suggest that local assemblages are not tightly organised and saturated with species but are open to recruitment from the regional species pool. Whether or not such a relationship holds in Antarctica is unknown. It is concluded that further studies of the Southern Ocean are likely to provide new findings fundamental to the "new" discipline of macroecology, which examines patterns and processes at the geographic scale.

Please note: When citing this article, please use the reference to the original source in the journal Polar Biology as given on the top of this page.

J.S. Gray
Biologisk Institutt, Universitetet i Oslo,
Pb 1064, 0316 Blindern, Norway
E-mail: j.s.gray@bio.uio.no

J.S. Gray
Department of Biology and Chemistry,
City University, Tat Chee Avenue,
Kowloon, Hong Kong

Introduction

Many terrestrial plant and animal taxa show an increase in the number of species from the poles to the tropics (Rosenzweig 1995). This is not surprising since there are few species capable of living in the harsh conditions of polar winters, and tropical environments such as rain forests are extremely rich in species. Yet in the marine environment, many species are capable of surviving in temperatures of a few degrees to around zero degrees, both in shallow polar seas and in the cold deep sea. Thorson (1957) reported a cline of increasing species numbers in epifauna from the Arctic to the tropics, and Stehli et al. (1967) showed for bivalves that there was a cline in both hemispheres. These findings led to the conclusion that clines of increasing species numbers from the poles to the tropics conformed to the terrestrial pattern. Samples of benthos from the deep sea also suggested that both in the North and South Atlantic there were similar patterns of increasing species numbers from poles to tropics (Rex 1981, 1983; Rex et al. 1997). These patterns are widely accepted and reported in textbooks, yet were these data appropriate to make such generalisations? Here I re-examine the data on which the paradigm of a marine latitudinal species gradient was raised.

Using new marine data from Antarctica, Clarke (1992) questioned the idea that there was a latitudinal gradient of increasing species numbers from the Antarctic to the tropics (see also Brey et al. 1994). In further papers, Clarke and Crame (1997) and Crame and Clarke (1997) showed that the terrestrial pattern probably does not apply in the marine environment in the southern hemisphere. In fact, recent data (Arntz and Rios 1999) show that for a number of taxa, species numbers increase from the Magellan area towards the Antarctic continent. The gradient of increasing species richness from Arctic to tropics probably does exist, however, in the northern hemisphere (Clarke and Crame 1997; Rex et al. 1997).

Measurements of species diversity are concerned not only with number of species but also include the distribution of individuals among the species (Magurran 1988). There are a variety of indices and methods for the measurement of diversity and I have recently reviewed these (Gray 2001). Whereas the latitudinal gradient in number of species is based on species lists, studies of gradients of diversity must involve quantitative sampling. This inevitably means that one has to take account of the spatial scale sampled. It is by no means clear that diversity measured at a small scale is linearly related to diversity at a large scale (but see Cornell and Karlson 1996). Few marine studies have explicitly taken this into account when comparing diversity at different latitudes. Here I examine the different spatial scales of diversity and how this may affect conclusions concerning a latitudinal gradient of species diversity.

The origins of the idea of the latitudinal gradient of marine species richness

Thorson (1957) recorded an increase in the number of species of some taxa of hard substratum epifauna from the Arctic to the tropics, but no increase in the number of species of soft sediments. Later, Stehli et al. (1967) showed that bivalve molluscs, at species, genus and family levels, increased in numbers towards the tropics in the Indo-Pacific. Thorson's data showed an average of 9 species of nudibranch in Arctic areas, 22 in the boreal and 128 in the tropics, and for Amphipoda, 150 species in the Arctic, 180 species in the boreal and 195 in the tropics. Stehli's data for bivalves showed ca. 50 species at 70°N, ca. 130 at 50°N, ca. 500 in the tropics, ca. 200 species at 30°S and ca. 70 species at 60°S. These are extremely small numbers of species on which to make general comparisons.

Recent analyses of species numbers show quite different data to those reported 30–40 years ago. In Southern Australia (Port Phillip Bay), Poore et al. (1975) found 197 species from an area of only 0.5 m^2 at a sandy site. In the Bass Strait, one 0.1 m^2 sample comprised 187 species, and 12 out of 38 taken on 1 occasion comprised more than 100 species (Coleman et al. 1997). Table 1 shows a summary of recent data from coastal areas of Australia, Norway and Scotland. Table 1 shows clearly that techniques have improved greatly since the 1960s and today we have some good quantitative data rather than the semi-quantitative data available at that time.

A compilation of all known species along the Norwegian coasts by Brattegard and Holthe (1997) gives a total of around 4,500 species of macrofauna and macroalgae. The data for 70°N show 90 species of Nudibranchia, 246 species of gammarid amphipods alone and 117 species of bivalves. These figures are much larger than those of Thorson and Stehli. Arntz et al. (1997) have compiled species richness data for the Southern Ocean (Table 2). The total number of species is approximately 5,200 species of invertebrate animals. Unfortunately, there are no recent compilations of species numbers from tropical areas. What is clear is that the species numbers used in comparative studies in the late 1950s and early 1960s are almost certainly not representative.

Sanders (1968) published an analysis of the diversity of marine sediments and showed two species richness gradients from the species-poor boreal region to the highly species-rich tropics and from low species-rich coasts to the high species-rich deep sea. Sanders's synthesis was based on data collected along a transect from

Table 2 Approximate species numbers in the Southern Ocean (from Arntz et al. 1997)

Taxon	Number of species
Porifera	ca. 300
Cnidaria	ca. 400
Polychaeta	ca. 700
Mollusca	ca. 900
Crustacea	ca. 1,800
Echinodermata	ca. 300
Lophophorata	ca. 500
Pisces	ca. 300

Table 1 Recent species richness data of some coastal areas

Area	Total number of species	Total number of individuals	Area (m^2)	Depth (m)
Snorre Norway[a]	343	20,668	20	270–330
Frigg, Norway[a]	592	290,401	75	70
Port Phillip Bay, Australia[a]	580	156,274	43	2–23
Bass Strait, Australia[a]	803	60,258	10.4	11–51
Lochs Linnhe and Eil, Scotland[b]	323	13,014	12	9–111
Firth of Lorne, Loch Etive, Loch Creran[c], Scotland	ca. 300	ca. 37,000	11	24–117

[a]Gray et al. (1997)
[b]Pearson (1970)
[c]Gage (1973)

Gayhead, United States to Bermuda (Hessler and Sanders 1967), and then from a variety of different geographical areas, from the boreal to the tropics and from estuaries to the deep-sea slope (Sanders 1968). Sanders used not the total species richness, but the polychaete-bivalve fraction of the fauna collected with a semi-quantitative anchor-dredge. Table 3 shows the data. Sanders (1968) claimed that the polychaete-bivalve fraction of the fauna usually represented over 80% of the total number of species. Abele and Walters (1979) reanalysed the data and showed that the polychaete-bivalve fraction varied from an average of 38% in estuaries to 60% on the shelf, around 70% in the deep sea, to over 80% in shallow waters. They argued that the data, therefore, were unlikely to be representative of total species richness. I pose another question: are the number of species of polychaetes and bivalves in Sanders's analyses adequate to postulate the existence of a latitudinal gradient? Sanders only sampled a few sites within each area and the maximum number of species was less than 100. The species numbers obtained cannot, therefore, be representative of tropical, boreal, estuarine or deep-sea areas but simply represent, at best, site species richness.

The scale of measurement of diversity

Whittaker (1960) suggested that the most basic level of species richness was the number of species in a sample, which he called alpha diversity. Later Whittaker (1972) suggested that diversity could be measured at four different scales: point diversity (a single sample), alpha diversity (samples within a habitat), gamma diversity (the diversity of a larger unit such as an island or landscape) and finally epsilon or regional diversity (the total diversity of a group of areas of gamma diversity). Magurran (1988) related these scales to sample, habitat, landscape and biogeographical province. Yet there is confusion about this terminology. For instance, Rosenzweig (1995) called alpha diversity "point" diversity whereas Pielou (1976) stated "alpha diversity pertains to a small area and is a property of a particular community; even though recognition of such an entity is nearly always subjective the risk of being seriously mistaken is negligible". This begs the question, as Underwood (1997) pointed out, of how a community should be defined.

Table 3 Expected number of species for 750 individuals (ES_{750}) (from Sanders 1968)

Tropical shallow	70–90 species
Boreal continental	57–63 species
Boreal maritime	44 species
Tropical estuary	19–22 species

There is a need to provide a uniform terminology for the different scales of species richness and diversity. Table 4 shows my attempt to provide such a scheme (Gray 2000).

Statisticians define the sampling unit as a single core, grab or other sampling device used in an area. In quantitative studies of marine benthic assemblages, one usually takes four or five replicate grab samples. In statistical terms these would be called the sampling units, and the five combined the sample. However, in terms of species richness and diversity measurements a single grab, covering only 0.1 m^2, may only sample a small fraction of the species and does not necessarily provide a point sample of species richness or diversity. In comparing species richness and diversity over different scales, I (Gray 2000) called the five replicate grabs the point sample species richness (or diversity) and the different sites sampled within a small area, the sample species richness (or diversity). However, tests of this assumption are needed.

It is important to note that in the above discussion of scales of species richness/diversity, there is no mention of beta diversity. There is considerable confusion in the marine literature about this term. Sanders (1968) and others (e.g. Gage 1996) claimed that studies of the species richness/diversity of soft sediments of similar grain size was of alpha (or within habitat) diversity, whereas a comparison between soft sediments of different grain sizes was of between habitat (beta diversity). Gage (1996) argued that it was invalid to compare coastal sediments with deep-sea sediments as Gray (1994) had done since one could not compare alpha and beta diversity. This view of alpha and beta diversity is mistaken. Beta diversity as defined by Whittaker (1972) refers to the change in species composition (or species turnover) along a gradient. Clarke and Crame (1997)

Table 4 Proposed unifying terminology for scales of diversity (from Gray, 2000)

Scale of species richness	Definition
Point species richness: SR_P	The species richness of a single sampling unit
Sample species richness: SR_S	The species richness of a number of sampling units from a site of defined area
Large area species richness: SR_L	The species richness of a large area which includes a variety of habitats and assemblages
Biogeographical province species richness: SR_B	The species richness of a biogeographical province
Type of species richness	
Habitat species richness: SR_H	The species richness of a defined habitat
Assemblage species richness: SR_A	The species richness of a defined assemblage of species

used this term correctly and Clarke and Lidgard (2000) made an interesting study of turnover diversity of marine bryozoans.

Most studies of marine species richness/diversity are at the scale of point or sample species richness (e.g. Sanders 1968). Rex et al. (1993) compared a series of samples taken in the deep sea in the Atlantic Ocean and claimed that there was a gradient of increased diversity from poles to tropics in both hemispheres. Both Sanders and Rex et al.'s studies used data obtained from semi-quantitative anchor-dredge samples covering unknown areas of seabed. In order to be able to make comparisons between samples, the rarefaction method was devised by Sanders (1968). This is based on the total number of individuals and species and does not consider species:area relationships, as the area sampled was not known. The method takes the distribution of individuals among species and calculates the expected number of species ES_n for a given number of individuals, n for smaller sample sizes. Gray (1997a, b) has shown that rarefaction can greatly overestimate the expected number of species for small samples.

Sanders's comparisons showed a maximum number of species in any site was less than 100 species. Rex et al. (1993) did not show the total number of species found, but the 97 anchor-dredge samples contained 214,508 individuals. This is a comparatively low number for such a large number of samples, when compared to recent data shown in Table 1. Furthermore, comparisons of ES_n were based on $n=50$ for Gastropoda, $n=75$ for Bivalvia and $n=200$ for Isopoda. These are extremely small sample sizes with which to make inferences on latitudinal patterns of species richness. Despite Rex et al.'s rebuttal (1997), I remain sceptical about Rex et al.'s conclusion that there is a latitudinal gradient in deep-sea benthos shown in their data (Rex et al. 1993). There does not appear to be a gradient in the southern hemisphere in any of the three taxa studied. Furthermore, there were very few samples taken from between 5°N and 5°S. Those that are shown have low variance whereas all the other samples show the high variance that is expected since sample species richness was measured. I believe it is essential to obtain more data points from tropical areas before one can assert that there is a latitudinal gradient in sample species richness. It is well known that the Norwegian Sea was recently glaciated and that species are still colonising the area (Crame and Clarke 1997), so the low figures for this area can be explained.

Kendall and Aschan (1993) made a comparison of the sediment-living macrofauna at a site at 123 m depth at 78°N off Svalbard, Norway, with a site off Northumberland, United Kingdom, 50°N at 80 m, with that at 30 m off the coast of Java, 7°N. Five replicate 0.1 m^2 grab samples were taken at each of the sites. Using ES_{200} they obtained respective species numbers of: 32.9 ± 1.4 for 78°N off Svalbard, 34.6 at 50°N in the North Sea, United Kingdom and 33.2 ± 1.9 at 7°N off the coast of Java, where the $\pm$ are 95% confidence intervals. The authors concluded that these data showed no trend in latitudinal species richness. However, Kendall and Aschan's data are from small sample sizes of very limited numbers of species and, in addition, depth varied considerably between sites. In coastal areas, small differences in depth are known to alter species composition and probably species richness (Olsgard and Gray 1995). Kendall and Aschan's paper simply shows that there was no difference in species richness between three very small areas sampled in different places, which happen to be polar, temperate and tropical.

Studies that aim to test whether or not there is a gradient in sample species richness with latitude need to be done at a large number of sites in polar, temperate and tropical areas before any general statements can be made. Merely sampling a few small areas does not provide evidence one way or another for a latitudinal gradient even in sample species richness. As long ago as the 1960s, MacArthur (1972) had suggested that in terrestrial systems there was no difference in sample species richness between temperate and tropical areas.

At the other extreme to studies of sample species richness are studies done using species lists that represent large areas such as 5° or 10°bands of latitude or biogeographical provinces. Species records of prosobranch molluscs along the coasts of N. America (1,907 species in the eastern Pacific and 2,009 in the western Atlantic) show clear gradients of increased species richness from the North Pole to the tropics, with a maximum richness between 20°N and 30°N (Roy et al. 1996, 1998). Yet although the study was done only in the northern hemisphere, Roy et al. (1998) stated that "latitudinal gradients, peaking in the tropics and tailing off toward the poles are the rule". Yet Roy et al.'s conclusions apply only to the northern hemisphere and the tropics did not have maximal species richness!

A recent analysis of Norwegian species lists shows clearly that there is a cline of increasing species richness from 70° to 60°N (F. Olsgard, unpublished work). It would be interesting to pool the data in Rex et al. (1993) and calculate species richness for large areas, as done by Clarke and Lidgard (2000) in their study of marine bryozoans.

Recent species-list data from the southern hemisphere support Clarke's (1992) contention that the latitudinal gradient of decreasing species richness towards Antarctica does not hold for the southern hemisphere. Lowry (personal communication) analysed the fauna of islands from Antarctica to New Zealand and found no clear gradient of increasing species richness, but recorded changes in the frequencies of various taxa. Data from Australia (Coleman et al. 1997; Gray et al. 1997) show extremely high species richness at 40°S, and there are no tropical data yet published showing as high a species richness. Clearly there is a need for more quantitative data from the southern hemisphere and from the tropics, but there is increasing evidence that there is not a cline of increasing diversity from Antarctica to the tropics. This pattern may also occur in terrestrial systems since the Fynbos of S. Africa (Cowling et al. 1992) and Kwongan of southwestern Australia have extremely high plant species diversity (Marchant 1991; Hopper 1992). Thus one can summarise that, in the marine domain, there is probably a cline in increasing species richness from the species-poor Norwegian Sea to the tropics but no clear evidence yet that there is a similar cline in the southern hemisphere.

The data that are available cover very small areas and are of sample species richness, or refer to species lists over extremely large areas. What are needed are studies that examine species:area relationships at intermediate scales covering areas larger than the sample, e.g. at the landscape scale, and for defined habitats and assemblages. Such studies for soft sediments must separate out the influence of depth and sediment-grain size. For other habitats, the relationship between habitat heterogeneity, area and species richness needs to be determined. Marine diversity studies are thus in their infancy!

What is special about Antarctic benthos?

Arntz et al. (1997) have reviewed Antarctic marine biodiversity. Based on their extensive experience of Antarctic benthos, they suggested that abundances per unit area are similar to those of other marine areas, and that both abundances and biomass are highly patchy. However, on average, biomass per unit area may be higher in the Antarctic largely due to the high biomass of suspension-feeding species. Arntz et al. made two important points: first, where ice-berg scour affects the assemblages, the immediate area affected is species poor, but due to successional processes over larger scales species richness may be increased, and second, that endemism is high in the Antarctic fauna and flora.

In one of the few quantitative studies done, Starmans (1997) compared the mega-benthos in an area off Greenland (Arctic) with that of the Bellinghausen Sea and the Weddell Sea, Antarctica. He found that the Weddell Sea had the greatest species richness followed by the Bellinghausen Sea, and Greenland had the smallest species richness. In comparing the Antarctic with the Arctic there are two main considerations. First, age: whereas, according to some authors the Antarctic has been isolated for 40 million years (e.g. Dunton 1992, but see Crame 2000 who suggests that separation started 30 million years ago and was completed 20 million years ago), the Arctic has only been isolated for 2–3 million years (Dayton et al. 1994). Thus age may play a dominant role in accounting for differences in species richness and diversity between the two poles. (I will not discuss the evolutionary aspects in detail since these have been well covered by Crame and Clarke 1997 and Crame 2000).

There are many physical barriers in the Southern Ocean and few in the Arctic. Thus allopatric speciation is likely to be more significant in the Antarctic and, due to the greater age, sympatric speciation also may be more prevalent than in the Arctic, although no detailed analysis is yet available. Clarke and Crame (1997) concluded that there was no strong evidence from marine taxa that speciation has occurred mainly in the tropics and that extinction has occurred at higher rates towards the poles. They suggested that speciation and extinction rates are similar in temperate and tropical marine species and that radiation rates were also similar. Clarke and Crame argue that tropical molluscan clades are 3 times as species rich as temperate ones simply because of the younger age of the latter. In this context the small species richness of the Norwegian Sea can be explained by its young geological age and that there has not been sufficient time for establishment of a full species complement.

Another argument for differences in species richness between the two poles is that species richness is known to increase with area. In relation to the Arctic and Antarctic, Dayton (1990) calculated that the area of the Arctic Ocean is less than half that of the Southern Ocean. Thus due to the larger area it is expected that species richness will be greater in the Antarctic. In one of the few published studies of the species:area relationship in relation to latitude, Abele and Walters (1979) reanalysed Sanders's (1968) data and showed convincingly that there was a linear trend between area and species richness. Thus, they argued that Sanders's data provided no evidence of a latitudinal gradient of species richness.

Rosenzweig (1995) proposed three arguments for the increase in number of species with area sampled. These are: (1) larger areas have more individuals, (2) larger areas have more habitats, and (3) larger areas contain more biogeographical provinces.

The analysis of the species:area relationship found in Sanders's data done by Abele and Walters (1979) did not set out to distinguish between hypothesis 1, that larger areas contain more individuals, and hypothesis 2, that the coast has more habitats than estuaries. Roy et al. (1998) tested the species:area relationship on their extensive data set on prosobranch molluscs. Their analysis was based on species lists of prosobranch molluscs from within latitudinal intervals for the east and west coasts of the United States. However, they plotted species richness for individual samples against the area of the sample and found no correlation. The problem is that the test was again of sample species richness against area, whereas they should have tested whether the relationship held over larger areas, such as the total area contained within a latitudinal boundary (5° or 10° intervals), that was the focus of their study.

Rohde (1997) analysed the surface areas of the major provinces in the various oceans and found no relationship between species richness and area. There is, however, no reason to expect that the species richness of sediment-living fauna (or the fauna and flora of other major marine habitats such as coral reefs, mangroves, segrass beds, kelp forests, etc.) will be related to the area of the sea surface within provinces. Data are needed on the areas covered by soft sediments and other important habitats within the various latitudinal boundaries. The species:area relationship, therefore, remains to be investigated in detail in polar, temperate and tropical areas.

In relation to Rosenzweig's second hypothesis that larger areas contain more habitats than smaller areas in the Southern Ocean, the Porifera and Lophophorata have large species richness and abundances (Table 2), quite unlike the Arctic. These taxa provide extensive three-dimensional substrata and greater structural heterogeneity in the Southern Ocean compared with the Arctic Ocean. It is well known in both terrestrial and marine habitats that greater structural heterogeneity leads to larger species richness (MacArthur and Wilson 1967). This again will result in greater numbers of species being found in the Southern Ocean than in the Arctic. Species lists for the various regions of the Southern Ocean have not been analysed in the context of quantitative relationships between structural heterogeneity and species richness/diversity.

Gaston (1998) analysed terrestrial data and suggested that numbers of individuals increased at a faster rate than the area over which they were distributed. More widely distributed species not only tend to have larger numbers of individuals but also tend to occur at higher local densities. Whether this is also true for polar systems remains to be tested. Gray et al. (1997) found no evidence that there were fewer individuals per species in the deep sea than in coastal areas. There are no published comparisons across latitudes. If this relationship is verified in further tests then there are unlikely to be differences in evenness between deep sea and coast.

At larger scales, the landscape and biogeographical province scale, there are no data for coastal systems where species richness across different habitats is treated in a comparative manner, a point made also by Arntz et al. (1997). Thus Rosenzweig's second and third hypotheses remain untested in the marine domain.

In addition to the species:area relationship mentioned above, Roy et al. (1998) tested two different hypotheses for the latitudinal gradient found in prosobranch molluscs: Rapoport's rule and the energy hypothesis (Wright 1983). Rapoport (1994) suggested that there is a decrease in species range lengths towards lower latitudes. Roy et al. showed that Rapoport's rule did not hold for prosobranch molluscs since median latitudinal ranges of species were greatest at low latitudes. Wright (1983) suggested that the total (or average) amount of available energy is greatest in the tropics and decreases towards the poles, thus supporting a larger number of species in the tropics. Roy et al. (1998) found a clear correlation with sea-surface temperature and thus favour an explanation based on energy input for the latitudinal gradient found in prosobranch molluscs in the northern hemisphere (but see the discussion below).

Discussion

In this final section I will try to put the points so far raised into a more general ecological context. I have emphasised that whilst there are data on sample species richness, there are rather few at larger spatial scales. In the general ecological literature there has been a renewed focus on research directed towards geographic scales, so-called macroecology. Gaston and Blackburn (1999) give four main reasons for this change in emphasis of ecological research. First, many of the issues facing mankind are at geographic rather than local scales. An example is the Convention on Biological Diversity requiring signatory nations to conserve biodiversity and use it sustainably. Second, there is a need to combine results from small-scale studies with large-scale patterns. Third, local assemblage composition is the result of both local scale and regional processes and, fourth, ecological patterns observed at one scale do not necessarily translate to another scale.

In reviewing the research agenda for macroecology, Lawton (1999) lists as key research areas (among others): (1) the relationships between species richness, population abundance, size of geographic range and population size; (2) species-area relationships; (3) the latitudinal pattern of species richness; (4) energy inputs and species richness. Many of these topics have been discussed in the foregoing and are already part of the research agenda in marine ecology. For example, Rex et al. (1997) have tested the relationship between local and regional species richness for deep-sea prosobranch molluscs and found a highly significant correlation. A similar pattern has been found in coral species (Cornell and Karlson 1996). Cornell and Karlson compared local species richness in quadrats and line transects at over 100 sites with the regional species pool obtained from species lists available in the primary literature and independent of the local richness data. Cornell and Karlson's analyses suggested that although coral species interact intensively at the local scale, the assemblages are open to recruitment from the regional species pool. Thus local assemblages are not tightly organised saturated assemblages that have an upper limit to species richness. Similar patterns occur in a wide variety of terrestrial ecosystems and habitats (Schluter and Ricklefs 1993). In the context of local and regional species pools, one might expect that with the greater age of marine systems compared with terrestrial, it is more likely that regional species pools reach saturation. Comparisons among sediment-living organisms between comparable assemblages in the "young" Arctic Ocean and the "old" Southern Ocean and between the species-rich Antarctic and tropics are likely to provide data of fundamental interest in relation to the issue of whether or not saturation occurs. Clearly more studies are needed from the marine environment and particularly from polar and tropical regions.

There are no published marine studies that have tested Lawton's ideas (Lawton 1999) on whether or not the plot of local versus regional species richness shows a direct proportionality (i.e. local species interactions are unimportant in understanding species richness), giving Type I curves. Alternatively, if species interactions are important in setting limits to the number of species present then the plot of local versus regional richness should reveal a negatively accelerating curve with an asymptote, Type II curves. As Lawton states, "the question of where real ecological assemblages lie on the continuum from Type 1 to Type II systems has major implications for community ecology, and the search for patterns and rules in nature". As mentioned earlier, it is not known whether species ranges in the Antarctic are small or whether most species have circumpolar distributions. Thus it is difficult to design tests of Lawton's Type I and Type II models when the regional species pools are so poorly known. However, the fundamental question of whether or not species interactions are important in limiting total species richness is clearly a central issue that urgently needs testing in Antarctica.

Studies of species richness/diversity to area relationships over different spatial scales are needed. As Lawton (1999) pointed out, the lognormal distribution of abundance of individuals among species and among body-size classes is a general pattern found across all assemblages, freshwater, marine and terrestrial. In relation to species abundances, the lognormal distribution implies that most species are rare and occur at low population abundances (Ugland and Gray 1982). Yet the pattern of species accumulation with increasing area has not been adequately studied. Models have been developed to predict total species richness based on accumulation rates of rare species (Chao 1984, 1987), but there are few examples where the models have been applied (but see Gray 2001).

Another problem that needs to be considered is that there are different types of rareness. Species that are restricted to a single site were termed "uniques" or spot endemics by Colwell and Coddington (1994) and species represented by a single individual, "singletons". In a study of the macrofauna inhabiting a coral lagoon, Schlacher et al. (1998) found that whereas the cumulative number of spot endemics showed a rapid approach to an asymptote, total species richness did not. This led them to the conclusions that species with small local densities dominated, that distribution ranges are highly compressed (spot endemics), that boundaries of assemblages overlap, and there is a large turnover and a positive interspecific abundance-range size relationship. For conservation purposes, they concluded that a strategy based on total species richness would lead to missing a significant fraction of rare species, and a strategy based on spot endemism was recommended. In a general ecological framework, it is important to determine the degree of endemism in the Antarctic fauna, and studies need to take a quantitative approach so that spot endemism can be distinguished from singleton species.

In relation to the distribution of body size, there is an important issue of whether or not body-size spectra are similar across latitudinal gradients (Brey and Gerdes 1997). The assumption often made is that such spectra are similar and, therefore, it is possible to compare samples using similar sampling methods (e.g. sieve mesh size for macrofauna). However, it is commonly known that the sediment-living tropical fauna is smaller than boreal and polar species, with the last being largest (Thorson 1957). Such differences may affect comparisons and so body-size spectra need to be studied across latitudinal gradients.

Although Roy et al. (1998) rejected Rapoport's rule of a decrease in species range at low latitudes, studies of species ranges in the Antarctic have not yet provided data to test Rapoport's rule. It is not clear whether species are distributed across the whole Southern Ocean or whether species are confined to restricted areas such as the Weddell, Ross and Davies Seas. Data on distributional records would provide key facts to test whether or not Rapoport's rule needs to be abandoned as some studies suggest (Rohde et al. 1993; Rohde 1996).

The energy input hypothesis as an explanation for the latitudinal gradient of species richness, as proposed by Roy et al. (1998) poses some problems. Roy et al. used sea-surface temperature as a surrogate for energy input and productivity. Yet as Lambshead (1993) pointed out there may not be a gradient of decreasing productivity from tropics to poles. Primary productivity is closely coupled to secondary production in the tropics so that there is little flux to the seabed. Moving polewards, the primary bloom becomes more decoupled from the grazing bloom so that much phytoplankton settles to the seabed. The settlement of material is, however, highly patchy and patchiness of food is greatest in polar areas (Grassle 1989). This patchiness should, over evolutionary time, lead to higher, not lower, species richness (Lambshead 1993). Rosenzweig (1995) showed clearly that the relationship between species richness and productivity is not linear but shows a maximum at intermediate values of productivity in many different ecological systems. Thus the energy hypothesis needs to be analysed more carefully, testing explicit hypotheses before it can be suggested as an explanation for the marine gradient in species richness in the northern hemisphere.

Finally, another important but neglected aspect of marine diversity studies is that of turnover (beta) diversity.

> Whittaker (1975) stated that alpha and beta diversity will be recognised as the consequences of niche diversification and habitat diversification of species, respectively. MacArthur found that alpha diversity for birds, which is closely related to vegetation structure, was not much higher in tropical than in temperate communities of similar structure, but beta diversity increased towards the tropics. Evolution in the tropics acts not to increase alpha diversity but to fit additional species in along environmental gradients by habitat differentiation and narrowed habitat distributions.

Harrison et al. (1992) reported that terrestrial tropical areas have turnover diversity almost an order of magnitude higher than in Britain. Gray (2000) reported turnover diversity values from the Norwegian continental shelf equivalent to those of terrestrial systems in Britain. Whether or not this is true for high-diversity systems such as in the Southern Ocean, or low-diversity systems such as the Arctic Ocean remains to be studied.

Clarke and Lidgard (2000) analysed turnover diversity in bryozoans in the North Atlantic and found no correlation with latitude. However, there was a higher turnover at low latitudes within biogeographical provinces if one excluded data from the Mediterranean Sea. It is important to analyse how turnover diversity varies among marine taxa (as done in a terrestrial system by Harrison et al. 1992). Further studies are needed on whether or not there are variations in turnover diversity with size (meiofauna compared with macrofauna) and whether or not tropical and Antarctic areas show turnover diversity at similar high values to terrestrial systems.

The focus of macroecology is the search for patterns and is a challenging task. As Gaston and Blackburn (1999) pointed out, "macroecology needs the establishment of a more balanced and better documented body of published patterns". I could not agree more. It is easy to shoehorn results into accepted patterns rather than having an open mind to alternatives; the marine latitudinal gradient is a good example. Whilst the gradient in the northern hemisphere seems well documented, more confirmatory evidence is needed. The explanatory hypotheses such as the species:area relationship or the energy-input hypothesis have not been adequately tested. Alternative hypotheses to explain this pattern must be erected and rigorously tested. We do not yet know if a latitudinal gradient exists in the southern hemisphere and exploring this pattern is an important task, where the species-rich Southern Ocean around Antarctica is central. I am convinced that major new patterns will emerge from more detailed studies of diversity in the Southern Ocean, which in turn will create challenges in erecting hypotheses to explain the observed patterns.

Acknowledgements I thank Wolf Arntz for inviting me to give this paper at the symposium and for many helpful comments during the preparation of the paper. Carlos Paredes, Tom Brey and Angelika Brandt are thanked for commenting on the manuscript. Any remaining faults are my own.

References

Abele L, Walters J (1979) Marine benthic diversity: a critique and alternative explanation. J Biogeogr 6:115–126

Arntz WE, Rios C (eds) (1999) Magellan-Antarctic: ecosystems that drifted apart. Sci Mar 63 [Suppl 1]:518

Arntz WE, Gutt J, Klages (1997) Antarctic marine biodiversity. In: Battaglia B, Valencia J, Walton DWH (eds) Antarctic communities: species, structure and survival. Cambridge University Press, Cambridge, pp 3–14

Brattegard T, Holthe T (1997) Distribution of marine, benthic macro-organisms in Norway. A tabulated catalogue. Preliminary edition. Research report no. 1991-1. Directorate for Nature Management, Trondheim, Norway

Brey T, Gerdes D (1997) Is Antarctic biomass really higher than elsewhere? Antarct Sci 9:266–267

Brey T, Klages M, Dahm C, Gorny M, Gutt J, Hain S, Stiller M, Arntz WE, Wägele J-W, Zimmermann A (1994) Antarctic benthic diversity. Nature 368:297

Chao A (1984) Non-parametric estimation of the number of classes in a population. Scand J Stat 11:265–270

Chao A (1987) Estimating the population size for capture-recapture data with unequal catchability. Biometrics 43:783–791

Clarke A (1992) Is there a latitudinal diversity cline in the sea? Trends Ecol Evol 7:286–287

Clarke A, Crame JA (1997) Diversity, latitude and time: patterns in the shallow sea. In: Ormond RJD, Gage JD, Angel MV (eds) Marine biodiversity: patterns and processes. Cambridge University Press, Cambridge, pp 122–147

Clarke A, Lidgard SM (2000) Spatial patterns of diversity in the sea: bryozoan species richness in the North Atlantic. J Anim Ecol

Coleman N, Gason A, Poore GCB (1997) High species richness in the shallow marine waters of south-east Australia. Mar Ecol Prog Ser 154:27–40

Colwell RK, Coddington JA (1994) Estimating terrestrial biodiversity through extrapolation. Philos Trans R Soc Ser B 345:101–118

Cornell HV, Karlson RH (1996) Species richness of reef-building corals determined by local and regional processes. J Anim Ecol 65:233–241

Cowling RM, Gibbs Russell GE, Hoffman MT, Hilton-Taylor C (1992) Patterns of plant species diversity in southern Africa. In: Huntley BJ (ed) Biotic diversity in southern Africa: concepts and conservation. Oxford University Press, Cape Town, pp 19–50

Crame JA (2000) Evolution of taxonomic diversity gradients in the marine realm: evidence from the composition of recent bivalve faunas. Paleobiology 26:188–214

Crame JA, Clarke A (1997) History and marine taxonomic diversity gradients. In: Ormond R, Gage JD, Angel MV (eds) Marine biodiversity: patterns and processes. Cambridge University Press, Cambridge, pp 258–273

Dayton PK (1990) Polar benthos. In: Smith WO (ed) Polar oceanography, Part B: chemistry, biology and geology. Academic Press, San Diego, pp 631–685

Dayton PK, Mordida BJ, Bacon F (1994) Polar marine communities. Am Zool 34:90–99

Dunton K (1992) Arctic biogeography: the paradox of the marine benthic fauna and flora. Trends Ecol Evol 7:183–189

Gage JD (1993) Community structure of the benthos in Scottish sealochs. I. Introduction and species diversity. Mar Biol 14:281–297

Gage JD (1996) Why are there so many species in deep-sea sediments? J Exp Mar Biol Ecol 200:257–286

Gaston KJ (1998) Rarity as double jeopardy. Nature 394:229–230

Gaston KJ, Blackburn TM (1999) A critique for macroecology. Oikos 84:353–368

Grassle JF (1989) Species diversity in deep-sea communities. Trends Ecol Evol 4:12–15

Gray JS (1994) Is deep-sea species diversity really so high: species diversity of the Norwegian continental shelf. Mar Ecol Prog Ser 112:205–209

Gray JS (1997a) Marine biodiversity: patterns, threats and conservation needs. Biodivers Conserv 6:153–175

Gray JS (1997b) Gradients of marine biodiversity. In: Ormond R, Gage JD, Grassle JF (eds) Marine biodiversity: patterns and processes. Cambridge University Press, Cambridge, pp 18–34

Gray JS (2000) The measurement of marine species diversity, with an application to the benthic fauna of the Norwegian continental shelf. J Exp Mar Biol Ecol 250:23–49

Gray JS (2001) Marine Diversity: the paradigms in patterns of species richness examined. Sci Mar 65:41–56

Gray JS, Poore GCB, Ugland KI, Wilson RS, Olsgard F, Johannessen Ø (1997) Coastal and deep-sea benthic diversities compared. Mar Ecol Prog Ser 159:97–103
Harrison S, Ross SJ, Lawton JH (1992) Beta diversity on geographic gradients in Britain. J Anim Ecol 61:151–158
Hessler RR, Sanders HL (1967) Faunal diversity in the deep sea. Deep Sea Res 14:65–78
Hopper SD (1992) Biogeographical aspects of speciation in the southwest Australian flora. Annu Rev Ecol Syst 10:399–422
Kendall MA, Aschan M (1993) Latitudinal gradients in the structure of macrobenthic communities: a comparison of Arctic, temperate and tropical sites. J Exp Mar Biol Ecol 172:157–179
Lambshead PJD (1993) Recent developments in marine benthic biodiversity research. Oceanis 19:15–24
Lawton JH (1999) Are there general laws in ecology? Oikos 84:177–192
MacArthur RH (1972) Geographical ecology. Harper & Row, New York
MacArthur RH, Wilson EO (1967) The theory of island biogeography. Princeton University Press, Princeton, pp 1–203
Magurran A (1988) Ecological diversity and its measurement. Croom Helm, London
Marchant A (1991) The vascular flora of South West Australia. Association of Societies for Growing Australian Plants, 16th Biennial Conference, Perth
Olsgard F, Gray JS (1995) A comprehensive analysis of the effects of offshore oil and gas exploration and production on the benthic communities of the Norwegian continental shelf. Mar Ecol Prog Ser 122:277–306
Pearson TH (1970) The benthic ecology of Loch Linnhe and Loch Eil, a sea-loch system on the west coast of Scotland. I. The physical environment and the distribution of the macrobenthic fauna. J Exp Mar Biol Ecol 5:1–34
Pielou EC (1976) Population and community ecology. Gordon & Breach
Poore GCB, Rainer SF, Spies RB, Ward E (1975) The zoobenthos program in Port Phillip Bay, 1969–73. Fish Wildl Pap Vict 7:1–78
Rapoport EH (1994) Marine and continental biogeography. Philos Trans R Soc Lond Ser B 343:71–78
Rex MA (1981) Community structure in the deep-sea benthos. Annu Rev Ecol Syst 12:331–353
Rex MA (1983) Geographic patterns of species diversity in deep-sea benthos. In: Rowe GT (ed) The seas, vol 8. Wiley, New York, pp 453–472
Rex MA, Stuart CT, Hessler RR, Allen JA, Sanders HL, Wilson GDF (1993) Global-scale patterns of species diversity in the deep-sea benthos. Nature 365:636–639
Rex MA, Etter RJ, Stuart CT (1997) Large-scale patterns of species diversity in the deep-sea benthos. In: Ormond R, Gage JD, Angel MV (eds) Marine biodiversity: patterns and processes. Cambridge University Press, Cambridge, pp 94–121
Rohde K (1996) Rapoport's Rule is a local phenomenon and cannot explain latitudinal gradients in species diversity. Biodivers Lett 3:10–13
Rohde K (1997) The larger area of the tropics does not explain latitudinal gradients in species diversity. Oikos 79:169–172
Rohde K, Hea M, Heap D (1993) Rapoport's Rule does not apply to marine teleosts and cannot explain latitudinal gradients in species richness. Am Nat 142:1–16
Rosenzweig ML (1995) Species diversity in space and time. Cambridge University Press, Cambridge
Roy K, Jablonski D, Valentine JW (1996) Higher taxa in biodiversity studies: patterns from eastern Pacific marine molluscs. Philos Trans R Soc Lond Ser B 351:1605–1613
Roy K, Jablonski D, Valentine JW, Rosenberg G (1998) Marine latitudinal diversity gradients: tests of causal hypotheses. Proc Natl Acad Sci USA 95:3699–3702
Sanders HL (1968) Marine benthic diversity: a comparative study. Am Nat 102:243–282
Schlacher TA, Newell P, Clavier J, Schlacher-Hoenlinger MA, Chevillon C, Britton J (1998) Soft-sediment benthic community structure in a coral reef lagoon – the preeminence of spatial heterogeneity and 'spot endemism'. Mar Ecol Prog Ser 174:159–174
Schluter D, Ricklefs RE (1993) Convergence and the regional component of species diversity. Species diversity. In: Ricklefs RE, Schluter D (eds) Ecological communities: historical and geographical perspectives. University of Chicago Press, pp 230–240
Starmans A (1997) A comparative study of the ecology and biodiversity of the Arctic and Antarctic mega-benthos. Ber Polarforsch 250:1–150
Stehli FG, McAlester AL, Helsey CE (1967) Taxonomic diversity of Recent bivalves and some implications for geology. Geol Soc Am Bull 78:455–466
Thorson G (1957) Treatise on marine ecology and palaeoecology. In: Hedgpeth JW (ed) Mem Geol Soc America. pp 461–534
Ugland KI, Gray JS (1982) Lognormal distributions and the concept of community equilibrium. Oikos 39:171–178
Underwood AJ (1997) Experiments in ecology: their logical design and interpretation using analysis of variance. Cambridge University Press, Cambridge
Whittaker RH (1960) Vegetation of the Siskiyou Mountains, Oregon and California. Ecol Monogr 30:279–338
Whittaker RH (1972) Evolution and measurement of species diversity. Taxon 21:213–251
Whittaker RH (1975) Communities and ecosystems, 2nd edn. Macmillan, London
Wright DH (1983) Species-energy theory: an extension of species-area theory. Oikos 41:496–506

Polar Biol (2001) 24: 839–845
DOI 10.1007/s003000100289

Jean Bouillon · Francesc Pagès · Josep-Maria Gili

New species of benthopelagic hydromedusae from the Weddell Sea

Abstract Four medusa species were collected by an epibenthic sledge during the "Polarstern" ANT XV/3 cruise carried out from January to March 1998 in the eastern Weddell Sea. The specimens were collected in the benthic boundary layer at depths ranging between 1,583 and 2,034 m; 2 of the species collected are new to science. The narcomedusa *Sigiweddellia bathypelagica* gen. nov. et sp. nov. is characterised by two types of marginal tentacles and closed marginal statocysts. The trachymedusa *Voragonema laciniata* sp. nov. (known only from the single holotype) is characterised by the number and irregular shape of the centripetal canals. These findings are the first to report benthopelagic hydromedusae in deep Antarctic waters. Examination of several specimens of *Benthocodon pedunculata* (Bigelow 1913) leads us to move it to the genus *Voragonema* Naumov 1971 because of the clear presence of centripetal expansions in the ring canal.

Introduction

In the oceans, the general pattern of biomass and species number decrease as depth increases is an aspect of marine biodiversity that has received much attention by the scientific community during the last decade (Gray 1997). Even though many examples of such patterns are known for both planktonic and benthic communities (Angel and Baker 1982; Lampitt et al. 1986), other results support a counter-trend that argues for high diversity in deep-sea benthic communities (Grassle 1989). The existence of a rich fauna and a high biomass of pelagic organisms near the deep-sea floor is generally accepted after the investigations conducted during the last decade, although these faunas are only well known in few selected marine areas (Angel 1990). These studies have pointed out that the deep-sea benthopelagic fauna is distinct and perhaps more diverse than the overlying bathy- and mesopelagic faunas (Childress et al. 1989).

One of the most neglected components of the fauna inhabiting the deep-sea benthic boundary layer appears to be the gelatinous zooplankton. The use of new observation techniques such as submersibles, remotely operated vehicles and underwater cameras has allowed documentation of the near-bottom gelatinous fauna in a way that was previously impossible (Larson et al. 1992). One of the most conspicuous organisms of the deep-sea fauna, the medusa, may develop dense concentrations near the bottom (Smith 1982), with quite different populations in comparison with the better-known mesopelagic ones (Mackie 1985).

Although the main scope of the EASIZ (Ecology of Antarctic Sea Ice Zone) Programme is the study of the coastal and shelf communities, the possibility of also investigating the slope benthopelagic fauna was considered an interesting contribution on cruise ANT XV/3, carried out in the Weddell Sea from January to March 1998. Several deep-water stations were sampled by an epibenthic sledge. A set of rare medusae was collected, which represents the first record of benthopelagic medusae in deep Antarctic waters, where the deep-sea medusa fauna is poorly known (Pagès 1997; Pugh et al. 1997). This report identifies the species collected and describes the new ones.

Please note: When citing this article, please use the reference to the original source in the journal Polar Biology as given on the top of this page.

F. Pagès (✉) · J.-M. Gili
Institut de Ciències del Mar (CSIC),
Plaça del Mar s/n, 08039 Barcelona,
Catalonia, Spain
E-mail: fpages@icm.csic.es
Fax: +34-93-2217340

J. Bouillon
Laboratoire de Biologie Marine,
Université Libre de Bruxelles,
Av. F.D. Roosevelt 50, Brussels 1050,
Belgium

Materials and methods

Four medusa species were collected during the "Polarstern" ANT XV/3 cruise, carried out from January to March 1998 in the eastern Weddell Sea. Benthopelagic samples were taken by an epibenthic sledge of 1 m mouth width and equipped with a net of 500 μm mesh size bearing a 300-μm mesh size cod-end. Each station was trawled for 10 min over the seafloor at 1 knot mean velocity (Brandt et al. 1999). The gelatinous specimens collected were preserved in 4% formalin immediately after the sledge arrived onboard.

Results

Anthomedusae, Family Bythotiaridae Maas 1905

Calycopsis borchgrevinki Browne 1910

Material. One specimen collected by an epibenthic sledge on 10 February 1998 at st. 48/142, 74°35.8′S 27°17.4′W; bottom depth 1650 m.

Narcomedusae, Family Cuninidae Bigelow 1913

New diagnosis

Narcomedusae with perradial and undivided manubrial pouches, with or without peripheral canal system, with tentacles leaving umbrella in the central edge of each manubrial pouch, equal in number to pouches; pouches not extending beyond points of origin of tentacles; **with** or without secondary tentacles on umbrella margin; with or without otoporpae; statocysts opened **or enclosed**.

Sigiweddellia gen. nov.

Diagnosis. Cuninidae without otoporpae; with secondary tentacles on the umbrella margin, with marginal sensory clubs enclosed in a polystratified ectodermal vesicle; with a peripheral canal system.

Sigiweddellia benthopelagica sp. nov.

Material. ANT XV/3 cruise: 11 specimens collected by an epibenthic sledge on 4 February 1998 at st. 48/89, 73°27.5′S 22°43.1′W; 1,583 m bottom depth. Five specimens collected by an epibenthic sledge on 8 February 1998 at st. 48/130, 73°23.6′S, 22°09.3′W, 2,034 m depth (Figs. 1, 2, 3).

Additional material examined. ANT XIII/3 cruise, 1 specimen collected by an amphipod trap on 1 March 1996 at st. T6, 70°09′S, 08°16′W; 2,008 m bottom depth. The holotype has been deposited in the Zoologisches Institut und Zoologisches Museum der Universität Hamburg (Germany), where it is entered as C 11654.

The rest of the collection is designated as paratypes, which have been deposited in the Cnidarian Collection of the Institut de Ciències del Mar of Barcelona with the catalogue number.

Etymology. The genus is dedicated to our colleague Sigi Schiel, for her friendship and outstanding contributions to the knowledge of the zooplankton communities of the Weddell Sea. The specific name is related to the apparent benthopelagic habitat, previously unknown in the Narcomedusae.

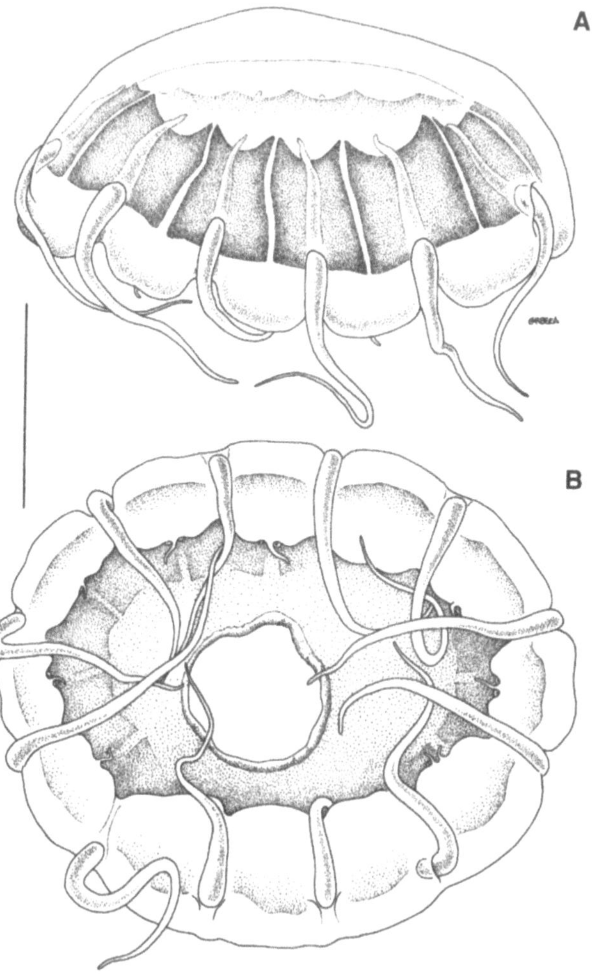

Fig. 1A, B *Sigiweddellia benthopelagica.* Lateral (**A**) and oral (**B**) views of a preserved specimen. Scale: 1 cm

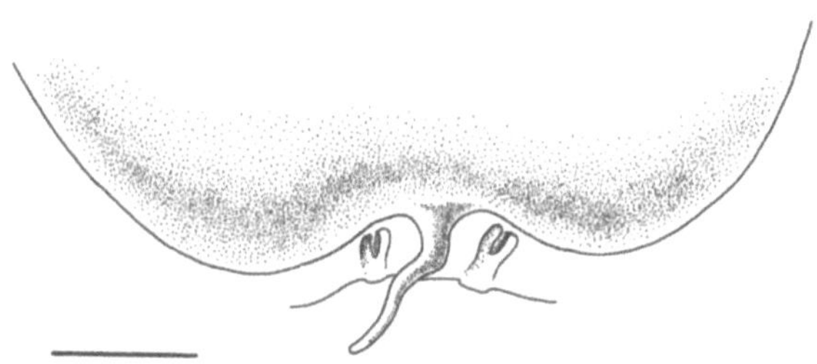

Fig. 2 *Sigiweddellia benthopelagica*: marginal lappet with one secondary tentacle and two multistratified ectodermal vesicles. Scale: 1 mm

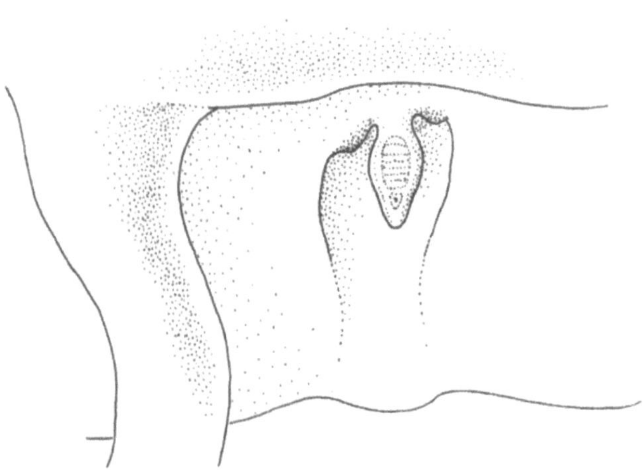

Fig. 3 *Sigiweddellia benthopelagica*: detail of a multistratified ectodermal vesicle that encloses a marginal sensory club. Scale: 0.5 mm

Description. Umbrella up to 30 mm wide and 13 mm high, somewhat flatter than hemispherical, sometimes almost dome-shaped in more contracted specimens (Fig. 1A); mesoglea quite thick and rigid in the centre, thinning slightly at the margin; 10–12 rectangular marginal lappets (Fig. 1B), slightly broader than long, with rounded angles; with a narrow peripheral canal system; manubrium largely conical, with 10–12 rectangular manubrial pouches with elevated, deep, lateral walls (parallepipedal) and lobed summit in mature specimens, the pouches very close together and only separated by deep, very narrow clefts; endoderm of the gastric pouches with digestive cells full of numerous excretory vacuoles with brown pigments and glandular cells; mouth circular, rather small, 7 mm in diameter, with a thick, dense edge formed by numerous glandular granular cells; velum well developed; 10–12 primary marginal tentacles issuing from umbrella at the upper end of peronia, in the central edge of each manubrial pouch, with root-like endodermal continuations in the apical mesoglea extending up to three-quarters length of marginal pouches; the ectoderm of the manubrial pouches is rich in cnidocysts of two types, apotrichous isorhizas and the other type not determined; 1 or 2 short secondary tentacles at the centre of the margin of each lappet (Figs. 1, 2), some of them sometimes missing, with conical base, sometimes coiled, each one on the central margin of each lappet, with numerous egg-shaped cnidocysts; gonads as irregular longitudinal folds of the manubrial pouches; no otoporpae; 2 marginal sensory clubs, 1 on each side of the secondary tentacles (Fig. 2), enclosed in a multistratified ectodermal vesicle (Fig. 3); manubrium and manubrial pouches coloured brownish by excretory vacuoles.
Diagnosis. The same as the genus.
Remarks. This species undoubtedly belongs to the family Cuninidae by its perradial manubrial pouches. It is nevertheless distinct from the genera *Cunina*, ?*Cunissa* and *Solmissus* by the presence of secondary tentacles and closed marginal sensory clubs; the latter morphological feature is unique among Narcomedusae and the creation of a new genus is thus proposed. This genus has another particular feature: all tissues show a very thick and meaty consistency in comparison with the softness and fragility of the rest of the known Cuninidae.

Trachymedusae, Family Halicreatidae

Botrynema brucei Browne 1908

Material. One specimen collected by an epibenthic sledge on 4 February 1998 at st. 48/089, 73°27.5′S, 22°43.1′W; bottom depth 1,583 m.

Family Rhopalonematidae Russell 1953, Genus *Voragonema* Naumov 1971

Voragonema laciniata sp. nov.

Material. One specimen collected by an epibenthic sledge on 4 February 1997 at st. 48/089, 73°27.5′S, 22°43.1′W; bottom depth 1,583 m. The holotype has been deposited in the Zoologisches Institut und Zoologisches Museum der Universität Hamburg (Germany), where it is entered as C 11653 (Figs. 4, 5).
Additional material examined. Holotype of *Benthocodon hyalinus* (Larson and Harbison 1990); USNM 87603, deposited at the Smithsonian Institution, National Museum of Natural History, Washington, D.C. Six specimens of *Benthocodon pedunculata* (Bigelow 1913) collected in Monterey Canyon, California, United States of America, and kindly supplied by George Matsumoto. Kodachrome photograph of *Benthocodon pedunculata* taken by Claudia E. Mills and collected in the Bahamas.
Etymology laciniata (Latin) inspired by the jagged edge of the ring canal.

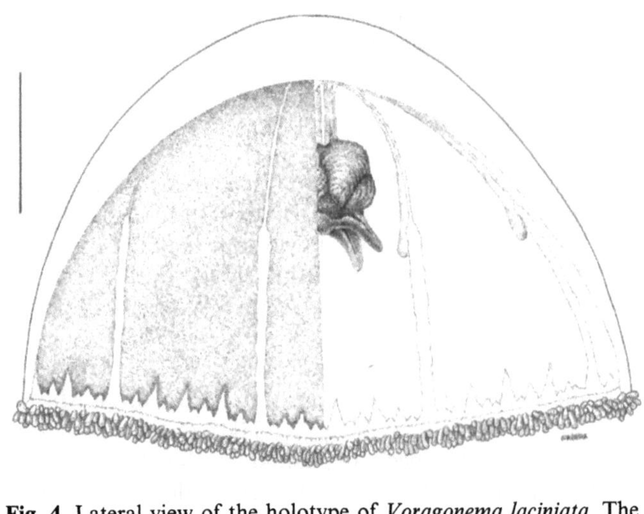

Fig. 4 Lateral view of the holotype of *Voragonema laciniata*. The *left half* shows the dark subumbrella, which has been deleted in the *right half* to illustrate the peduncle, the manubrium and the gonads. Scale: 1 cm

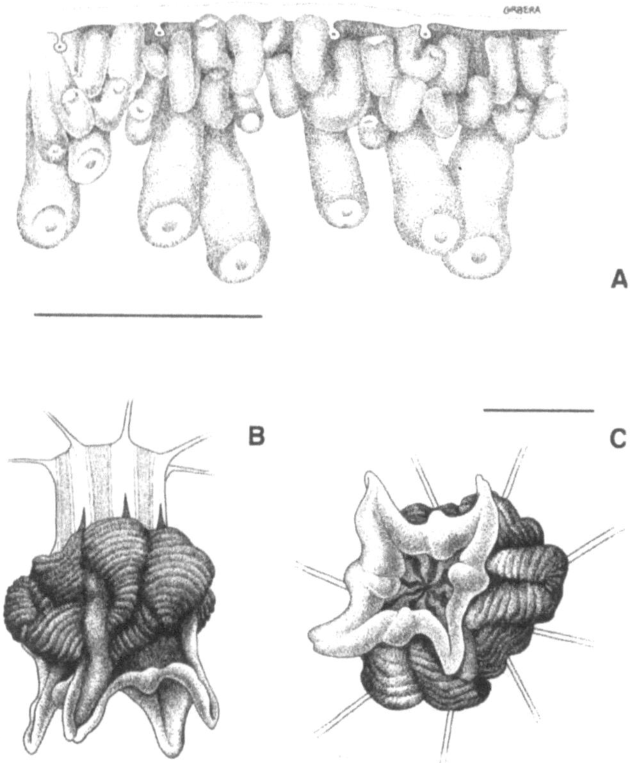

Fig. 5A–C *Voragonema laciniata*. **A** detail of the bell margin showing the proximal part of marginal tentacles superimposed in several rows together with the free statocysts placed among them; drawn from the inside, looking out. Scale: 5 mm. **B** Infralateral view of the peduncle and manubrium on which the gastric pouches stand out. Scale: 5 mm. **C** Oral view of the manubrium showing the perradial lips. Scale: 5 mm

Diagnosis. *Voragonema* with subumbrella brownish-coloured, circular canal with irregular conical centripetal expansions; about 1,200 marginal tentacles.

Description. Umbrella almost as high (30 mm) as wide (35 mm) (Fig. 4); mesoglea 4 mm thick in the apex and tapering towards the umbrella margin; exumbrella transparent with fine ridges, subumbrella dark-red-brownish, with circular and crossed oblique-radial muscles in the subumbrella; dark-red-brownish colour due to pigmented cells.

Gastric peduncle large (Figs. 4, 5B), circular from above and slightly octogonal from the subumbrellar cavity (Fig. 5B), 4.2 mm in diameter and 5.0 mm in height; ectodermal walls of the gastric peduncle dark-red.

Radial canals 8, white-cream coloured, straight; peduncular part 5.0 mm long and 1.5 mm wide; proximal half of the subumbrellar part narrow, 0.5 mm wide, distal half broader, 1.3 mm wide.

Manubrium short, broad, square, dark-red, slightly longer than the peduncle, with thick muscular walls; mouth with four prominent and elongated perradial lips and swollen interradial edges; manubrial cavity divided into eight longitudinally folded gastric pouches that open into the radial canals; peduncle and manubrium height is half that of the subumbrella (Fig. 5B, C).

Gonads extending along the proximal half of the radial canals, flattened, with distal part slightly pendent.

Circular canal whitish and wide, with numerous centripetal, conical and irregular expansions; bell margin thickened by a conspicuous whitish ring of cnidocysts; velum dark-red, 6.0 mm wide, with marked circular muscles.

Tentacles with reddish ectodermal pigment, superimposed in 5–6 rows, each with an adnate base, about 150 in number per octant; size increases from the adaxial to the abaxial side; younger tentacles are 2.0 mm in length and 0.15 mm in diameter; maximum length unknown; numerous free statocysts with a single large statolith, placed among the marginal tentacles (Fig. 5A).

Remarks. This new species belongs to the genus *Voragonema* Naumov 1971 because of the presence of centripetal expansions arising from the ring canal. *V. laciniata* differs markedly from *V. profundicola* by the coloured sumbumbrellar surface, a higher number of marginal tentacles and by the number and irregular shape of the centripetal canals (Table 1). The description of *V. profundicola* was rather poor (Naumov 1971) and some of the characters (e.g. the shape of the centripetal canals) are only known from the holotype illustration, which contains little detail.

In this new species, the presence of septal pouches in the manubrial cavity presumably increases the absorption surface of the endoderm. This morphological feature is physiologically similar to the gastric digitations observed by Bouillon (1988) in the anthomedusan genus *Koellikerina*. Likewise, the length of larger tentacles is unknown since none of them were recovered. However, all tentacles are cut off at the same level and their bases show identical structure; the ectoderm seems to cover the endodermal layer of the hollow tentacle. This feature suggests that *V. laciniata* is able to autotomise the marginal tentacles during adverse conditions, as has been observed in other Trachymedusae like *Colobonema sericeum* and *Benthocodon pedunculata* (Larson and Harbison 1990; Matsumoto et al. 1997), as well as in most other trawl-collected Trachymedusae.

The strong musculature of the wide velum suggests that *V. laciniata* is at least periodically a powerful swimmer. Matsumoto et al. (1997) observed in situ that *Benthocodon pedunculata*, a closely related species, performs upwards-swimming and sinking bouts. The genus *Benthocodon* was created for the new Antarctic species *Benthocodon hyalinus* (Larson and Harbison 1990). This genus differs mostly from the closely related genus *Crossota* by the presence of a gastric mesoglean peduncle non-existent in the genus *Crossota*, with the exception of the species *Crossota pedunculata*. This latter species was therefore moved to the genus *Benthocodon*. The strong morphological similarities between *Benthocodon* and *Voragonema*, already noted by Larson and Harbison (1990), caused us to investigate the main differences between the genera.

Microscopical examination of six specimens of *Benthocodon pedunculata*, collected in Monterey Bay (Cali-

Table 1 Main morphological features of the three species of the genus *Voragonema* and *Benthocodon hyalinus*. Key interspecific differences are in **bold**

	Exumbrella	Subumbrella	Radial canals	Peduncle	Manubrium	Mouth	Gonads	Centripetal canals	Tentacles	Statocysts	Velum	Distribution and depth	Main references
Voragonema profundicola Naumov 1971	10 mm high, 15 mm wide; transparent	**Transparent,** hemispherical	Eight, slightly undulated along the peduncle, uniformly narrow along the subumbrella	Short, solid, cylindrical in section	Narrow	With four simple oral lips	Unknown	**Eight per octant, triangle-shape, regular in length and width**	About 500	Unknown	Rather wide	Kurile-Kamchatka Trench, 8,700–6,800 m	Naumov (1971)
V. pedunculata (Bigelow 1913)	Up to 31 mm high and 41 mm in diameter; transparent with a little rusty pigment, with many fine meridional furrows	**Dark-reddish-brown,** more conical than hemispherical	Eight, white-cream, proximal half narrow, distal half broader	Short, small	Short, broad, square, dark-red, folded gastric pouches	With four flared lips	White-cream, thin, proximal half attached along the radial canal, distal half free, pendent	**Eight per octant, triangle-shape, regular in length and width**	Superimposed in 4–5 rows; 1,000–2,000 in number; 2 types: outermost, solid and heavily pigmented; inner, hollow, shorter and thinner	Up to 30 per octant	6.0 mm wide	Northwest USA, 90 m; Bahamas, 830–910 m; Virgin Is. 870–900 m; Monterey Canyon, California, 175–975 m	Bigelow (1913); Larson et al. (1992); Matsumoto et al. (1997); Wrobel and Mills (1998)
V. laciniata Bouillon, Pagès and Gili	Almost as high (30 mm) as wide (35 mm in diameter); transparent with numerous fine meridional ridges	Dark-red-brownish, circular and crossed oblique-radial muscles	Eight, white-cream, proximal half narrow, distal half broader	Large, circular from above and slightly octogonal from below	Short, broad, square, dark-red, thick muscular walls; cavity divided by eight longitudinal and folded gastric pouches	With four prominent elongated, perradial lips and with prominent interradial swellings	Whitish, thin, proximal half attached along the radial canal, distal half free, pendent	**11–13, usually 12 per octant, triangle-shape, irregular in length and width**	Superimposed in 5–6 rows, about 1,200 in number, 2 types: outermost row, solid, with adnate base; shorter and thinner in the innermost rows	Free, numerous, with a single and large statolith	6.0 mm wide	Weddell Sea, 1,583 m	This study
Benthocodon pedunculata Larson and Harbison 1990	Dome-shaped, 40 mm in diameter, 25 mm high; thick, **smooth**	Transparent, hemispherical	Eight, proximal half narrow, distal half broader	Cylindrical, 1/4 the length of the manubrium or longer	Tubular, extends to the margin in living specimens; reddish-brown	With four flared oral lips with smooth margin	White, as wavy bands along most of the radial canals	**Absent**	Superimposed in at least 5 rows, more than 800 in number, 1 type	Not seen	Thin, fairly broad	McMurdo Sound, Ross Sea, 20–0 m depth	Larson and Harbison (1990)

fornia) and kindly supplied by Dr. George Matsumoto, showed the existence of many short centripetal expansions in the ring canal, a conspicuous feature apparently not detected in previous studies (Larson and Harbison 1990; Larson et al. 1992; Matsumoto et al. 1997; Wrobel and Mills 1998), although visible in one photograph of a specimen collected in Bahamas waters and produced by Claudia E. Mills (Fig. 6). This finding demonstrates that this species actually belongs to the genus *Voragonema* and we propose to move the species *pedunculata* to this genus. Unfortunately, the holotype of *V. profundicola* could not be examined for comparison.

The diagnosis of the genus *Benthocodon* mentions that centripetal canals are absent (Larson and Harbison 1990). The holotype of *Benthocodon hyalinus*, which is quite damaged, showed no tracks of centripetal expansions of the ring canal when examined under the microscope. The lack of centripetal expansions suggests keeping *Benthocodon* as a valid genus because *Voragonema* is characterised by this uncommon feature, not shown in the rest of the Rhopalonematidae. An excellent photograph of a living specimen of *Benthocodon hyalinus* published under the name *Arctapodema* (Curtsinger 1986) makes evident its strong morphological similarity in comparison to *V. pedunculata* (Fig. 6, Larson et al. 1992). However, the presence of centripetal expansions in the ring canal is an outstanding feature, unknown in other Trachymedusae except the family Geryoniidae, which warrants the maintenance of both genera. *Voragonema* and *Benthocodon* are quite different in comparison to other genera in the family Rhopalonematidae for the very large number of tentacles and the separation of small downward-pointing tentacles and larger outward-pointing ones, seemingly an adaptation to a benthopelagic lifestyle. *Crossota* and *Vampyrocrossota* are two genera of Rhopalonematidae morphologically similar to *Voragonema* and *Benthocodon*. However, *Crossota* has pendent gonads (Wrobel and Mills 1998) and *Vampyrocrossota* shows the tentacles arranged in a single row (Thuesen 1993).

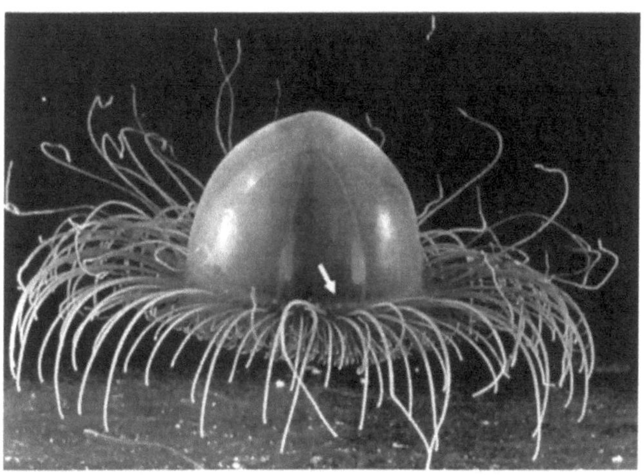

Fig. 6 Specimen of *Voragonema* (*Benthocodon*) *pedunculata* collected in Bahamas waters on 5 November 1989 and photographed in an aquarium on board ship. Note the regularly spaced triangular centripetal expansions on the ring canal. Photograph by Claudia E. Mills

In summary, the genus *Voragonema* comprises three species at present: *V. profundicola* Naumov 1971, *V. pedunculata* (Bigelow 1913) and the new species *V. laciniata*. The main morphological features that distinguish them are the colour of the subumbrella and the number of centripetal canals (Table 1).

Outlook

The bathypelagic fauna has a close relationship with the benthic communities on which many planktonic crustaceans (Cartes 1998) and gelatinous organisms (Matsumoto et al. 1997) feed. Interest in the role of this bathy-benthopelagic fauna in plankton-benthos coupling processes is increasing. The study of these processes is one of the main objectives of the EASIZ-SCAR Programme. However, while pelagic-benthic interactions are increasingly well studied on the high-Antarctic shelf (Gili et al. 2001), their importance in the deep sea is as yet unknown. This contribution is a first step towards understanding these processes in deep water.

Acknowledgements We are greatly indebted to Claudia E. Mills for valuable comments on the content, style and language of the manuscript, permission to reproduce her photograph of *Voragonema (Benthocodon) pedunculata*, as well as her field notes on this species. Valuable comments by Nando Boero and Peter Schuchert are acknowledged. We thank the officers and crew of the FS "Polarstern" and many colleagues, especially Covadonga Orejas, Pablo López-González, Wolf Arntz and Angelika Brandt, for their help during the EASIZ XV/3 (EASIZ II) cruise. We are grateful to George Matsumoto and Jenny Paduan (Monterey Bay Aquarium Research Institute) for the gift of several specimens of *Voragonema pedunculata*. We thank Steve Cairns and the Smithsonian Institution for the loan of the holotype of *Benthocodon hyalinus*. Financial support for this study was provided by a CICYT Spanish grant (ANT97-1533-E). J.B. acknowledges the "Fondation David et Alice Van Buuren". The figures were drawn by Jordi Corbera.

References

Angel MV (1990) Life in the benthic boundary layer: connections to the mid-water and sea floor. Philos Trans R Soc Lond A 331:15–28

Angel MV, Baker AC (1982) Vertical distribution of the standing crop of plankton and micronekton at three stations in the northeast Atlantic. Biol Oceanogr 2:1–30

Bigelow HB (1913) Medusae and Siphonophorae collected by the U.S. Fisheries Steamer "Albatros" in the North Western Pacific, 1906. Proc US Nat Mus 44:1–119

Bouillon J (1988) Notes additionelles sur les hydroméduses de la mer de Bismarck (Hydrozoa-Cnidaria) II. Indo-Malay Zool 5:87–99

Brandt A, Hilbig B, Linse K, Wegener G (1999) Biodiversity and zoogeography of Mollusca, Polychaeta, and Crustacea Peracarida. In: Arntz WE, Gutt J (eds) The Expedition ANT XV/3 (EASIZ II) of RV "Polarstern". Ber Polarforsch 301:135–149

Cartes JE (1998) Dynamics of the bathyal benthic boundary layer in the northwestern Mediterranean: depth and temporal variations in macrofaunal-megafaunal communities and their possible connections within deep-sea trophic webs. Prog Oceanogr 41:111–139

Childress JJ, Gluck DL, Carney RS, Gowing MM (1989) Benthopelagic biomass distribution in a deep-sea benthic boundary layer dominated by gelatinous organisms. Limnol Oceanogr 34:913–930

Curtsinger B (1986) Under Antarctic ice. Natl Geogr 169:494–511

Gili J-M, Coma R, Orejas C, López-González P, Zabala M (2001) Are Antarctic suspension-feeding communities different from those elswhere in the world? Polar Biol 24:473–485

Grassle JF (1989) Species diversity in deep-sea communities. Trend Ecol Evol 4:12–15

Gray JS (1997) Marine biodiversity: patterns, threats and conservation needs. Biodivers Conserv 6:153–175

Lampitt RS, Billett DSM, Rice AL (1986) Biomass of the invertebrate megabenthos from 500 to 4100 m in the northeast Atlantic Ocean. Mar Biol 93:69–81

Larson RJ, Harbison GR (1990) Medusae from McMurdo Sound, Ross Sea, including the descriptions of two new species, *Leuckartiara brownei* and *Benthocodon hyalinus*. Polar Biol 11:19–25

Larson RJ, Matsumoto GI, Madin LP, Lewis LM (1992) Deep-sea benthic and benthopelagic medusae, recent observations from submersibles and a remotely operated vehicle. Bull Mar Sci 51:277–286

Mackie GO (1985) Midwater macroplankton of British Columbia studied by submersible PISCES IV. J Plankton Res 7:753–777

Matsumoto G, Baxter C, Chen EH (1997) Observations of the deep-sea trachymedusa *Benthocodon pedunculata*. Invertebr Biol 116:17–25

Naumov DV (1971) Hydromedusae and scyphomedusae from the Kurile-Kamchatka trench. Tr Inst Okeanol 92:9–17

Pagès F (1997) The gelatinous zooplankton in the pelagic system of the Southern Ocean: a review. Ann Inst Océanogr Paris 73:139–158

Pugh PR, Pagès F, Boorman B (1997) Vertical distribution and abundance of pelagic cnidarians in the eastern Weddell Sea, Antarctica. J Mar Biol Assoc UK 77:341–360

Smith KL (1982) Zooplankton at a bathyal benthic boundary layer: in situ rates of oxygen consumption and ammonium excretion. Limnol Oceanogr 27:461–471

Thuesen EV (1993) *Vampyrocrossota childressi*, a new genus and species of black medusa from the bathypelagic zone off California (Cnidaria Trachymedusae: Rhopalonematidae). Proc Biol Soc Wash 106:190–194

Wrobel D, Mills CE (1998) Pacific coast pelagic invertebrates: a guide to the common gelatinous animals. Sea Challengers and Monterey Bay Aquarium

Polar Biol (2001) 24: 832–838
DOI 10.1007/s003000100288

A. Louise Allcock · Uwe Piatkowski
Paul G.K. Rodhouse · John P. Thorpe

A study on octopodids from the eastern Weddell Sea, Antarctica

Abstract A comprehensive study of octopodids in the Weddell Sea was conducted during cruise ANT XIII/3 of RV "Polarstern". The study yielded eight species of incirrate octopodids, three of which were undescribed. There appeared to be no differences in octopodid species assemblage in the two sampling areas (Kapp Norvegia and south of Drescher Inlet), supporting the theory that there is a single macrobenthic assemblage on the eastern shelf of the Weddell Sea. Six of the octopodid species belonged to the genus *Pareledone*, illustrating the extensive radiation of this endemic genus within the Southern Ocean. The fragmented nature of suitable habitats and disturbance caused by glaciation cycles are proposed as mechanisms for this radiation. Comparative data illustrate that both the number of octopodid species present and their abundance are greater at higher latitudes than close to the Antarctic Polar Frontal Zone.

Introduction

Although 16 species of octopodids are described from the Southern Ocean (Table 1), they have received little scientific attention that is not taxonomic in nature. Indeed, because of confusion in the taxonomy, it has been almost impossible to make the accurate identifications necessary for ecological, physiological and behavioural studies. In addition, research has been concentrated in particular areas of the Southern Ocean: the eastern sector of Antarctica near Prydz Bay, the Antarctic Peninsula and the Scotia Sea. The Scotia Sea is defined as the body of water bounded by the Scotia Ridge and the extension of the Shackleton Fracture Zone to the southern tip of South America (Locarnini et al. 1993). It is separated from the Weddell Sea by the Weddell-Scotia Confluence whose waters lack the prominent subsurface temperature and salinity maxima that characterise the Circumpolar Deep Water found on either side (Whitworth et al. 1994). Prior to the present study, there were few records of octopodids in the Weddell Sea, and identification was hampered by poor preservation (Voß 1988).

Off South Georgia, an island on the Scotia Arc just south of the Antarctic Polar Frontal Zone (APFZ), octopodid beaks are present in the stomach contents of, for example, southern elephant seals, *Mirounga leonina* Linnaeus 1758 (Rodhouse et al. 1992), black-browed albatrosses, *Diomedea melanophris* (Temmink 1928), Rodhouse 1990) and blue-eyed shags, *Phalacrocorax atriceps nivalis* Falla 1937 (Wanless et al. 1992). Although these beaks were not identified to species level, they suggest that octopodids may be of ecological importance to some of the large Antarctic predators. Should the Weddell Sea cephalopod fauna prove to be abundant and diverse, then an important component of the ecosystem has, up until now, been overlooked.

The dominant genera in other shallow-water areas of the Southern Ocean appear to be *Pareledone* and *Megaleledone* (Allcock et al. 1997; Kuehl 1988; Lu and Stranks 1994). Both genera are found only south of the APFZ. Specimens that have been recorded farther north, [e.g. *Pareledone turqueti* (Joubin 1905) recorded off Brazil by the Terra Nova Expedition; Massy 1916] appear to have been misidentified. A taxonomic review

Please note: When citing this article, please use the reference to the original source in the journal Polar Biology as given on the top of this page.

A.L. Allcock (✉) · J.P. Thorpe
Port Erin Marine Laboratory, University of Liverpool,
Port Erin, Isle of Man IM9 6JA, UK

U. Piatkowski
Institut für Meereskunde, Düsternbrooker Weg 20,
24105 Kiel, Germany

P.G.K. Rodhouse
British Antarctic Survey, High Cross, Madingley Road,
Cambridge CB3 0ET, UK

Present address: A.L. Allcock
National Museums of Scotland,
Chambers Street, Edinburgh EH1 1JF, UK
e-mail: louise@nms.ac.uk,
Fax: +44-131-2204819

Table 1 Octopodidae currently described from the Southern Ocean

Subfamily	Genus	Species	Junior synonym	Weddell Sea records
Eledoninae	*Pareledone*	*charcoti* (Joubin 1905)	*aurorae* (Berry 1917)	Voß 1988, present study
		harrissoni (Berry 1917)	*antarctica* (Thiele 1920)	
		turqueti (Joubin 1905)		Voß 1988, present study
		prydzensis Lu and Stranks 1994		
		framensis Lu and Stranks 1994		
		adeliana (Berry 1917)	*umitakae* Taki 1961	
		polymorpha (Robson 1930)		Voß 1988, present study
	Megaleledone	*senoi* Taki 1961		Present study
Graneledoninae	*Graneledone*	*setebos* Robson 1932		
		antarctica Voss 1976		
	Bentheledone	*albida* (Berry 1917)		
		rotunda (Hoyle 1885)		
	Thaumeledone	*gunteri* Robson 1930		
		brevis (Hoyle 1885)		
Bathypolypodinae	*Benthoctopus*	*levis* (Hoyle 1885)		Present study
		thielei Robson 1932		

of *Pareledone* was published by Lu and Stranks (1994) following extensive fishing in eastern Antarctic waters. Of the 12 *Pareledone* species that have been described, 3 have been synonymised (Table 1). Two others, *Pareledone carlgreni* (Thore 1945) and *Pareledone nigra* (Hoyle 1910) (neither of which is found in the Southern Ocean), are being removed from the genus (C.F.E. Roper, personal communication). Seven valid species remain.

Pareledone framensis (Lu and Stranks 1994), *Pareledone prydzensis* (Lu and Stranks 1994), *Pareledone harrissoni* (Berry 1917) and *Pareledone adeliana* (Berry 1917) appear to be restricted to eastern Antarctica. *Pareledone turqueti* and *Pareledone polymorpha* (Robson 1930) are likewise thought to be restricted to western Antarctica, whilst *Pareledone charcoti* (Joubin 1905) is thought to be the only circumpolar species (Lu and Stranks 1994).

The other prominent octopodid species in shallow Antarctic waters, *Megaleledone senoi* (Taki 1961), is distributed in the Indo-Atlantic sector of the Southern Ocean. Although additional species are recognised from the area (Table 1), these are restricted to deeper waters and their taxonomy is, in many cases, still extremely confused. The aim of this paper is to present basic ecological data on species composition and abundance in shallower waters (< 1,000 m) in the Weddell Sea.

Materials and methods

Octopods were collected during the ANT XIII/3 cruise of RV "Polarstern" to the Weddell Sea in February 1996. The sampling programme was planned with the knowledge that most of the faunal components of the area were familiar (Arntz and Gutt 1997), and investigations concentrated on two main areas, using a variety of gear types. A combination of up to 17 different gear types was launched at 33 stations (see Arntz and Gutt 1997 for full details). The gear types included three trawl nets that yielded octopodids: a standard Agassiz trawl (AGT); a benthopelagic net (BPN); and a bottom trawl (GSN). The GSN was a commercial-scale 47-m headline bottom trawl with an effective mouth opening of 22.5 m by 3 m and a cod-end mesh size of 20 mm. It was the same net as that used 10 months later in a sampling programme off the Antarctic Peninsula (ANT XIV/2), and hence comparison between these two data sets was facilitated.

Octopodids were collected regardless of the gear that yielded them. They were identified to species and measured. Catch per unit effort (CPUE) was calculated as the number of octopodids caught in each haul divided by the duration of the trawl in minutes. The cumulative number of species captured was plotted against total trawl time to assess whether it was likely that all the species present were captured. Trawls were plotted in sequential order.

Multi-dimensional scaling (MDS) using the PRIMER statistical package (Clarke and Warwick 1994) was performed to assess whether the depth bands chosen were appropriate. Similarities between the compositions of a pair of samples (i and l) were calculated using the Bray-Curtis coefficient, Sil, where $Sil = 100^*(1 - (\mathrm{sum}|yij - ylj|/(\mathrm{sum}(yij) + \mathrm{sum}(ylj))))$ and where yij and ylj are the counts for species j in samples i and l, respectively and n is the total number of species recorded. The Bray-Curtis coefficient is equal to 100 if the samples are identical, but equal to zero if there are no species in common. Joint absences do not affect the similarity between samples. Given a similarity matrix defined by Bray-Curtis coefficients, we used the subroutine ANOSIM to test if communities were different. This procedure compares the average rank similarity within predefined groups with the average similarity between groups. Values close to 1 indicate that there is some discrimination between the groups. The significance of the between-group to within-group comparison can be tested using a randomisation test (Clarke and Warwick 1994). This test arbitrarily assigns samples to the predefined groups and generates a test statistic distribution, which indicates whether the measured discrimination statistic is likely to have risen by chance alone.

Results

The sampling programme yielded 446 octopodids: 294 from off Kapp Norvegia and 152 from southwest of the Drescher Inlet (Fig. 1). A total of 84 specimens were caught in a single benthopelagic net that unintentionally touched the sea floor; 22 specimens were caught in a total of 5 Agassiz trawls, and 1 specimen was caught in each of the epibenthic sled and multibox corer. The remaining specimens (338) were taken in a series of 10 bottom trawls, the majority of which were in the more southerly sampling area (Table 2).

A single cirrate species, *Cirroctopus glacialis* (Robson 1930), and eight species of incirrate octopodids were

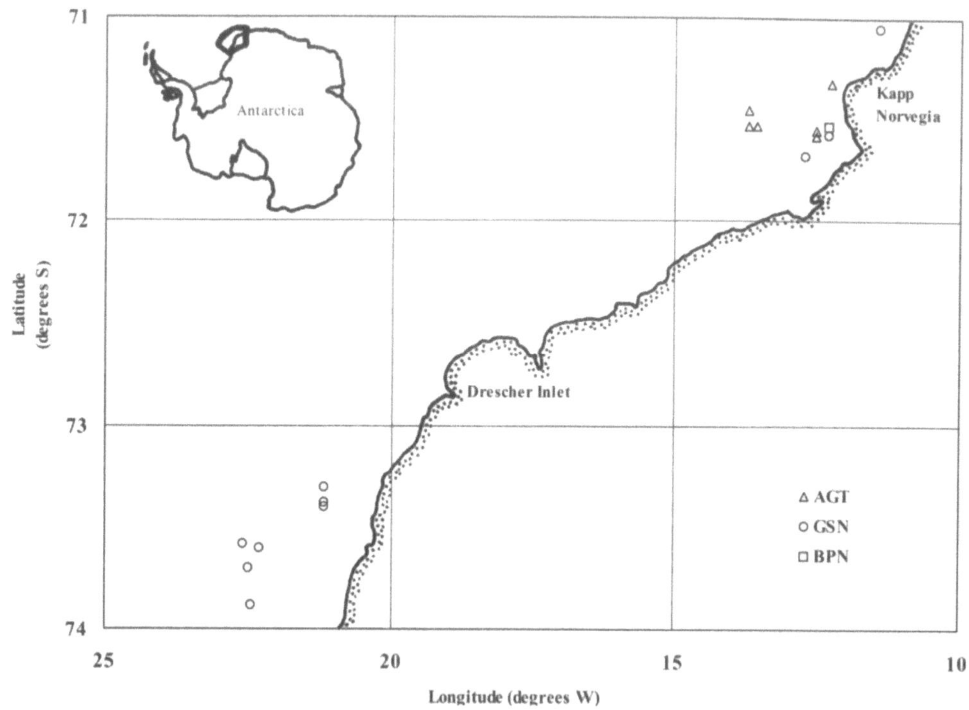

Fig. 1 Sampling locations in the Weddell Sea during cruise ANT XIII/3 of RV "Polarstern"

identified from the catches. Six species belonged to the endemic genus *Pareledone* (*Pareledone turqueti*, *Pareledone charcoti*, *Pareledone polymorpha* and three undescribed species); one species belonged to the endemic monotypic genus *Megaleledone* and one belonged to the taxonomically confused genus *Benthoctopus*.

No major differences in community assemblage between the two main sampling areas were obvious. All the species were found in both areas with the exception of *Benthoctopus* cf. *levis*. The genus *Benthoctopus* is generally associated with increased bottom depth and the species' apparent absence from Kapp Norvegia may be an artifact caused by lack of fishing at appropriate depths.

Because it is well known that community composition is affected by depth (Brey and Clarke 1993), the trawls were split into arbitrary depth bands in an attempt to eliminate this confounding factor from the data (Table 3). Trawls that spanned the boundary between two depth ranges were assigned to the deeper band. The necessity for grouping data into broad ranges is further illustrated by the fact that three of the eight species appear to be restricted in their distribution to two of the three depth bands, while a fourth (*Pareledone* sp. B) is restricted to a single depth band.

Table 2 Number of gear deployments in each area. Figures in *parentheses* indicate how many deployments captured octopodids

Gear	Off Kapp Norvegia	South of Drescher Inlet
Bottom trawl	3 (3)	7 (7)
Agassiz trawl	10 (5)	1 (0)
Benthopelagic trawl	1 (1)	3 (0)
Epibenthic sled	5 (1)	0 (0)
Multibox corer	31 (1)	3 (0)

The catch data were root transformed and an MDS plot (Fig. 2) was generated. The discrete groupings of the components of the three depth bands suggest that these bands are appropriate, as does the ANOSIM statistic $R = 0.808$. The latter equates to a probability of there being a difference in community composition between the depth bands of $P = 0.001$, which is significant.

Three of the ten bottom trawls were located at the more northerly sampling area off Kapp Norvegia, and these are indicated by filled-in symbols on the MDS plot (Fig. 2). The wide dispersal of these points on the MDS plot supports the suggestion that there is no difference in community assemblage between Kapp Norvegia and the more southerly sampling area. Furthermore, when an ANOSIM statistic was calculated to examine any differences that might exist between these two areas, the value of $R = 0.190$ was not significant ($P = 0.167$). A significant result would be unlikely considering the lack of replication at each depth and location.

The grouping into depth bands allowed the data to be compared with those from a more extensive sampling programme off the Antarctic Peninsula (Allcock 1997; Piatkowski et al. 1998) which used the same commercial bottom trawl. At the two shallower depths, the Antarctic Peninsula appears more speciose than the Weddell Sea (Fig. 3). Five species were captured in the Weddell Sea at 200–400 m depth compared with 10 species off the Antarctic Peninsula. Seven species were captured in the Weddell Sea at 400–600 m depth compared with 11 species off the Antarctic Peninsula. In the deeper band (600–800 m) both curves are similar in shape, suggesting that at

Table 3 Numbers of each species of octopodid caught during each of ten trawls with a commercial bottom trawl in the Weddell Sea

Depth band	200–400 m			400–600 m				600–1000 m			Total
Depth of trawl (m)	227–232	242–246	333–338	428–446	457–459	465–468	462–481	574–604	620–640	850–859	
Pareledone charcoti	35	8	8	6	2	1	33	2	1	2	98
Pareledone sp. A	10	19	1			1	2				33
Pareledone sp. B				6	2	3	5				16
Pareledone turqueti	16	17	9	4	2	5	8		2		63
Pareledone sp. C				10	2	8	2	1	7	2	32
Pareledone polymorpha	10		5	4	3	6	52				80
Megaleledone senoi	2	3			1		5	3			14
Benthoctopus cf. levis										2	2
Total number of specimens	73	47	23	30	12	24	107	6	10	6	338
Trawl duration (mins)	16	15	15	16	15	16	26	14	15	15	163

these depths the Weddell Sea and the Antarctic Peninsula may support a similar number of octopodid species, although further survey work is required in the Weddell Sea to verify this.

The overall abundance of octopodids in the Weddell Sea and off the Antarctic Peninsula (in terms of catch per unit effort to standardise for different trawl durations) was also compared for each depth band using t-tests. No significant differences ($P < 0.05$) in CPUE between the Weddell Sea and the Antarctic Peninsula were found at any of the depth ranges (Fig. 4). Multiple t-testing was used in preference to ANOVA to avoid the problems inherent in an unbalanced design, as we were not interested in the interaction term. Whilst multiple hypothesis testing increases the probability of a type I error, the latter effect is only important when the null hypothesis is rejected. Levene's test for equality of variances was insignificant in each case and therefore equal variances were assumed throughout.

Size-frequency analysis showed that the most abundant size class was 21–30 mm mantle length (Fig. 5).

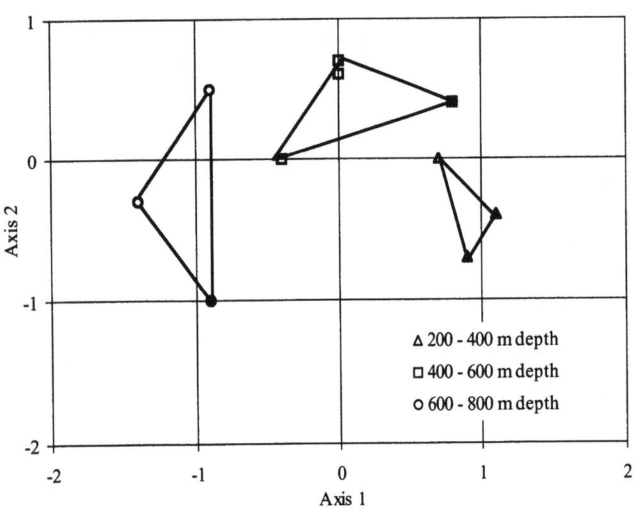

Fig. 2 MDS plot for ten GSN hauls taken at three depth bands in the Weddell Sea. *Filled symbols* represent hauls from the region off Kapp Norvegia. *Unfilled symbols* represent hauls from the area south of the Drescher Inlet

Discussion

Of the eight incirrate octopod species captured in the Weddell Sea, seven were from genera endemic to the Southern Ocean and, of the six species of *Pareledone*, at least two of the undescribed species may be endemic to the Weddell Sea. *Megaleledone senoi* has been reported from both eastern Antarctica (Lu and Stranks 1994) and from the Antarctic Peninsula (Kubodera and Okutani 1994; Piatkowski et al. 1998) so its presence in the Weddell Sea is not unexpected. It is possible that this species may be circumpolar although its presence in the Ross Sea is yet to be confirmed.

Only the genus *Benthoctopus* is non-endemic. *Benthoctopus* has two representatives in the Southern Ocean: *B. levis* (Hoyle 1885) described from off Heard Island, and *B. thielei* (Robson 1932) described from the shores of Kerguelen. Both descriptions are poor, and even the diagnosis of the genus is unclear (Grimpe 1921; Robson 1927). *Benthoctopus* specimens were also reported from Elephant Island, South Shetlands (Kuehl 1988) but these were not identified to species level. A recent cruise to the Antarctic Peninsula yielded specimens also diagnosed as *Benthoctopus* cf. *levis* (Piatkowski et al. 1998).

The endemic Antarctic genus *Pareledone* not only accounted for six of the species captured but also for over 95% of the specimens collected by the GSN. A survey of eastern Antarctica by Lu and Stranks (1994) yielded two new species of this genus whilst a survey of the Antarctic Peninsula by RV "Polarstern" in 1996 also yielded several undescribed species of *Pareledone* (Piatkowski et al. 1998). It is evident that this genus has undergone widespread radiation in the Southern Ocean. Like many families of endemic fishes (e.g. Artedidraconidae, Bathydidraconidae, Channichthyidae), *Pareledone* was probably associated in its earlier history with the waters of the Antarctic Plate prior to the formation of the Southern Ocean. The genus dominates the octopodid fauna and has probably had little competition from other octopodid genera during its radiation, as all other octopodids in the Southern Ocean are associated with deeper water.

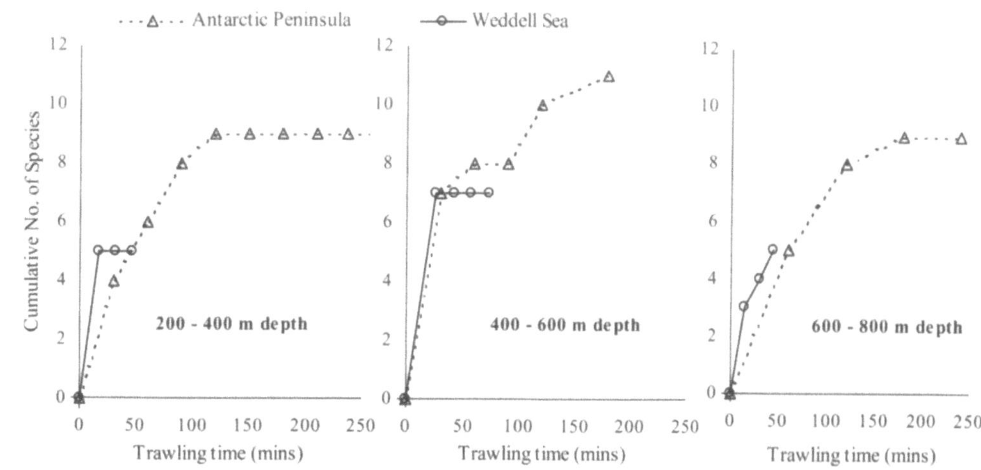

Fig. 3 Total number of species present at each depth band in the Weddell Sea and off the Antarctic Peninsula. Plots of cumulative number of species captured against trawl time. Trawls plotted sequentially

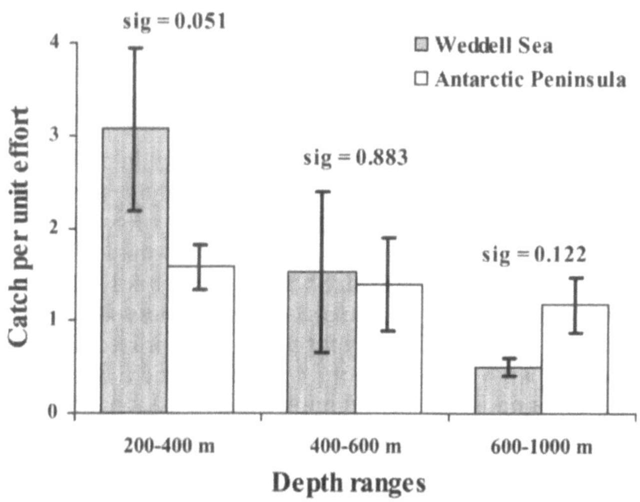

Fig. 4 The non-significant differences in catch per unit effort between the Weddell Sea and the Antarctic Peninsula at each depth band. *Error bars* represent 1 standard error of the mean

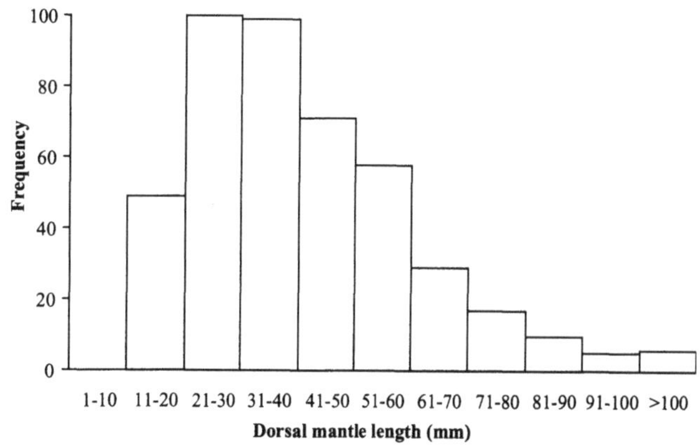

Fig. 5 Size-frequency analysis of octopodids captured during cruise ANT XIII/3 of RV "Polarstern" to the Weddell Sea

Radiation of octopodid genera in this way is most unusual. Of the 23 genera that are currently recognised in the family Octopodidae (Sweeney and Roper 1998), 8 are monotypic, and a further 5 comprise only 2 species each. When the three new species from this survey are included in the genus *Pareledone*, only two octopodid genera are more speciose. These are *Octopus* and *Benthoctopus*, both of which are considered to be taxonomically confused, and are likely to be split into several less speciose genera. Why then has *Pareledone* shown a degree of radiation that appears to be untypical of its family?

There are two factors that may be pertinent. Firstly, the habitat occupied by *Pareledone* in the Southern Ocean is extremely fragmented. As well as inhabiting the shelf and upper slope of the Antarctic continent, *Pareledone* is also present on the shelf and slope areas provided by the numerous islands in the vicinity of the Antarctic Peninsula and the Scotia Arc. Many of these islands are separated by tracts of ocean where the depth of the seafloor far exceeds that which may be colonised by *Pareledone*, and many populations of *Pareledone* are therefore isolated. The ovaries of mature females of *Pareledone* contain eggs up to 20 mm in length (A.L. Allcock, unpublished data). Eggs of such large size give rise to benthic, crawl-away young (Hochberg et al. 1992). Since fertilisation is internal and adult movement is probably limited, the lack of a planktonic stage may have severe consequences for dispersal and hence the maintenance of genetic panmixia. Indeed, populations of *Pareledone turqueti* from South Georgia and Shag Rocks show wide genetic divergence (Allcock et al. 1997) even though these islands of the Scotia Arc are separated by only 90 miles of ocean. Nonetheless, some genetic exchange must take place and, at some point, octopodids crossed these barriers and colonised the islands of the Scotia Arc. Rafting, particularly of small juveniles, is the most likely explanation for this. There is extensive literature on organisms rafting on floating and submerged drift algae (Holmquist 1994; Johannesson 1988; Worcester 1994), including long-distance dispersal in the Southern Ocean (Helmuth et al. 1994).

The second factor that may be pertinent to the radiation of *Pareledone* is the glacial nature of Antarctica. It is possible that the early *Pareledone* underwent vicariant speciation due to events on a geological time scale, such as warming in the Pliocene, which led to the formation of interglacial fjords (Webb 1990). These fjords may have provided environments where populations of *Pareledone* flourished. Such populations would also be isolated because of the limited dispersal capabilities of the benthic hatchlings, and conditions would be ideal for allopatric speciation. Although this situation is hypothetical, it is true that high levels of diversity are often found in regions that are climatically stable over years or hundreds of years, but are geologically or climatically unstable on a time scale of 10^5–10^6 years (Cook 1991). The glaciation cycles of Antarctica appear to fulfil these criteria. A superficially similar radiation is seen in the shallow-water octopuses of the Philippines (another fragmented habitat), where more than 20 species are recognised (Norman and Sweeney 1997). This is one of the highest octopodid diversities in the world. Although 18 of these are currently placed in the genus *Octopus*, they comprise 5 "species groups", which are likely to be split into genera at a later date. Until the taxonomic revision of *Octopus* is complete, it is not possible to determine whether this radiation is comparable.

The total number of species captured will increase as the time devoted to trawling increases, until finally all the species susceptible to that gear in the area have been captured. However, without greater fishing effort in the Weddell Sea, we can only speculate on whether all the octopodid species present have been captured. However, the octopodid fauna of the high Antarctic does appear to be richer than that found close to or just outside the Antarctic Polar Frontal Zone. Around the Falkland Islands, to the north of the APFZ, only three species of octopodids are recorded from depths shallower than 400 m (Allcock 1997; Robson 1932), while demersal fish surveys around South Georgia have yielded only two species of *Pareledone* and only three octopodid species in total in these depths (Allcock 1997; Daly and Rodhouse 1994). The third species, *Thaumeledone gunteri* Robson 1930, is in fact a deep-water species, whose range marginally extends to this depth band; just three specimens were caught in the deepest trawl of the fish survey, which reached 394 m depth.

Although on rare occasions the Falkland Islands and South Georgia may also have suffered disturbance due to glaciation effects (for example, the last glacial maximum that occurred approximately 20,000 years ago is thought to have led to fully glacial conditions in southern South America; Quilty 1990), it is likely that the impact of glacial cycles has been much greater at higher latitudes, and perhaps this is partly responsible for the latitudinal differences seen in species richness.

The similarities in the octopodid species assemblage between Kapp Norvegia and the area southwest of the Drescher Inlet are consistent with the theory that there is a single macrobenthic assemblage in these areas, termed the "Eastern Shelf Community" (Voß 1988). The fragmented nature of habitats of shallower depths around the Antarctic Peninsula suggests that more than one community assemblage was sampled in that area. This may explain why octopodid species richness off the Antarctic Peninsula exceeds that in the Weddell Sea. This inference is supported by the projected similarity of the actual numbers of octopodid species in the Weddell Sea and off the Antarctic Peninsula in the deepest waters sampled (600–800 m) (Fig. 3) as deep habitats off the Antarctic Peninsula are less fragmented than those in shallow waters.

Octopodid abundance also appears to be reduced around South Georgia. Eighty-two 30-min hauls by an FP 120 commercial bottom trawl yielded just 359 specimens of *Pareledone turqueti* (Allcock et al. 1997). *Pareledone turqueti* was the most abundant of the three species captured around South Georgia (Allcock 1997). The catch figures suggest an order of magnitude lower abundance compared with the Weddell Sea. This supports the view of Brey and Clarke (1993) who suggested that high standing stocks on the Antarctic shelf and slope may be the result of adaptations to low and oscillating food levels.

The standing stock in the Weddell Sea in February 1996 might actually have been higher than that suggested by catch per unit effort. The majority of individuals captured by the GSN were quite small (Fig. 5). Since the cod-end mesh size was 20 mm, it is likely that many individuals with mantle lengths less than this escaped the net. This hypothesis is supported by the size-frequency distribution (Fig. 5), which shows a considerable lack of animals below 20 mm mantle length. There is no information on the reproductive cycle of *Pareledone* although most species of octopus are semelparous brooders (Hochberg et al. 1992). We are unaware at what age these animals reach maturity and during which season spawning occurs, if indeed spawning is seasonal. The existence of all maturity stages of *Pareledone* in the catch suggests that seasonal spawning may not occur in this genus. It is, however, necessary to extend our sampling window from just a few months in the Antarctic summer if we are to gain a deeper insight into the life-cycle of these animals.

Clearly these data raise many questions and there is ample room for further study. Although Antarctic science is costly, ecological studies of the Southern Ocean octopodid fauna could be amalgamated with the extensive surveys of fish stocks at little extra cost. The commercial bottom trawls used to capture demersal fishes have proved to be by far the most efficacious gear for sampling octopodids, and the data yielded by rigorously designed fishing strategies would be of considerable scientific merit, especially for ecological studies.

Acknowledgements This work was funded by a Natural Environment Research Council studentship awarded to A.L.A. British Antarctic Survey were the CASE partners. The Alfred-Wegener-Institut für Polar- und Meeresforschung, Bremerhaven provided sea time aboard RV "Polarstern". Mike Vecchione and Silke Steimer helped with data collection. Mark Johnson assisted with statistical analysis.

References

Allcock AL (1997) The genetics and taxonomy of Southern Ocean Octopodidae, with special reference to the genus *Pareledone*. PhD Thesis, Liverpool University

Allcock AL, Brierley AS, Thorpe JP, Rodhouse PGK (1997) Restricted geneflow and evolutionary divergence between geographically separated populations of the Antarctic octopodid *Pareledone turqueti*. Mar Biol 129:97–102

Arntz W, Gutt J (eds) (1997) The Expedition ANTARKTIS XIII/3 (EASIZ 1) of "Polarstern" to the eastern Weddell Sea in 1996. Ber Polarforsch 249:1–148

Brey T, Clarke A (1993) Population dynamics of marine benthic invertebrates in Antarctica and Sub-Antarctic environments: are there unique adaptions? Antarct Sci 5:253–266

Clarke KR, Warwick RM (1994) Change in marine communities: an approach to statistical analysis and interpretation. National Environment Research Council, Plymouth Marine Laboratory

Cook LM (1991) Genetic and ecological diversity. Chapman & Hall, London

Daly HI, Rodhouse PG (1994) Comparative morphology of two sympatric *Pareledone* species from South Georgia. Antarct Sci 6:163–169

Grimpe G (1921) Teuthologische Mitteilungen. VII. Systematische Übersicht der Nordsee-Cephalopoden. Zool Anz 52:297–305

Helmuth B, Veit RR, Holberton R (1994) Long-distance dispersal of a sub-Antarctic brooding bivalve (*Gaimardia trapesina*) by kelp-rafting. Mar Biol 120:421–426

Hochberg FG, Nixon M, Toll RB (1992) Octopoda. In: Sweeney MJ, Roper CFE, Mangold KM, Clarke MR, Boletzky SV (eds) "Larval" and juvenile cephalopods: a manual for their identification. Smithson Contrib Zool 513:213–280

Holmquist JG (1994) Benthic macroalgae as a dispersal mechanism for fauna – influence of a marine tumbleweed. J Exp Mar Biol Ecol 180:235–251

Johannesson K (1988) The paradox of Rockall – why is a brooding gastropod (*Littorina saxatilis*) more widespread than one having a planktonic larval dispersal stage (*Littorina littorea*). Mar Biol 99:507–513

Kubodera T, Okutani T (1994) Eledonine octopods from the Southern Ocean: systematics and distribution. Antarct Sci 6:205–214

Kuehl S (1988) A contribution to the reproductive biology and geographical distribution of Antarctic Octopodidae (Cephalopoda). Malacologia 29:89–100

Locarnini RA, Whitworth T, Nowlin WD (1993) The importance of the Scotia Sea on the outflow of Weddell Sea deep water. J Mar Res 51:135–153

Lu CC, Stranks TN (1994) Synopsis of *Pareledone* and *Megaledone* species, with description of two new species from East Antarctica (Cephalopoda: Octopodidae). Mem Natl Mus Victoria Melbourne 54:221–242

Massy AL (1916) Mollusca. Part II. Cephalopoda. Br Antarct *Terra Nova* Exped 1910 Zool 2:141–175

Norman MD, Sweeney MJ (1997) The shallow-water octopuses (Cephalopoda: Octopodidae) of the Philippines. Invertebr Taxon 11:89–140

Piatkowski U, Allcock L, Hevia M, Steimer S, Vecchione M (1998) Cephalopod ecology. In: Kattner G (ed) The expedition ANTARKTIS XIV/2 of RV "Polarstern" in 1996/1997. Ber Polarforsch 274:41–47

Quilty PG (1990) Significance of evidence for changes in the Antarctic marine environment over the last 5 million years. In: Kerry KR, Hempel G (eds) Antarctic ecosystems – ecological change and conservation. Springer, Berlin Heidelberg New York, pp 4–8

Robson GC (1927) Notes on the Cephalopoda. III. On the anatomy and classification of the North Atlantic species of *Bathpolypus* and *Benthoctopus*. Ann Mag Nat Hist Ser 9 vol XX:249–263

Robson GC (1932) Monograph of the recent Cephalopoda. Part II. Octopoda (excluding Octopodinae). British Museum (Natural History), London

Rodhouse PG (1990) Cephalopod fauna of the Scotia Sea at South Georgia: potential for commercial exploitation and possible consequences. In: Kerry KR, Hempel G (eds) Antarctic ecosystems. Ecological change and conservation. Springer, Berlin Heidelberg New York, pp 289–298

Rodhouse PG, Arnbom TR, Fedak MA, Yeatman J, Murray AWA (1992) Cephalopod prey of the southern elephant seal *Mirounga leonina* L. Can J Zool 70:1007–1015

Sweeney MJ, Roper CFE (1998) Classification, type localities, and type repositories of recent Cephalopoda. In: Voss NA, Vecchione M, Toll RB, Sweeney MJ (eds) Systematics and biogeography of cephalopods, vol II. Smithson Contrib Zool 586:561–599

Voß J (1988) Zoogeographie und Gemeinschaftsanalyse des Makrozoobenthos des Weddellmeeres (Antarktis). Ber Polarforsch 45:1–145

Wanless S, Harris MP, Morris JA (1992) Diving behaviour and diet of the blue-eyed shag at South Georgia. Polar Biol 12:713–719

Webb P-N (1990) The Cenozoic history of Antarctica and its global impact. Antarct Sci 2:3–21

Whitworth T, Nowlin WD, Orst AH, Locarnini RA, Smith SG (1994) Weddell Sea shelf water in the Bransfield Strait and Weddell-Scotia confluence. Deep Sea Res 41:629–641

Worcester SE (1994) Adult rafting versus larval swimming – dispersal and recruitment of a botryllid ascidian on eelgrass. Mar Biol 121:309–311

Polar Biol (2001) 24: 883–891
DOI 10.1007/s003000100287

M.C. Gambi · F.P. Patti · G. Micaletto · A. Giangrande

Diversity of reproductive features in some Antarctic polynoid and sabellid polychaetes, with a description of *Demonax polarsterni* sp. n. (Polychaeta, Sabellidae)

Abstract The Polynoidae and Sabellidae represent two polychaete families with contrasting life habits and different phylogenetic constraints. Species of both families studied in this work were collected on the shelf of the Eastern Weddell Sea (Antarctica) during the EASIZ-I (1996) and EASIZ-II (1998) expeditions. Among the Polynoidae, three different species of the genus *Harmothoe* were observed to brood eggs under the dorsal elytra, which is an uncommon feature in this family. The egg size of all brooding taxa ranged between 120 and 216 μm.

Three Sabellidae species were studied: *Myxicola* cf *sulcata* is gonochoric with eggs among the largest (up to 1000 μm) recorded for the sabellids. Eggs, as well as larvae, are incubated within the branchial crown, a feature newly discovered for the genus *Myxicola*. *Demonax polarsterni* n. sp. (a newly discovered taxon) and *Euchone pallida* are gonochoric and possess relatively large eggs (350 μm and 270 μm, respectively). In comparison with cogeneric species of temperate areas, the Antarctic Sabellidae have larger egg dimensions and egg brooding that seems unconstrained by adult size.

Introduction

The unique and extreme climatic conditions of the Antarctic marine environment seem to favour K-selected traits in the life history of marine invertebrates, including offspring brooding (Duchene 1985). However, more recent analyses of single and homogeneous taxonomic groups have pointed out that phylogenetic and morpho-functional (e.g., adult size) constraints also play an important role and have to be considered in the evaluation of the diversity of adaptations to the polar marine environment (Pearse et al. 1991; Giangrande et al. 1994). In this respect, the presence of planktotrophic larvae (an exception to "Thorson's rule"), or egg and larval brooding have assumed a different ecological significance.

In order to evaluate the adaptive significance of the main traits of the reproductive biology of the Antarctic marine polychaetes, some species belonging to the families Polynoidae and Sabellidae, which represent different and contrasting life habits and reproductive strategies (Giangrande 1997), were examined and compared in the light of phylogenetic constraints existing within each family.

The Polynoidae seem to be more constrained in their reproductive features, and possess a general common pattern in having planktotrophic development and relatively small eggs (Giangrande 1997). They also display a relatively wide variety of life-habits and behaviour, being epibenthic vagile forms or symbiotic, and carnivorous, parasitic or scavengers. The Sabellidae are more canalized in life-habits, with almost all species being tubicolous and filter-feeding. The representatives of this family, despite a relative anatomical simplicity and uniformity, and a synapomorphy in larval development which is always lecithotrophic, show a high diversification of reproductive habits (Rouse and Fitzhugh 1994; Gambi et al., 2000).

The Polynoidae and Sabellidae are well represented in Antarctic waters, in respect to species richness and abundance (Hartman 1978; Stiller 1996). The taxonomy of the Polynoidae is better understood than that of other Antarctic polychaete families (Stiller 1996), including the Sabellidae (Giangrande and Gambi

Please note: When citing this article, please use the reference to the original source in the journal Polar Biology as given on the top of this page.

M.C. Gambi (✉) · F.P. Patti
Stazione Zoologica "A. Dohrn",
Laboratorio di Ecologia del Benthos,
80077 Ischia, Naples, Italy
E-mail: gambimc@alpha.szn.it
Fax: +39-81-984201

G. Micaletto · A. Giangrande
Dipartimento di Biologia, Università di Lecce, Lecce, Italy

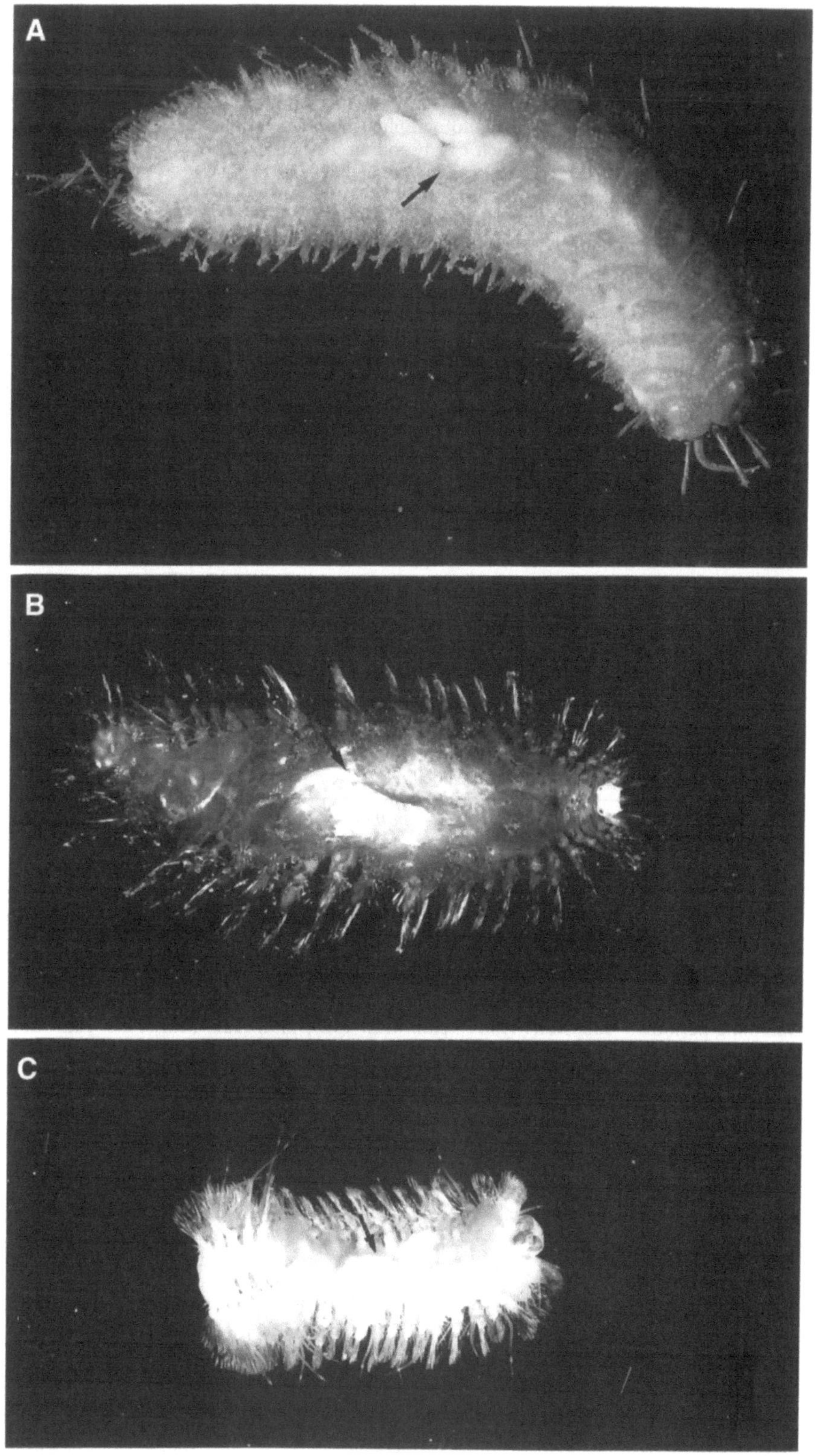

Fig. 1 **A** Living specimen of the polynoid *Harmothoe spinosa* (light form) brooding the eggs (grouped in single oval masses) under the dorsal elytra (*arrow*); **B** living specimen of *Harmothoe* sp., brooding egg masses under the dorsal elytra (*arrow*); **C** living specimen of *Harmothoe* cf *magellanica* brooding egg masses under the elytra (*arrow*). The photos were taken and kindly provided by Dr. Martin Rauschert (AWI, Berlin, Germany)

1997). However, the reproductive biology of the Polynoidae is poorly understood (Stiller 1996; Giangrande 1997), and although the Sabellidae are the best known family among polychaetes (Giangrande 1997), this does not include the Antarctic species (Gambi et al., 2000).

Materials and methods

The specimens of Polynoidae and Sabellidae analyzed here were collected by one of us (M.C.G.), during two Antarctic cruises on board of the R/V "Polarstern": EASIZ-I (ANT-XIII/3) and EASIZ-II (ANT-XV/3) (Arntz and Gutt 1997, 1999). The worms were collected by means of Agassiz (AGT) and bottom (GSN) trawls, and during EASIZ-II also with a large grab which was equipped with a TV camera (TVG) (Gambi 1997, 1999a, b). After collection, all worms were allowed to relax in order to avoid or limit body damage and autotomy, and were then fixed in 4% formalin. Some specimens in good condition were photographed before fixation. For the Polynoidae species in which egg brooding was observed, the egg size was measured, whereas for the studied species of Sabellidae, the size-frequency distribution was studied by measuring at least 50 oocytes for each of the females analyzed (3–6 specimens per species). The population structure of the newly described species *Demonax polarsterni*, which was particularly abundant, was studied by measuring the total body length. Some male specimens of the studied species of Sabellidae, and larvae of *Myxicola* cf. *sulcata*, were preserved in 2.5% glutaraldehyde for electron microscope (SEM) analysis of the larvae and of the mature spermatozoa. The structure of the mature sperm could provide some insight into fertilization and the fate of eggs (Jamieson and Rouse 1989). The material prepared for SEM was observed with a Philips 505 microscope, after being dehydrated in ethanol, critical-point-dried with CO_2 and coated with carbon and gold.

Results

Polynoidae

A total of about 29 species of Polynoidae were collected during the two cruises (Gambi 1997, 1999b). Analysis of the reproductive biology of the most common and widespread taxa is still in progress. Here we report the first and most interesting observations on the external brooding of eggs, observed under the elytra of a few specimens of three species belonging to the genus *Harmothoe*: *Harmothoe spinosa* Kinberg (Fig. 1A) (one specimen), *Harmothoe* sp. (one specimen, Fig. 1B) and *H.* cf *magellanica* Mc Intosh (four specimens, Fig. 1C). Among the specimens bearing eggs, one specimen each of *H.* cf *magellanica* and *Harmothoe* sp. were collected during the EASIZ-I cruise, and the others were sampled during the EASIZ-II cruise. The sizes and mass shape of the egg broods varied in the different species (Table 1). Regarding *H. spinosa*, which is the most common polynoid in Antarctic waters (Hartman 1978), Stiller (1996) distinguished two differently coloured forms in the Weddell Sea, but which were morphologically very similar: a light form (pale brown, as the specimen in Fig. 1A) and a dark form (black) characterized in living animals by iridescent scales (Gambi, personal observation). These two morphotypes were also recorded in our samples (Gambi 1997, 1999b), with the light form being more abundant than the dark one. Egg broods were observed in a single specimen belonging to the light form. The eggs were included in a thin, transparent membrane, and

Table 1 Summary of data on sampling methods and sites, distribution and ecology, and on the principal reproductive features of the studied specimens of Polynoidae and Sabellidae

	Harmothoe spinosa (light form)	*Harmothoe* cf *magellanica*	*Harmothoe* sp.	*Euchone pallida*	*Myxicola* cf *sulcata*	*Demonax polarsterni*
Distribution	Circumantarctic	Antarctic-Subantarctic	–	Circumantarctic	Circumantarctic	Endemic Weddell Sea (?)
Sampling site	Eastern Weddell Sea	Eastern Weddell Sea	Eastern Weddell Sea	Eastern Weddell Sea	Eastern Weddell Sea	Four Seasons Inlet (East. W. S.)
Sampling period	Jan.-Mar. 1998	Jan.-Mar. 1996, 1998	Jan.-Mar. 1996	Jan.-Mar. 1998	Jan.-Mar. 1998	Jan.-Mar. 1998
Sampling gear	AGT, GSN	AGT, GSN	AGT, GSN	AGT, GSN	AGT, GSN, TVG	TVG
Depth of occurrence (m)	250	120–400	120	75–400	200	75–180
No. of specimens	1	4	1	14	10	111
Examined bottom type	3-D sponge assemblage	3-D sponge assemblage	3-D sponge assemblage	Sand-muddy sand	Sand-muddy sand	Coarse gravel/pebbles
Body length (range, cm)	3	2.5–3	2.5	3–7	1–4	1–5
Reproductive habit	Gonochoric	Gonochoric	Gonochoric	Gonochoric	Gonochoric	Gonochoric
Egg size (range μm)	120–180	120–216	144–192	30–270	75–1000	50–350
Egg (modal class μm)	132	180	156	110	125 and 675	142 and 256
Egg (mean μm)	143	174	160	133	500	251
Egg fate	External brooding	External brooding	External brooding	Free-spawning (?)	External brooding	Free-spawning (?)
Development type	Planktotrophic	Planktotrophic	Planktotrophic	Lecitotrophic	Lecitotrophic	Lecitotrophic
Sperm morphology	–	–	–	Ent-aquasperm	Ent-aquasperm	Ect-aquasperm
Acrosome morphology	–	–	–	Pointed (nipple-like)	pointless cone	Rounded, globose

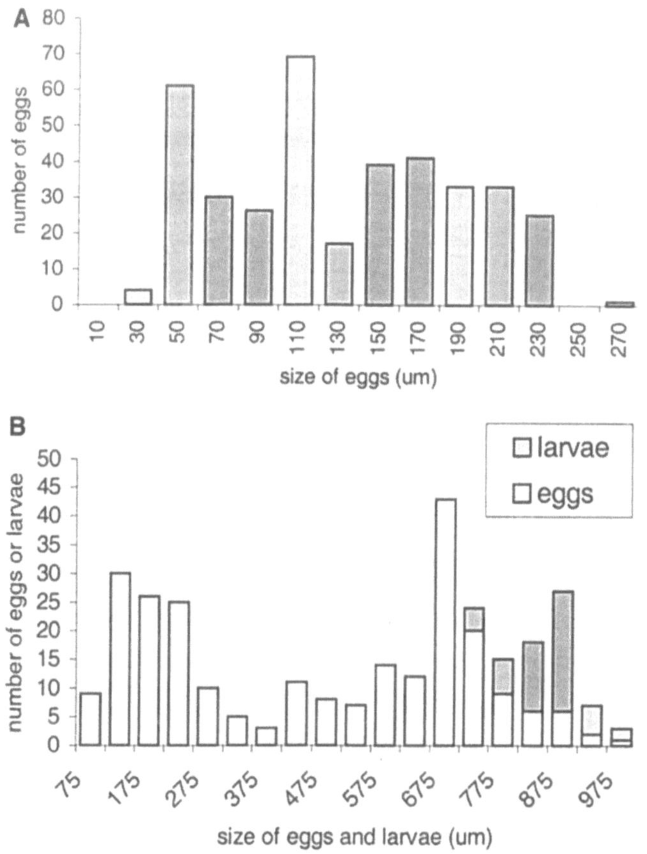

Fig. 2 A Size-frequency distribution in the sabellid *Euchone pallida*. **B** Size-frequency distribution of oocytes and larvae in the sabellid *Myxicola* cf *sulcata*

grouped in single oval masses located under the dorsal elytra (only three of such oval masses occurred in our specimen, Fig. 1A). The eggs were whitish, rounded and with a modal class at 132 μm (mean = 120 ± 17 μm). The eggs of *H*. cf *magellanica* were also whitish and elongated-lens shaped, with a shorter axis of 120 μm, a longer axis ranging from 120 to 216 μm, and a modal class of 180 μm (mean = 174 ± 22 μm). The eggs were included in a thin membrane, and were packed in flattened masses located on both dorsal sides of the worm, under the elytra (Fig. 1B). Finally, the eggs of *Harmothoe* sp. were whitish, rounded and had an irregularly bumped surface. They were also packed in flattened masses, and located under the elytra as in *H*. cf. *magellanica* (Fig. 1C); their modal size was 156 μm (mean = 160 ± 19.2 μm).

The number of brooded eggs was visually estimated at several hundreds for *H. spinosa*, and at several thousands for the other two species studied.

Sabellidae

Five different species of Sabellidae were collected, among which the two most common were *Perkinsiana littoralis* Hartman and *P. borsibrunoi* Giangrande & Gambi (Giangrande and Gambi 1997; Gambi 1999a). Data on their reproductive biology, as well as on the other two cogeneric taxa found in different Antarctic and Subantarctic areas (*P. milae* and *P. antarctica*), have been reported previously (Gambi and Patti 1998, 1999; Gambi et al. 2000). The other species collected and studied here were *Euchone pallida* Ehlers, *Myxicola* cf. *sulcata* Ehlers and *Demonax polarsterni*; the last-mentioned is a new taxon and represents the first record of the genus *Demonax* for Antarctica. Data on the distribution, size and main features of the reproductive biology of these sabellids are summarized in Table 1.

Euchone pallida is one of the most common sabellids recorded in Antarctic waters and is also well known from the Weddell Sea (Hartman 1978; Hartmann-Schroeder and Rosenfeldt 1989, 1991). In our samples, the species showed a wide depth distribution (from 75 m to

Fig. 3 A SEM micrographs of the structure of the mature spermatozoa in *Myxicola* cf *sulcata*; **B** SEM micrographs of the structure of the mature spermatozoa in *Euchone pallida*; **C** SEM micrographs of the structure of the mature spermatozoa in *Demonax polarsterni* n. sp

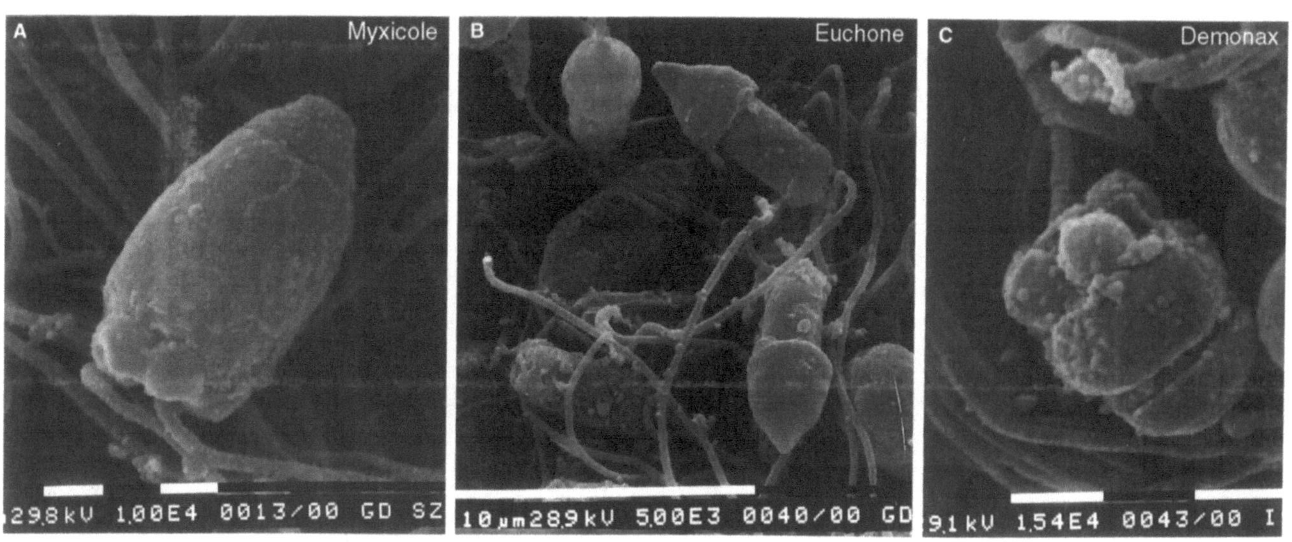

about 400 m). *E. pallida* is gonochoric (Table 1): 6 males and 8 females were found. Eggs and sperms are present only in the abdomen, and had a maximum diameter of up to 270 μm. The egg-size distribution showed a pattern that can be interpreted as unimodal, with a modal class at 110 μm (Fig. 2A), and a mean of 133 ± 3 μm. The morphology of mature sperm is very peculiar (Fig. 3B): the nucleous is cylindrical and the acrosome is cup-like and pointed, resembling the "nipple" of a feeding-bottle.

The species *Myxicola* cf. *sulcata* has been sampled at three stations, all located at 200 m depth (Table 1). The species is also relatively common in Antarctic waters. The genus *Myxicola* is unique among Sabellidae in having a typical mucous tube and in feeding at the sediment surface, with the inferior side of the crown lying on the bottom. The species is gonochoric, and 4 males and 6 females were found. A bimodal egg-size distribution (Fig. 2B) was evident, with modal classes at 125 and 675 μm, and with a maximum egg diameter up to 1,000 μm. Two females were observed to brood the offspring in the branchial crown. One specimen was bearing mature eggs, with a brilliant orange-reddish colour. The other specimen was brooding larvae (of which we counted 946). Larvae were also a brilliant orange colour, and ranged from 750 μm to 1 mm in length (Fig. 2B). They had a round-pointed prostomium (Fig. 4) with a well-developed prototroch and neurotroch (Fig. 4B, C) and two eye-spots. The four body segments bore two small capillary, slightly limbate chaetae each (Fig. 4B). No uncini were present. The pygidium was rounded and with two very small eye-spots. The morphology of the mature sperm in this species is also relatively peculiar (Fig. 3A), with an oval nucleus and a point-less conical acrosome. This structure resembles an ent-aquasperm type of spermatozoa (Jamieson and Rouse 1989; Rouse 1995).

Demonax polarsterni sp.n.

Description

Genus: *Demonax* Kinberg, 1867.

Type species: *Demonax krusensterni* Kinberg, 1867; subsequent designation by Bush (1905) *Demonax polarsterni* sp.n.

Material examined: 111 specimens from the eastern Weddell Sea, Antarctica, 75–180 m depth.

Holotype (Australian Museum no. W25914): station ANT-XV/3 no. 48/212 (75 m depth) (coordinates: 71° 07.3' S, 11° 27.8' W), March 1998. Collected by M.C. Gambi.

Paratypes (Australian Museum no. W25915): 2 specimens from the same locality and sample of the holotype.

Other material: 81 specimens from the same locality and sample of the holotype and paratypes, and 27 specimens from station ANT-XV/3 no. 48/277 (180 m depth) (coordinates: 71° 18.0' S, 012° 15.3' W), March 1998 (all specimens are at the Biology Department of the University of Lecce in the A. Giangrande collection).

Etymology: the species was named after the German R/V "Polarstern", the mythical icebreaker devoted to research in polar waters.

Fig. 4A–D SEM micrographs of larvae of *Myxicola* cf *sulcata*, brooded inside the branchial crown. **A** General view of three larvae; **B** lateral view of a larva showing the segments bearing the notochaete; **C** ventral view of a larva showing both the prototroch and the neurotroch; **D** dorsal view of a larva showing the pointed prostomium, the prototroch and the dorsal ciliary spots

Holotype: complete with 8 thoracic and about 50 abdominal chaetigers. Total length 53 mm of which 11 mm is the branchial crown; maximum body width 2 mm (Fig. 5A). Thoracic chaetigers wider than long. Each half of branchial crown with 15 radioles. Palmate membrane and radiolar flanges absent. End of radioles with tapering terminal filament (Fig. 5G), radioles subquadrangular in cross-section, fused at the base. Radiolar skeleton rounded in cross section (Fig. 5I). Radiolar appendage of dorsal lip elongate, pinnular appendages small (Fig. 5H). Ventral lips paired, well developed, beginning ventrally between ventral lappets of the collar, with margin extending dorsally to near dorsal lips. Ventral sacs absent. Collar oblique bilobed, ending ventrally in triangular or rounded lappets (Fig. 5B), with a wide gap mid-dorsally (Fig. 5C), dorsolateral margin covering the junction between the thorax and the base of the branchial crown. Thoracic tori short, not indenting the ventral shields (Fig. 5A, B). Thoracic uncini avicular about 35 per torus, with handle length about twice the distance between crest and breast (Fig. 5D). Upper thoracic notochaetae slender with narrow hood, inferior thoracic notochaetae limbate

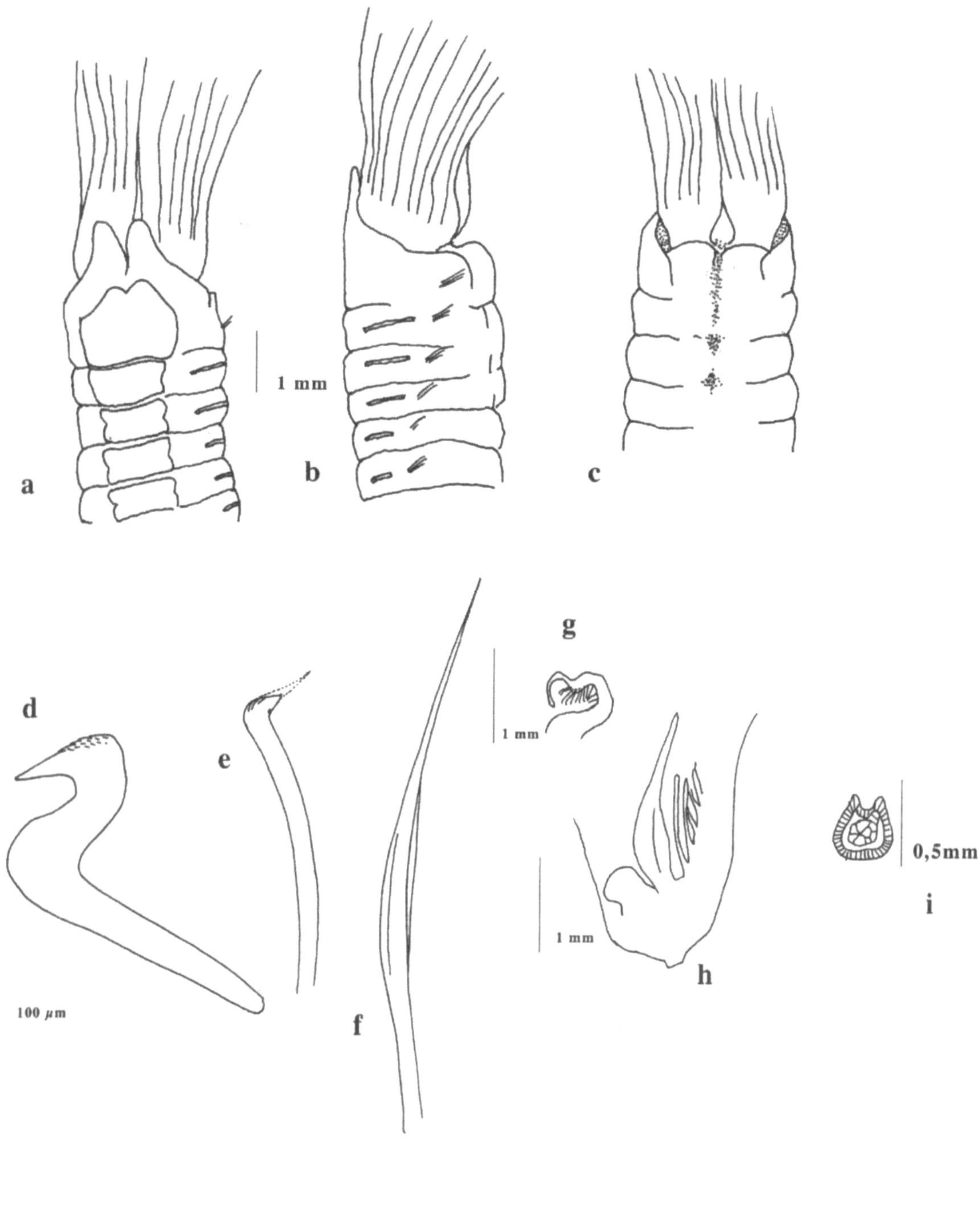

Fig. 5A–I Drawings of the sabellid *Demonax polarsterni* sp.n. **A** Ventral view of the anterior part and of the collar and lappets; **B** lateral view of the anterior part and of the gap of the collar; **C** dorsal view of the anterior part and of the gap of the collar; **D** thoracic avicular uncinus; **E** companion chaeta; **F** limbate inferior thoracic notochaeta; **G** tapering terminal filament of the radioles; **H** radiolar and pinnular appendages of dorsal lips; **I** cross section of the radiolar skeleton of the branchial crown

(Fig. 5F). Thoracic uncini with teeth above the main fang of equal size, hood absent, handles of medium size. Companion chaetae typical for the genus with a bulbous tooth-head, hooked in side view and with a filamentous blade arising from the centre, at right angles to the shaft (Fig. 5E). Abdominal uncini similar to the thoracic ones, abdominal neurochaetae arranged in two transverse rows, an anterior and a posterior one: both rows have elongate, narrowly hooded chaetae in anterior abdominal chaetigers, whereas in posterior segments the posterior rows have modified elongated narrowly hooded chaetae.

The tubes of all specimens collected at the shallowest stations (75 m depth) were heavily encrusted with coarse sand, gravel and small pebbles, whereas those sampled at 180 m depth were encrusted with fine sand.

Remarks

D. polarsterni belongs to a group of species within the genus having thoracic tori that do not indent the ventral shields, a character which seems to be linked to the length of the handles of the uncini, as also observed in the genus *Perkinsiana*. The species with those features are *D. japonicus* (Moore & Bush 1904), *D. lacunosus* (Perkins 1984), and *D. torulis* (Knight-Jones 1983). All these species are smaller, and have differently shaped inferior thoracic notochaetae. The most similar species to *D. polarsterni* seems to be *D. japonicus*, which has, however, more broadly hooded inferior thoracic chaetae.

Concerning reproductive biology, *D. polarsterni* is gonochoric, with a sex ratio close to 1:1. The population structure (based on total length) in sample 48/212 (75 m depth) is shown in Fig. 6A. Most of the individuals, belonging to all size ranges observed, appeared to be sexually mature, as indicated by the distribution of females among sizes (Fig. 6A). Females possessed relatively large eggs, up to 350 µm in diameter (Table 1), and of a bright pink colour. In each of the females examined, regardless of their dimension, the egg-size distribution showed a typical bimodal pattern (Fig. 6B), with modal classes at 256 and 142 µm, and a mean of 251.1 ± 3.4 µm.

Finally, the morphology of the mature sperm (Fig. 3C) showed the typical ect-aquasperm structure (Jamieson and Rouse 1989), with a rounded nucleus and a globose acrosome.

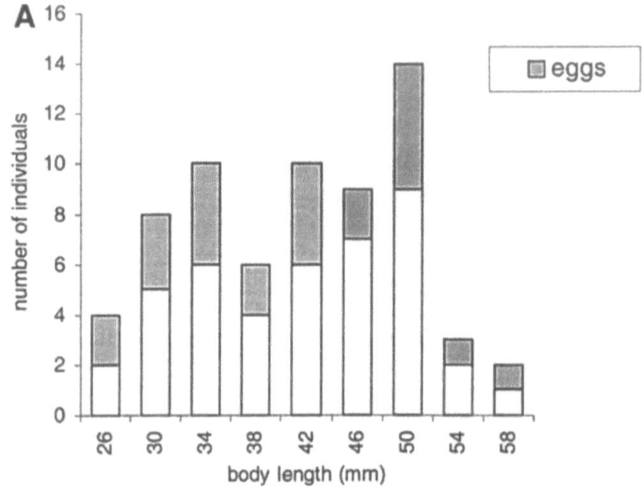

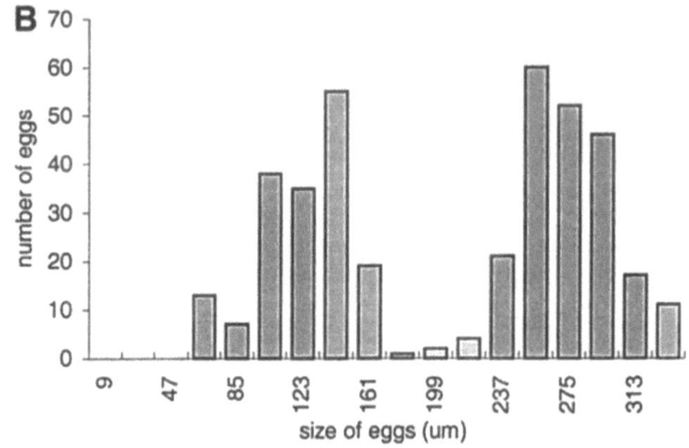

Fig. 6 **A** Size-frequency distribution (based on total body length) of the population of *Demonax polarsterni* sp.n. sampled in station 48/212 (ANT-XV/3). The specimens with eggs are shown. **B** Size-frequency distribution of oocytes in *D. polarsterni* sp.n

Discussion

The most interesting feature of the studied polynoids is the record of external brooding of the eggs under the dorsal elitra. Literature data on the reproductive biology of the Polynoidae, even though still relatively limited (Giangrande 1997), showed that brooding of the eggs is very rare within this family, being recorded only in a population of *Harmothoe imbricata*, suggesting a certain degree of intraspecific variability, at least within the genus *Harmothoe*. However, the taxonomy of this group is still scarcely known and the possibility of sibling species cannot be excluded. In the classic paper by Knox (1977) on Antarctic polychaete fauna characteristics and distribution, the author indicated in an anecdotal sentence that "Many polynoids, such as *Austrolaenilla pelagica*, *Harmothoe lagiscoides*, *Polynoe antarctica* and *Harmothoe crosetensis* brood their eggs under the elytra", but he did not give any other information, data or reference about this finding.

Therefore, the brooding habit of polynoids in Antarctic waters is probably a more common reproductive feature than previously considered.

In the Polynoidae studied by us, the fact that the eggs were easily removable from the elytra, and the egg size suggests that the external brooding is probably short and that the hatched larvae are pelagic planktotrophics. This hypothesis is supported by the fact that all known Polynoidae have a planktotrophic development (Giangrande 1997), and that the egg size of the studied

Antarctic species is similar to that of the temperate ones. In addition, Knox's personal observation on plankton from inshore waters off McMurdo Sound (Ross Sea) (Knox 1977) indicate that the most common pelagic larvae were those of Polynoidae. However, as for the brooding habit, the scarcity of observation on larval release from adult specimens in polynoids does not allow to exclude the occurrence of lecitotrophic larvae or even of direct development in this group of polychaetes.

Comparison of our data with those of Stiller (1996) on specimens of *H. spinosa* and *H.* cf *magellanica* collected during different cruises in the Weddell Sea, showed that the maximum and mean egg size for both species is comparable. However, Stiller (1996) never observed brooding specimens in his samples, and we have recorded this habit only in a few individuals. The scarcity of brooding specimens could be due mainly to two factors: the sampling season, and the sampling gear and fixation procedure.

According to the data of Stiller (1996), a constant increase in egg size was observed from January to May for both *H. spinosa* and *H.* cf. *magellanica*. Observations are lacking from May to October, while in October maximum oocyte diameter was observed, together with free gametes into the coelom. This pattern, observed by Stiller (1996) also in other polynoids species, suggested a seasonal reproductive cycle in this group, with egg maturation starting in autumn and lasting until October, and spawning between late spring and early summer. This is consistent with the probable planktotrophic development of larvae that are released in the water column during the summer season, when phytoplankton availability reaches its potential maximum. The external brooding observed in our specimens, even though relatively rare, question the above-described cycle, since mature individuals were observed in January–March. It is possible, however, that the few brooding specimens found were early mature individuals and that the population is not so well synchronized. As regards to the sampling procedure, polynoids are very delicate animals and they generally loose very easily the elytra and break their body (autotomy). The benthic sampling tools used on board of Polarstern were generally the relatively rough Agassiz and bottom trawls. It is therefore possible that the brooded eggs are lost by most of the specimens during sampling and fixation not preceded by relaxation of the worms (Gambi 1997).

As for the Antarctic Sabellidae, the present data confirm the high variability and plasticity in the reproductive features of this family. This was already evidenced from previous research on the genus *Perkinsiana* (Gambi and Patti 1999; Gambi et al., 2000), in which brooding, observed in *P. antarctica*, was not linked to the size of the worm, and a probable asexually reproducing species (*P. milae*) was also found.

Among the presently reported species, *Myxicola* cf. *sulcata* was particularly interesting, because it was found incubating embryos within the crown, a strategy usually adopted by very small sabellid species. As in *P. antarctica*, the species reaches a larger size than that reported for brooding. Moreover, until now the incubation habit for the genus *Myxicola* has not been reported (Rouse and Fitzhugh 1994). Larvae inside the branchial crown measured up to 1 mm in length and were all at the same developmental stage. Because this was the only stage observed, it cannot be excluded that the larvae might have a short pelagic phase before metamorphosis. Moreover, the eggs of this species reach the largest diameter recorded for Sabellidae (up to 1,000 μm), a size which is also one of the largest found among polychaetes (Giangrande 1997). Egg size distribution in this species indicates two egg cohorts characterized by large oocytes ready to spawn, and small ones in development. This could indicate that the species have different spawning events in a single reproductive season, or that the batch of small oocytes will be mature for the next reproductive season. The first case implies a relatively fast oogenesis process, whereas the other hypothesis is consistent with a relatively slow process which in the cold waters of the Antarctic seems to be a more common feature.

The spermatozoan morphology of this species could suggest an ent-aquasperm structure (Jamieson and Rouse 1989; Rouse 1995), according to a probable presence of in situ fertilization. As observed for *P. antarctica*, brooding in the branchial crown in *M.* cf. *sulcata* seems to be related to environmental conditions, rather than to size or other phylogenetic contraints.

Euchone pallida shows a different egg size distribution suggesting a more synchronized oogenesis with a single reproductive episode during the reproductive season. This species possessed the smallest egg size among the sabellids investigated here, and the morphology of the spermatozoa might suggest the presence of in situ fertilization.

Finally, *Demonax polarsterni* shows an intermediate egg size, and a spermatozoan morphology more readily defined as an ect-aquasperm type, suggesting a free spawning strategy, whereas the egg size distribution was very similar to that found in *M.* cf. *sulcata*, in turn suggesting similar hypotheses about spawning events and oogenesis process.

Some general conclusions can be extrapolated from current evidence collected on Antarctic species belonging to Polynoaidae and Sabellidae families. Even though the two families have different phylogenetic constraints concerning reproduction, both show some traits, such as brooding and fertilization, which can be explained more easily in terms of environmental adaptation than in terms of genealogy.

In comparison to previous literature data, the patterns recorded in the two investigated families suggest an interplay between ecological and phylogenetic factors that enhances the variability of the traits of the life cycle, limiting the generalizations possible.

Acknowledgements Thanks are due to Prof. Wolf Arntz (AWI, Germany) for inviting M.C.G. to the EASIZ-I and EASIZ-II cruises to the Weddell Sea, and to the Captains and crew members of the R/V "Polarstern" for logistic support and comfort on board.

We also wish to thank Dr. Martin Rauschert (AWI, Germany) for the pictures of living polynoids and sabellids, and Dr. Sabine Cochrane (Akvaplan-Niva, Norway) for checking the identification of *Euchone pallida*. Electron Microscopy Service of the Stazione Zoologica provided the technical assistance for SEM analyses.

References

Arntz WE, Gutt J (eds) (1997) The expedition ANT XIII/3 (EASIZ I) of "Polarstern" to the eastern Weddell Sea. Ber Polarforsch 249:1–148

Arntz WE, Gutt J (eds) (1999) The expedition ANT XV/3 (EASIZ II) of "Polarstern" to the eastern Weddell Sea. Ber Polarforsch 301:1–158

Bush JK (1905) Tubicolous annelids of the tribes Sabellides and Serpulides from the Pacific Ocean. Harriman Alaska Exped 12:167–346

Duchene JC (1985) Adaptation de la reproduction dans les eaux froides en zone Subantarctique. Oceanis 11:87–100

Gambi MC (1997) Autecology of Aphroditidae and Polynoidae on the Continental Shelf and Slope of the Eastern Weddell Sea. In: Arntz WE, Gutt J (eds) The expedition ANT XIII/3 (EASIZ I) of R/V "Polarstern" in 1996. Ber Polarforsch 249:66–73

Gambi MC (1999a) Preliminary observations on distribution, biology and feeding ecology of some suspension feeding polychaetes in the Eastern Weddell Sea. In: Arntz WE, Gutt J (eds) The expedition ANT XV/3 (EASIZ II) of R/V "Polarstern" in 1998. Ber Polarforsch 301:83–94

Gambi MC (1999b) Preliminary observations on species composition and distribution of Aphroditidae and Polynoidae (Polychaeta) in the Eastern Weddell Sea. In: Arntz WE, Gutt J (eds) The expedition ANT XV/3 (EASIZ II) of R/V "Polarstern" in 1998. Ber Polarforsch 301:161–163

Gambi MC, Patti FP (1998) Observations on reproductive biology of *Perkinsiana antarctica* (Kinberg) (Polychaeta, Sabellidae). Biol Mar Mediterr 5:336–338

Gambi MC, Patti FP (1999) Reproductive biology of *Perkinsiana antarctica* (Kinberg) (Polychaeta, Sabellidae) in the Straits of Magellan (Chile): systematics and ecological considerations. In: Arntz WE, Rios C (eds) Magellan-Antarctic: ecosystems that drifted apart. Sci Mar 63 [Suppl 1]:253–259

Gambi MC, Giangrande A, Patti FP (2000) Comparative observations on reproductive biology of four species of *Perkinsiana* (Polychaeta, Sabellidae). In: Reish D, Lana P (eds) Sixth International Polychaete Conference, 2–7 August 1998, Curitiba (Paranà, Brasil). Bull Mar Sci 67(1):299–309

Giangrande A (1997) Polychaete reproductive patterns, life cycles and life histories: an overview. Oceanogr. Mar Biol Annu Rev 35:323–386

Giangrande A, Gambi MC (1997) The genus *Perkinsiana* (Polychaeta, Sabellidae) in Antarctica, with the description of the new species, *P. milae* and *P. borsibrunoi*. Zool Scr 26:267–278

Giangrande A, Belmonte G, Geraci S (1994) Life cycle and life history traits diversity in marine invertebrates and implications in community dynamics. Oceangr Mar Biol Annu Rev 32:305–333

Hartman O (1978) Polychaeta from the Weddell Sea quadrant, Antarctica. Biology of the Antarctic Seas VI. Antarctic Res Ser 26:125–223

Hartmann-Schroeder G, Rosenfeldt P (1989) Die Polychaeten der "Polarstern" – Reise ANT-II/2 in die Antarktis 1984. Teil 2: Cirratulidae bis Serpulidae. Mitt Hamb Mus Inst 86:65–106

Hartmann-Schroeder G, Rosenfeldt P (1991) Die Polychaeten der "Walther Herwig" – Reise 68/1 nach Elephant Island (Antarktis) 1985. Teil 2: Acrocirridae bis Sabellidae. Mitt Hamb Zool Mus Inst 88:73–96

Jamieson BGM, Rouse GW (1989) The spermatozoa of the Polychaeta (Annelida): an ultrastructural review. Biol Rev 64:93–157

Kinberg JGH (1867) Annulata nova. Ofversigt af Svenska Vetenskapsakademien Forhandlingar, Stockholm 23:337–357

Knox GA (1977) The Antarctic polychaete fauna: its characteristics, distribution patterns, and evolution. In: Llano GA (ed) Adaptation within Antarctic Ecosystem, Proc. III SCAR Symp. Antarct. Biol. Smithsonian Institution, Washington D.C.: 1111–1127

Pearse JS, McClintock JB, Bosch I (1991) Reproduction of Antarctic marine invertebrates: tempos, modes, and timing. Am Zool 31:65–80

Rouse GW (1995) Is sperm ultrastructure useful in polychaete systematics? An example using 20 species of the Fabriciinae (Polychaeta: Sabellidae). Acta Zool (Stockh) 76:57–74

Rouse GW, Fitzhugh K (1994) Broadcasting fables: is external fertilization really primitive? Sex, size and larvae in sabellid polychaetes. Zool Scr 23:271–312

Stiller M (1996) Distribution and biology of the Aphroditidae and Polynoidae (Polychaeta) in the eastern Weddell Sea and Lazarev Sea (Antarctica) (in German). Ber Polarforsch 185: 1–200

Polar Biol (2001) 24: 538–544
DOI 10.1007/s003000100259

Brigitte Hilbig

Deep-sea polychaetes in the Weddell Sea and Drake Passage: first quantitative results

Abstract Based on samples obtained with a multibox corer during the "Polarstern" expedition EASIZ II (ANT XV/3) to the Weddell Sea and King George Island, the polychaete communities from deep water (1,000–2,400 m) are described. The results presented are the first quantitative data for the area. Species richness and abundances of polychaetes from four different downslope transects are presented, and a first preliminary description is provided of the polychaete community structure. Comparisons are made with polychaete communities from northern, temperate deep-sea areas in the Atlantic and Pacific. While the density of polychaetes was on average about a tenth of that in other slope areas, the species richness was very high if the low polychaete abundances and the small area sampled are considered. The community structure differed from that in the northern, temperate slope areas in the absence or poor representation of usually abundant and species-rich families such as spionids, capitellids and dorvilleids and the predominance of ampharetids, terebellids and maldanids.

Introduction

Although polychaetes have been collected and identified, along with other benthic infauna, during many expeditions to Antarctic seas since the end of the last century (for example, McIntosh 1885; Benham 1921; Monro 1930, 1936; Augener 1932; Fauvel 1936; Hartman 1964, 1966, 1967; Hartmann-Schröder and Rosenfeldt 1988, 1989), no quantitative investigations of the benthic polychaete fauna below 1,000 m in Antarctic waters have been published to this date. The present paper is the first description of deep-sea polychaete assemblages, based on quantitative data from the EASIZ II (ANT XV/3) campaign on board R/V "Polarstern" in 1998. Species richness (i.e. number of species per station), abundance and faunal composition are presented in this paper. A first interpretation of the data is attempted with regard to the mode of life and feeding type of the most abundant species.

A preliminary comparison with polychaete data from a different area of the continental slope in the northeast Pacific is made to help answer the question of whether there is a specifically Antarctic deep-sea community in the Weddell Sea, and if so, what the typical characteristics of this assemblage might be.

B. Hilbig
Zoological Institute and Zoological Museum,
Martin-Luther-King-Platz 3, 20146 Hamburg, Germany
E-mail: bhilbig@zoologie.uni-hamburg.de

Materials and methods

During the expedition EASIZ II (ANT XV-3) aboard RV Polarstern in 1998, 36 samples were collected with a multibox corer (Gerdes et al. 1992) in the Weddell Sea and off King George Island (Bransfield Strait and Drake Passage). The samples analysed for this investigation were collected at ten stations located on three downslope transects in the Weddell Sea and one off King George Island in the Drake Passage (Table 1) in depths between 1,000 and 2,400 m. Up to seven cores, each sampling an area of 0.024 m^2, from each haul were sieved through 300-μm (three cores) and 500-μm screens (one to four cores). The smaller mesh size was intended to produce data comparable to those collected in other deep-sea areas during several American projects by means of a Sandia box core (Hessler and Jumars 1974; Blake and Grassle 1994). The analysis of the polychaetes from EASIZ II, however, showed that there was no noticeable difference between samples sieved through 300-μm mesh and 500-μm mesh, respectively, as far as the percentage of small animals is concerned, although there was no statistical test of this observation. The reported high number of very small deep-sea polychaetes, which would have been lost using the coarser screens, in the temperate Atlantic and Pacific Oceans did not seem to be present in the studied areas. Consequently, the data from both sieve sizes were pooled and are considered comparable with the above-mentioned data.

The samples were processed using standard methods, i.e. fixed in a buffered 4% formaldehyde solution in seawater and later preserved in 70% alcohol. The polychaetes were separated from the sediment, sorted to family, identified to species and enumerated. The raw data were logged into a spreadsheet for any later statistical treatment.

Table 1 Deep-sea stations sampled with the multibox corer during EASIZ II

Station	Transect	Latitude	Longitude	Depth (m)
48-091	Southwest of Vestkapp (western transect)	73°28.4′S	22°48.8′W	1,510
48-092		73°34.2′S	22°38.0′W	994
48-093		73°28.3′S	22°54.5′W	1,988
48-131	Southwest of Vestkapp (eastern transect)	73°23.7′S	22°09.1′W	1,985
48-132		73°20.7′S	22°18.4′W	2,387
48-136	Halley Bay	74°33.0′S	27°13.1′W	2,012
48-137		74°36.2′S	27°12.7′W	1,521
48-146		74°39.4′S	27°08.0′W	1,021
48-330	Drake Passage	61°20.6′S	58°15.1′W	2,009
48-334		61°26.7′S	58°06.6′W	1,028

Results

The 10 deep-sea samples yielded 822 individuals of polychaetes (including 39 not identifiable to species) belonging to 115 species in 28 families (Table 2). The species list includes 28 species that are new to science; most of them belong to the Cirratulidae, and they may have been collected in the past and misidentified. Until the recent generic review by Blake (1996), genera and species in this family had been ill-defined, resulting in erroneous identifications. Some new species may prove to be typical for the deep sea, rather than exhibiting a eurybathic distribution characteristic for many Antarctic marine organisms.

Both density and species richness showed a depth-dependent pattern which was not equally distinct in all areas (Figs. 1, 2). Among the stations on the two transects off Vestkapp in 1,000–2,400 m depth, there was a continuous gentle decline in polychaete density from more than 1,300 to about 250 individuals per m^2, whereas off Halley Bay, densities were already quite low at 1,000 m (about 500 individuals per m^2), and basically did not decrease down to 2,000 m. At the stations in the Drake Passage in 1,000 and 2,000 m, respectively, there was a distinct decrease of density from 800 to 300 individuals per m^2 (Fig. 1).

The species richness generally followed the density (Fig. 2), with a gentle decline from shallow to deep water present at the Vestkapp transects, an almost unchanged number of species at the Halley Bay transect, and a rather sharp decline between the two stations in the Drake Passage. Generally, the number of species per station ranged from 13 (Vestkapp, 2,400 m) to nearly 50 (Drake Passage, 1,000 m). However, the seemingly obvious interdependence of species richness and depth becomes less straightforward if samples from shallower water are included; for example, among 7 stations in 200–300 m off Kapp Norvegia, the range in species richness was almost as wide, with less than 20 to 55 species per station (B. Hilbig, unpublished work). It is possible that parameters other than depth are influencing species richness, including sedimentary characteristics.

For a first analysis of the deep-sea infaunal polychaete community structure, the species richness of the polychaete families was determined (Fig. 3). Of the 28 families found, the most speciose was the cirratulids with 13 species, followed by ampharetids (11 species), terebellids (9 species), paraonids (8 species) and maldanids (8 species). Seven families were represented by only one species each. All of these species-rich families are sedentary or discretely motile surface or subsurface deposit feeders that live either at the sediment-water interface (cirratulids, ampharetids, terebellids and paraonids) or deeper in the sediment (maldanids). The only vagile family with a relatively large number of species was the syllids. Spionids and capitellids were poorly represented (six and two species, respectively), as well as epibenthic and highly vagile forms, such as polynoids, and motile carnivores, such as nephtyids and glycerids.

Discussion

Densities of polychaetes were generally low in the deep Weddell Sea (about an order of magnitude lower than in temperate Atlantic and Pacific deep waters), but at the same time the species richness was relatively high. For example, at the deepest stations off Vestkapp (48-131 and 48-132), there were only 40 and 33 specimens, respectively, but they belonged to 21 and 13 species, respectively. The total area sampled during EASIZ II in depths between 1,000 and 2,400 m was 1.4 m^2, harbouring more than 100 polychaete species; Grassle et al. (1990) reported nearly 4 times as many polychaete species (385) from the slope off the American east coast, but the sampled area was about 15 times larger (21 m^2). It can therefore be expected that additional samples will yield many more species, resulting in high species richness and diversity values. The polychaete fauna of the deep Weddell Sea can by no means be considered depauperate compared to other deep-sea areas.

As far as depth-dependence of polychaete abundance and species richness is concerned, any conclusions drawn from a single sampling event are to be viewed with great caution. While there appears to be a decline of both density and species richness with increasing depth, the ranges were almost as wide among shallow stations, e.g. off Kapp Norvegia. Differences in density were generally much less pronounced in depths greater

Table 2 List of species collected during EASIZ II (ANT XV/3) at stations in depths of 1,000 m and below

Family	Species
Acrocirridae	*Flabelligella* sp. 2
Ampharetidae	*Ampharete kerguelensis* McIntosh, 1885
	Amphicteis gunneri (Sars, 1835)
	Amphisamytha sp. 1
	Anobothrella antarctica (Monro, 1939)
	Anobothrus gracilis (Malmgren, 1866)
	Anobothrus sp. 1
	Anobothrus sp. 2
	Ecamphicteis sp. 1
	Glyphanostomum pallescens (Théel, 1878)
	Melinantipoda antarctica Hartman, 1967
	Melinna elisabethae McIntosh, 1914
Amphinomidae	*Paramphinome australis* Monro, 1930
Aphroditidae	*Hermionopsis levisetosa* Seidler, 1923
Cirratulidae	*Aphelochaeta cincinnata* (Ehlers, 1908)
	Aphelochaeta cf. *epitoca* (Monro, 1930)
	Aphelochaeta sp. 1
	Aphelochaeta sp. 2
	Aphelochaeta sp. 3
	Chaetozone sp. 1
	Chaetozone sp. 2
	Chaetozone sp. 3
	Cirratulidae sp. 1
	Cirratulus parafiliformis (Hartmann-Schröder & Rosenfeldt, 1989)
	Monticellina sp. 1
	Monticellina sp. 2
	Protocirrineris antarctica (Monro, 1930)
Capitellidae	*Capitella capitata complex* (Fabricius, 1780)
	Notomastus latericeus Sars, 1850
Eunicidae	*Eunice pennata* (O.F.Müller, 1776)
Euphrosinidae	*Euphrosine antarctica* Hartmann-Schröder & Rosenfeldt, 1992
	Euphrosine cirrata Sars, 1862
Glyceridae	*Glycera kerguelensis* McIntosh, 1885
Lumbrineridae	*Augeneria tentaculata* Monro, 1930
	Lumbrineris cf. *kerguelensis* (Grube, 1878)
	Paraninoe brevipes (McIntosh, 1903)
Maldanidae	*Asychis amphiglyptus* (Ehlers, 1897)
	Maldane sarsi antarctica Arwidsson, 1911
	Nicomache sp. 1
	Notoproctus oculatus antarcticus Arwidsson, 1911
	Praxillella kerguelensis (McIntosh, 1885)
	Rhodine intermedia Arwidsson, 1911
	Rhodine loveni Malmgren, 1865
Nephtyidae	*Aglaophamus posterobranchus* Hartman, 1967
	Aglaophamus trissophyllus (Grube, 1877)
	Micronephtys sp. 1
Nereididae	*Ceratocephale* sp. 1
Onuphidae	*Kinbergonuphis notialis* (Monro, 1930)
	Kinbergonuphis sp. 1
	Notuphis antarctica (Monro, 1930)
Opheliidae	*Ophelina cylindricaudata* (Hansen, 1878)
	Ophelina nematoides (Ehlers, 1913)
	Travisia kerguelensis McIntosh, 1885
Orbiniidae	*Leitoscoloplos kerguelensis* (McIntosh, 1885)
	Orbinia sp. 1
	Scoloplos marginatus (Ehlers, 1897)
Oweniidae	*Galathowenia wilsoni* Blake, 1984
	Myriochele joinvillensis Hartmann-Schröder & Rosenfeldt, 1989
	Myriochele sp. 1
	Myrioglobula antarctica Hartman, 1967
Paraonidae	*Aricidea antarctica* Hartmann-Schröder & Rosenfeldt, 1988
	Aricidea cf. *antennata* Annenkova, 1934
	Aricidea oculata Hartmann-Schröder & Rosenfeldt, 1990
	Aricidea simplex Day, 1963
	Aricidea strelzovi Hartmann-Schröder & Rosenfeldt, 1990
	Aricidea nr *trilobata* Imajima, 1973
	Cirrophorus brevicirratus Strelzov, 1973
	Levinsenia antarctica (Strelzov, 1973)
Pholoidae	*Pholoe* sp. 1

Table 2 (Contd.)

Family	Species
Phyllodocidae	*Mystides borealis* Théel, 1879
	Protomystides cf. *occidentalis* (Ditlevsen, 1917)
	Sige sp. 1
Polynoidae	*Eunoe anderssoni* (Bergström, 1916)
	Harmothoe nr. *crosetensis* (McIntosh, 1885)
Sabellidae	*Euchone pallida* Ehlers, 1908
	Euchone undulocincta Hartmann-Schröder & Rosenfeldt, 1989
	Fabricia stellaris stellaris Blainville, 1828
	Perkinsiana borsibrunoi Giangrande & Gambi, 1997
	Perkinsiana milae Giangrande & Gambi, 1997
Scalibregmatidae	*Asclerocheilus ashworthi* Blake, 1981
	Asclerocheilus nr *tropicus* Blake, 1981
	Hyboscolex dicranochaetus (Schmarda, 1861)
	Oligobregma sp. 1
	Pseudoscalibregma bransfieldium (Hartman, 1967)
Sphaerodoridae	*Ephesiella antarctica* (McIntosh, 1885)
	Sphaerodoropsis arctowskyensis Hartmann-Schröder & Rosenfeldt, 1988
	Sphaerodoropsis parva (Ehlers, 1913)
	Sphaerodoropsis sp. 1
Spionidae	*Laonice cirrata* (Sars, 1851)
	Laonice weddellia Hartman, 1978
	Microspio moorei (Gravier, 1911)
	Prionospio cf. *dubia* Monro, 1930
	Prionospio cf. *ehlersi* Fauvel, 1928
	Spiophanes sp. 1
Syllidae	*Eusyllis kerguelensis* McIntosh, 1885
	Exogone heterosetoides australis Hartmann-Schröder & Rosenfeldt, 1988
	Exogone obtusa Hartmann-Schröder & Rosenfeldt, 1988
	Exogone sp. 1
	Sphaerosyllis antarctica Gravier, 1807
	Sphaerosyllis joinvillensis Hartmann-Schröder & Rosenfeldt, 1988
	Typosyllis pennelli Knox & Cameron, 1998
	Typosyllis variegata (Grube, 1860)
Terebellidae	*Amphitrite cirrata* Müller, 1771
	Lanicides sp. 1
	Laphania cf. *boecki* Malmgren, 1866
	Leaena antarctica McIntosh, 1885
	Leaena collaris Hessle, 1917
	Leaena pseudobranchia Levenstein, 1964
	Leaena sp. 1
	Pista patriciae Hartmann-Schröder & Rosenfeldt, 1989
	Thelepides koehleri Gravier, 1911
Trichobranchidae	*Octobranchus antarcticus* Monro, 1936
	Octobranchus sexlobatus Hartmann-Schröder & Rosenfeldt, 1989
	Octobranchus sp. 1
	Terebellides sp. 1

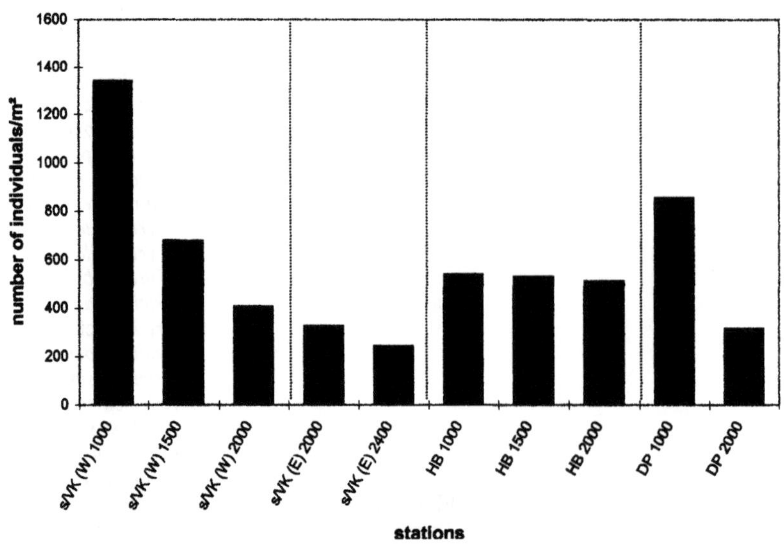

Fig. 1 Polychaete density (individuals per m^2) in the studied areas [*s/VK* south of Vestkapp, *HB* Halley Bay, *DP* Drake Passage, (*W*) western transect, (*E*) eastern transect]; *numbers* are depths in metres

Fig. 2 Polychaete species richness (number of species) in the studied areas. Abbreviations as in Fig. 1

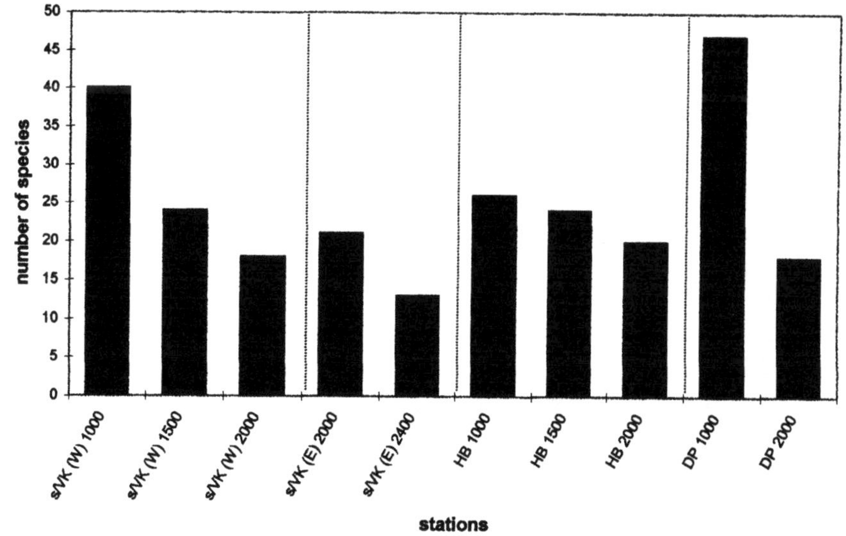

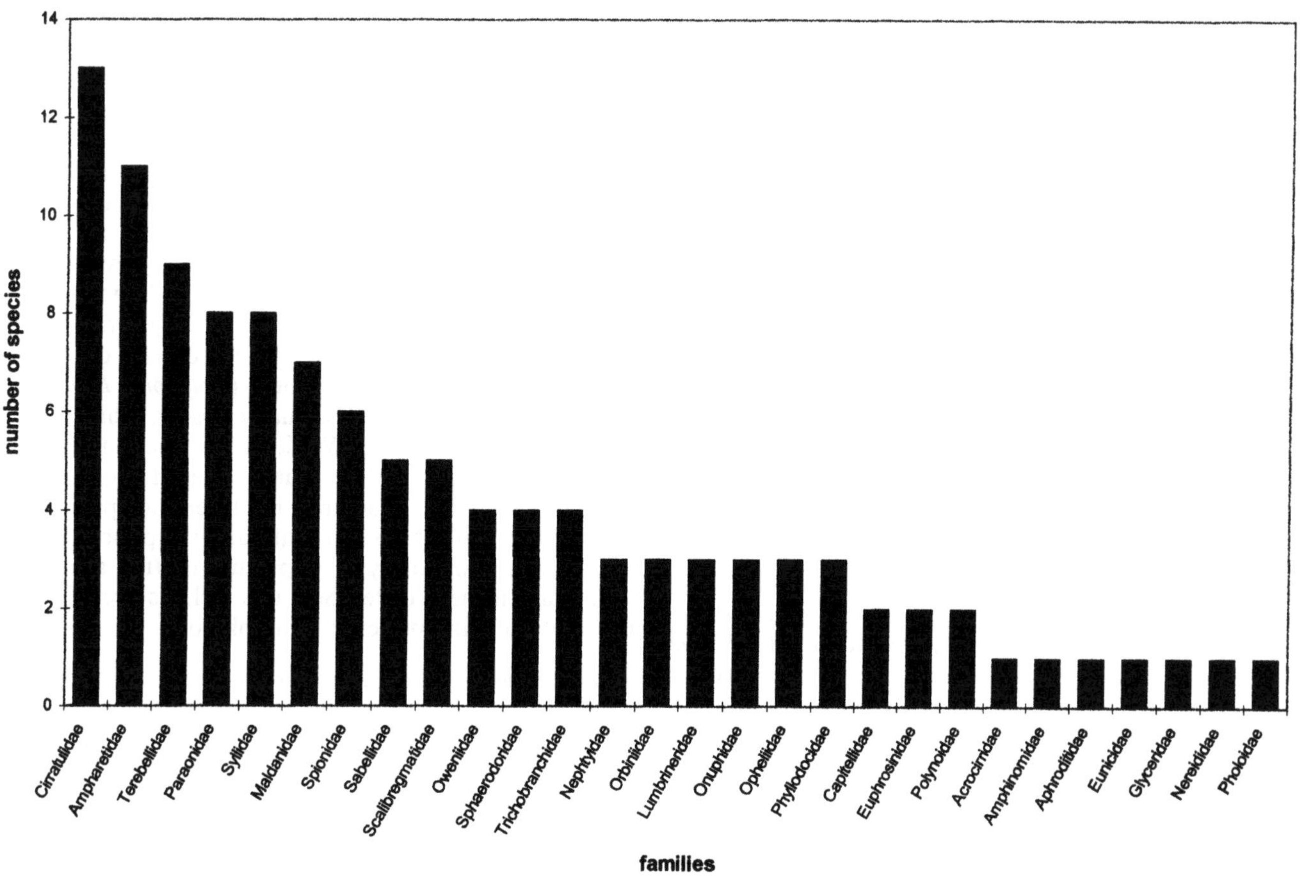

Fig. 3 Species richness of individual polychaete families in the studied areas; pooled data of all stations

than 1,000 m than in shallower water, where values differed by as much as an order of magnitude between stations at similar depth but in different locations (B. Hilbig, unpublished work).

If the decrease in abundance and especially species richness with depth does indeed prove to be true, this would represent a substantial difference to continental slopes of the temperate Atlantic and Pacific Oceans where a peak in species richness around 2,000 m seems to be a widespread and frequent phenomenon (Rex 1981, 1983; Paterson et al. 1985, 1998; SAIC 1992). Regular collection of environmental data, such as sediment grain size, organic carbon concentration in the sediment, and C/N ratio, would help to elucidate the importance of environmental factors other than depth

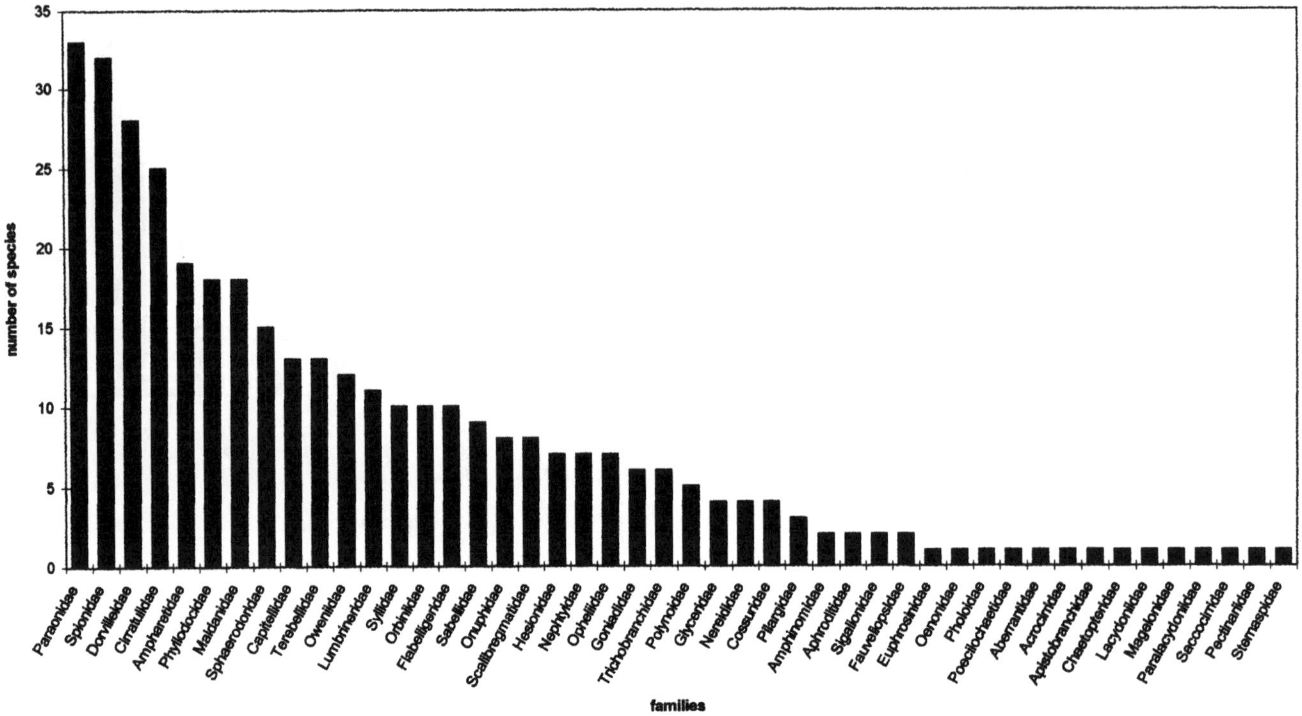

Fig. 4 Species richness of individual polychaete families in the deep American Pacific off California (after SAIC 1992)

for the structure of infaunal communities in the Weddell Sea.

The composition of the polychaete assemblages in the studied area proved to be markedly different from other slope areas in the northern hemisphere, for example, the slope off the United States Pacific coast. Samples collected there in 800–2,000 m revealed that the most speciose families in that area were paraonids and spionids, with more than 30 species each, followed by dorvilleids with 28 species (Fig. 4). The last were virtually absent in the deep Weddell Sea samples analysed to date; members of this family are small, vagile organisms, some of which are apparently opportunists that can undergo rapid mass development when a large food fall becomes available. At peak density, they can form a turf covering the entire area (Smith, personal communication). The tube-dwelling forms that were so predominant in the Weddell Sea were much less important in the eastern Pacific, even though the number of species was still fairly high, with 19 ampharetids, 18 maldanids and 15 terebellids. In sharp contrast to the Weddell Sea, there were also 15 species of capitellids. Analyses of benthic communities on the continental slope in the northwest Atlantic (Hilbig 1994) show that the polychaete assemblages there are very similar to the Pacific.

The absence or poor representation of polychaete families in the deep Weddell Sea and Drake Passage, which are quite speciose in northern temperate deep-sea areas, are due to different reasons. Dorvilleids are known to form a group of organisms specially adapted to opportunism under deep-sea conditions in the Atlantic and Pacific. Spionids and some capitellids are also regarded as opportunists in the deep sea and were observed to be pioneer colonisers in recolonisation tray experiments (Grassle 1977; Desbruyères et al. 1980, 1985); they are typical members of nearly every soft-bottom community and represented by a large number of species in the deep sea, but seem to be generally rarer in Antarctic waters, with an entire group of otherwise widespread, closely related genera (*Polydora*, *Dipolydora*, *Boccardia*, etc.) being virtually absent. This is particularly true for the deep Weddell Sea where no typical deep-sea spionids could be found, except for one undescribed *Spiophanes*. There may not be enough patches of organic enrichment on the seafloor to support a large and diverse community of opportunists. It is also possible that, in the Weddell Sea, this niche is occupied by the overwhelming presence of amphipods.

Acknowledgements Special thanks are due to Dieter Gerdes who saved my project by sharing his material, to crew and shipmates aboard R/V "Polarstern" who helped to obtain and process the samples, to Günther Sach and Olgierd Wrzesinski for sorting the samples, and to Claude de Broyer for the bottomless supply of espresso when it was needed. Dieter Fiege, Volker Storch and an anonymous reviewer are thanked for their helpful comments on an earlier version of the manuscript. This publication is part of a study funded by the Deutsche Forschungsgemeinschaft (DFG) under the project number BR1121-7.

References

Augener H (1932) Antarktische und antiboreale Polychaeten nebst einer Hirudinee. Sci Res Norwegian Antarctic Exped, 1927–28. Nor Videns Akad Oslo 9:86

Benham WB (1921) Polychaeta. Australasian Antarctic Expedition 1911–1914. Under the leadership of Sir Douglas Mawson. Sci Rep Ser C Zool Bot 6:128

Blake JA (1996) Family Cirratulidae Ryckholdt, 1851. Including a revision of the genera and species from the eastern North Pacific. In: Blake JA, Hilbig B, Scott PH (eds) Taxonomic atlas of the benthic fauna of the Santa Maria Basin and the western Santa Barbara Channel, vol 6. The Annelida Part 3 – Polychaeta: Orbiniidae to Cossuridae. Santa Barbara Museum of Natural History, pp 263–376

Blake JA, Grassle JF (1994) Benthic community structure on the U.S. South Atlantic slope off the Carolinas: spatial heterogeneity in the current dominated system. Deep Sea Res II 41:835–874

Desbruyères D, Bevas JY, Khripounoff A (1980) Un cas de colonisation rapide d'un sédiment profond. Oceanol Acta 3:285–291

Desbruyères D, Deming J, Dinet A, Khripounoff A (1985) Réactions de l'écosystème benthique profond aux perturbations: nouveaux résultats expérimentaux. In: Laubier L, Monniot C (eds) Peuplements profonds du Golfe de Gascogne. IFREMER, Brest, pp 121–142

Fauvel P (1936) Polychètes Expédition antarctique Belgica. Résultats du voyage de la Belgica en 1897–1899, sur le commandement de A. de Gerlache de Gomery

Gerdes D, Klages M, Arntz WE, Herman RL, Galéron J, Hain S (1992) Quantitative investigations on macrobenthos communities of the southeastern Weddell Sea shelf based on multi box corer samples. Polar Biol 12:291–301

Grassle JF (1977) Slow recolonization of deep-sea sediment. Nature 265:618–619

Grassle JF, Maciolek NJ, Blake JA (1990) Are deep-sea communities resilient? In: Woodwell G (ed) The earth in transition. Cambridge University Press, New York, pp 385–393

Hartman O (1964) Polychaeta Errantia of Antarctica. Antarct Res Ser 3:1–131

Hartman O (1966) Polychaeta Myzostomidae and Sedentaria of Antarctica. Antarct Res Ser 7:1–158

Hartman O (1967) Polychaetous annelids collected by the USNS Eltanin and Staten Island cruises, chiefly from Antarctic seas. Allan Hancock Monogr Mar Biol 2:1–387

Hartmann-Schröder G, Rosenfeldt P (1988) Die Polychaeten der "Polarstern"-Reise ANT III/2 in die Antarktis 1984. Teil 1: Euphrosinidae bis Chaetopteridae. Mitt Hamb Zool Mus Inst 85:25–72

Hartmann-Schröder G, Rosenfeldt P (1989) Die Polychaeten der "Polarstern"-Reise ANT III/2 in die Antarktis 1984. Teil 2: Cirratulidae bis Serpulidae. Mitt Hamb Zool Mus Inst 86:65–106

Hessler RR, Jumars PA (1974) Abyssal community analysis from replicate box cores in the central North Pacific. Deep Sea Res 21:185–209

Hilbig B (1994) Faunistic and zoogeographical characterization of the benthic infauna on the Carolina continental slope. Deep Sea Res II 41:929–950

McIntosh WC (1885) Report on the Annelida Polychaeta collected by H.M.S. Challenger during the years 1873–76. Challenger Rep 12:1–554

Monro CCA (1930) Polychaete worms. Discovery Rep 2:1–222

Monro CCA (1936) Polychaete worms. II. Discovery Rep 12:59–198

Paterson GLJ, Lambshead PJD, Sibuet M (1985) The Ophiuroidea fauna of the Bay of Biscay. In: Laubier L, Monniot C (eds) Peuplements profonds du Golfe de Gascogne. IFREMER, Brest, pp 491–507

Paterson GLJ, Wilson GDF, Cosson N, Lamont PA (1998) Hessler and Jumars (1974) revisited: abyssal polychaete assemblages from the Atlantic and Pacific. Deep Sea Res II 45:225–251

Rex MA (1981) Community structure in the deep-sea benthos. Annu Rev Ecol Syst 12:331–353

Rex MA (1983) Geographic patterns of species diversity in the deep-sea benthos. In: Rowe (ed) The sea, vol 8. Wiley, New York, pp 453–472

SAIC (1992) Benthic ecology and sediment characterization ocean studies report. Detailed physical and biological oceanographic studies for an ocean site designation effort under the Marine Protection, Research and Sanctuaries Act of 1972 (MPRSA). Final report prepared by Science Applications International Corporation for EPA Region IX under EPA contract no. 68-C8–0062

Polar Biol (2001) 24: 785–789
DOI 10.1007/s003000100290

Angelika Brandt

Great differences in peracarid crustacean density between the Arctic and Antarctic deep sea

Abstract Density of peracarid crustaceans (Amphipoda, Cumacea, Isopoda, Mysidacea and Tanaidacea) from epibenthic sledge samples is presented from Weddell Sea stations (Vestkapp and Halley Bay ANT XV-3). These data from both the supranet (1–1.3 m above the ground) and epinet (27–60 cm above the ground) of the epibenthic sledge are compared with samples taken at comparable depths in the Arctic Ocean (ARK X-1 and ARK XI-2) using the same gear. The latter showed 1 magnitude higher density values in similar depth zones in the Arctic Ocean, possibly reflecting the age of the environment and maturity of the benthic communities. Densities of Peracarida increased slightly with increasing depth in the Arctic, whereas they decreased with increasing depth in the Weddell Sea, especially at around 1,000 m in the epinet samples. Species richness of Peracarida is presented from the supranet samples for the first time in the Weddell Sea. In the Antarctic samples, Mysidacea with their hyperbenthic mode of life were generally collected more frequently in the supranet than in the epinet, while the other taxa were more abundant in the epinet, even at deeper stations at the shelf break.

Introduction

Few investigations on the suprabenthic community have been carried out in polar areas (e.g. Brandt 1995, 1997; Brandt et al. 1996; Sirenko et al. 1996; San Vicente et al. 1997; Arnaud et al. 1998; Saiz-Salinas et al. 1998). Except for Euphausiacea and Copepoda, the Peracarida are the most successful taxon of Crustacea in the Southern Ocean, with a high percentage of endemic species (≈80%) (Brandt 1999). One goal of the EASIZ II (ANT XV/3) cruise was to fill gaps in the taxonomic inventory of many groups (Arntz and Gutt 1999). In order to collect smaller species and especially the vagile, hyperbenthic fauna, an epibenthic sledge (EBS) (Brandt and Barthel 1995; Rothlisberg and Pearcy 1977) was employed. Densities of Peracarida sampled in the Weddell Sea were compared with results obtained in the same way from the Arctic Ocean, and an explanation for differences sought.

Please note: When citing this article, please use the reference to the original source in the journal Polar Biology as given on the top of this page.

A. Brandt
Zoological Institute and Zoological Museum,
University of Hamburg, Martin-Luther-King-Platz 3,
20146 Hamburg, Germany
E-mail: Abrandt@zoologie.uni-hamburg.de

Materials and methods

The present investigation is based on material collected with an epibenthic sledge in February and March 1998 in the Weddell Sea and Bransfield Strait aboard RV "Polarstern" during ANT XV-3 (Table 1). Two vertical transects were sampled in the Weddell Sea, southwest of Vestkapp and off Halley Bay (one station each at ca. 200, 400, 750, 1,000, 1,500, 2,000 m), and one horizontal deep-sea transect of 3 stations at about 2,000 m depth off Kapp Norvegia, southwest of Vestkapp and off Halley Bay. Another transect was sampled off King George Island towards the Bransfield Strait. Seventeen successful EBS hauls were made.

Data from two expeditions from the Arctic Ocean using the same EBS, which was employed in a standardised way in ARK X/1 (≈75°N) and ARK XI/2 (≈79°N) (station data in Brandt 1997; Brandt and Schnack 1999), are compared with these preliminary results from the EASIZ II expedition (Brandt et al. 1999).

The EBS carries two sampling nets: an epinet, 100 cm wide and 33 cm high, extending from 27 to 60 cm above the bottom, and a supranet from 100 to 133 cm above the ground. A net of 0.5-mm mesh size with a cod end of 0.3-mm mesh was used. The mouth of the nets opens mechanically on bottom contact and closes when the EBS leaves the seafloor.

The sledge was hauled over the ground for 10 min at a mean velocity of about 1 knot. The haul distances were calculated on the basis of the GPS-derived positions at start and end of each haul. The area swept was calculated by multiplying the distance travelled by the net width. As the haul distances varied (Table 1), densities were calculated for a standardised 1,000-m haul (compare Basford et al. 1989; Svavarsson et al. 1990; Brandt 1993, 1995; Brandt and Piepenburg 1994).

The samples were sieved through a 0.3-mm screen and preserved in 4% buffered formalin solution. After 3 days, the samples were washed and transferred into 70% ethanol. For faunistic comparison the complete samples were analysed; no subsamples were taken.

While the material from the supranets was sorted to species level but is not yet completely determined, not all of the samples from the epinet were sorted and counted.

Results

Species richness of supranet samples analysed so far (Table 1) is high. For example, at station 48-114 in 755 m depth, 19 amphipod species were identified from 32 specimens (Andres, personal communication). Numbers of species were highest for the suprabenthic Amphipoda, followed by Isopoda and Cumacea. At station 48-130 (1,645 m) 34 peracarid species, and at station 48-089 (almost 2,000 m) 29 peracarid species were collected. Only one mysid species was sampled at any station with the supranet. Numbers of species of Amphipoda from the supranet samples did not decrease with depth (Table 1). In about 2,000 m depth 14 species were still identified from these supranet samples. In the supranet, most peracarid specimens and species occurred on the transects southwest of Vestkapp and in the Bransfield Strait off King George Island. Lysianassidae were the most abundant amphipod family. Munnidae, Paramunnidae and Jaeropsididae were most abundant among the isopods, followed by the Munnopsididae. Most of the species from other families only occurred in small numbers of specimens. Preliminary results on the larger species of isopods from the Weddell Sea samples have been published (Brandt et al. 1999). Mysidacea were most abundant southwest of Vestkapp, where besides Peracarida, copepods were the most abundant taxon (A. Brandt, unpublished work). Within the cumaceans, the Nannastacidae were the most speciose family.

Numbers of specimens generally decreased with depth in both supra- and epinet. Epinet samples contained more specimens than supranet samples (Table 2). The exception was Mysidacea, which were more frequent in the supranet. Scaled abundances from the supranet samples were low for almost all peracarid taxa, except at station 48-310 (204 m) in the Bransfield Strait. At this shallowest station, Cumacea and Amphipoda dominated, whereas at some mid-depth stations (e.g. 231–1,054 m) Mysidacea were most abundant. In the epinet samples, numbers of individuals were higher at stations at less than 1,000 m depth. As in the supranet samples, density was highest for the Amphipoda, especially at the shallowest stations, followed by Cumacea and Isopoda. Density values decreased at depths greater than 938 m (Table 2).

In the Arctic Ocean, densities of all peracarids were more than 1 order of magnitude higher at similar depths than in the Antarctic, especially deeper than 1,000 m (Table 3, Fig. 1). In the Arctic at 75°N (ARK X-1), peracarid densities increased with depth to about 1,900 m and then slightly decreased again. At 79°N (ARK XI-2: Table 3, Fig. 1) numbers of peracarids were also higher at deeper stations (Brandt and Schnack 1999). The lower peracarid densities, especially at depths greater than 1,000 m in the samples from our EASIZ II

Table 1 Station list of epibenthic sledge stations from the expedition ANT XV/3 with RV "Polarstern" ordered by depth (m) (*swVK* southwest of Vestkapp; *HB* Halley Bay; *KN* Kapp Norvegia; *BS* Bransfield Strait; *A* Amphipoda; *C* Cumacea; *I* Isopoda; *M* Mysidacea; *T* Tanaidacea; * no supranet employed). Values = numbers of species identified from supranet samples; haul (m) = trawled distance

Station	Date	Time	Region	Latitude (S)	Longitude (W)	Depth (m)	Haul (m)	A	C	I	M	T	Sediment
48-310	16.03.98	13:30	BS	62°15	58°22	204	336	12	4	10	1	0	Fine clay and silt
48-171	12.02.98	17:56	HB	75°26	26°39	231	392	3	1	1	0	0	Small pebbles, sponges
48-111	06.02.98	14:12	swVK	73°38	22°11	397	382	0	0	1	0	0	Sponge spicules, coarse gravel, sand
48-165	12.02.98	08:28	HB	75°00	27°13	397	482	13	1	3	1	1	Coarse sand, some silt
48-311	16.03.98	14:30	BS	62°16	58°23	413	295	3	1	1	1	0	No sediment in net
48-316	16.03.98	19:19	BS	62°16	58°16	742	207	0	0	0	1	0	Very fine clay and silt
48-114	06.02.98	18:43	swVK	73°36	22°24	755	397	19	8	1	1	3	Two large and some small stones
48-156	11.02.98	14:28	HB	74°37	26°58	782	328	10	1	3	1	1	Coarse sand
48-107	06.02.98	08:24	swVK	73°34	22°38	938	420	14	8	14	1	3	Fine gravel
48-145	10.02.98	17:15	HB	74°38	27°10	1054	317	4	4	4	1	0	Coarse sand, sponge spicules
48-320	17.03.98	07:09	BS	62°22	58°41	1064	243	8	0	1	1	0	Very fine clay and silt
48-323	17.03.98	13:16	BS	62°25	58°42	1488	215	0	0	0	1	0	Pebble-sized clay balls
48-142	10.02.98	10:28	HB	74°36	27°16	1574	442	12	4	3	1	0	Fine sand
48-089	04.02.98	10:24	swVK	73°27	22°45	1645	452	14	4	7	1	3	No sediment in net
48-130	08.02.98	14:40	swVK	73°23	22°08	1983	367	16	6	8	1	3	Coarse sand, fine silt, mucus
48-272	26.02.98	16:09	KN	71°28	15°10	2077	466	*	*	*	*	*	Foraminifera, coarse sand with clay
48-133	09.02.98	10:19	HB	74°31	27°12	2119	335	0	0	1	1	0	Very fine, clay and silt

Table 2 Comparison of scaled densities (per 1,000 m^2) of Peracarida from supra- and epinet (*five right-hand columns*) samples; stations ordered by depth (*A* Amphipoda; *C* Cumacea; *I* Isopoda; *M* Mysidacea; *T* Tanaidacea; * epinet samples not processed yet)

Station	Depth (m)	A	C	I	M	T	A	C	I	M	T
48-310	204	500	1580	467	9	51	*	*	*	*	*
48-171	231	9	0	6	90	0	1699	648	538	69	43
48-111	397	108	22	55	489	15	1246	233	508	115	50
48-165	397	70	11	18	50	0	*	*	*	*	*
48-311	413	17	7	3	14	0	*	*	*	*	*
48-316	742	10	0	5	0	0	*	*	*	*	*
48-114	755	162	124	124	424	12	*	*	*	*	*
48-156	782	57	13	19	54	0	*	*	*	*	*
48-107	938	0	0	3	0	0	738	833	640	24	190
48-145	1054	85	3	24	168	3	*	*	*	*	*
48-320	1064	53	0	8	4	0	*	*	*	*	*
48-323	1488	0	0	14	0	0	*	*	*	*	*
48-142	1574	89	6	23	10	15	317	102	624	9	25
48-089	1645	128	60	63	10	15	442	246	~600	24	~100
48-130	1983	95	112	38	19	22	~950	267	95	35	5
48-272	2077	0	0	0	0	0	219	197	73	0	67
48-133	2119	18	5	43	0	0	167	27	119	24	6

Table 3 Comparison of density data of peracarid taxa from ANT XV-3 with the two Arctic expeditions ARK X-1 and ARK XI-2

	Peracarida	Amphipoda	Cumacea	Isopoda	Mysidacea	Tanaidacea
ANT XV-3						
48-171 (231 m)	2997	1699	648	538	69	43
48-111 (397 m)	2152	1246	233	508	115	50
48-107 (938 m)	2426	738	833	640	24	190
48-142 (1,574 m)	1077	317	102	624	9	25
48-89 (1,645 m)	1412	442	246	~ 600	24	~ 100
48-130 (1,983 m)	1353	~ 950	267	95	35	5
48-272 (2,077 m)	556	219	197	73	0	67
48-133 (2,119 m)	343	167	27	119	24	6
ARK X-1						
31-017 (198 m)	5312	1753	2270	455	744	9
31-014 (321 m)	6577	1711	1079	3106	333	74
31-016 (774 m)	4008	1719	144	1797	116	117
31-020 (817 m)	300	106	35	136	12	6
31-009 (1,525 m)	23098	4475	3590	14307	4	859
31-006 (2,681 m)	13557	2900	234	9238	63	1171
ARK XI-2						
37-016 (188 m)	5999	1701	1313	716	2220	49
37-019 (299 m)	4943	1723	1299	1140	781	0
37-018 (311 m)	19316	13273	1558	2235	2087	163
37-014 (800 m)	10149	1435	665	5589	1224	1236
37-008 (1,150 m)	21326	1352	2695	11673	4803	803
37-021 (1,809 m)	35738	1416	24593	5561	680	3488

expedition (ANT XV/3), differ markedly from the Arctic Ocean results.

Discussion

Numbers of specimens sampled with the epibenthic sledge were about 1 order of magnitude lower in the Antarctic than in the Arctic samples (Table 3, Fig. 1) at comparable depths. This contrasts with the findings of Knox and Lowry (1977), who demonstrated a higher biomass for the benthos in the Antarctic than in the Arctic at similar depths. According to Knox (1994), highest densities in the Antarctic (at Arthur Harbour) occur in shallow-shelf areas in depths not exceeding 100 m, whereas densities showed a reduction of 1 order of magnitude for sites below 300 m.

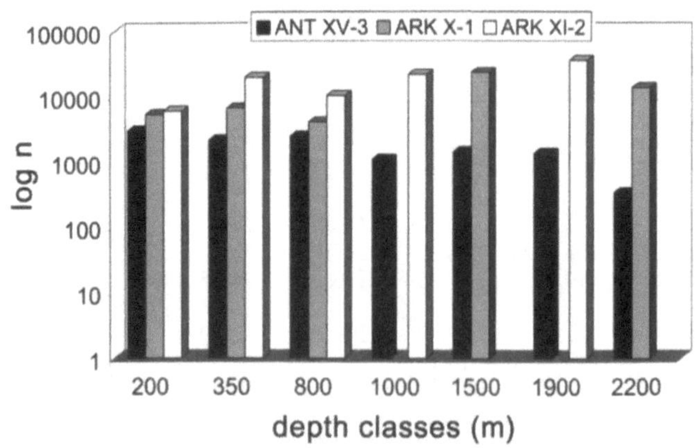

Fig. 1 Comparison of densities of all Peracarida sampled at different depth levels (for comparison only depth classes are illustrated) in the Antarctic (ANT XV-3) and in the Arctic (ARK X-1 and ARK XI-2) (logarithmic scale)

For logistic reasons it was not possible to take the Antarctic samples at exactly the same depths where the Arctic EBS samples were taken; therefore, only depth classes can be compared in the present paper. Supranet samples of ANT XV-3 were sorted to species level and numbers are presented from deep Weddell Sea stations in order to demonstrate that species richness is high in a depth of about 2,000 m (34 species from 369 individuals), even in a small sample size compared to the Arctic (e.g. 50 species in 5,120 individuals; Brandt 1997). The higher densities of Peracarida in the Arctic compared to the Antarctic could possibly reflect the relative ages of the two environments. The Arctic is the more recent polar environment and its peracarid community might therefore be more immature and dominated by fewer colonising species which occur in higher densities (e.g. MacArthur and Wilson 1963; MacArthur 1972; Cornell 1993). The Antarctic, however, has been glaciated for at least 22–25 million years (Clarke 1990; Crame 1997; Brandt 1999) and isolated by deep sea, while the Arctic is much younger (2–2.5 million years). Therefore it is likely that Antarctic communities generally show a higher maturity and diversification. Watling and Thurston (1989) and Brandt (1991) reported adaptive radiation processes for Antarctic amphipod and isopod families.

In a response to Poore and Wilson (1993) and Rex et al. (1993), Brey et al. (1994) showed that the benthic diversity in the Antarctic deep sea was much higher for Isopoda than in the Arctic. Sedimentation of organic carbon in the southeastern Weddell Sea lies in the same range and is of similar variability as in north polar seas (e.g. Hebbeln and Wefer 1991). Therefore, differences in the age of the environments are considered a probable reason for the differences in densities.

The question of whether we will find more species within a much lower density of Peracarida in the Antarctic than in the Arctic Ocean, however, remains unresolved until the epinet material from all stations can be sorted and identified to species level. Those results, including a species list, will be published in a subsequent paper.

Acknowledgements The author is grateful to the crew of RV "Polarstern" for their help and logistic support during the expedition ANT XV/3. My sincere thanks go to Dr. B. Hilbig and Ms. K. Linse for discussions, as well as to Ms. C. Stark and Ms. U. Weber for sorting. Dr. B. Hilbig kindly read, corrected and commented on an earlier version of the manuscript. Dr. G. Poore and two anonymous reviewers kindly commented on the manuscript, and their critiques improved this paper.

References

Arnaud PM, López CM, Olaso L, Ramil F, Ramos-Esplá, Ramos A (1998) Semi-quantitative study of macrobenthic fauna in the region of the South Shetland Islands and the Antarctic Peninsula. Polar Biol 19:160–166

Arntz WE, Gutt J (1999) The expedition ANTARKTIS XV/3 (EASIZ II) of RV "Polarstern" in 1998. Ber Polarforsch 301: 1–229

Basford DJ, Eleftheriou A, Raffaelli D (1989) The epifauna of the northern North Sea (56°–61°N). J Mar Biol Ass UK 69:387–407

Brandt A (1991) Zur Besiedlungsgeschichte des antarktischen Schelfes am Beispiel der Isopoda (Crustacea, Malacostraca). Ber Polarforsch 98:1–240

Brandt A (1993) Composition, abundance, and diversity of peracarid crustaceans on a transect of the Kolbeinsey Ridge, north of Iceland. Polar Biol 13:565–576

Brandt A (1995) Peracarid fauna (Crustacea, Malacostraca) of the NE Water Polynya off Greenland: documenting close benthic-pelagic coupling in the Westwind Trough. Mar Ecol Prog Ser 121:39–51

Brandt A (1997) Abundance, diversity, and community patterns of epi- and benthic-boundary layer Crustacea Peracarida at 75°N of East Greenland. Polar Biol 17:337–349

Brandt A (1999) On the origin and evolution of Antarctic Peracarida (Crustacea, Malacostraca). In: Arntz WE, Ríos C (eds) Magellan-Antarctic: ecosystems that drifted apart. Sci Mar 63 [Suppl 1]:261–274

Brandt A, Barthel D (1995) An improved supra- and epibenthic sledge for catching Peracarida (Crustacea, Malacostraca). Ophelia 43:15–23

Brandt A, Piepenburg D (1994) Peracarid crustacean assemblages of the Kolbeinsey Ridge, north of Island. Polar Biol 14:97–105

Brandt A, Schnack K (1999) Macrofaunal abundance at 79°N off East Greenland: focussing on Crustacea, Peracarida and Polychaeta. Polar Biol 22:75–81

Brandt A, Vassilenko S, Piepenburg D, Thurston M (1996) The species composition of the peracarid fauna (Crustacea, Malacostraca) of the Northeast Water Polynya (Greenland). Medr Groenland, Biosci 44:1–30

Brandt A, Hilbig B, Linse K, Wegener G (1999) Biodiversity and zoogeography of Mollusca, Polychaeta, and Crustacea Peracarida. In: Arntz WE, Gutt J (eds) The expedition ANTARKTIS XV/3 (EASIZ II) of RV "Polarstern" in 1998. Ber Polarforsch 301:1–229

Brey T, Klages M, Dahm C, Gorny M, Gutt J, Hain S, Stiller M, Arntz WE, Wägele JW, Zimmermann A (1994) Antarctic benthic diversity. Nature 368:297

Clarke A (1990) Southern Ocean cooling and the Antarctic marine fauna. In: Kerry KR, Hempel G (eds) Temperature and evolution. Springer, Berlin Heidelberg New York, pp 9–22

Cornell HV (1993) Unsaturated patterns in species assemblages: the role of regional processes in setting local species richness. In: Ricklefs RE, Schluter D (eds) Species diversity in ecological communities. Historical and geographical perspectives. University of Chicago Press, Chicago, pp 243–252

Crame JA (1997) An evolutionary framework for the polar regions. J Biogeogr 24:1–9

Hebbeln D, Wefer G (1991) Effects of ice coverage and ice-rafted material on sedimentation in the Fram Strait. Nature Lond 350:409–411

Knox GA (1994) The biology of the Southern Ocean. Cambridge University Press, Cambridge

Knox GA, Lowry JK (1977) A comparison between the benthos of the Southern Ocean and the North Polar Ocean with special reference to the Amphipoda and Polychaeta. In: Dunbar MJ (ed) Polar oceans. Proceedings of the Polar Oceans Conference, McGill University, Montreal. Arctic Institute of North America, Calgary, pp 423–462

MacArthur RH (1972) Geographical ecology: patterns in the distribution of species. Harper & Row, New York

MacArthur RH, Wilson EO (1963) The theory of island biogeography. Princeton University Press, Princeton, NJ

Poore GCB, Wilson GDF (1993) Marine species richness. Nature 361:597–598

Rex MA, Stuart CT, Hessler RR, Allen JA, Sanders HL, Wilson GDF (1993) Global-scale latitudinal patterns of species diversity in the deep-sea benthos. Nature 365:636–639

Rothlisberg PC, Pearcy WG (1977) An epibenthic sampler to study the ontogeny of vertical migration of *Pandalus jordani* (Decapoda, Caridea). Fish Bull 74:994–997

Saiz-Salinas JI, Ramos A, Munilla T, Rauschert M (1998) Changes in the biomass and dominant feeding mode of benthic assemblages with depth off Livingston Island (Antarctica). Polar Biol 19:424–428

San Vicente C, Ramos A, Jimeno A, Sorbe JC (1997) Suprabenthic assemblages from South Shetland Islands and Bransfield Strait (Antarctica): preliminary observations on faunistical composition, bathymetric and near-bottom distribution. Polar Biol 18:415–422

Sirenko BJ, Markhaseva EL, Buzhinskaya GN, Golikov AA, Menshutkina TV, Petryashov VV, Semenova TN, Stepanjants SD, Vassilenko SV (1996) Preliminary data on suprabenthic invertebrates collected during the RV "Polarstern" cruise in the Laptev Sea. Polar Biol 16:345–352

Svavarsson J, Brattegard T, Strömberg JO (1990) Distribution and diversity pattern of asellote isopods (Crustacea) in the deep Norwegian and Greenland Seas. Prog Oceanogr 24: 297–310

Watling L, Thurston M (1989) Antarctica as an evolutionary incubator: evidence from the cladistic biogeography of the amphipod family Iphimediidae. In: Crame JA (ed) Origin and evolution of Antarctic biota. Geol Soc Spec Publ 47: 297–313

Polar Biol (2001) 24: 657–662
DOI 10.1007/s003000100265

P. Dauby · Y. Scailteur · G. Chapelle · C. De Broyer

Potential impact of the main benthic amphipods on the eastern Weddell Sea shelf ecosystem (Antarctica)

Abstract As they represent one of the most diversified taxonomic groups on Antarctic bottoms, amphipods are likely to play a complex role in biogeochemical fluxes that occur within benthic ecosystems. The aim of this paper is to present, using both digestive tract analyses and relative species abundance data, the impact of amphipod feeding on different potential preys of the Weddell Sea. The study is based on data obtained for 29 representative amphipod species collected at 130 stations distributed along the eastern shelf of the Weddell Sea (depth range: 60–2,000 m) during three summer cruises, from 1989 to 1998. Sedimenting plankton particles (10–27%), crustaceans (22–32%) and fish carrion (5–18%) are the main food resources. Other abundant potential preys, such as molluscs or tunicates, do not seem to be consumed. Variations in the proportions of the different preyed food items are observed, mainly related to differences in relative amphipod species composition in samples. Presented results will help in refining ecological models of the prospected area, but also underline the need for accurate and reliable measurements of the feeding rates of Antarctic benthic organisms.

Introduction

In a recent attempt to model the organic matter flows within the benthic communities of the Weddell Sea continental shelf (Jarre-Teichmann et al. 1997), biologists of the Alfred-Wegener Institute (Bremerhaven, Germany) underlined the need for a better understanding and for an accurate measurement of several key processes that govern biogeochemical fluxes in these ecosystems. Among these processes, those related to nutrition of benthic organisms (like food intake or assimilation rates) are of primary interest.

Within the Southern Ocean benthic ecosystems, crustaceans are by far the most speciose taxon (Arntz et al. 1997) and, among crustaceans, amphipods represent the richest group, with more than 820 recorded species (De Broyer and Jazdzewski 1993, 1996), about 230 of which inhabit Weddell Sea waters. These peracarids have colonised a wide variety of ecological niches, from epontic to below-ground biotopes (De Broyer et al. 2001). They achieved a successful eco-ethological diversification, occupying apparently almost all the possible micro-habitats and developing various feeding strategies, from suspension-feeding to scavenging on vertebrate carcasses. Indeed, in a recent study (Dauby et al. 2001), we showed that at least eight different feeding types occurred among Weddell Sea gammaridean amphipods (despite the absence of macro-herbivory owing to the lack of benthic macroalgae in that area), including some specialised modes such as micropredatory browsing on invertebrate colonies.

In this study, we try to answer the questions: what do amphipods eat (what is the share of every potential food in their diet), and also, what is their impact on the different available food sources? We then analyse the role of the Weddell Sea amphipod taxocoenosis in consuming these food sources, taking into account their food preferences and their in situ relative occurrence. The presented results are, however, limited to the austral summer season and are thus not likely to represent a year-round situation.

Please note: When citing this article, please use the reference to the original source in the journal Polar Biology as given on the top of this page.

P. Dauby (✉) · Y. Scailteur · G. Chapelle · C. De Broyer
Laboratoire de Carcinologie,
Département des Invertébrés,
Institut Royal des Sciences Naturelles de Belgique (IRScNB),
rue Vautier, 29, 1000 Brussels, Belgium
E-mail: patrick.dauby@naturalsciences.be

Materials and methods

Amphipods were collected from benthic and suprabenthic samples taken in the eastern Weddell Sea during three Antarctic summer RV "Polarstern" cruises: EPOS leg 3 (1989; Arntz et al. 1990),

Table 1 Mean relative proportions (in %) of the different food items in the digestive tract of the most common amphipod species from the eastern Weddell Sea. Corresponding trophic types from Dauby et al. (2001)

Species	*n*	Family	Inorganic matter	Unidentified organic matter	Sedimenting plankton	Porifera	Cnidaria	Polychaeta	Crustacea	Bryozoa	Ophiuroidea	Holo-thuroidea	Pisces	Feeding type
Ampelisca richardsoni	20	Ampeliscidae		36	54	7			3					Suspension feeder
Epimeria georgiana	31	Epimeriidae	13		30	7	14	10	26					Deposit feeder
Epimeria macrodonta	10	Epimeriidae	5	18	8	14	30		15			10		Opportunistic predator
Epimeria robusta	10	Epimeriidae	7	25	8	12	19	7	19			3		Opportunistic predator
Epimeria rubrieques	12	Epimeriidae		8	18	7	18	8	27			14		Opportunistic predator
Epimeria similis	27	Epimeriidae		5	9	14	63	9						Micropredatory browser
Epimeriella walkeri	17	Epimeriidae			20	10	6		2		36	1	25	Macropredator/ scavenger
Eusirus perdentatus	20	Eusiridae	27	25				4	44					Macropredator/ scavenger
Rhachotropis antarctica	12	Eusiridae			17		9	11	63					Macropredator/ scavenger
Antamelita sp. nov.	27	Melitidae	10	18	1	4	1	3	57	1	4	1		Deposit feeder
Paraceradocus gibber	20	*Ceradocus* group	18	39	7	12		6	14		4			Deposit feeder
Echiniphimedia hodgsoni	20	Iphimediidae	8	20	9	62						2		Micropredatory browser
Gnathiphimedia mandibularis	15	Iphimediidae		5	2	15				78				Micropredatory browser
Iphimediella cyclogena	31	Iphimediidae			10			20				70		Micropredatory browser
Maxilliphimedia longipes	6	Iphimediidae			6	12	73	9						Micropredatory browser
Jassa goniamera	10	Ischyroceridae			50				50					Suspension feeder
Liljeborgia georgiana	22	Liljeborgiidae		13		13	13	24	37					Deposit feeder/ predator
Abyssorchomene nodimanus	14	Lysianassoidea		5									95	Carrion feeder
Abyssorchomene rossi	21	Lysianassoidea		5	5	5			5				80	Opportunistic carrion feeder
Hirondellea antarctica	18	Lysianassoidea					50		25				25	Micropredatory browser
Orchomenella pinguides	16	Lysianassoidea			50				50					Deposit feeder macrophage

Tryphosella murrayi	35	Lysianassoidea		5				5	43			47	Opportunistic carrion feeder
Uristes gigas	42	Lysianassoidea	14		11	13			27			35	Deposit feeder/ carrion feeder
Waldeckia obesa	30	Lysianassoidea	5		5	5						85	Carrion feeder
Melphidippa antarctica	10	Melphidippidae			71	6	2		21				Suspension feeder
Oediceroides calmani	24	Oedicerotidae	22	19	21	8		4	26				Deposit feeder/ predator
Oediceroides emarginatus	26	Oedicerotidae	25	3	2	7		12	51				Deposit feeder/ predator
Heterophoxus videns	17	Phoxocephalidae		16	20	8		37	19				Macropredator/ scavenger
Bathypanoploea schellenbergi	12	Stilipedidae			9		81		5	1	4		Micropredatory browser

EASIZ I (1996; Arntz and Gutt 1997) and EASIZ II (1998; Arntz and Gutt 1999). In total, 130 catches provided amphipod material in water depths from 60 m down to 2,000 m. Gears used included Agassiz, benthopelagic and bottom trawls, dredges, epibenthic sledges, TV grabs, giant and multibox corers, and baited traps.

Ethological observations (food detection and capture, mobility patterns) were performed with living specimens of about 40 species kept in cool containers on board and afterwards in a cool laboratory at IRScNB, Brussels. Amphipods (between 25 and >100 individuals, depending on species) were maintained at a temperature of −1°C (±1°C) in 2- to 30-l aquaria. Feeding experiments were performed in these aquaria, using as potential food source different living organisms (like crustaceans, echinoderms or plankton) or dead material (such as pieces of amphipods, fishes or squid) laid on the bottom or offered with forceps. These experiments were usually conducted over periods of 6–8 weeks.

Amphipod gut content analyses were done mainly on specimens fixed just after sampling in 4% formaldehyde. The amount of food in the stomach (C_s) and gut (C_g), respectively, was coded with arbitrary scores (4, more than 75% of the volume is filled; 3, from 50 up to 75%; 2, from 25 up to 50%; 1, less than 25%). Every item present in the digestive tract was determined to the group or species level when possible, and their proportion was coded using a similar coefficient (P_s, $P_g = 1$, 2, 3 or 4). A semi-quantitative approach, inspired from the "percentage points" method (Hynes 1950; Williams 1981), was used, with the following equations:

$$I(i) = \sum_{n=1}^{x} C_s(n) * P_s(n) + C_g(n) * P_g(n) \qquad (1)$$

where $I(i)$, dimensionless, is the importance of item i in the diet of a given species, and x the number of specimens dissected;

$$R(i) = \frac{I(i)}{\sum_{n=1}^{y} I(n)} * 100 \qquad (2)$$

where $R(i)$, in %, represents the relative importance of item i in the total diet of a given species, and y the number of different items.

The trophic impact of these amphipod species on Weddell Sea ecosystems has been evaluated by coupling feeding preferences and relative species abundances with the basic formula:

$$T(i) = \sum_{sp=1}^{x} \frac{\bar{N}_{sp}}{\bar{N}_{tot}} * R(i)_{sp} \qquad (3)$$

where $T(i)$ is the trophic impact on food item i (in %), x the number of analysed species, and $\bar{N}$ the mean number of individuals of a defined species (sp) and of all the x analysed species (tot) for all the samples of a cruise. Only classical benthic sampling devices (trawls and box corers) were taken into account for evaluating $\bar{N}$, as baited traps, for instance, do not reflect the actual instantaneous abundance of a species in a defined sampling area.

Results

The mean relative abundance [$R(i)$, in %] of the different food items in the digestive tract of the most important Weddell Sea amphipod species is presented in Table 1, with information about their principal feeding types. The latter, determined from both digestive tract analyses and behavioural observations, are extensively described in Dauby et al. (2001).

The 29 amphipod species selected for the present evaluation were chosen on the basis of their relative abundance, each one representing at least 0.4%, on average, of the total amphipod population collected during each cruise. For the EPOS cruise (1989), these 29 species represent 52.8% of total amphipods (58% of all the samples were analysed) while, for the EASIZ I cruise (1996), they represent 70.9% of total amphipods (82% of the samples were investigated).

The amphipod trophic impact on the different available food items [*T*(*i*)] of the Weddell sea is presented in Table 2, for the EPOS and the EASIZ I cruises, respectively.

For the EPOS samples, crustaceans appear to be the most frequently consumed items (32%), followed by fish carcasses (18%). If we do not take into account the unidentified organic matter (unrecognisable organic bodies without any specific features like cnidocysts, or without any hard structures like chitinous plates or spicules), particles of planktonic origin represent the third most common item. Sponges, cnidarians and worms (polychaetes and nematodes) are about equally consumed (5–7%), while bryozoans and echinoderms represent only a tiny fraction of the amphipod diet.

Results from the EASIZ I cruise are rather different. Plankton-originating cells (together with the unidentified matter) are the main dietary component (27%), before crustaceans (22%). Fish flesh constitutes only a small fraction (5%), and is of the same order of occurrence as worms, bryozoans or holothuroids. Sponges, with 7%, are the third item in importance in the amphipod diet.

Discussion

In their recent effort to summarise into a conceptual flow diagram the trophic interactions that occur on Weddell Sea bottoms, Jarre-Teichmann et al. (1997) admitted that crustaceans form a "very heterogenous box". Literature on high-Antarctic amphipod feeding strategies is indeed scarce (Oliver and Slattery 1985; Slattery and Oliver 1986; Coleman 1989a, b, c, 1990a, b; Klages and Gutt 1990a, b), and is concerned with less than 5% of known species. Owing to this lack of knowledge, Jarre-Teichmann et al. (1997) arbitrarily attributed the following predation impacts to crustaceans in their model:

Table 2 Mean relative trophic impacts [*T*(*i*)] (in %) of the eastern Weddell Sea benthic amphipod community on different potential food sources, during the EPOS (1989) and EASIZ I (1996) cruises. Inorganic material (sediment grains) were omitted from calculations

	EPOS	EASIZ I
Unidentified organic matter	17	25
Sedimenting plankton	10	27
Porifera	7	7
Cnidaria	7	3
Worms	5	4
Crustacea	32	22
Bryozoa	2	3
Ophiuroidea	1	1
Holothuroidea	1	3
Pisces	18	5

10% polychaetes, 3% crustaceans, 4% echinoderms, 1% bryozoans and 82% detritus. From Tables 1 and 2, it clearly appears that such a *T*(*i*) distribution is inappropriate. "Detritus"-feeders represent only one-half of the Weddell Sea amphipod species (Dauby et al. 2001) even when gathering together the deposit-feeders and all the occasional opportunistic scavengers of all types from micro-feeders to necrophages. The species used in Jarre-Teichmann et al.'s (1997) paper (*Paraceradocus gibber*, *Gnathiphimedia mandibularis*, *Epimeria robusta*, (*Abyss-*) *Orchomene plebs* and *Bovallia gigantea*) are not at all representative of the whole benthic amphipod community, and thus induce a bias in the model.

The present study shows that, when averaging over site and year variations, crustaceans, planktonic particles and carcasses form the bulk of the Weddell Sea amphipod diet. The other sources of feeding (worms, echinoderms, bryozoans or sponges) count for a minor part and are mainly consumed by some specialised predatory feeders.

The share between living and non-living food items is, however, sometimes difficult to establish. If it is relatively evident for, for example, fish muscles (although we have in very few cases observed cohorts of Lysianassoidea species attacking living fishes in the aquarium), it is not so obvious for items such as crustaceans. The presence of euphausiid remains in amphipod digestive tracts, for instance, would suggest a scavenging deposit-feeding mode, but the occurrence of amphipod remains can be attributed to both a predatory and a scavenging feeding mode. Complementary in vivo observations of feeding behaviour are thus necessary.

Two groups of benthic animals which are well represented in the high Antarctic are apparently absent from the diet of amphipods: molluscs (in terms of biodiversity, the second phylum after crustaceans; Arntz et al. 1997) and ascidians. The absence of these items could be related to their "protective" shell (although it is not much calcified in cold waters) or tunic. Such an argument is, however, not consistent, as some amphipod species are able to browse on holothuroid podia, on sponges or on bryzoan colonies. Some molluscs and ascidians may have been consumed, and may form part of the unidentified organic matter, which represents about one-fifth of the total ingested food, but no radula was found in amphipod digestive tracts.

The difference between the *T*(*i*) distributions for the two cruises (Table 2) is likely to be attributed to differences in sampling areas, and concurrently in the sampled amphipod communities (De Broyer et al. 2001). For instance, in the EPOS cruise, *Uristes gigas*, a scavenging lysianassoid, was abundant (>7% of total amphipods) in the samples, while in EASIZ I, one *Ampelisca richardsoni*, a typical suspension-feeder, was dominant (22%). Such differences emphasise the need for multiple sampling with devices that give a better overview of the benthic community composition, such as corers.

In the same way, the results presented in Table 2 should be regarded cautiously, as the analysed amphipod species, albeit dominant in samples, are not necessarily the dominant species of the prospected areas. Indeed, owing to the gears used, larger species are significantly more numerous in the samples. Except for one species (*Heterophoxus videns*), all the amphipods we analysed are longer than the mean size (17 mm) for all known East Antarctic species. Moreover, all of them were adult specimens, and the diet of juveniles may be different.

The conversion of the measured trophic impacts into actual food consumptions requires knowledge of both biomass and feeding rate of the amphipods. Reliable biomass data (based on multibox corer sampling effort) for eastern Weddell Sea amphipods were only published by Gerdes et al. (1992) for cruises performed from 1987 to 1989. The authors found a mean biomass of 669 mg WW m^{-2}, with values ranging from 8 to 6,396 mg WW m^{-2} depending on the sampling station. Amphipods counted for about 5‰ of total benthic animal biomass. Feeding rate measurements, however, were never published for high-Antarctic amphipods, maybe because of the experimental complexity in feeding most of these crustaceans in culture conditions. Some attempts, however, have been made to indirectly estimate this feeding rate (for example, using P/B ratios; Brey and Clarke 1993; Jarre-Teichmann et al. 1997), but these should be cautiously regarded. We have carried out some experimental series in aquaria on different necrophagous Lysianassoidea (De Broyer and Klages 1990; Y. Scailteur, G. Chapelle, C. De Broyer, unpublished work), and measured feeding rate values ranging from 14 to 40 mg g^{-1} day^{-1}, depending on species and prey. These values, however, are likely not applicable to the whole amphipod community, as necrophages are known to display a discrete method of feeding, depending greatly on food availability.

The present paper, hopefully shedding better light on the role of the amphipod taxocoenosis within the trophic flows which occur on the Weddell Sea bottoms, emphasises the necessity to amplify the research effort in at least two directions: a better knowledge of the benthic community structure (mainly species biomass estimates), and the development of techniques to accurately measure the organic matter fluxes at the individual (or population) level. Without such additional information, the elaboration of consistent ecosystem models, such as those presented, for example, by Rakusa-Suszczewski (1993) or Knox (1994), may remain at a standstill for a long time.

Acknowledgements The present research was performed under the auspices of the Scientific Research Programme on Antarctic (phase IV) from the Belgian State Prime Minister's Federal Office for Scientific, Technical and Cultural Affairs (OSTC contract no. A4/DD/B02). Samples were collected during the European "Polarstern" study (EPOS), sponsored by the European Science Foundation, and during the EASIZ I and II campaigns, sponsored by the Alfred-Wegener-Institut für Polar- und Meeresforschung (A.W.I.), Bremerhaven, Germany. We are indebted to the officers and crews of R/V "Polarstern" for the attention they paid in deploying sampling devices. We would also like to thank Professor W. Arntz and Drs. D. Gerdes, M. Klages, T. Brey, J. Gutt (A.W.I.) and all the people who helped in collecting, sorting and analysing the samples.

References

Arntz WE, Gutt J (eds) (1997) The expedition ANTARKTIS XIII/3 (EASIZ I) of RV "Polarstern" to the Weddell Sea in 1996. Ber Polarforsch 249:148

Arntz WE, Gutt J (eds) (1999) The expedition ANTARKTIS XV/3 (EASIZ II) of RV "Polarstern" in 1998. Ber Polarforsch 301:229

Arntz WE, Ernst W, Hempel I (eds) (1990) The expedition ANTARKTIS VII/4 (EPOS Leg 3) and VII/5 of the R/V "Polarstern" in 1989. Ber Polarforsch 68:214

Arntz WE, Gutt J, Klages M (1997) Antarctic marine biodiversity: an overview. In: Battaglia B, Valentia J, Walton DWH (eds) Antarctic communities: species, structure and survival. Cambridge University Press, Cambridge, pp 3–14

Brey T, Clarke A (1993) Population dynamics of marine benthic invertebrates in Antarctic and Sub-Antarctic environments: are there unique adaptations? Antarct Sci 5:253–266

Coleman CO (1989a) On the nutrition of two Antarctic Acanthonotozomatidae (Crustacea: Amphipoda). Polar Biol 9:287–294

Coleman CO (1989b) Burrowing, grooming and feeding behaviour of *Paraceradocus*, an Antarctic amphipod genus (Crustacea). Polar Biol 10:43–48

Coleman CO (1989c) *Gnathiphimedia mandibularis* (KH Barnard 1930), an Antarctic amphipod (Acanthonotozomatidae, Crustacea) feeding on Bryozoa. Antarct Sci 1:343–344

Coleman CO (1990a) Two new Antarctic species of the genus *Epimeria* (Crustacea: Amphipoda: Paramphithoidae), with description of juveniles. J R Soc N Z 20:151–178

Coleman CO (1990b) *Bathypanoploea schellenbergi* (Holman & Watling, 1983), an Antarctic amphipod (Crustacea) feeding on Holothuroidea. Ophelia 31:197–205

Dauby P, Scailteur Y, De Broyer C (2001) Trophic diversity within the eastern Weddell Sea amphipod community. Hydrobiologia 443:69–86

De Broyer C, Jazdzewski K (1993) Contribution to the marine biodiversity inventory. A checklist of the Amphipoda (Crustacea) of the Southern Ocean. Doc Trav Inst R Sci Nat Belg 73:1–155

De Broyer C, Jazdzewski K (1996) Biodiversity of the Southern Ocean: towards a new synthesis for the Amphipoda (Crustacea). Boll Mus Civ Stor Nat Verona 20:547–568

De Broyer C, Klages M (1990) Studies on amphipod biology. In: Arntz WE, Ernst W, Hempel I (eds) The expedition ANTARKTIS VII/4 (EPOS Leg 3) and VII/5 of the R/V "Polarstern" in 1989. Ber Polarforsch 68:113–115

De Broyer C, Scailteur Y, Chapelle G, Rauschert M (2001) Diversity of epibenthic habitats of gammaridean amphipods in the eastern Weddell Sea. Polar Biol 24:744–753

Gerdes D, Klages M, Arntz WE, Herman RL, Galéron J, Hain S (1992) Quantitative investigations on macrobenthos communities of the southeastern Weddell Sea shelf based on multibox corer samples. Polar Biol 12:291–301

Hynes HBN (1950) The food of the freshwater sticklebacks *Gasterosteus aculeatus* and *Pygosteus pungitius*, with a review of the methods used in studies of the food of fishes. J Anim Ecol 19:36–58

Jarre-Teichmann A, Brey T, Bathmann UV, Dahm C, Dieckmann GS, Gorny M, Klages M, Pages F, Plötz J, Schnack-Schiel SB, Stiller M, Arntz WE (1997) Trophic flows in the benthic community of the eastern Weddell Sea, Antarctica. In: Battaglia B, Valentia J, Walton DWH (eds) Antarctic communities: species, structure and survival. Cambridge University Press, Cambridge, pp 118–134

Klages M, Gutt J (1990a) Observations on the feeding behaviour of the Antarctic gammarid *Eusirus perdentatus* (Chevreux, 1912) (Crustacea: Amphipoda) in aquaria. Polar Biol 10:359–364

Klages M, Gutt J (1990b) Comparative studies on the feeding behaviour of high Antarctic amphipods (Crustacea) in laboratory. Polar Biol 11:73–79

Knox GA (1994) The biology of the Southern Ocean. Cambridge University Press, Cambridge

Oliver JS, Slattery PN (1985) Effects of crustacean predators on species composition and population structure of soft-bodied infauna from McMurdo Sound, Antarctica. Ophelia 24:155–175

Rakusa-Suszczewski S (1993) Matter transport and budgets in the Admiralty Bay area. In: Rakusa-Suszczewski S (ed) The maritime Antarctic coastal ecosystem of Admiralty Bay. Department of Antarctic Biology, Polish Academy of Science, Warsaw, pp 199–212

Slattery PN, Oliver JS (1986) Scavenging and other feeding habits of lysianassid amphipods (*Orchomene* spp.) from McMurdo Sound, Antarctica. Polar Biol 6:171–177

Williams MJ (1981) Methods for analysis of natural diet in portunid crabs (Crustacea: Decapoda: Portunidae). J Exp Mar Biol Ecol 52:103–113

Polar Biol (2001) 24: 744–753
DOI 10.1007/s003000100276

Claude De Broyer · Yves Scailteur · Gauthier Chapelle
Martin Rauschert

Diversity of epibenthic habitats of gammaridean amphipods in the eastern Weddell Sea

Abstract Gammaridean amphipods appear ubiquitous in the shelf communities of the eastern Weddell Sea where more than 230 species have been recorded. Comparison of catches of the 40 most common species taken by different gears and ethological observations in the aquarium allowed us to distinguish 6 groups: endobenthic, epibenthic, hyperbenthic (or suprabenthic), benthopelagic, pelagic and cryopelagic. Only epibenthic species, which form the bulk of the fauna, are dealt with in the present paper. Aquarium observations indicated three different epibenthic strata occupied by amphipods: the sediment surface, and a lower and upper level on the substrata formed by sessile suspension feeders. In addition, four symbiotic microhabitats were detected in or on sponges, ascidians, hydrozoans and gorgonians.

Introduction

Recent observations by still and video underwater cameras, coupled with analyses of benthos samples, have allowed a precise description of the variety of benthic assemblages from the eastern Weddell Sea shelf (i.e. Galéron et al. 1992; Gutt and Schickan 1998; Gutt and Starmans 1998). The continental shelf, to a depth of more than 600 m, is colonised in many places by species-rich assemblages of abundant, diverse and multistratified suspension feeders, such as sponges, bryozoans, cnidarians, hydrozoans, holothurians and crinoids. However, there is a gradient of species-rich assemblages of suspension feeders, mostly in the Kapp Norvegia region, to extremely poor detritus-feeder assemblages, in the southernmost part of the Weddell Sea. The shelf-bottom cover appears mostly patchy and ranges from a few percent to 100% of the bottom surface. The patchy, diverse and multistratified sessile benthos offers a high diversity of potential microhabitats to small vagile invertebrates.

Gammaridean amphipods, often collected in benthic samples, seem to be ubiquitous in the bottom communities of the eastern Weddell Sea, where they constitute an abundant and diverse group (Voss 1988; De Broyer and Klages 1990; Klages 1991; De Broyer et al. 1997, 1999). Their precise habitats, however, are usually unknown. Some recent "Polarstern" campaigns provided new data on amphipod distribution, which are partly analysed in this paper, in an attempt to document the habitats (and as far as possible the microhabitats) of the 40 most common epibenthic species of the eastern Weddell Sea.

Materials and methods

Sampling

The amphipod material analysed here was collected in the eastern Weddell Sea during the Antarctic summer cruises EPOS 3 (1989), EASIZ I (1996) and EASIZ II (1998) of the RV "Polarstern" (Arntz et al. 1990; Arntz and Gutt 1997, 1999). Benthic samples were taken along different transects, mainly off Kapp Norvegia, southwest of Vestkapp, in Drescher Inlet, and off Halley Bay, at depths from 60 m down to 2,500 m (Fig. 1). In total, 130 catches provided about 80,000 specimens of amphipods belonging to more than 230 species. Collecting gears included Agassiz trawl, bottom trawl, benthopelagic trawl, Rauschert's dredge, epibenthic sledge, TV grab, giant box corer, multibox corer and "autonomous" baited traps. Voucher specimens are kept at the "Antarctic Marine Biodiversity Reference Centre", devoted to amphipods, at the Royal Belgian Institute of Natural Sciences in Brussels, Belgium.

Please note: When citing this article, please use the reference to the original source in the journal Polar Biology as given on the top of this page.

C. De Broyer (✉) · Y. Scailteur · G. Chapelle
Institut Royal des Sciences Naturelles de Belgique,
Département des Invertébrés, rue Vautier 29,
1000 Brussels, Belgium
E-mail: claude.debroyer@naturalsciences.be

M. Rauschert
Bruno-H.-Bürgel-Allee 31, 15732 Eichwalde,
Germany

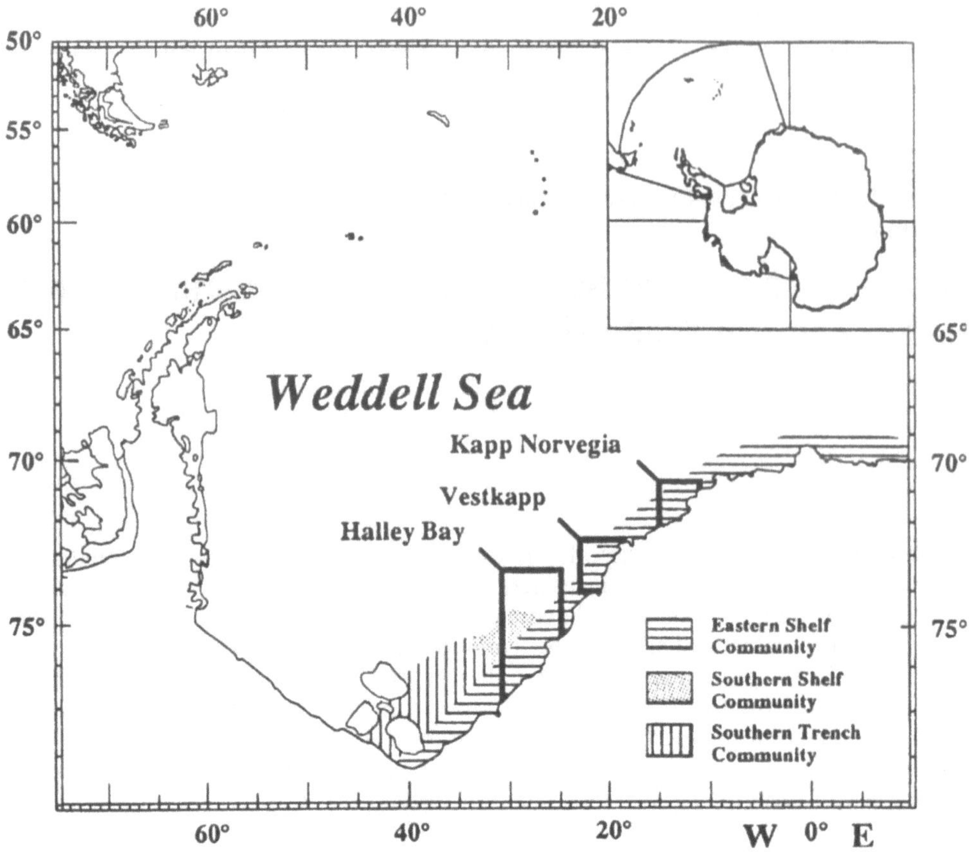

Fig. 1 Investigated areas and major macrobenthic assemblages (after Voss 1988; Galeron et al. 1992) on the eastern Weddell Sea shelf

Habitat characterisation

Identification of the amphipod macrohabitats was based on a comparative analysis of catches taken by different collecting gears. Description of particular microhabitats relied on aquarium and incidental catch observations during the cruises, as well as on published records (Kunzmann 1996). Bathymetric distribution (Table 1) refers to collection data from the cited Weddell Sea cruises.

Aquarium observations

Ethological observations (habitat choice and mobility patterns: swimming, crawling, walking, climbing or burrowing) were performed on living specimens kept in a cool container on board (mostly during the "Polarstern" EASIZ II cruise), and afterwards in a cool laboratory at IRScNB, Brussels. The amphipods were maintained at $-1°C$ ($\pm 1°C$) in aquaria with volumes of 2–30 l. Aquaria were provided with different kind of substrates, according to the known or suspected life-style of studied species. A large aquarium with a "reconstituted natural bottom" (30 cm high) was used to study the species behaviour. The "reconstituted" bottom was composed of a mosaic of mixed (fine/coarse) sediment, sponge spicule mat, stones and different common sessile organisms such as sponges, cnidarians, hemichordates and bryozoans. Observations were qualitative; movements and position of amphipods in aquaria were checked at least twice a day for a period of 10–56 days and typical behaviours were video-recorded.

Results

Comparative analysis of catches allowed to distinguish six major amphipod habitats in the neritic zone (Fig. 2). The *endobenthic* habitat is composed of the first centimetres of the sediment and is occupied by sedentary tube- or "cell"-dwellers of the family Ampeliscidae and by sedentary and temporary burrowers belonging to the Oedicerotidae, Phoxocephalidae and Lysianassoidea. The *epibenthic* habitat with three strata, described hereafter, is colonised by numerous free-living species, the most abundant being the Epimeriidae, Eusiridae s.l., Iphimediidae and Lysianassoidea. Among the epibenthos, several *symbiotic* and *inquilinous* microhabitats are occupied mainly by some Colomastigidae, Dexaminidae, Leucothoidae, Lysianassoidea, Sebidae, Stenothoidae and Stilipedidae. The Benthic Boundary Layer forms the *hyperbenthic* (or suprabenthic) habitat of some swimming Eusiridae s.l. and Lysianassoidea. Few Epimeriidae, Eusiridae s.l., Lysianassoidea and Stegocephalidae occupy the neritic water column, showing either a *benthopelagic* or a purely *pelagic* life-style. In the *cryopelagic* habitat, i.e. the under-surface of the sea ice, three species (one lysianassoid and, surprisingly, two stenothoids) have been incidentally recorded so far. Only the epibenthic level is analysed in detail here; the other compartments will be dealt with elsewhere.

The epibenthic habitats

The epibenthic layer of the eastern Weddell Sea shelf includes all habitats from the sediment surface level up to the top of sessile organisms, which offer secondary

Table 1 Epibenthic habitats and motility patterns of gammaridean amphipods from the eastern Weddell Sea (*according to Kunzmann 1996)

Species	Epibenthic macrohabitat	Substrate/ microhabitat	E. Weddell Sea depth range (m)	Motility pattern	Aquarium observations		Feeding type (according to Y. Scailteur, unpublished work)
					N (days)	*N* (ind.)	
Melitidae sp. nov.	Surface	Soft	246–836	Weakly motile	45	84	Macrophagous deposit feeder
Melphidippa antarctica	Surface	Soft	232–672	Weakly motile	50	28	Suspension feeder (passive)
Epimeria georgiana	Surface	Soft + mixed	184–928	Weakly motile	10	70	Macrophagous deposit feeder
	Lower level	Sessile benthos			51	52	
Liljeborgia georgiana	Surface	Detritic + mixed	196–1030	Weakly motile	50	28	Macropredator/deposit feeder
Paraceradocus gibber	Surface	Detritic + mixed	118–793	Weakly motile	56	34	Macrophagous deposit feeder
Waldeckia obesa	Surface + ?	Detritic?	80–1030	Highly motile	50	500	Scavenger
Abyssorchomene scotianensis	Surface + ?	Detritic?	600–1500	Highly motile			Scavenger
Orchomenella acanthura	Surface + ?	Detritic?	170–261	Highly motile			Scavenger? Detritivore?
O. pinguides	Surface + ?	Detritic?	234–301	Highly motile			Scavenger
Parschisturella carinata	Surface + ?	Detritic?	219–2081	Highly motile	53	100	Scavenger
Tryphosella murrayi	Surface + ?	Detritic?	80–920	Highly motile	55	71	Scavenger/macro-predator
Uristes adarei	Surface + ?	Detritic?	233–690	Highly motile			Macropredator/deposit feeder
U. gigas	Surface + ?	Detritic?	236–810	Highly motile	40	32	Macrophagous deposit feeder
Echiniphimedia hodgsoni	Lower level	Sponges + bryozoans?	188–1130	Weakly motile	53	40	Micropredatory browser
Echiniphimedia scotti	Lower level	Sponges + bryozoans?	118–571	Weakly motile	50	25	Micropredator
Gnathiphimedia mandibularis	Lower level	Bryozoans + sponges	170–773	Weakly motile	55	55	Micropredatory browser
Iphimediella cyclogena	Lower level	Sessile benthos	178–889	Weakly motile			Micropredator
Maxilliphimedia longipes	Lower level	Sessile benthos	186–617	Weakly motile			Micropredatory browser
Epimeria rubrieques	Lower level	Sessile benthos	352–1030	Weakly motile	46	30	Predator (opportunistic)
Eusirus perdentatus	Surface + lower level	Mixed + sessile benthos	100–928	Moderately motile	55	120	Macropredator
Jassa goniamera	Lower level	Sessile benthos	170–584	Weakly motile	55	20	Suspension feeder (active)
Bathypanoploea schellenbergi	Lower/upper level	Gorgonians?	396–672	Weakly motile			Micropredatory browser
Epimeria macrodonta	Upper level	Sessile benthos	182–1030	Weakly motile	52	62	Predator (opportunistic)
Epimeria robusta	Upper level	Hydrozoans?	233–793	Weakly motile	55	15	Predator (opportunistic)
Epimeria similis	Upper level	Sessile benthos	242–1030	Weakly motile	46	10	Micropredatory browser
Epimeriella walkeri	Upper level	Sessile benthos	184–928	Highly motile	53	28	Macropredator
Hirondellea antarctica	Upper level	Cnidarians?	65–1136	Highly motile			Micropredator
Alexandrella mixta	Upper level	Gorgonians?	242–1543	Highly motile	50	6	Macropredator
*Andaniotes linearis**	Symbiotic	Sponges*	405*	Sedentary?			
*Colomastix simplicicauda**	Symbiotic	Sponges*	228*	Sedentary?			
*Leucothoe spinicarpa**	Symbiotic?	Sponges*	332–771*	Sedentary?			
Scaphodactylus sp. nov. 1	Symbiotic	Sponges	105	Sedentary?			

Table 1 (Contd.)

Species	Epibenthic macrohabitat	Substrate/ microhabitat	E. Weddell Sea depth range (m)	Motility pattern	Aquarium observations N (days)	N (ind.)	Feeding type (according to Y. Scailteur, unpublished work)
Polycheria antarctica	Symbiotic	Sponges* + ascidians	118–611	Sedentary			Suspension feeder (passive) parasitic predator*
Seba antarctica	Symbiotic	Sponges* + ascidians	253–399	Sedentary			Suspension feeder (passive) parasitic predator*
Leucothoe sp.	Symbiotic	Ascidians	399	Sedentary?			
Orchomenyx sp.	Symbiotic	Ascidians	186	Sedentary?			
Metopoides sp. nov. 2	Symbiotic	Ascidians	600–800	Sedentary?			
Torometopa sp. nov. 1	Symbiotic	Hydrozoans	230	Sedentary?			
Torometopa sp. nov. 6	Symbiotic	Hydrozoans	165	Sedentary?			
Thaumatelson sp. Z	Symbiotic	Hydrozoans?	214	Sedentary?			
Torometopa sp. nov. 5	Symbiotic	Gorgonians	400	Sedentary?			
Polycheria sp.	Symbiotic	Gorgonians?	64–110	Sedentary			

substrates to colonisers. Its thickness can reach about 1 m, as shown by bottom pictures and the size of the largest sessile invertebrates collected. Amphipods appear roughly distributed on the bottom in three different strata: the sediment surface and the lower and upper strata of the sessile epibenthos (Table 1 and Fig. 3).

The sediment surface

Heterogeneous in its composition, structure and thickness, the sediment surface varies from the soft type such as sand and mud to the detritic type such as coarse bryozoan debris or sponge spicule mats. Dropstones from icebergs can be mixed with this soft or coarse sediment. This heterogeneous sediment surface, often with cracks, interstices and holes, offers a great variety of microhabitats to a number of small vagile invertebrates such as amphipods (Fig. 4). Some amphipod species live only temporarily in this biotope, looking for transitory substrate and protection, while others appear sedentary, finding shelter and food there.

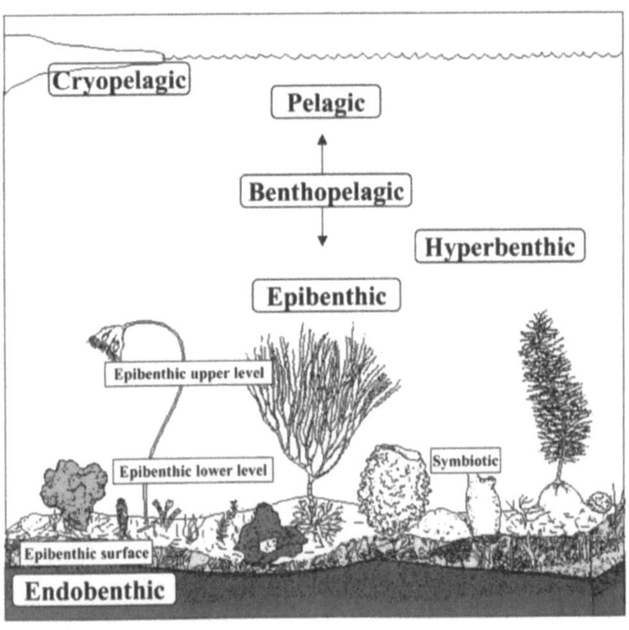

Fig. 2 Scheme of the amphipod macrohabitats in the eastern Weddell Sea neritic zone

Representative species from soft-sediment surface

Melitidae gen nov., sp. nov. Position in aquarium with "reconstituted bottom": stays on the soft-sediment surface, without attempting to burrow or to climb on sessile animals or stones; always keeps contact with the substrate (thigmotropy). Mobility: moderately active; usually walking upright on pereopods; not side-crawling like *Paraceradocus gibber* (Coleman 1989a) despite the similarity of morphology, but can rest on its side; seems unable to swim.

Melphidippa antarctica Schellenberg, 1926 (*Melphidippidae*) Usual position in aquarium is upside down, with dorsum on the bottom resting on the three posterior pereopods with the setose pereopods 3 and 4, as well as antenna 1, directed towards the water column (quite like *Melphidippella macra* Norman, 1869, described by Enequist 1949). Mobility: stays mostly motionless; rarely moving; swimming was not observed; probably sedentary.

Epimeria georgiana Schellenberg, 1931 (*Epimeriidae*) Collected in catches from different kinds of bottoms: bryozoan debris, sponge spicule mats, rich sessile fauna of hydrozoans, sponges, bryozoans and also from soft

Fig. 3 Scheme of presumed epibenthic habitats of some representative species of gammaridean amphipods on the eastern Weddell Sea shelf

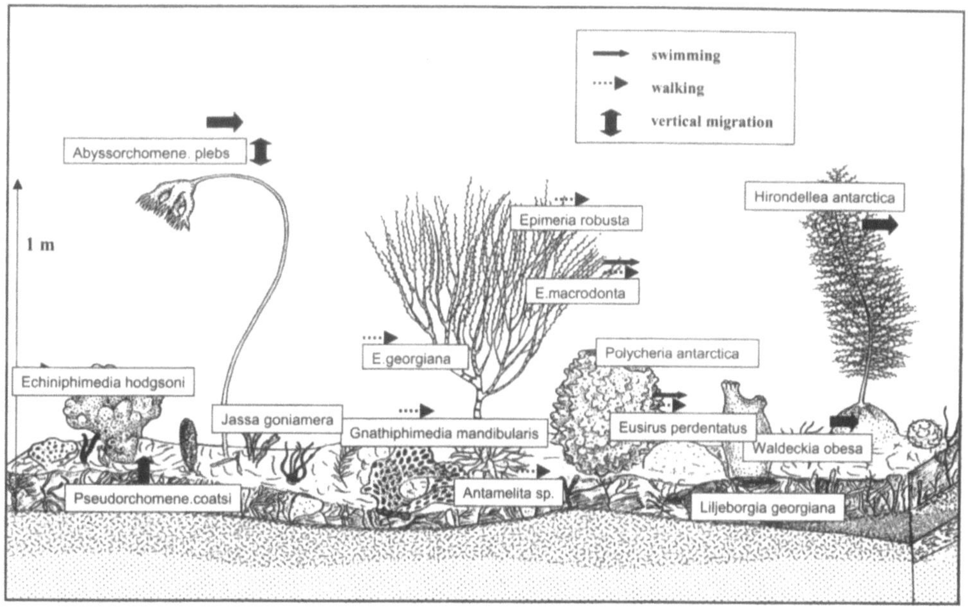

muddy bottom very poor in sessile fauna. Usual position in aquarium: apparently not substrate selective; can stay on the sediment surface, without burrowing, or climb and sit on the lower level of erected sessile invertebrates in the aquarium with "reconstituted natural bottom", e.g. at the base of hydrozoan colonies. Mobility: stays usually motionless on the bottom; never observed swimming spontaneously; when induced by direct stimulation with forceps, swimming is always short and clumsy, the amphipod quickly returning to the substrate.

Representative species from coarse or mixed sediment surface

Liljeborgia georgiana Schellenberg, 1931 (*Liljeborgiidae*). Present, always in low number, in mixed-bottom samples, with bryozoan debris,mud or sand. In aquarium, on "reconstituted bottom", usually stays motionless on the mixed sediment or below the pieces of gauze provided as artificial substrate. Living treelike hemichordates can be used as shelter or substrate. Mobility: able to walk slowly on the substrate in an upright position; the long last pereopods are used for jumping when disturbed by other amphipods or by forceps; swimming, very rarely observed, is jerky and very slow.

Fig. 4 Undisturbed sample of detritic bryozoan bottom taken by the multibox corer (ANT XV/3 sta. 02 MG4, northwest of Kapp Norvegia, 174 m)

Paraceradocus gibber Andres, 1984 (*Melitidae*) Found in diverse catches often containing a variety of biogenic debris with stones, sand and mud. In a large aquarium with "reconstituted bottom" and in few small aquaria with different bottom types, *P. gibber* shows a strong thigmotropy toward the sediment and does not attempt to climb stones or epibenthic animals. It usually tries to hide the entire body, except the antennae, in any available shelter. On soft sediment, with or without stones, it stays at the surface and was not observed trying to dig. Coleman (1989a), however, observed the species digging a burrow under stones. *P. gibber* seems to be sedentary and able to defend its shelter against other amphipods. Mobility: in all cases, specimens newly introduced into aquarium walk around, mostly crawling on one side. Never observed swimming, which confirms the previous observations by Coleman (1989a) and Klages and Gutt (1990b).

In addition to the two former species probably restricted to this type of bottom, some other species seem to use the sediment surface as a temporary substrate from which they can move, swim and forage in a much larger area, according to their motility patterns. The species of this numerous group all belong to the Lysianassoidea and show common behavioural traits in the aquarium: they usually remain motionless on the substrate and can live close together in high densities; they do not appear to be substrate selective and do not burrow in soft sediment. Most are able to swim fast. Representative species include the following.

Waldeckia obesa (*Chevreux*, 1905) From catch observations, they seem to occur on nearly all types of bottoms; always in low numbers in trawl samples but can be collected in thousands in baited traps. In the aquarium, sits at the surface of diverse bottom substrates (except pure soft sediments). Mobility: a sporadic swimmer leaving the bottom substrate to swim sub-vertically to the aquarium surface. Can swim very fast. Also undertakes horizontal swimming trips very close to the bottom. Can walk slowly upright on the substrate. Never seen digging in any kind of sediment.

Uristes gigas Dana, 1949 and *U. adarei* (*Walker*, 1903). On the aquarium bottom, usually seen lying on one side, antennae slowly moving. No observed attempt to dig in the soft sediment or hide in the coarse detritic sediment.

Abyssorchomene scotianensis (*Andres*, 1983), *Lepidecreella* sp A, *Orchomenella acanthura* (*Schellenberg*, 1931), *Orchomenella pinguides* (*Walker*, 1903), *Parschisturella carinata* (*Schellenberg*, 1926), *Tryphosella murrayi* (*Walker*, 1903). All have been observed in the aquarium staying upright (or lying on side: *T. murrayi*) on diverse bottom surfaces. *P. carinata* and *T. murrayi* sometimes hide part of the body in the coarse detritic sediment. Mobility: usually seen motionless but able to swim fast (except *T. murrayi* which swims rather slowly).

The lower level of sessile epibenthos

Covering the bottom in patches or in continuous layers, the diverse assemblages of hexactinellids and demosponges (Barthel and Gutt 1992), the various bryozoans and the multiple cnidarians and hydrozoans form highly heterogeneous structures. Due to its diversity of shapes and structures (including multiple cavities), this sessile epibenthos provides numerous potential substrates for amphipods. According to their feeding type, many amphipods also find food in abundance there (Coleman 1989b, c, 1990; Klages and Gutt 1990b; Dauby et al. 2001).

Aquarium observations showed that some species seem to prefer to occupy the lower part of the erected substrates, whereas others occur on the upper part. But other species show no apparent preference and can occur in both levels and sometimes also on the sediment surface.

Representative species from the lower epibenthic level

The Iphimediidae family is probably the most representative of this level, with at least five common species. All these species are typical walker-climbers, walking on the three last pairs of pereopods and rarely swimming (in the aquarium). Different microhabitats can be distinguished depending on the nature of the "living substrate".

Echiniphimedia hodgsoni Walker, 1906 and *Echiniphimedia scotti K. H. Barnard*, 1930 (*Iphimediidae*). Usually found in trawl samples with abundant stones and sponges. Position in aquarium with "reconstituted bottom": stay at the external surface of sponges, sometimes on bryozoans. Often gathering together. Mobility: usually motionless with only antenna 2 moving in the water, or walking slowly on the substrate. Swimming is very rarely observed; when disturbed *Echiniphimedia hodgsoni* can swim with jerky movements up to the aquarium surface.

Gnathiphimedia mandibularis K.H. Barnard, 1930 (*Iphimediidae*). In the aquarium, often seen sitting on pieces of bryozoan colony or, more rarely, on sponges, with strong antenna 1 erected and more flexible antenna 2 slowly whipping the water. Mobility: walks very slowly on the surface of calcareous bryozoans; swimming very rarely observed. The real habitat of this species (exceptionally recorded in situ) is shown in a bottom picture (Fig. 5).

Iphimediella cyclogena K.H. Barnard, 1930 and *Maxilliphimedia longipes Walker*, 1906 (*Iphimediidae*). Their presence in trawl samples seems to be linked to the abundance of cnidarians. Position in aquarium: observed motionless at the surface of diverse substrates. Mobility very similar to *Gnathiphimedia mandibularis*.

Epimeria rubrieques De Broyer & Klages, 1991 (*Epimeriidae*) Occurrence in catches: very few trawls contained this species, which seems to have a more restricted habitat than the other *Epimeria*. The most important catch during the EASIZ II cruise (sta. ANT XV 206 AGT15) contained 25 specimens among boulders and colonies of cnidarians. Usual position in aquarium: stays on the surface of stones, on diverse colonies of cnidarians and bryozoans. Mobility: walks very slowly. Swimming, which was observed only after stimulation, is jerky, heavy and powerful.

Fig. 5 Habitat of *Gnathiphimedia mandibularis* (*arrow*): bryozoan colonies and sponges (ANT VII/4 frame 262). Picture courtesy of J. Gutt, A.W.I.

Eusirus perdentatus Chevreux, 1912 (*Eusiridae*) In the aquarium, usually observed staying motionless on the substrate surface or on stones. Mobility: seems influenced by the nature of the substrate and by the aquarium population density. On a "reconstituted bottom" and with few (two or three) specimens in a 30-l aquarium, *Eusirus perdentatus* stays mostly motionless but can be seen walking slowly on the bottom; swimming is very rarely observed. With higher density conditions (unlikely in nature for this predator) and on artificial substrate, *Eusirus perdentatus* was observed swimming upwards, especially at night. Swimming is jerky and always short.
Jassa goniamera Walker, 1903 (*Ischyroceridae*) Position in aquarium: usually fixed by pereopods 5–7 on substrates like bryozoans, hydrozoans or sponges. The powerful first gnathopods can be used to grip substrate. Pereopodal glands produce secretion, building a kind of translucent, protecting cocoon. Mobility: usually motionless; can very briefly swim vertically with jerky movements. Probably sedentary.

The upper level of sessile epibenthos

ROV video records of the sessile suspension-feeder community showed a succession of ball-like, urn- or finger- shaped sponges, tree-like hydrozoans or flower-shaped gorgonians or bryozoans. The "canopy" of these assemblages, composed of a mixture of delicate and strong organisms, constitutes a secondary bottom colonised by other invertebrates, the most obvious being echinoderms (e.g. the crinoid *Promachocrinus kerguelensis* Carpenter, 1888, several *Ophiurolepis*), and by fish (*Artedidraco skottsbergi* Lönnberg, 1905, *Trematomus scotti* Boulenger, 1907. Amphipods are suspected of being also present in this layer but, due to their size, are usually not visible on pictures.

Representative species from the upper epibenthic level

The family Epimeriidae comprises typical representatives of this habitat with at least four common species: *Epimeria macrodonta* Walker, 1906, *Epimeria robusta* K.H. Barnard, 1930, *Epimeria similis* Chevreux, 1912 and *Epimeriella walkeri* K.H. Barnard, 1930. Some differences in usual position and motility were noticed.
Epimeria macrodonta In the aquarium, stays at different levels; it was observed walking slowly or sitting on the sediment surface, on sponges or, more frequently, on cnidarian colonies (*Primnoisis formosa* Gravier, 1913, *Dasystanella* sp., *Thorella* sp.). Mobility: swimming is rarely observed but can be spontaneous and fast; swims in an upright position, usually turns two or three times around the aquarium and returns to the bottom.
Epimeria robusta Occurs in trawl catches with coarse detritic sediment (bryozoan fragments, sponges and gorgonians), stones and diverse sessile benthos. In the aquarium with "reconstituted bottom", does not select a particular substrate but never seen on bare sediment surface; usually sits near the top of hydrozoan colonies. Mobility: usually motionless for long periods; can swim actively and fast when disturbed.
Epimeria similis This species has very similar habits to *Epimeria macrodonta*.
Epimeriella walkeri In aquarium with "reconstituted bottom", usually stays motionless at the sediment surface or more frequently on the top of gorgonian or hydrozoan colonies. Mobility: swims very fast when disturbed.
Hirondellea antarctica (*Schellenberg*, 1926) (*Lysianassoidea*) Frequently collected in low numbers by trawls; also caught in baited traps. In the aquarium, stays motionless at the sediment surface or more frequently on the top of gorgonian colonies. Mobility: although usually very quiet, can spontaneously swim upwards very fast.
Alexandrella mixta (*Nicholls*, 1938) (*Stilipedidae*) In the aquarium, usually stays on sessile animals like gorgonians. Mobility: like *Hirondellea*, usually motionless, but able to swim fast.

The symbiotic and inquilinous habitats

Among the amphipods living on the sessile invertebrates, some species, usually sedentary, have established different symbiotic relationships with their hosts. Hereafter are recorded the amphipod-benthos associations observed or suspected among the eastern Weddell Sea bottom communities.

Species associated with sponges

No less than 16 species of amphipods were recorded on common hexactinellids and demosponges by Kunzmann (1996). On the walls or at the bottom of the central cavity of some hexactinellids (*Rossella antarctica* Cater, 1872, *R. nuda* Topsent, 1901, *R. racovitzae* Topsent, 1901, *Scolymastra joubini* Topsent, 1910), she found, in order of frequency, *Leucothoe spinicarpa* Abildgaard, 1789 s.l., *Eusirus microps* Walker, 1906, *Eusirus* cf *antarcticus* Thomson, 1880, *Seba antarctica* Walker, 1906, *Alexandrella* sp. and *A. mixta*. In addition, *Polycheria antarctica* Stebbing, 1875 s.l. was found on the external surface of *Rossella* spp. and among their basal scleres, and *Seba antarctica*, as well as some unidentified Lysianassoidea, on the external surface and in the surface tissue lesions of *Rossella nuda* and *Scolymastra joubini*.

The internal cavities of eight common demosponges (mostly *Mycale acerata* and *Tedania trirhapis*) examined by Kunzmann (1996) hosted *Seba antarctica*, *Andaniotes linearis* K.H. Barnard, 1932, *Leucothoe spinicarpa*, *Ampelisca* sp., *Euandania gigantea* Stebbing, 1883 and some unidentified Stegocephalidae and Stenothoidae (Kunzmann's identification of *Euandania gigantea* appears doubtful; the species is known as a purely pelagic one, usually recorded in the bathypelagic zone). *L. spinicarpa*, *Eusirus* cf *antarcticus*, and some Lysianassoidea were found on the external surface, in addition to *Polycheria antarctica*, which live – mostly lying on the back – in self-excavated oval pits in the host tissues, and *Colomastix simplicicauda* Nicholls, 1938, which form ball-shaped groups at the sponge surface.

Kunzmann (1996) established no host-specific relationships. Among her material, 4 species appear frequently and in relatively high numbers in some sponges (48–898 individuals per sponge): *Seba antarctica*, *Polycheria antarctica* s.l., *Colomastix simplicicauda* and *Andaniotes linearis*. They can be considered preferential spongicolous species, but not exclusive, as *S. antarctica* and *P. antarctica* s.l. have also been recorded on ascidians (see below). According to Kunzmann (1996), *S. antarctica* and *P. antarctica* s.l are ectoparasites eating the host tissues and using the sponge as a shelter from predators.

The following additions are made to Kunzmann's records.

Polycheria antarctica s.s.(Dexaminidae). Occurrence/habitat: 15 specimens (sta. ANT VII/4 259 BPN 4) were found in holes in the surface tissues of the demosponge *Crella crassa* Hentschel, 1914 (det. O. Tendal, Copenhagen). Stomach content analysis (12 ind.) revealed no sponge spicules but only small particles (less than 100 μm) of unidentifiable organic matter, diatom fragments and mineral grains (Dauby et al. 2001), which does not confirm Kunzmann's observation of ectoparasitism.

Scaphodactylus sp. nov. 1 (Stenothoidae) was found on an unidentified demosponge at sta. ANT XV/3 37 D, 105 m.

Species associated with ascidians

Species of the families Leucothoidae, Lysianassoidea, Stegocephalidae or Stenothoidae (still in process of identification) have been found in the branchial cavity of different ascidians. *Leucothoe* sp. was found in *Corella eumyota* Traustedt, 1882 (sta. ANT VII/4 AGT 15). The lysianassoid, *Orchomenyx* sp., was found in *Ascidia challengeri* Herdman, 1842 (sta. ANT VII/4 224 GSN4) and an unidentified species in a "large red ascidian" (sta. ANT XV/3 49 AGT). Stegocephalids have been found in *Ascidia challengeri* at sta. ANT XV/3 120 GSN and in *Eugyrioides polyducta* Monniot & Monniot, 1983 (sta. ANT VII/4 AGT 8). *Metopoides* sp. nov. 2 (Stenothoidae) was recorded in *Ascidia challengeri* at depths of 600 m (sta. ANT XV/3 120 GSN) and 710 m (sta. ANT XV/3 150 GSN). Preliminary examination indicated no apparent host specificity.

Species associated with hydrozoans

Two new stenothoid species of the genus *Torometopa* were found on *Oswaldella billardi* Briggs,1938: *Torometopa* sp. nov. 1 (sta. ANT XV/3 62 AGT) and *Torometopa* sp. nov. 6 (sta. ANT XV/3 5 TVG). *Thaumatelson* sp. Z (Stenothoidae) was found on a unidentified hydrozoan (sta. ANT XV/3 62 AGT). The stolons produced by *Tubularia ralphii* Bale, 1884 and *Oswaldella antarctica* on several stones from the underwater hilltop of Four Seasons Inlet (northeast of Kapp Norwegia; sta. ANT XV/3 212) host a few hundreds of stenothoids from two spp, which could not, however, be associated with the hydrozoans but simply shelter in this tri-dimensional substrate (De Broyer et al. 1999; Gili et al. 1999).

Species associated with gorgonians

One stenothoid species of the genus *Torometopa* (sp. nov. 5) was found on *Primnoella* sp. at 400 m (sta. ANT XV/3 62 AGT). *Polycheria* sp (Dexaminidae) occurred on an unidentified gorgonian host at sta. ANT XV/3 214 D, 110–64 m.

Discussion

About the methods

Contrary to the coastal zone where detailed investigations of amphipod habitats have been done successfully by scuba-diving (e.g. Bregazzi 1972; Richardson and Whitaker 1979), shelf habitat determination has to rely on indirect approaches. Bottom pictures and video records have been particularly useful for characterising the habitats of fishes (Ekau and Gutt 1991) or conspicuous macrobenthos (e.g. holothurians, Gutt 1991; shrimps, Gutt et al. 1991; sponges, Barthel and Gutt 1992) but are of little help for the small and often hidden amphipods. Analysis of trawl-catch contents is of limited or no value for indicating the potential habitat of collected amphipods because of the usually disturbed state of the catch (e.g. a mixture of sediments, stones and diverse fauna), and also the usual high patchiness of sampled assemblages (see, for example, Gutt and Koltun 1995). They can, however, be informative in the case of homogeneous bottom catches or symbiosis on well-preserved hosts, for instance. However, undisturbed bottom samples from corers and large grabs (which should be more systematically checked) can provide useful epibenthic habitat indications (Fig. 3). Aquarium observations can provide information on general behaviour (see, for example, Enequist 1949; Klages and Gutt 1990a, b) and on the species' ability to select a particular habitat (e.g. Coleman 1989a). However, extrapolations on the basis of similar morphologies to infer similar habitats can be hazardous, as shown for instance by the *Eusirus* case: *Eusirus perdentatus* is a typically epibenthic animal, walker and poor swimmer (see Klages 1993) although its sister species *Eusirus propeperdentatus* Andres, 1979 is a purely pelagic animal (Andres 1979; De Broyer and Jazdzewski 1993).

The habitat diversity

From a preliminary comparison with the other amphipod macrohabitats in the eastern Weddell Sea, the epibenthic zone, here subdivided into three different levels, appears the most heterogeneous and the richest in species. The presumed habitats of some representative epibenthic species are presented schematically in Fig. 5. Each epibenthic stratum from the heterogeneous sediment surface to the top of the erect sessile benthos offers to amphipods different habitats, characterised by some physical parameters and by nature and availability of food. Some species find there a temporary substrate or a shelter; others forage in this habitat, in some cases on the living substrate itself. The comparatively high number of "walker-climber" species (mainly belonging to the Iphimedioidea, with more than 50 species), mostly found on the different levels of the rich suspension-feeder assemblages, seems unique to the eastern Weddell Sea. It is most probably linked to the diversity of microhabitats and the abundance of food offered by the rich epibenthos to these specialised micropredatory grazers or unspecialised predators (feeding types according to Coleman 1989b, c; Klages and Gutt 1990b; Dauby et al., 2001).

Aquarium observations have indicated two possible levels of amphipod distribution on the sessile epibenthos substrates, which require confirmation. It seems, nevertheless, possible to differentiate the environmental conditions of the two levels. At the top of the epibenthic substrates, the upper level stratum can be more exposed to strong currents (a current of 40 cm/s was recorded in Kapp Norvegia at 5 m above the bottom at a depth of 676 m; Fahrbach et al. 1992). This seems a priori a favourable position for the free-living suspension-feeders (Ischyroceriidae) to collect the organic rain from the water column above and from lateral advection. Currents can also carry carrion smell and this area could be advantageous for scavengers such as Lysianassoidea (see Ingram and Hessler 1983). These trophic advantages are balanced, however, by a greater exposure to predators. Demersal fish stomach contents (*Trematomus* spp., *Pogonophryne* spp., *Artedidraco orianae*), for instance, frequently revealed *Epimeria* species supposed to stay at this level (Olaso 1999; Olaso et al. 2000).

Are the results representative of all habitats in this taxocoenosis?

The selected species undoubtedly represent the most conspicuous and most common ones in the epibenthic catches so far analysed. But the majority of specimens have been caught by trawls with a 15-mm mesh size, which do not always allow adequate collection of small species, which can be numerous judging from the preliminary analysis of material from small-mesh-sized dredge (M. Rauschert, unpublished work). The diversity of symbiotic and inquilinous habitats is probably highly underestimated. Only sponge habitats have been systematically investigated (Kunzmann 1996) so far. Ascidians remain to be more systematically checked for their inquilinous fauna. Potential associations with cnidarians and hydrozoans should attract more attention. A number of small species, among them the numerous Stenothoidae detected by M. Rauschert, might have developed preferential relationships with some hosts. They could represent an important part of the specific diversity of eastern Weddell Sea amphipods.

Acknowledgements The authors are very grateful to Professor Wolf Arntz and the Alfred-Wegener-Institut for the invitation to the EPOS and EASIZ I and II campaigns of RV "Polarstern". Wim Vader (Tromsoe) and Oliver Coleman (Berlin) made helpful criticisms on the manuscript. Josep-Maria Gili (Barcelona) kindly identified the cnidarians. This work was performed under the contract A4/DD/B02 of the "Belgian Antarctic Research Programme-Phase IV" sponsored by the Federal Office for Scientific, Technical and Cultural Affairs.

References

Andres HG (1979) Gammaridea (Amphipoda, Crustacea) der Antarktis-Expedition 1975/76. Auswertung der Dauerstation südlich von Elephant Island. Meeresforschung 27:88–102

Arntz WE, Gutt J (eds) (1997) The expedition ANT XIII/3 (EASIZ I) of "Polarstern" to the eastern Weddell Sea in 1996. Ber Polarforsch 249:1–148

Arntz WE, Gutt J (eds) (1999) The expedition ANT XV/3 (EASIZ II) of RV "Polarstern" in 1998. Ber Polarforsch 301:1–229

Arntz WE, Ernst W, Hempel G (eds) (1990) The expedition ANT VII/4 (Epos leg 3) and VII/5 of RV "Polarstern" in 1989. Ber Polarforsch 68:1–214

Barthel D, Gutt J (1992) Sponge associations in the eastern Weddell Sea. Antarct Sci 4:137–150

Bregazzi PK (1972) Habitat selection by *Cheirimedon femoratus* (Pfeffer) and *Tryphosella kergueleni* (Miers) (Crustacea, Amphipoda). Br Antarct Surv Bull 31:21–31

Coleman CO (1989a) Burrowing, grooming, and feeding behaviour of *Paraceradocus*, an Antarctic amphipod genus (Crustacea). Polar Biol 10:43–48

Coleman CO (1989b) On the nutrition of two Acanthonotozomatidae (Crustacea: Amphipoda). Gut contents and functional morphology of mouthparts. Polar Biol 9:287–294

Coleman CO (1989c) *Gnathiphimedia mandibularis* (KH Barnard 1830), an Antarctic amphipod (Acanthonotozomatidae, Crustacea) feeding on Bryozoa. Antarct Sci 1:343–344

Coleman CO (1990) *Bathypanoploea schellenbergi* (Holman & Watling 1983), an Antarctic amphipod (Crustacea) feeding on Holothuroidea. Ophelia 31:197–205

Dauby Y, Scailteur Y, De Broyer C (2001) Trophic type diversity within the eastern Weddell Sea amphipod community. Hydrobiologia 443:69–86

De Broyer C, Jazdzewski K (1993) Contribution to the Marine Biodiversity Inventory: a checklist of the Amphipoda (Crustacea) of the Southern Ocean. Doc Trav IRSNB 73:1–154

De Broyer C, Klages M (1990) Studies on amphipod biology. In: Arntz WE, Ernst W, Hempel I (eds) The expedition ANT VII/4 (EPOS Leg 3) and VII/5 of RV "Polarstern" in 1989. Ber Polarforsch 68:113–115

De Broyer C, Rauschert M, Chapelle G (1997) Trophodynamics, biodiversity and gigantism of the amphipod Crustacea taxocoenosis. In: Arntz W, Gutt J (eds) The expedition ANT XIII/3 (EASIZ I) of "Polarstern" to the eastern Weddell Sea in 1996. Ber Polarforsch 249:76–79

De Broyer C, Rauschert M, Scailteur Y (1999) Structural and ecofunctional biodiversity of the benthic amphipod taxocoenoses. In: Arntz WE, Gutt J (eds) The expedition ANT XV/3 (EASIZ II) of RV "Polarstern" in 1998. Ber Polarforsch 301:163–174

Ekau W, Gutt J (1991) Notothenioid fishes from the Weddell Sea and their habitat, observed by underwater photography and television. Proc NIPR Symp Polar Biol 4:36–49

Enequist P (1949) Studies on the soft-bottom amphipods of the Skagerak. Zool Bijdr 28:297–492

Fahrbach E, Rohardt G, Krause G (1992) The Antarctic Coastal Current in the southeastern Weddell Sea. Polar Biol 12:171–182

Galeron J, Herman RL, Arnaud PM, Arntz WE, Hain S, Klages M (1992) Macrofaunal communities on the continental shelf and slope of the southeastern Weddell Sea, Antarctica. Polar Biol 12:283–290

Gili JM, Arntz WE, Filipe P, Lopez P, Orejas C, Ros J, Teixido N (1999) The role of benthic suspension feeders in Antarctic communities. In: Arntz WE, Gutt J (eds) The expedition ANT XV/3 (EASIZ II) of RV "Polarstern" in 1998. Ber Polarforsch 301:30–83

Gutt J (1991) On the distribution and ecology of holothurians in the Weddell Sea (Antarctica). Polar Biol 11:145–155

Gutt J, Koltun (1995) Sponges of the Lazarev and Weddell Sea, Antarctica: explanations for their patchy occurrence. Antarct Sci 7:227–234

Gutt J, Schickan T (1998) Epibiotic relationships in the Antarctic benthos. Antarct Sci 10:398–405

Gutt J, Starmans A (1998) Structure and biodiversity of megabenthos in the Weddell and Lazarev Seas (Antarctica): ecological role of physical parameters and biological interactions. Polar Biol 20:229–247

Gutt J, Gorny M, Arntz WE (1991) Spatial distribution of Antarctic shrimps (Crustacea: Decapoda) by underwater photography. Antarct Sci 3:363–369

Ingram C, Hessler RH (1983) Distribution and behavior of scavenging amphipods from the central North Pacific. Deep Sea Res 30:683–706

Klages M (1991) Biologische und populationsdynamische Untersuchungen an ausgewählten Gammariden (Crustacea; Amphipoda) des südöstlichen Weddellmeeres, Antarktis. Thesis, University of Bremen

Klages M (1993) Distribution, reproduction and population dynamics of the Antarctic gammaridean amphipod *Eusirus perdentatus* (Chevreux 1912) (Crustacea). Antarct Sci 5:349–359

Klages M, Gutt J (1990a) Observations on the feeding behaviour of the Antarctic gammarid *Eusirus perdentatus* (Chevreux 1912) (Crustacea: Amphipoda) in aquaria. Polar Biol 10:359–364

Klages M, Gutt J (1990b) Comparative studies on the feeding behaviour of high Antarctic amphipods (Crustacea) in the laboratory. Polar Biol 11:73–79

Kunzmann K (1996) Die mit ausgewählten Schwämmen (Hexactinellida und Demospongiae) aus dem Weddellmeer, Antarktis, vergesellschaftete Fauna. Ber Polarforsch 210:1–93

Olaso I (1999) The pelagic fish food web. In: Arntz WE, Gutt J (eds) The expedition ANT XV/3 (EASIZ II) of RV "Polarstern" in 1998. Ber Polarforsch 301:110–118

Olaso I, Rauschert M, De Broyer C (2000) Impact of the family Artetidraconidae (Pisces) on the eastern Weddell Sea benthic communities. Mar Ecol Prog Ser 194:143–158

Richardson MG, Whitaker TM (1979) An Antarctic fast-ice food chain: observations on the interaction of the amphipod *Pontogeneia antarctica* (Chevreux) with ice-associated micro-algae. Br Antarct Surv Bull 14:107–115

Voss J (1988) Zoogeographie und Gemeinschaftsanalyse des Makrozoobenthos des Weddellmeeres (Antarktis). Ber Polarforsch 45:135–144

Polar Biol (2001) 24: 497–501
DOI 10.1007/s003000100245

Christoph Held

No evidence for slow-down of molecular substitution rates at subzero temperatures in Antarctic serolid isopods (Crustacea, Isopoda, Serolidae)

Abstract The molecular slow-down hypothesis suggests a temperature-controlled slow-down of molecular substitution rates to be a common characteristic of polar poikilotherm organisms. There is evidence that metabolic rate may be controlling the rate of nucleotide substitution in many organisms. In this study, absolute rates of nucleotide change based on a geologically calibrated molecular clock from the 16S ribosomal RNA (LSU) gene of 12 species of serolid isopods living in Antarctic waters are calculated and compared with data from sesarmid crabs from tropical latitudes. Contrary to the prediction, no difference of molecular substitution rates between Crustacea from different latitudes can be detected. This result is corroborated by testing the relative rate in serolid isopods against other Crustacea from tropical, temperate and Antarctic waters. While an influence of temperature on the rate of spontaneous mutation cannot be dismissed, the results show that temperature is not the key determinant of the rate of nucleotide change as has been suggested by the molecular slow-down hypothesis.

Introduction

The rate of nucleotide substitution varies between organisms. It is currently the subject of debate whether: (1) oxidative by-products lower the fidelity of DNA replication ("metabolic rate hypothesis"), (2) the number of germ line replication events ("generation time hypothesis") or (3) differential efficiency of DNA repair mechanisms exert the strongest influence on the rate of nucleotide change (see Martin 1999 for a review). Martin and Palumbi (1993) studied genetic change in the mitochondrial genome and found consistently slower rates in ectotherms than in endotherms of comparable size. They interpreted this as the consequence of a temperature-dependent speed of molecular change. Slowinski and Arbogast (1999) presented an alternative explanation for that particular data set, but it is a generally accepted view that metabolic rate and the rate of molecular change are correlated, at least for mitochondrial genes (Kocher et al. 1989; Rand 1994).

When Bargelloni et al. (1994) found much lower genetic diversities among species of notothenioid fish than expected from previous age estimates of that group (Eastman 1993), they suggested these animals to be subject to slower rates of molecular substitution in response to the extremely low temperatures of the Southern Ocean.

For this, they coined the term "molecular slow-down". According to this hypothesis, polar poikilotherm organisms are expected to show a slow-down of molecular rates of change relative to poikilotherms from tropical and temperate latitudes. In addition to low metabolic rate, the long generation times of polar ectotherms also support the expectation of slower rate of change against absolute time. In a more recent study, the authors point out that a higher rate may be more appropriate for their data set (Bargelloni et al. 1997). Nevertheless, the concept deserves attention, because recent studies from various organisms highlight the importance of oxidative damage to the DNA as an important source of nucleotide change (Martin 1999 and references therein).

This study uses molecular data from Held (2000) and the literature to address the validity of the assumption that low ambient temperatures will lead to a reduced rate of molecular substitution in poikilotherm organisms from polar waters. Extant serolid isopods from

Please note: When citing this article, please use the reference to the original source in the journal Polar Biology as given on the top of this page.

C. Held
Ruhr-Universität Bochum, Lehrstuhl für Spezielle Zoologie, Gebäude ND05-755, 44780 Bochum, Germany
E-mail: Christoph.Held@ruhr-uni-bochum.de
Tel.: +49-0234-3225577
Fax: +49-0234-3214114

Antarctic waters are particulary well suited to test this hypothesis since it is likely that they evolved *in situ* when the Southern Ocean was already cold (Held 2000), their metabolic rate is low (Luxmoore 1984) and their generation times are long (Wägele 1987).

Materials and methods

Absolute rates of molecular change per time

Using data from the mitochondrial large ribosomal subunit (16S rRNA) gene and the V4 hypervariable region of the nuclear small ribosomal subunit (18S rRNA) gene, Held (2000) analysed the phylogeny of serolid isopods from Antarctica, South America and Australia. In this study, all Antarctic species grouped into a single monophyletic clade, which most likely resulted from a single colonisation event after the opening of the Drake Passage isolated shallow-water habitats in Antarctica from similar ones in South America (Fig. 1).

The molecular data used in this study are accessible at the EMBL database (accession numbers AJ2697799–AJ2697828) and specimens, unless destroyed during DNA extraction, have been deposited at the Museum für Naturkunde zu Berlin (ZMB: collection numbers in Held 2000). The first 16 bases and also approximately 50 nucleotides corresponding to a secondary structure stem and loop from base 911 to 964 in the *Drosophila melanogaster* 16S rRNA gene (Gutell et al. 1993) are hypervariable in serolid isopods and were therefore excluded from the analysis. For details on sample collection, molecular genetic protocols, secondary structure alignment and phylogenetic analysis, see Held (2000).

Also using the mitochondrial 16S rRNA gene, Schubart et al. (1998) presented a molecular phylogeny for brachyuran decapod crabs (Sesarminae), which became geographically separated by the rising of the Panama landbridge. These relatively well-dated geological events (Ehrmann 1994; Schubart et al. 1998) were then used to calibrate the local molecular clocks for the mitochondrial 16S rRNA gene for serolid isopods and sesarmid decapods (Fig. 2).

The average branch length from the root of the Antarctic monophylum of species to the tips was divided by the geological age estimate of the Drake Passage (23.5 my), which equals the maximal age of the Antarctic radiation of serolid species. In order to avoid too high estimates of serolid substitution rates because of overcompensation for saturation effects, two relatively simple models of nucleotide substitution were chosen: uncorrected *p*-distances and the Kimura 2 parameter (K2P) model. The rate of sesarmid decapods was calculated using the same models of molecular change following the procedure in Schubart et al. (1998).

The second molecular data set for serolid isopods in Held (2000) cannot be used for calculation of rates because the 18S ribosomal V4 segment is probably unique for peracarid crustaceans, and homology of this region can be difficult to establish even among closely related species.

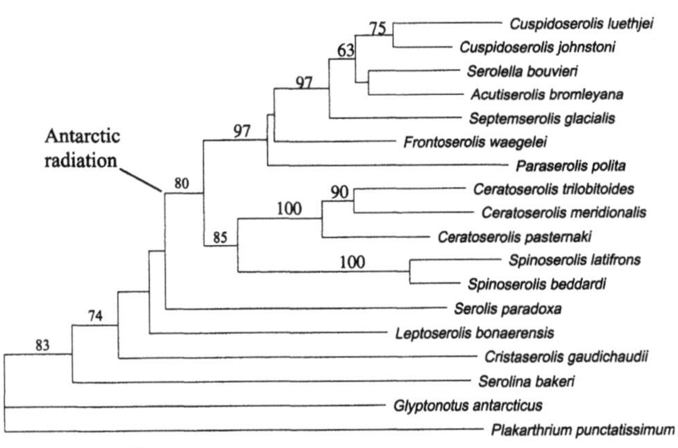

Fig. 1 Neighbour-joining dendrogram based on the 3′-terminus of the mitochondrial large subunit rRNA gene (16S) of 16 species of serolid isopods as used for the two-cluster test. *Numbers* above branches represent bootstrap support values of 1,000 replications; no values lower than 60 given, Kimura 2 parameter model, transitions and transversions. For a phylogenetic hypothesis, see Held (2000). Parsimony, likelihood and distance-based methods reconstruct very similar topologies which perform almost identically in the two-cluster test. *Glyptonotus* and *Plakarthrium* serve as the outgroup

Relative rate tests

The Antarctic serolid species in this study were tested against Crustacea of different geographic origin and taxonomic affiliation. The null hypothesis in the relative rate test is equality of rates. The rationale is that the amount of molecular differences which can be observed today between two taxa must have evolved in either the stemline of one group or the other. By comparing the two groups against a common outgroup, it is determined which of the two lineages or groups of species is more likely to be subject to higher substitution rates. The "two-cluster test" in the program package lintre is an extension of the relative rate test to more than two sequences (Takezaki et al. 1995). The Antarctic serolids are compared with sesarmid crabs from tropical latitudes (Schubart et al. 1998), hermit crabs from temperate waters (Cunningham et al. 1992), Antarctic euphausiids (Patarnello et al. 1996) and non-Antarctic Serolidae (Held 2000). Three crustacean species (*Clibanarius vittatus*, *Coenobita* sp. and *Artemia salina*; Cunningham et al. 1992) and two insect species from Genbank (*Omus californicus*, L4925; *Zonosemata electa*, U39415) served as outgroups for the relative rate tests.

If the molecular slow-down hypothesis holds true, the expectation is that Antarctic serolids should display a gradual decrease of substitution rate relative to the other taxa, with the difference being greatest between tropical decapods and polar serolids.

Results

Calibration of the mitochondrial 16S rRNA gene molecular clock

The geologically calibrated molecular clocks for Serolidae and Sesarminae indicate a higher substitution rate when both substitution types are considered compared to rates based on transversions alone. In contrast to the Serolidae, however, the inferred rates for the decapods are not strongly influenced by the choice of the model of substitutions (Table 1). The rate estimates of the same gene fragment from the Serolidae, however, increase considerably when more realistic assumptions concerning the process of molecular substitution are made; in this case, when transitions and transversions are allowed to take place at different rates (K2P model). This indicates that the K2P model at least partly corrects the number of substitutions occurring repeatedly at the same sites (multiple hits), which go undetected when only the observed transition and transversion events are counted. Most of these multiple substitutions are presumably transitions, because the increase in rate estimates between the two models is negligible when only

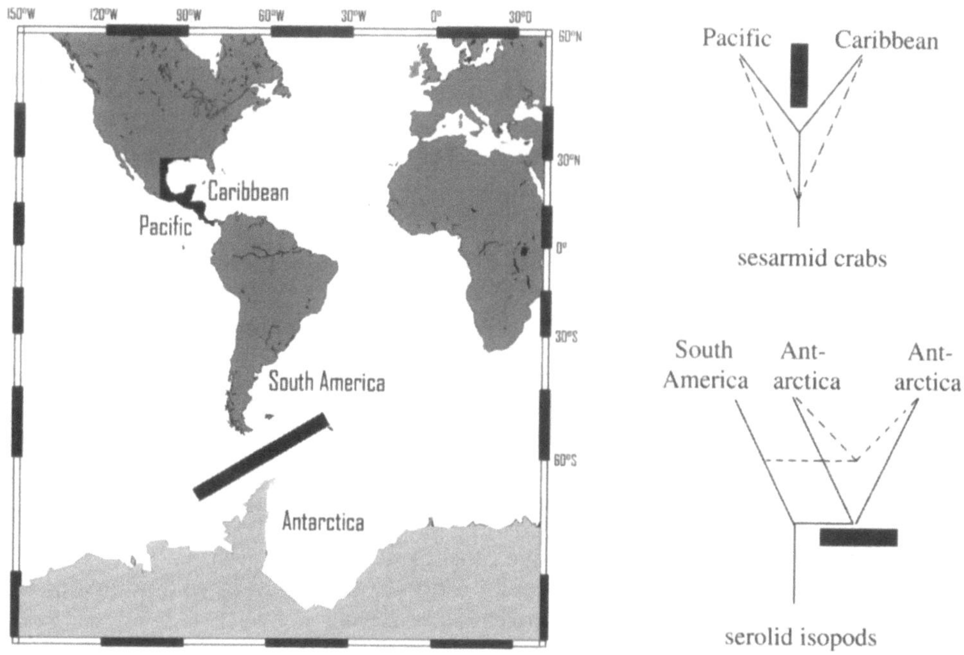

Fig. 2 Biogeographical inference and possible errors in dating speciation and radiation events. In the calculations, separation (*black area*) and speciation were assumed to be occurring simultaneously (*solid lines*), but could also have occurred before (decapods, *above*) or later (*serolids*, below). The Isthmus of Panama today separates groups of Pacific and Caribbean intertidal decapods (Schubart et al. 1998); the Drake Passage separates Antarctica from South America, from where serolid isopods presumably colonised Antarctic shallow waters (Held 2000)

transversions are considered (Table 1, see also Held 2000).

The stronger effects of mutational saturation among the Serolidae can be explained by the fact that there are large differences in the timespan during which substitutions could accumulate. Whereas the Sesarminae became geographically separated around 3.1 my ago, the Antarctic radiation of serolid isopods is around 23.5 my old. The relatively fast rate in serolid isopods becomes even more prominent when only the portion of the sesarmid 16S rRNA gene is analysed which is also used for serolid isopods (Table 1, numbers in parentheses).

This effect is only detectable amongst transitions; the more slowly accumulating transversions did not preferentially accumulate in the more variable portions of the gene in the Sesarminae. The relative constancy of inferred transversion rates in serolid isopods is also prominent when even more complex models are evaluated (not shown). This is consistent with the findings from other studies that transversions accumulate more linearly over time than transitions (Holmquist 1983).

The relative rate tests indicate significantly faster rates in Antarctic Serolidae than in all three comparisons with Crustacea from different latitudes (tropical Sesarminae, temperate Decapoda, Antarctic krill; two-cluster test, $P < 0.05$); this is true for a wide range of substitution models. The only group which does not differ significantly from the Antarctic Serolidae are Serolidae from South American and Australian waters ($P > 0.05$).

It should be emphasised that the results of the first three relative rate tests have to be viewed with caution. The rather distant relationship between the two groups in these comparisons necessarily entails higher genetic divergence between the groups than within, thus making mutational saturation more likely. Furthermore, the differences in temperature whose influence on substitution rates are to be tested have lasted a relatively shorter period of time since the last common ancestor of the two groups. Neither condition, however, would necessarily favour higher substitution rates to be postulated for the Serolidae. The fourth relative rate test, which does not suffer from these potential shortcomings, fails to detect an elevated substitution rate in non-Antarctic serolids compared to Antarctic serolids, despite the fact that they have lived at higher ambient temperatures.

Table 1 Geologically calibrated rates of molecular change of the mitochondrial 16S rRNA gene in Antarctic serolid isopods and tropical sesarmid crabs (Schubart et al. 1998). *Numbers in parentheses* indicate decapod rates excluding the more variable 50-bp region, which had to be discarded from the analysis of the serolid data due to alignment difficulties (*Ti* transition substitution, *Tv* transversion substitution)

Substitution type considered	Taxon	% base substitutions per my (*p*-distance)	% base substitutions per my (Kimura 2 parameter model)
Ti + Tv	Serolidae	0.37	0.43
	Sesarminae	0.38 (0.34)	0.38 (0.34)
Tv	Serolidae	0.14	0.15
	Sesarminae	0.11	0.11

Discussion

Robustness of rate estimates

Although the same gene fragment was amplified in all studies that have been presented here, differences in sequence readability near the primers and differential exclusion of data due to alignment difficulties often lead to slightly different portions of the gene fragment being compared (see Schubart et al. 2000). Nevertheless, a general pattern emerges from the results of the calculation of absolute and relative rates. Even when conditions are extremely conservative (no correction for saturation, exclusion of variable segments only in serolid isopods; Table 1), the deduced rates of nucleotide change in serolid isopods do not support the predictions of a molecular slow-down correlated with low temperatures. The calculation of absolute and relative rates of nucleotide change shows that rates in Antarctic isopods are equally high or higher than those of tropical Sesarminae, temperate Decapoda and Antarctic krill.

It is noteworthy that minimum rates for serolid isopods and maximum rates for the brachyuran decapods have been calculated. The Pacific and Caribbean clades in the study by Schubart et al. (1998) may already have diverged long before the Isthmus of Panama formed a physical barrier between them (Fig. 2; see Knowlton et al. 1993 for an example). However, if indeed the cooling of the Southern Ocean has been responsible for the monophyly of serolids in Antarctic waters, then this radiation could also have taken place a long time after the opening of the Drake Passage, requiring an even higher rate than calculated here for the observed genetic divergence to evolve.

Very little or no correction for multiple substitution has been employed in order to avoid too-high estimates of serolid rates due to overcompensation of the models of the substitution process. This increases the likelihood of some substitutions becoming masked by multiple hits in the Serolidae more strongly because of the longer timespan available for accumulation of substitutions.

These circumstances all result in underestimation of substitution rates in serolid isopods rather than overestimation. The rates for Antarctic serolids given in Table 1 are therefore conservative estimates.

The absence of detectable rate differences between Antarctic and non-Antarctic Serolidae in the fourth relative rate test suggests that temperature is not the principal component determining the rate of nucleotide change. The rate differences between non-Antarctic Serolidae and tropical, as well as temperate, decapods on the one hand and Antarctic Serolidae and Antarctic krill on the other, emphasise the reality of lineage-specific differences of substitution rates in spite of similar ambient temperatures.

Interestingly, the transversion rates of Sesarminae and Serolidae are very similar (0.11–0.14% per my). Equally high transversion rates for the mitochondrial 16S rRNA gene have been independently calculated in a number of other studies investigating a variety of invertebrate and vertebrate taxa (Allard et al. 1992; Sturmbauer et al. 1996; Caccone et al. 1997; Schubart et al. 1998). The transversion rate of 0.14% per my for serolid isopods can therefore be regarded as the most reliable clock available for this gene at present, and will be used for rate comparisons and to obtain at least approximate time estimates of speciation events.

There is also evidence that a normal or even slightly elevated rate of molecular substitution in polar poikilotherms is not confined to serolid isopods alone. A relative rate test of serolid isopods against 15 Antarctic species of the rather distantly related isopod family Arcturidae shows no evidence of differences in substitution rates between serolid and arcturid isopods (C. Held, unpublished data).

Ecological implications

One reason why the possibility of direct effects of temperature on the rate of molecular substitution receives some attention in the ecological literature is the fact that one could hypothesise a direct connection between the metabolic rate and the rate of molecular change itself (Clarke and Johnston 1996). If there were a general slow-down of molecular substitution rates at low temperatures, this might also help explain the popular view of the tropics as the sources of taxonomic diversity which are characterised by rapid speciation and low extinction rates whereas polar regions, in contrast, display slow speciation and high extinction rates and act as sinks of taxonomic diversity instead (see Crame 1997 for a review). While this pattern is relatively well established for terrestrial ecosystems, it is still unclear whether it also applies to marine systems, at least in the southern hemisphere (Clarke 1992). Taxonomic work already suggests that the Southern Ocean may be far less depauperate than previously thought and significantly richer in species than the younger Arctic Ocean (Dunton 1992; Arntz et al. 1997).

An answer to these ecological questions, however, would imply a direct connection between the rates of molecular substitution and the rate of evolution itself. In view of the highly integrated, pleiotropic nature of the genome such a proportional response to molecular substitution is doubtful. More importantly, molecular studies like the present one cannot provide a straightforward answer to these questions because the term "molecular evolution" is often used when "molecular change over time" would be more appropriate.

Conclusions

Currently, there is insufficient molecular data available from high-latitude poikilotherms to permit a conclusive evaluation of temperature-dependent effects on substitution rates such as those suggested by the molecular slow-down hypothesis. Similarly, the rates observed here in serolid and perhaps arcturid isopods do not prove that high-latitude poikilotherms generally show increased substitution rates either. Until more specific information concerning DNA repair efficiency, intensity of oxidative damage to the DNA and number of germ-line replication events becomes available, it will remain difficult to uncover the most important principles driving molecular evolution in extreme environments.

It should be stressed that, if variation of substitution rates is the expectation (Swofford et al. 1996), we are running the risk of a perception (and publication) bias. Low rates at low temperatures might be attributed to a causal link between these two parameters whereas findings that do not meet our expectations are viewed as what they probably are: variation around a mean. This is not to say, however, that temperature has no influence whatsoever on the speed of molecular evolution. It might well be that, all other things being equal, major differences in metabolic rates may cause an increased frequency of oxidative damage to the DNA by an increased amount of released free oxygen radicals. This, again, can be expected to play a greater role during relatively short-term changes, because there are a number of adaptations which can protect the organism from such damage on an evolutionary timescale (Rand 1994; Hoffmann and Parsons 1997). The data presented here suggest that the direct effects of temperature on the frequency of nucleotide change, if present, are probably subtle and frequently overridden by other circumstances such as lineage-specific differences of molecular substitution rates.

Acknowledgements The author is indebted to J.-W. Wägele at the Ruhr-Universität Bochum for supervision and support during the author's PhD Thesis. Many thanks are due to the officers, crew and fellow scientists onboard the "RV Polarstern". G. Poore (Melbourne) and G. Duhamel (Paris) provided additional material. C. Rios, E. Mutschke, J. Ramirez and E. Morrison (Universidad de Magallanes, Punta Arenas, Chile) provided logistical support during collection in the Magellan Strait. This work was supported by a grant from the Deutsche Forschungsgemeinschaft to J.-W. Wägele.

References

Allard MW, Miyamoto MM, Jarecki L, Kraus F, Tennant MR (1992) DNA systematics and evolution of the artiodactyl family Bovidae. Proc Natl Acad Sci USA 89:3972–3976

Arntz WE, Gutt J, Klages M (1997) Antarctic marine benthos: an overview. In: Battaglia B, Valencia J, Walton DHW (eds) Antarctic communities: species, structure and survival. Cambridge University Press, Cambridge, pp 3–14

Bargelloni L, Ritchie PA, Patarnello T, Battaglia B, Lambert DM, Meyer A (1994) Molecular evolution at subzero temperatures: mitochondrial and nuclear phylogenies of fishes from Antarctica (Suborder Notothenioidei), and the evolution of antifreeze glycopeptides. Mol Biol Evol 11:854–863

Bargelloni L, Patarnello T, Ritchie PA, Battaglia B, Meyer A (1997) Molecular phylogeny and evolution of notothenioid fish based on partial sequences of 12S and 16S ribosomal RNA mitochondrial genes. In: Battaglia B, Valencia J, Walton DWH (eds) Antarctic communities: species, structure and survival. Cambridge University Press, Cambridge, pp 45–50

Caccone A, Milinkovitc MC, Sbordoni V, Powell JR (1997) Mitochondrial DNA rates and biogeography in european newts (genus *Euproctus*). Syst Biol 46:126–144

Clarke A (1992) Is there a latitudinal diversity cline in the sea? Trends Ecol Evol 7:286–287

Clarke A, Johnston IA (1996) Evolution and adaptive radiation of Antarctic fishes. Trends Ecol Evol 11:212–218

Crame JA (1997) An evolutionary framework for the polar regions. J Biogeogr 24:1–9

Cunningham CW, Blackstone NW, Buss LW (1992) Evolution of king crabs from hermit crab ancestors. Nature 355:539–542

Dunton K (1992) Arctic biogeography: the paradox of the marine benthic fauna and flora. Trends Ecol Evol 7:183–189

Eastman JT (1993) Antarctic fish biology. Academic Press, San Diego

Ehrmann WU (1994) Die känozoische Vereisungsgeschichte der Antarktis. Ber Polarforsch 137:1–152

Gutell RR, Gray MW, Schnare MN (1993) A compilation of large subunit (23S and 23S-like) ribosomal RNA structures. Nucleic Acids Res 21:3055–3074

Held C (2000) On the phylogeny and biogeography of serolid isopods (Crustacea, Isopoda, Serolidae) and the use of ribosomal expansion segments in molecular systematics. Mol Phylogenet Evol 15:165–178

Hoffmann AA, Parsons PA (1997) Extreme environmental change and evolution, 2nd edn. Cambridge University Press, Cambridge

Holmquist R (1983) Transitions and transversions in evolutionary descent: an approach to understanding. J Mol Evol 19:134–144

Knowlton N, Weigt LA, Solórzano LA, Mills DK, Bermingham E (1993) Divergence in proteins, mitochondrial DNA, and reproductive compatibility across the Isthmus of Panama. Science 260:1629–1632

Kocher TD, Thomas WK, Meyer A, Edwards SV, Pääbo S, Villablanca FX, Wilson AC (1989) Dynamics of mitochondrial DNA evolution in animals. Proc Natl Acad Sci USA 86:6196–6200

Luxmoore RA (1984) A comparison of the respiration rate of some Antarctic isopods with species from lower latitude. Br Antarct Surv Bull 62:53–65

Martin AP (1999) Substitution rates of organelle and nuclear genes in sharks: implicating metabolic rate (again). Mol Biol Evol 16:996–1002

Martin AP, Palumbi SR (1993) Body size, metabolic rate, generation time, and the molecular clock. Proc Natl Acad Sci USA 90:4087–4091

Patarnello T, Bargelloni L, Varotto V, Battaglia B (1996) Krill evolution and the antarctic ocean currents: evidence of vicariant speciation as inferred by molecular data. Mar Biol 126:603–608

Rand DM (1994) Thermal habit, metabolic rate and the evolution of mitochondrial DNA. Trends Ecol Evol 9:125–131

Schubart CD, Diesel R, Hedges SB (1998) Rapid evolution to terrestrial life in Jamaican crabs. Nature 393:363–365

Schubart CD, Neigel JE, FelderDL (2000) The use of the mitochondrial 16S rRNA gene for phylogenetic and biogeographic studies of brachyuran Crustacea. Crust Issues 12:817–830

Slowinski JB, Arbogast BS (1999) Is the rate of molecular evolution inversely related to body size? Syst Biol 48:396–399

Sturmbauer C, Levinton JS, Christy J (1996) Molecular phylogeny analyis of fiddler crabs: test of the hypothesis of increased behavioural complexity. Proc Natl Acad Sci USA 93:10855–10857

Swofford DL, Olsen GJ, Waddell PJ, Hillis DM (1996) Phylogenetic inference. In: Hillis DM, Moritz C, Mable BK (eds) Molecular systematics, 2nd edn. Sinauer, Sunderland, Mass, pp 407–514

Takezaki N, Rzhetsky A, Nei M (1995) Phylogenetic test of the molecular clock and linearized trees. Mol Biol Evol 12:823–833

Wägele J-W (1987) On the reproductive biology of *Ceratoserolis trilobitoides* (Crustacea: Isopoda): latitudinal variation of fecundity and embryonic development. Polar Biol 7:11–24

Polar Biol (2001) 24: 604–614
DOI 10.1007/s003000100261

Volker Storch · Bodil A. Bluhm · Wolf E. Arntz

Microscopic anatomy and ultrastructure of the digestive system of three Antarctic shrimps (Crustacea: Decapoda: Caridea)

Abstract Mandibles and stomachs of three Antarctic shrimp species (*Nematocarcinus lanceopes*, *Notocrangon antarcticus*, and *Chorismus antarcticus*) were investigated by means of scanning electron microscopy. Transmission electron microscopy of the midgut glands was applied to find out the nutritional status of the animals, which contained a broad variety of food items in their stomachs. In contrast to the Antarctic krill, the three carideans possess a dual filter system; primary filters in the cardia and secondary filters in the pyloric chamber. Morphologically, the mandibles and stomachs of the three species vary considerably; however, their food items are similar. We conclude that stomach contents do not really reflect the different modes of life or habitat preferences of the investigated species.

Introduction

The scarcity of decapod species in the Southern Ocean has attracted attention since the beginning of modern Antarctic research (e.g. Yaldwyn 1965; Maxwell 1977; Kirkwood 1984). Only 11 benthic and 13 pelagic species have been found hitherto on the Antarctic continental shelves and 1 in the adjacent deep sea. In addition, 13 decapod species, all caridean and penaeid shrimps, have been recorded in the pelagial within the Antarctic Convergence (Gorny 2000).

Please note: When citing this article, please use the reference to the original source in the journal Polar Biology as given on the top of this page.

B.A. Bluhm (✉) · W.E. Arntz
Alfred-Wegener-Institut für Polar- und Meeresforschung, Columbusstrasse, 27515 Bremerhaven, Germany
E-mail: fsbab@uaf.edu

V. Storch
Zoologisches Institut, Universität Heidelberg, Im Neuenheimer Feld 230, 69120 Heidelberg, Germany

Scarcity of species does not necessarily mean low abundance. Research during the past decade has shown that some species of benthic caridean shrimps occur regularly, or are even dominant, in many areas of the Southern Ocean. In the southeastern Weddell Sea and the Lazarev Sea, which have been extensively sampled by RV "Polarstern", the three most abundant benthic shrimp species are distributed along an overlapping depth gradient: *Chorismus antarcticus* is most common in shallower waters (200–500 m), *Notocrangon antarcticus* in intermediate depths (300–700 m), and *Nematocarcinus lanceopes* occurs from 600 to > 2,000 m (Arntz and Gorny 1991; Sirenko et al. 1997). A substantial amount of information has been collected on abundance, distribution, reproductive ecology and growth of these species (e.g. Maxwell 1977; Clarke and Lakhani 1979; Gutt et al. 1991; Gorny et al. 1992, 1993). However, knowledge is still lacking on their role in the Antarctic food web, their feeding habits, as well as mechanical and biochemical degradation of their food in the course of its passage through the alimentary canal.

One of the most obvious characters of the alimentary canal of Decapoda is the complicated stomach or proventriculus, which is composed of an anterior cardiac and a posterior pyloric chamber. It serves several purposes: storage of ingested food, its trituration, its chemical degradation by means of the enzymes of the midgut glands (= hepatopancreas), and eventually straining of the resulting chymus through two sets of filters, the primary filter (= ventrolateral setose plates) for coarse material and the secondary filter (= gland filter) for fine material. The stomach of Decapoda and all other Malacostraca is part of the foregut and, as such, is lined by a cuticle which is differentiated according to its main tasks: thick and often calcified in trituration devices, extremely delicate in the filters. The filtrate resulting from the intragastric digestion process is conveyed into the midgut glands, the supernatant into the post-gastric section of the alimentary canal, the midgut. Before being processed in the stomach, food has to pass a series of mouthparts. Among these, the mandibles play a crucial rôle because they can crush the food items. They are located close to and in front of the mouth opening between the upper lip (labrum). The paired lower lips (labia) or their distal portions may be situated in the mouth cavity. Mouth parts, especially mandibles, and stomach morphology give an indication of feeding habits. R-cells of the midgut glands may give hints as to the quality of the absorbed food.

In this paper, we analyse mandibles and stomach filters of the three dominant Antarctic shrimps, *Nematocarcinus lanceopes* Bate (Nematocarcinidae), *Notocrangon antarcticus* Pfeffer (Crangonidae) and *Chorismus antarcticus* (Pfeffer 1887) (Hippolytidae), to estimate their efficiency in food degradation. The findings are discussed in an ecological context.

Materials and methods

The specimens investigated were caught with various types of bottom trawls in depths of 220 m to 2,100 m in the eastern Weddell Sea between 71°S 10°W and 76°S 30°W during ANT XV/3 of RV "Polarstern" in January to March 1998 (Arntz and Gutt 1998). Dissection was undertaken on board using a Wild stereomicroscope. The species were identified according to Kirkwood (1984) and Tiefenbacher (1990). The investigated specimens measured 24–30 mm carapace length (CL) in *Nematocarcinus lanceopes*, 20–23 mm CL in *Notocrangon antarcticus* and 17–19 mm CL in *Chorismus antarcticus*. These sizes lie well within the known adult size ranges of the respective species (Arntz and Gorny 1991; Gorny et al. 1992). Tissue samples of ten individuals of each species were fixed in a solution containing 1.5% glutaraldehyde and 1.5% paraformaldehyde in phosphate buffer (pH 7.4), adjusted to an osmolarity of 900 mOsm by the addition of sucrose and sodium chloride. After postfixation in a 1% solution of OsO_4 containing 1.5% potassium hexacyanoferrate, tissue samples were rinsed in 100 mM cacodylate buffer (pH 7.4), incubated (5 min) in a 1% solution of uranyl acetate (in 50 mM maleate buffer at pH 5.2), dehydrated in ethanol and embedded in Araldite. Thin sections were stained for 5 min with uranyl acetate (saturated solution in 70% methanol) and for another 5 min in lead citrate. The sections were examined under a Zeiss EM 10C (TEM) electron microscope. For scanning electron microscopy, specimens were critical-point dried, coated with gold, and examined with a Cambridge SEM S4-10 microscope. For light microscopic analysis of the stomach contents and filters, we used a Zeiss Axioskop equipped with an MC 100 camera. Stomach contents were analysed from 15 specimens per species.

Results

Mandibles

The mandibles of the three species investigated differ considerably in their position, size and architecture. Those of *Nematocarcinus lanceopes* (Figs. 1a, b, 2) consist of a base carrying two processes and a palp. The molar process is located orally (or may be accommodated in the mouth cavity); the incisor process is in an aboral position (or may lie immediately anterior to the mouth opening). The mandibles reach nearly the length of the stomach, are highly asymmetrical in their molar processes, and are deep red. Their incisor processes are flanked by an upper lip and paired lower lips and are equipped with acute teeth. The molar processes are massive and fitted with a complicated profile composed of protruberances and several rows of bristles (Figs. 1a, b, 2). The following structures can be distinguished: molar-like protruberances (Fig. 2a), conical-shaped teeth (Fig. 2b) and bristles (Fig. 2c). The differences between left and right molar process are documented in Fig. 1a, b. Each mandible bears a three-jointed palp with long setae on the two distal joints.

The mandibles of *Chorismus antarcticus* also almost reach the length of the stomach (Figs. 1e, f, 3). Their incisor processes are located behind the lips, have the shape of an acute blade with a cutting edge, and are fitted with several pointed subterminal teeth on their medial face. The yellow molar process stands at an angle of > 90° towards the incisor process and is partly fringed by a row of brown bristles. Its shape is similar to a cylinder with the surface of the right being smaller than the left (Fig. 1e, f) and fitting into each other so as to grind material. The complicated microstructure of the molar process (Fig. 1e, f) is characterised by rows of stout and blunt teeth, irregular and pointed teeth, and bristles (Fig. 3a–d). The dissimilarity between right and left pars molaris is documented in Fig. 1e, f. As in *Nematocarcinus lanceopes*, the palp is three-jointed and bears setae on all joints.

Regarding the mandibles, *Notocrangon antarcticus* stands apart; its mandibles bear neither incisor processes nor palps. They are long, curved, slightly asymmetrical structures, the distal ends of which, meeting in the oral cavity, are equipped with acute yellow teeth that mesh together (Fig. 1c, d). They reach barely half the length of the stomach. There are neither setae nor scales on their surfaces. Fig. 4 summarises the findings in a diagrammatic way.

Stomach

In *Nematocarcinus lanceopes*, the stomach reaches almost half of the carapace length; in *Notocrangon antarcticus* it is slightly longer than half the carapace length, and in *Chorismus antarcticus* it measures one-third of the carapace lenth. The stomach wall of *Nematocarcinus lanceopes* is very thin and transparent, whereas the other two species have muscular, thick stomach walls, especially *Notocrangon antarcticus*.

The primary filters are nearly rectangular, roof-shaped structures in *Nematocarcinus lanceopes* and wedge-shaped in the other two species (Fig. 5a, c, e) with the tip anteriorly directed. In adult specimens, they are 1.4–2.8 mm long and are composed of two rows of parallel feather-like setae covering one ventro-lateral channel each, which conveys the filtrate to the secondary filters. The distance between the two rows of filter setae varies considerably in the three species investigated. Each row of filter setae is composed of 400–500 setae, which have different lengths in the three species (Fig. 5b,

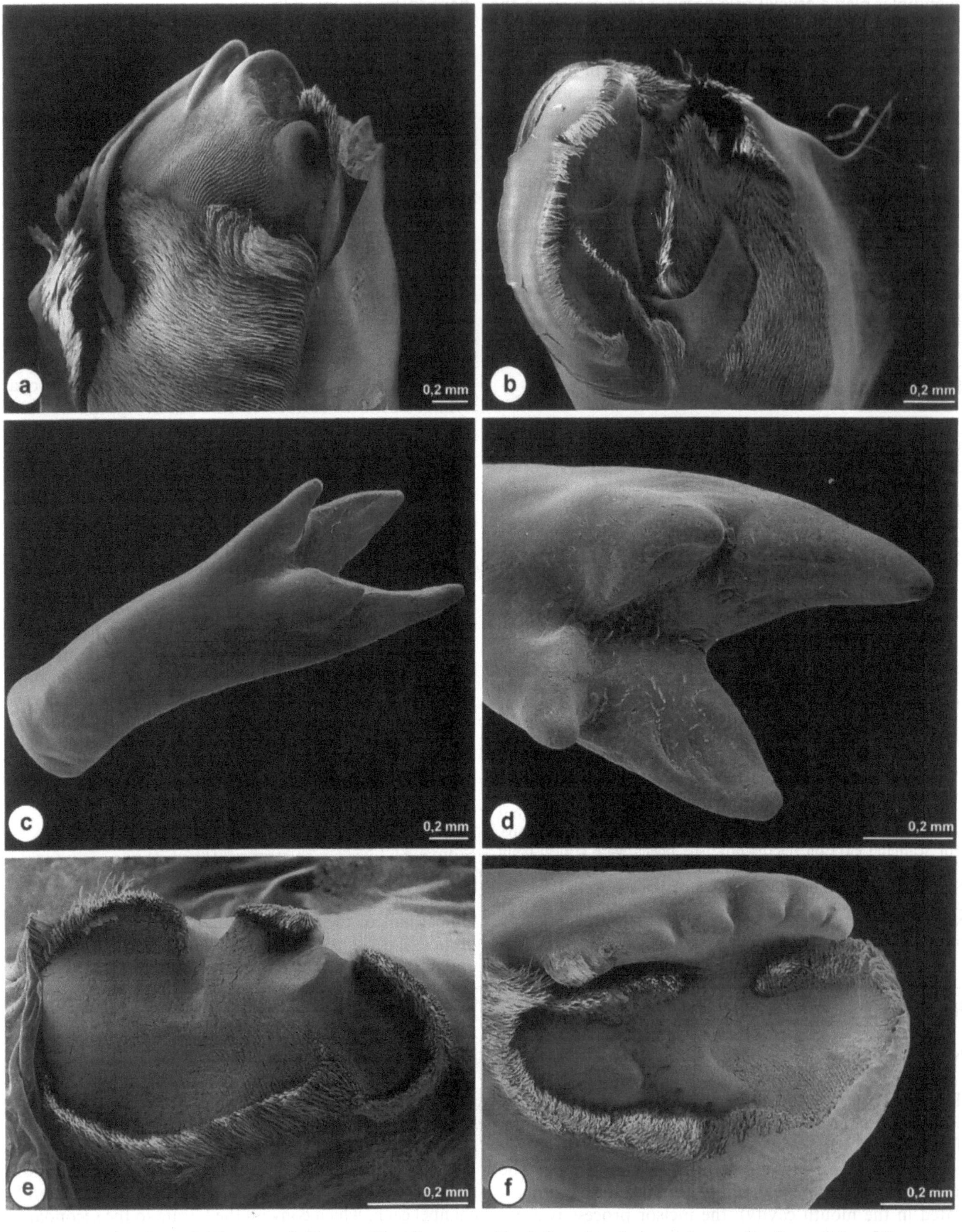

Fig. 1 Scanning electron micrographs of mandibles of *Nematocarcinus lanceopes*, pars molaris (**a**, **b**), *Notocrangon antarcticus* (**c**, **d**), and *Chorismus antarcticus*, pars molaris (**e**, **f**)

d, f), measuring from 0.17 mm to 0.48 mm. The distance between two setae is about 3 μm. In *Notocrangon antarcticus*, several folds of the lateral muscular stomach

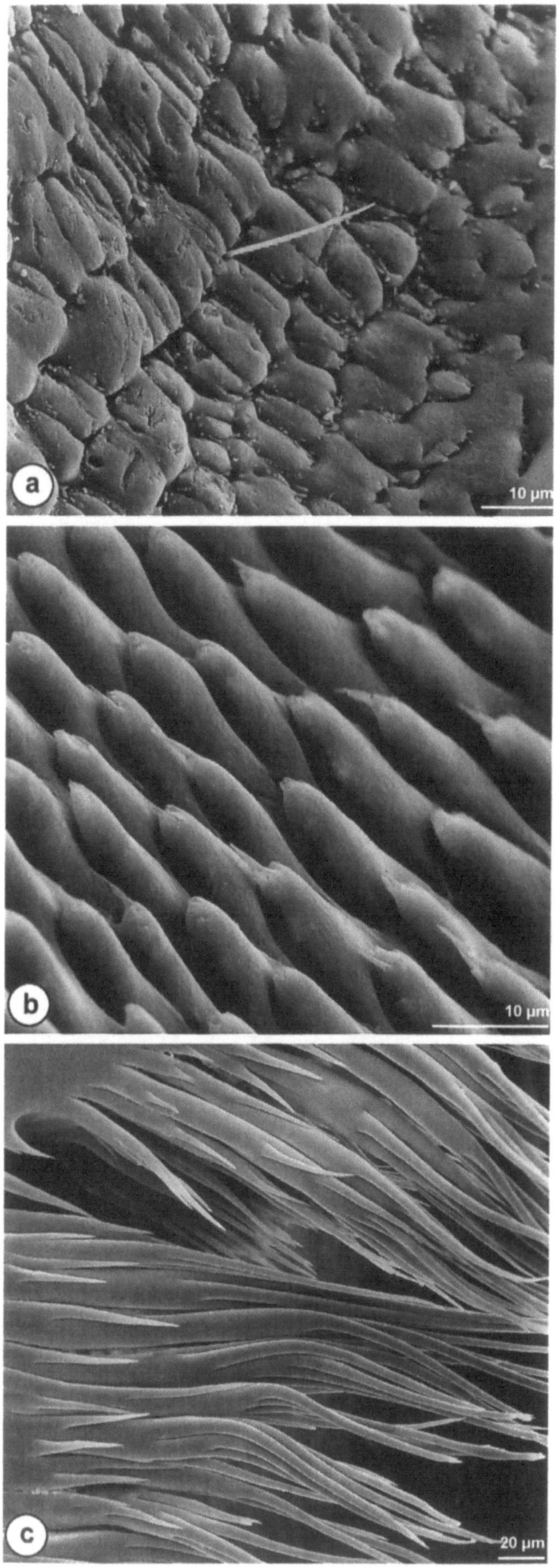

◄
Fig. 2a–c *Nematocarcinus lanceopes*: scanning electron micrographs of the pars molaris of the mandibles. **a** Molar-like protuberances, **b** conical-shaped teeth, **c** bristles

wall with anteriorly orientated setae may cover the primary filters in empty stomachs. In this species the primary filter is framed by a clear-cut and light lateral rim without bristles (Fig. 5c). In *Chorismus antarcticus*, too, folds of the lateral stomach wall may cover the filter setae. *Nematocarcinus lanceopes* stands apart from the other two species; the filter setae are partly covered by rows of long setae of the lateral stomach wall (Fig. 5a).

Primary filters are connected at their posterior ends to the secondary filters by a frame that is more strongly sclerotised than most other parts of the stomach (cardiopyloric valve). Between the two filters there is a constant angle of approximately 80° in *Nematocarcinus lanceopes*, 100° in *Notocrangon antarcticus* and 120° in *Chorismus antarcticus*. The secondary filters also differ considerably in the three species (Fig. 6a–c). They reach over 2 mm in length and 1.3–1.7 mm in height in *Nematocarcinus lanceopes*, 1.4–1.6 mm and 0.9–1.1 mm in *Chorismus antarcticus*, and 1–1.5 mm and 0.4–0.5 mm in *Notocrangon antarcticus*, respectively. They are composed of an inferomedianum (or gland filter), a midventrally located roof-shaped ridge, and two inferolateralia (or filter presses) lying adjacent in a lateral position (Fig. 7a, b). Both structures are equipped with rows of setae. The lateral surfaces of the inferomedianum are concave, thus fitting into the convex medial sides of the inferolateralia. This arrangement makes a second filtration process possible. In *Nematocarcinus lanceopes*, the setae of the inferolateralia are 4 times longer than the setae of the inferomedianum, giving the inferolateralia the appearance of brushes. In height, the inferolaterale measures 1.7–1.9 mm in *Nematocarcinus lanceopes*, 1.2–1.5 mm in *Notocrangon antarcticus* and 1.3–1.4 mm in *Chorismus antarcticus*, with 50–60 rows of setae in *Nematocarcinus lanceopes*, 20 in *Notocrangon antarcticus* and 50–60 in *Chorismus antarcticus*. In all three species, the number of rows of setae differs in the inferomedianum and inferolateralia. In the inferomedianum, there are about 20–22 rows of filter setae in *Nematocarcinus lanceopes*, 24–35 in *Chorismus antarcticus* and 9–16 in *Notocrangon antarcticus*. The distance between two rows of filter setae is 40–120 µm in *Nematocarcinus lanceopes* (the more dorsally located rows are nearer to each other than the more ventrally located); in *Chorismus antarcticus* the distance is 20–35 µm, and in *Notocrangon antarcticus* 20–30 µm. The distance between the filter setae varies; in *Nematocarcinus lanceopes*, there are 360 setae per millimetre, whereas *Chorismus antarcticus* and *Notocrangon antarcticus* have 600 setae per millimetre. The scanning electron microscope reveals differences in the ultrastructure of the filter bars of the secondary filter; however, they are pectinate in all three species. The setae of the inferolateralia, too, are different in the three species investigated (an

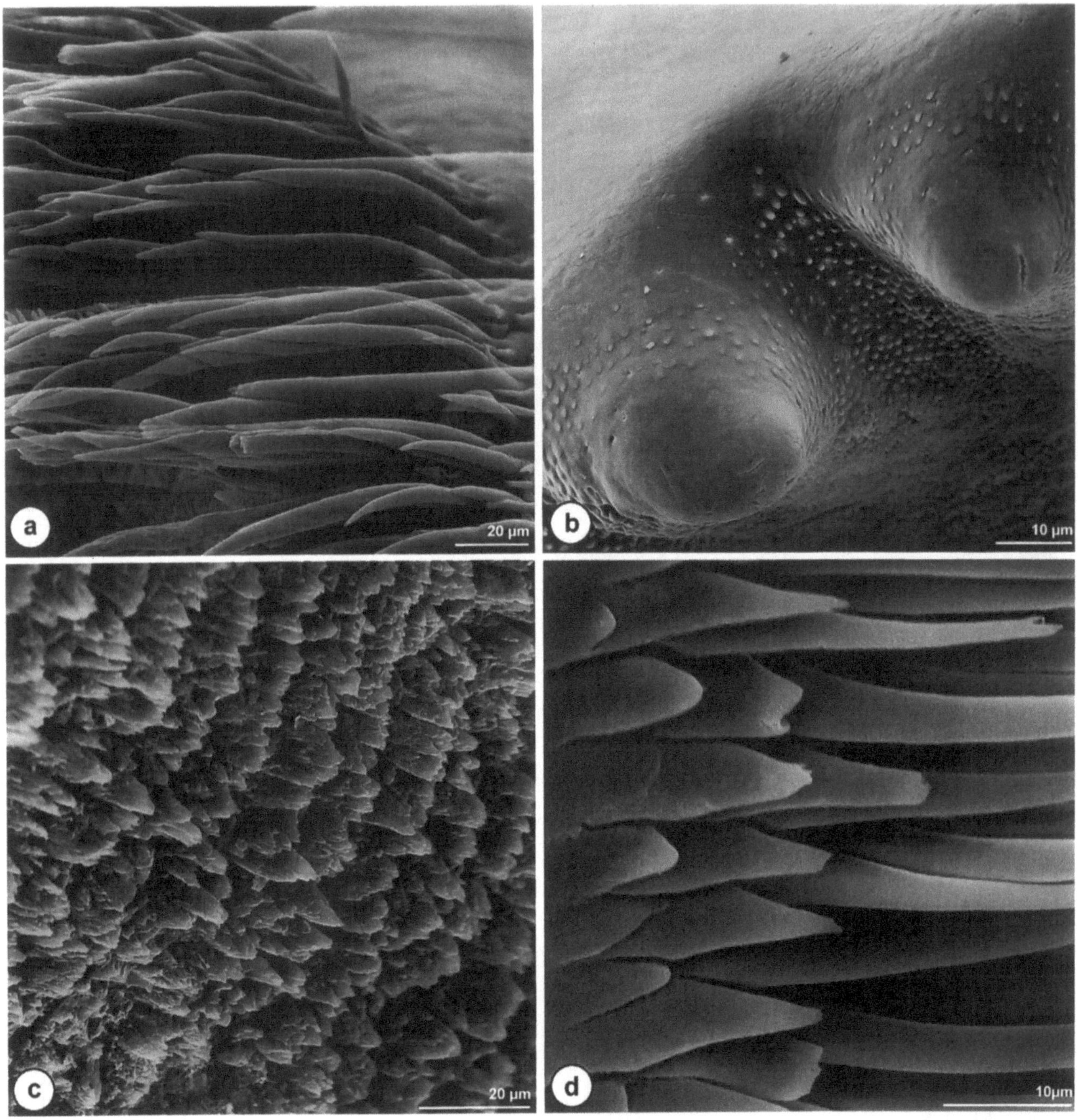

Fig. 3a–d *Chorismus antarcticus*: scanning electron micrographs of the pars molaris of the mandibles. **a** Bristles, **b** blunt teeth, **c** irregular teeth, **d** bristles

example in Fig. 7a, b). Generally, they are short in the periphery and longer (up to 60 µm) near the centre.

As regards the dorsal wall of the stomach, it is characterised by a system of furrows and outpocketings into the gastric lumen in *Nematocarcinus lanceopes*. Two oval protrusions are, at their posterior ends, connected to a middorsally situated bifid tooth bearing a series of sclerotised hooks at its tip (Fig. 8a, b). All three structures are equipped with long anteriorly directed hairy bristles (Fig. 8c). In *Notocrangon antarcticus*, the roof of the stomach bears a fishbone pattern with very long anteriorly directed bristles but no distinct protrusions. In *Chorismus antarcticus*, the dorsal side of the stomach is plain, flat and provided with short, again anteriorly directed bristles. It has a thinner layer of musculature than the lateral parts.

The stomachs of all three species investigated contained a large variety of food organisms (Table 1), consisting predominantly of Crustacea, Polychaeta and Mollusca. The bulk was found in the dorsal part of the cardiac chamber. Some hard and indigestable food items were of considerable size, such as a specimen of the bivalve *Yoldia sabrina*, measuring slightly over 5 mm. In the stomachs of several specimens of *Chorismus antarc-*

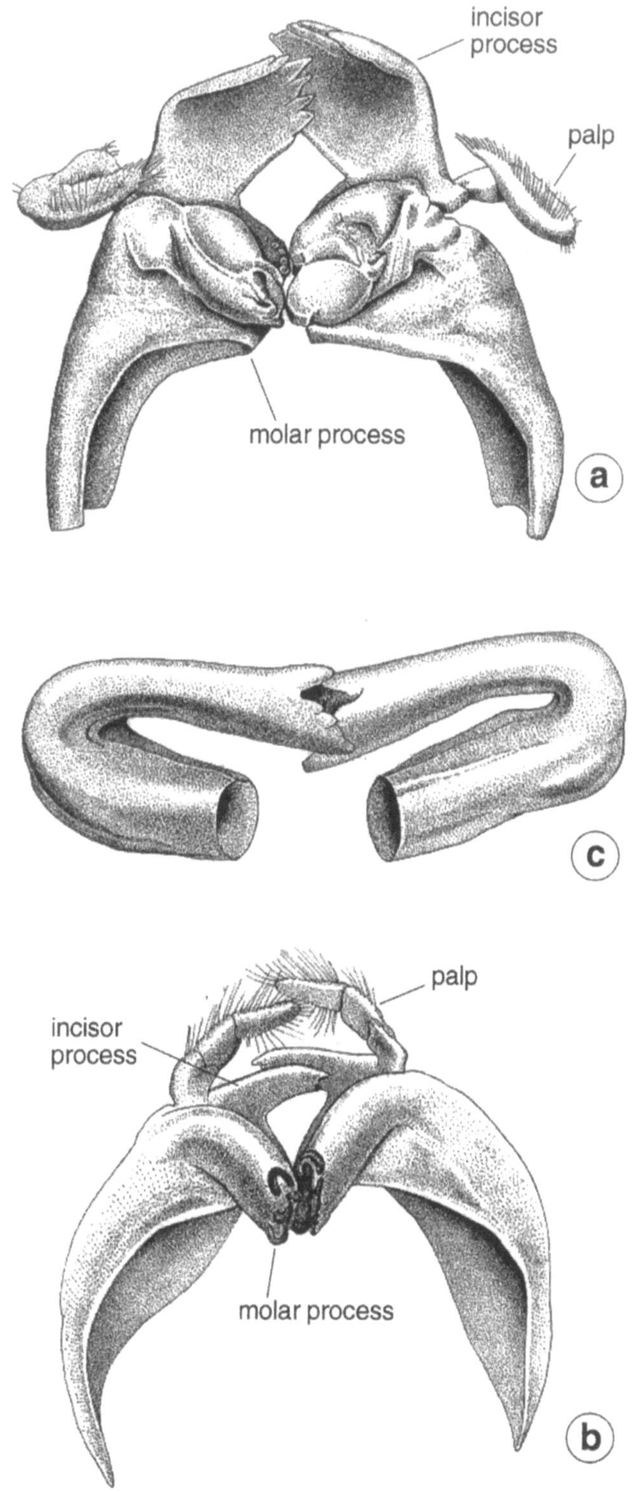

Fig. 4 Diagrammatic presentation of the mandibles of **a** *Nematocarcinus lanceopes*, **b** *Chorismus antarcticus*, and **c** *Notocrangon antarcticus*

ticus and *Notocrangon antarcticus*, polychaetes were conspicuous and were found in various stages of degradation. In some cases there were only bristles left (which, however, still allowed identification of the genus). In the same species, nearly complete specimens of Mysidacea occurred.

Midgut glands

The midgut glands of the three species investigated are surrounded by a thin layer of transparent tissue containing pigment cells. In *Nematocarcinus lanceopes*, they are yellow, orange and red in different parts of the gland and vary in size from 50 to 150 μm. In *Notocrangon antarcticus*, they are bright and dark red and have a diameter of 500 μm and 300 μm, respectively. In *Chorismus antarcticus*, their colour is dark red and they measure 300 μm in diameter. Ultrastructurally, we found the R-cells stuffed with lipids and glycogen in all species. The specimens showed no indication of starvation. There was a very large number of mitochondria.

Discussion

This investigation has widened our knowledge of the digestive system and food of common Antarctic shrimps and is, as far as we know, the first to study this topic in any of the three genera. Irrespective of their very different architecture, the mandibles can grab a variety of food items, which are conveyed via the mouth cavity into the short pharynx. Usually, the presence of a sharp incisor and a massive molar process, as found in *Nematocarcinus lanceopes* and *Chorismus antarcticus*, indicates the need for slicing and crushing, as well as grinding large food items (Patwardhan 1935; Felgenhauer and Abele 1985). Although *Notocrangon antarcticus* lacks such structures, large food items were found in all three species and can obviously also be manipulated with "chop stick"-like mandibles. The connecting pharynx is a straight tube whose extrinsic muscle cells may extend its lumen to allow comparatively large objects (e.g. complete polychaetes, gastropods, bivalves and various crustaceans) to be transported into the stomach.

The stomach is the most complicated part of the alimentary canal of Decapoda (and other Malacostraca). There is a vast literature on its architecture (for references, see Storch and Strus 1989; Ullrich and Storch 1993; Lin 1996; Stamhuis et al. 1998). However, in most publications, the filter systems have not been adequately described. In all three investigated shrimp species, primary and secondary filters are well developed, which is in concordance with most Malacostraca but not, however, with the Antarctic krill (*Euphausia superba*) and other Euphausiacea. In the last, the primary filter is one of the most effective cardiac filter systems in the Crustacea. It serves as a fine filter and is nearly as complex as the secondary filter of other Malacostraca, which are lacking in *E. superba* (Ullrich et al. 1991). Although the primary filters in *Nematocarcinus lanceopes*, *Notocrangon antarcticus* and *Chorismus antarcticus* differ markedly, all

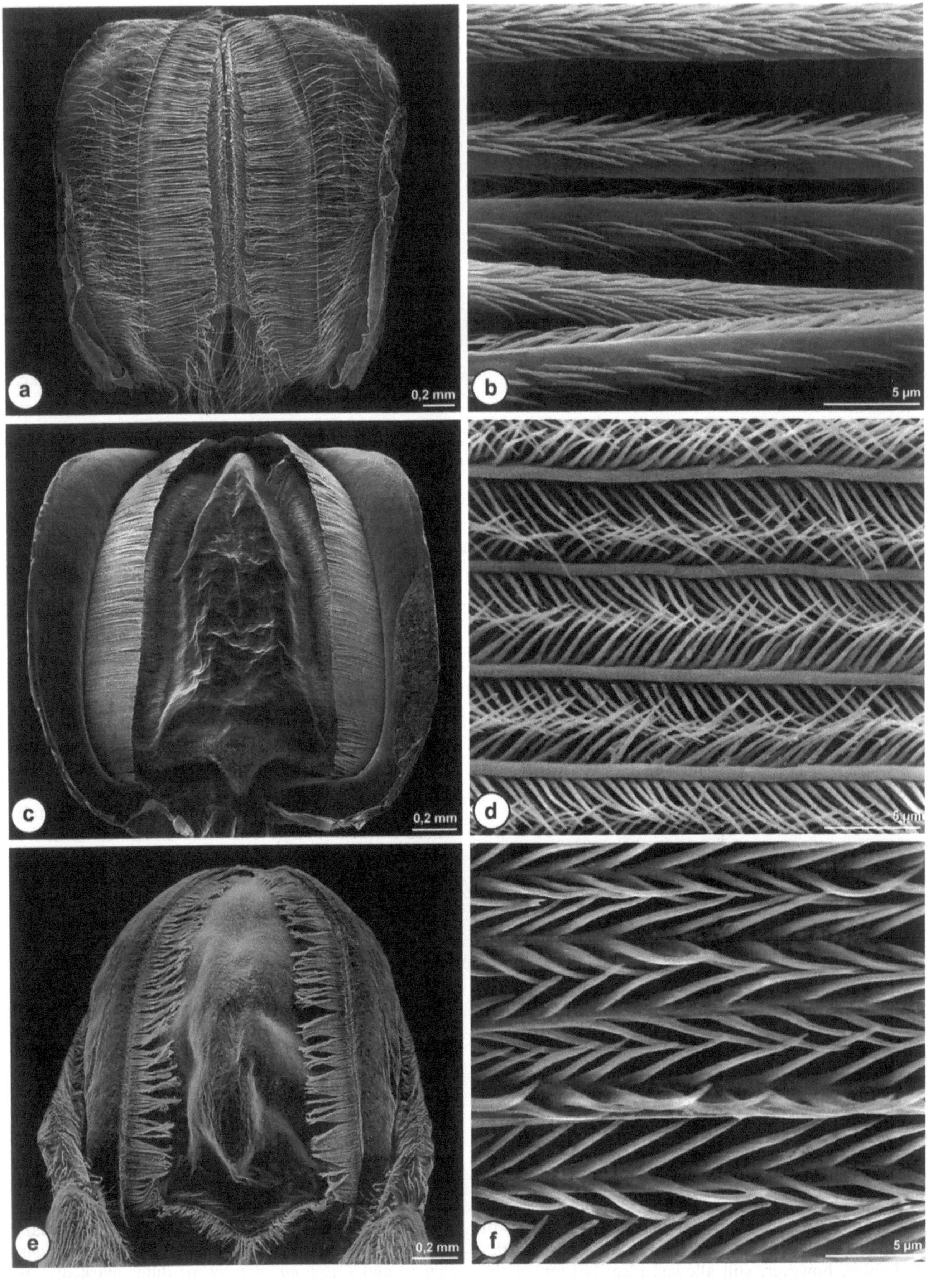
a
0,2 mm
b
5 µm
c
0,2 mm
d
e
0,2 mm
f
5 µm

◄
Fig. 5a–f Ventral face of the stomach. *Left column*: overview. *Right column*: setae of primary filters. **a**, **b** *Nematocarcinus lanceopes*, **c**, **d** *Notocrangon antarcticus*, **e**, **f** *Chorismus antarcticus*

three of them allow fluid and fine material to enter ventral channels which are connected to the secondary filters of the pyloric region. Coarse material can only be found in the cardiac chamber above the primary filters.

The setae of the secondary filters are arranged in a multi-storied system on either side of the inferomedianum, as opposed to those of Peracarida. In Amphipoda there is a two-storied filter; in other Peracarida, there is only one set of filter setae on each side (for literature, see Storch 1989). The mesh size of the secondary filters in the investigated species is certainly less than 1 µm, although exact data cannot be given as the mechanical properties of the pectinate setae are not known. The mesh size is smaller in *Chorismus antarcticus* and *Notocrangon antarcticus* than in *Nematocarcinus lanceopes*.

Chemical degradation occurs in the gastric fluid and both chambers of the stomach, which means that the stomach contents and the stomach's activity changes rhythmically (for details, see Vogt et al. 1989). Chemical degradation by enzymes synthesised in the F-cells of the midgut glands is followed by filtration through the secondary filters (conveying the resulting filtrate into the midgut glands) and absorption of their R-cells. Vogt (1994) suggested that coarse undigestible material is evacuated through the pharynx and mouth. This hypothesis is supported by the structure of the median dorsal bifid tooth of *Nematocarcinus lanceopes* and the anteriorly directed setae at the roof of the stomach, which might facilitate a retrograde transport of indigestible large particles. Smaller particles are conveyed into the very thin midgut.

Some areas on the Antarctic continental shelf impress by their richness in terms of diversity and biomass, consisting mainly of epifaunal sessile suspension feeders (e.g. Dayton et al. 1974; White 1984; Arntz and Gutt 1998). This biotic environment is, apart from carbon input, which via primary production and sedimentation is extremely seasonal in this area (Clarke 1988; Arntz et al. 1994), one important factor determining the aggregation, occurrence and abundance of shrimp species. *Chorismus antarcticus* appears to require sponges or similar structures as substratum. Gutt and Schickan (1998) even considered this relationship a transitional stage between commensalism and protocooperation, which they defined as feeding on organic material, e.g. deposited between the branches of bryozoans, and on motile prey organisms on the surface of sponges, which in turn keep the substratum organisms clean. Our findings from the stomachs confirm the predatory/omnivorous mode of feeding of *Chorismus antarcticus*. *Notocrangon antarcticus*, in contrast, seems to prefer a substratum that is not covered by such high concentrations of epifauna, possibly due to its digging behaviour (Gutt et al. 1991). In accordance with the observation that this species takes its food from the sediment (Gorny et al. 1992), we found fewer epibenthic taxa in the stomachs of *Notocrangon antarcticus* than in *Chorismus antarcticus*. The long antennae and legs of Nematocarcinidae, a family widely distributed over the deep sea of the world ocean from the North Atlantic to the Indo-Pacific and Southern Ocean, are interpreted as

Fig. 6a–c Secondary filters, inferomedianum. **a** *Nematocarcinus lanceopes*, **b** *Notocrangon antarcticus*, **c** *Chorismus antarcticus*

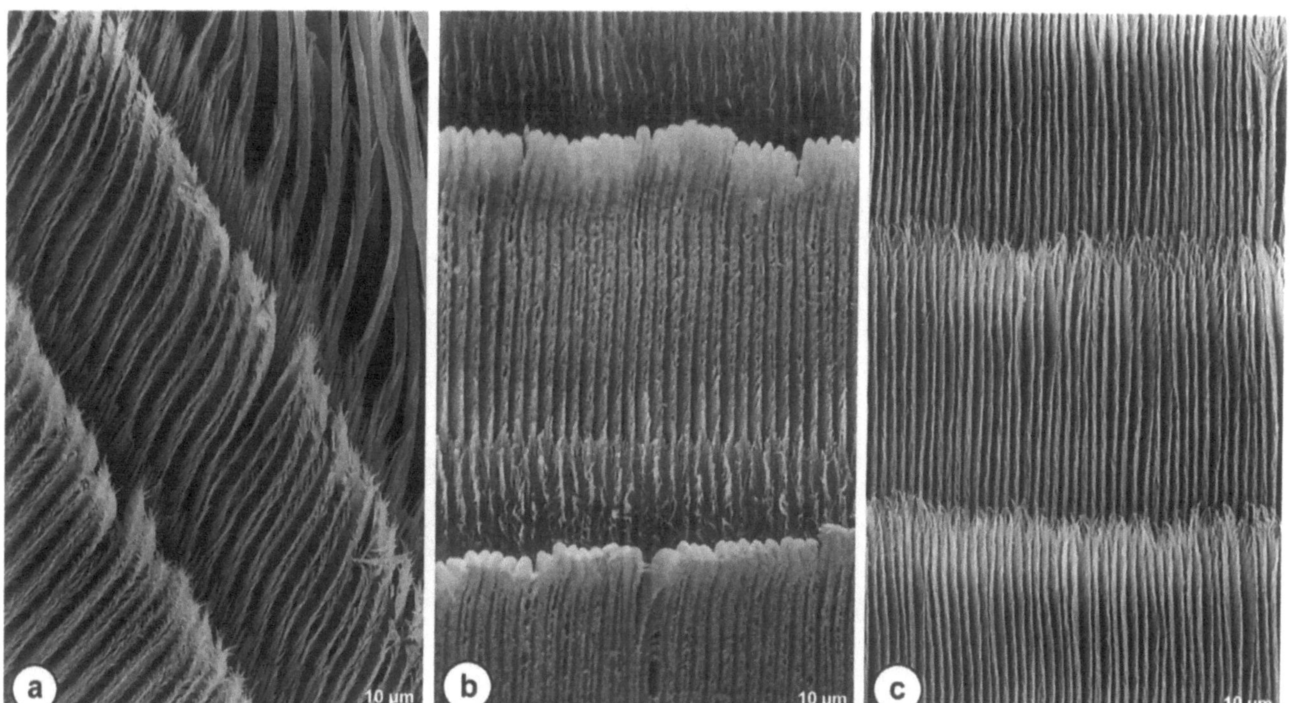

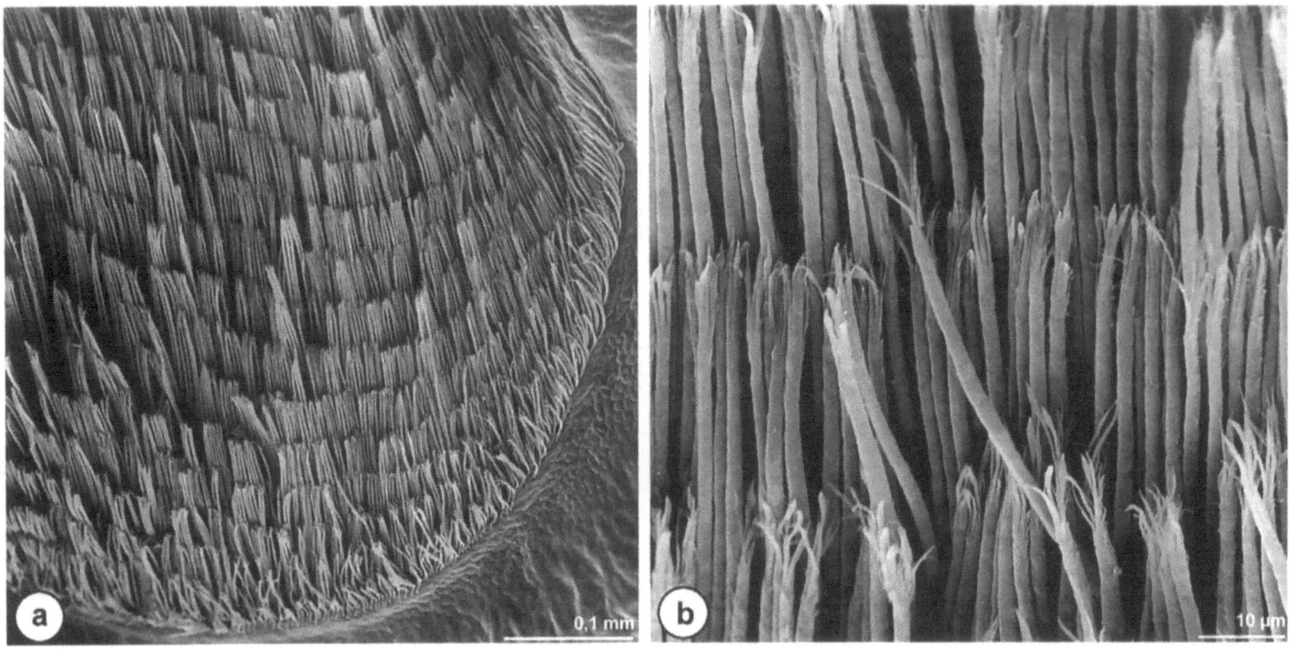

Fig. 7a, b Secondary filters, inferolaterale. **a** *Nematocarcinus lanceopes*, overview, **b** *Nematocarcinus lanceopes*

adaptations to walking across flocculent deep-sea muds (Gore 1985). Occasionally, Nematocarcinidae are described as facultatively benthic/hyperbenthic (Gore 1985) rather than as truly benthic. Stomach contents of species of this family indicate a generalised opportunistic feeding mode (Wenner 1979; Gore 1985) which, concluding from our observations, also holds true for *Nematocarcinus lanceopes*, the southernmost member of the genus.

It can be concluded from their mandible and stomach structure, as well as their stomach contents that all three species investigated have digestive systems adapted to an omnivorous diet. Stomach contents analyses, however, are difficult, as reliable identification of food items depends on their degradative state. Thanks to the com-

Fig. 8a–c Dorsal stomach wall of *Nematocarcinus lanceopes*. **a** Bifid tooth, **b** sclerotised hooks at the tip of the bifid tooth, **c** anteriorly directed bristles

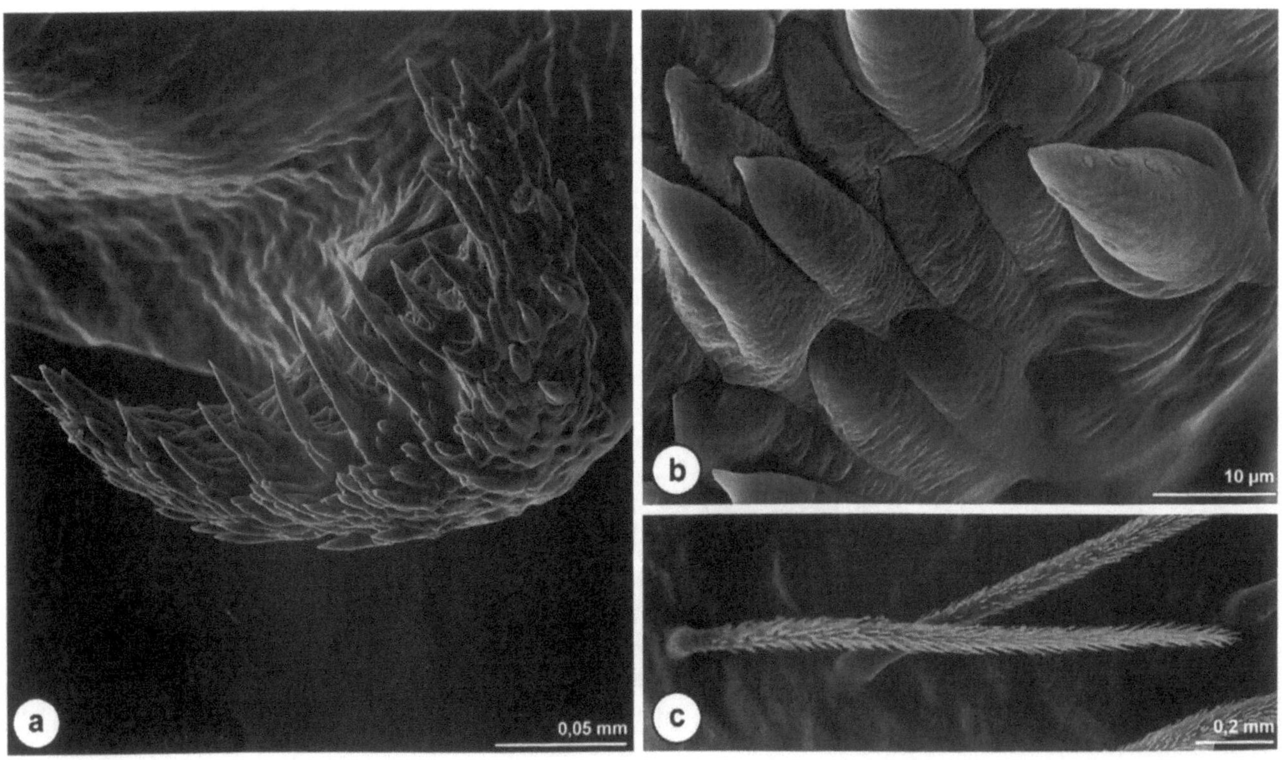

Table 1 Stomach contents of *Nematocarcinus lanceopes*, *Chorismus antarcticus* and *Notocrangon antarcticus* found in 15 stomachs per species

	Nematocarcinus lanceopes	*Chorismus antarcticus*	*Notocrangon antarcticus*
Diatomeae	x	x	x
-*Corethron* sp.		x	
Dinoflagellata		x	
Foraminifera	x	x	x
Radiolaria	x		
Tintinnid Ciliata	x		
Hexactinellid spicules	x	x	x
Cnidaria	x		
Bryozoa	x		
Nematoda		x	
-*Desmodoa* sp.	x		
Gastropoda		x	x
-Rissoidae		x[a]	
Bivalvia			
-*Yoldiella sabrina*	x		x[b]
Polychaeta			
-Polynoidae		x	
-Maldanidae		x	
-Glyceridae		x	
-*Nephtys* sp.		x	
-*Barukia* sp.		x[c]	
-Spionidae			x
-Sabellidae		x	x
Harpacticoid copepod	x		
Mysidacea		x	x
Amphipoda	x	x	
-*Orchomenella* sp.		x	
Isopoda			
-*Antarcturus* sp.		x	
Bristles and filter structures of Crustacea	x	x	x
Holothuroidea, sclerites		x	x
-*Taeniogyrus contortus*			
Fish scales		x	
Cross-striated musculature		x	
Lipid droplets		x	

[a]1.8 mm long, 0.8 mm wide
[b]5.1 mm long, 2.9 mm wide
[c]Large number of setae

bined knowledge of taxonomic specialists on board "Polarstern", a variety of food items could be determined to genus, though some remained unidentified. We conclude that stomach contents or morphological structures do not really reflect the different modes of life or habitat preferences of the species. To interpret the morphological structures, it is therefore also noteworthy that our findings are conclusive in the light of phylogeny. They agree with the general trend of increasing reduction of the gastric mill and ossicles among caridean shrimps (Coombs and Allen 1978).

Acknowledgements The help and support of the "Polarstern" crew and colleagues on board is greatly acknowledged. Special thanks are due to C. Kempendorf for technical assistance. Two anonymous referees made useful suggestions on the draft manuscript.

References

Arntz WE, Gorny M (1991) Shrimp (Decapoda, Natantia) occurrence and distribution in the eastern Weddell Sea, Antarctica. Polar Biol 11:169–177

Arntz WE, Gutt J (eds) (1998) The expedition Antarktis XV/3 (EASIZ II) of RV "Polarstern" in 1998. Ber Polarforsch 301:229

Arntz WE, Brey T, Gallardo VA (1994) Antarctic zoobenthos. Oceanogr Mar Biol 32:241–304

Clarke A (1988) Seasonality in the Antarctic marine environment. Comp Biochem Physiol 90B:461–473

Clarke A, Lakhani KH (1979) Measures of biomass, moulting behaviour and the pattern of early growth in *Chorismus antarcticus* (Pfeffer). Br Antarct Surv Bull 47:61–88

Coombs EF, Allen JA (1978) The functional morphology of the feeding appendages and gut of *Hippolyte varians* (Crustacea: Natantia). Zool J Linn Soc 64:261–282

Dayton PK, Robilliard GA, Paine RT, Dayton LB (1974) Biological accommodation in the benthic community at McMurdo Sound, Antarctica. Ecol Monogr 44:105–128

Felgenhauer BE, Abele LG (1985) Feeding structures of two atyid shrimps, with comments on caridean phylogeny. J Crust Biol 5:397–419

Gore RH (1985) Some rare species of abyssobenthic shrimp (families Crangonidae, Glyphocrangonidae, and Nematocarcinidae) from the Venezuela Basin, Caribbean Sea (Decapoda, Caridea). Crustaceana 48:269–285

Gorny M (2000) On the biogeography and ecology of the Southern Ocean decapod fauna. In: Arntz WE, Ríos C (eds) Magellan – Antarctic: Ecosystems that drifted apart. Sci Mar 63 [Suppl 1]: 367–382

Gorny M, Arntz WE, Clarke A, Gore DJ (1992) Reproductive biology of caridean decapods from the Weddell Sea. Polar Biol 12:111–120

Gorny M, Brey T, Arntz WE, Bruns T (1993) Growth, development and productivity of *Chorismus antarcticus* (Pfeffer) (Crustacea: Decapoda: Natantia) in the eastern Weddell Sea, Antarctica. J Exp Mar Biol Ecol 174:261–275

Gutt J, Schickan T (1998) Epibiotic relationships in the Antarctic benthos. Antarct Sci 10:398–405

Gutt J, Gorny M, Arntz W (1991) Spatial distribution of Antarctic shrimps (Crustacea: Decapoda) by underwater photography. Antarct Sci 3:363–369

Kirkwood JM (1984) A guide to the Decapoda of the Southern Ocean. ANARE Res Notes 11:1–47

Lin FY (1996) Structure of the gland filters in the pyloric stomach of *Penaeus japonicus* (Decapoda: Penaeidae). J Crust Biol 16:515–521

Maxwell JGH (1977) The breeding biology of *Chorismus antarcticus* (Pfeffer) and *Notocrangon antarcticus* (Pfeffer) (Crustacea, Decapoda) and its bearing on the problems of the impoverished Antarctic decapod fauna. In: Llano GA (ed) Adaptations within Antarctic ecosystems. Proceedings 3rd SCAR Symposium Antarctic Biology. Gulf, Houston, Tex, pp 335–342

Patwardhan SS (1935) On the structure and mechanism of the gastric mill in Decapoda. V. The structure of the gastric mill in natantous Macrura-Caridea. Proc Indian Acad Sci Sect B1:414–422

Sirenko B, Smirnov I, Arntz WE (1997) Taxonomic biodiversity of bottom invertebrates in the eastern Weddell Sea. In: Arntz WE, Gutt J (eds) The expedition Antarktis XIII/3 (EASIZ I) of "Polarstern" to the eastern Weddell Sea in 1996. Ber Polarforsch 249:25–31

Stamhuis EJ, Dauwe B, Videler JJ (1998) How to bite the dust: morphology, motion pattern and function of the feeding appendages of the deposit-feeding thalassinid shrimp *Callianassa subterranea*. Mar Biol 132:43–58

Storch V (1989) Scanning and transmission electron microscopic observations on the stomach of three mysid species. J Morphol 200:17–27

Storch V, Strus J (1989) Microscopic anatomy and ultrastructure in the alimentary canal of terrestrial isopods. Biol Monit Zool Ital NS 4:105–126

Tiefenbacher L (1990) Beiträge zur Taxonomie von *Nematocarcinus longirostris* Bate, 1888, und *Nematocarcinus lanceopes* Bate, 1888, neu für die westliche Antarktis (Crustacea, Decapoda, Natantia). Spixiana 13:229–235

Ullrich B, Storch V (1993) Development of the stomach in *Euphausia superba* Dana (Euphausiacea). J Crust Biol 13:423–431

Ullrich B, Storch V, Marschall MP (1991) Microscopic anatomy, functional morphology, and ultrastructure of the stomach of *Euphausia superba* Dana (Crustacea, Euphausiacea). Polar Biol 11:203–211

Vogt G (1994) Funktionsmorphologie des Garnelenmagens. Verh Dtsch Zool Ges 87:117

Vogt G, Stöcker W, Storch V, Zwilling R (1989) Biosynthesis of *Astacus* protease, a digestive enzyme from crayfish. Histochemistry 91:373–381

Wenner EL (1979) Distribution and reproduction of nematocarcinid shrimp (Decapoda: Caridea) from the northwestern North Atlantic. Bull Mar Sci 29:380–393

White MG (1984) Marine benthos. In: Laws RM (ed) Antarctic ecology, vol 2. Academic Press, London, pp 421–461

Yaldwyn JC (1965) Antarctic and subantarctic decapod Crustacea. In: Mieghem J van, Oye P van (eds) Biogeography and ecology in the Antarctic. Junk, The Hague, pp 324–330

Polar Biol (2001) 24: 941–945
DOI 10.1007/s003000100305

Tomás Munilla León

Synopsis of the pycnogonids from Antarctic and Subantarctic waters

Abstract This work summarizes existing knowledge on the faunal history, biodiversity and biogeography of the Antarctic and Subantarctic pycnogonids. It refers to material sampled between 1829 and 1999 from more than 40 expeditions run by 14 countries, and published in 75 papers. Up to now, 31 genera and 251 species have been recorded from a total of about 38,000 specimens captured at 2,000 stations. They constitute 38.75% and 21.5%, respectively, of the 80 genera and 1,164 species recorded worldwide. One hundred and twenty species have been found in Antarctic, 71 in Subantarctic waters, and 60 species in both zones, yielding an endemicity for the entire area of over 80%. Fifty five species revealed a circumpolar distribution and 161 were found mainly on the shelf. The Antarctic Ocean is suggested to be a centre of pycnogonid speciation and dispersion.

Introduction

The pycnogonids (Cheliceriformes, Arthropoda) from Antarctic and Subantarctic waters have been studied more extensively than those from any other world ocean of similar size. Exploration of this area began with the American Expedition of Nathaniel Palmer (Palmer Archipelago, 1829–1830). The naturalist of this cruise, James Eights, described (Eights 1835) and drew the first Antarctic pycnogonid *Decolopoda australis*, on a serolid and some fossils. Later on, many countries were to send their own "National Antarctic Expedition", which produced numerous monographs, the main ones being the works of Hoek (1881), Möbius (1902), Hodgson (1907, 1927), Bouvier (1913), Calman (1915), Gordon (1932, 1938, 1944), Fry and Hedgpeth (1969), Pushkin (1993) and Child (1994a, b, 1995a, b, c). The Child papers mentioned contain more detailed information about the historical background of distinct families.

The goal of this work is to extract inferences (about historical research, biodiversity, zoogeography and ecology) on the Austral pycnogonids by analysing the consulted literature.

Please note: When citing this article, please use the reference to the original source in the journal Polar Biology as given on the top of this page.

T. M. León
Unidad de Zoología, Universidad Autónoma de Barcelona, 08193 Bellaterra, Barcelona, Spain
E-mail: Tomas.Munilla@blues.uab.es

Materials and methods

This is a bibliographic paper. On the biodiversity and history I have consulted the above mentioned works, the Müller catalogue (Müller 1993) and 42 others. The main ecological and zoogeographical works examined are those of Fry (1964), Fry and Hedgpeth (1969), Hedgpeth (1969a, b, 1971), Arntz et al. (1994), Arntz (1997), Clarke (1990) and Clarke and Crame (1989, 1997). In total, 75 papers have been consulted.

Results and discussion

Historical research and biodiversity

The chronological and synoptic faunal histories of the Austral pycnogonids captured on different expeditions and in different decades are shown in Tables 1 and 2, respectively. This research involves 165 years, 14 countries and more than 40 ships or expeditions. The main results are that at present 38,000 specimens have been found at about 2,000 stations in the Southern Ocean.

These individuals belong to 31 genera and 251 species of pycnogonids, 1,164 being the total number of species in the world. They represent 21.5% of the actual world species, which were captured in Austral waters, that is to say, in 21% of the ocean areas (Jacques and Tréguer 1986). Nymphonidae is the most abundant family (69 species), the first genus is *Nymphon* (66 species), and *N. australe* the most frequently recorded species. In short, 101 species are endemic in the Antarctic area, 60 are present only in the Subantarctic zone and 42 species

Table 1 Chronological history of discovery of pycnogonids from Antarctic and Subantarctic waters. *N* Number of specimens, *STS*. number of stations, *S* number of species, *N. SP*. number of new species

Author	Country	Ships or expedition	Years of cruises	N	STS.	S	N. SP.
Eights 1835	USA	Palmer Exp.	1829–1830	4	3	1	1
Hoek 1881, 1898	UK	Challenger	1872–1876	78+	17	24	17
Möbius 1902	Germany	Valdivia	1898–1899	85	17	26	14
Hodgson 1902, 1904, 1907, 1908, 1914, 1915, 1927	UK, Germany	Southern Cross Discovery, Siboga Gauss	1898–1904	?	?	94	43
Bouvier 1905, 1906, 1913	France	Francais, Pourquoi-pas	1903–1910	500 approx.	41	34	18
Calman 1915, 1933	UK	Terranova	1915–1933	600	16	44	11
Loman 1923	Sweden	Swedish Antarct.	1901–1903	1,000 approx.	31	32	5
Gordon 1932, 1938, 1944	UK, Australia New Zealand	Discovery I, II Australian Antarct. Banzare	1911–1931	3,820 approx.	158	105	28
Stephensen 1947	Norway	Norwegian Antarct. Exp.	1927–1928	109	37	17	0
Hedgpeth 1950	USA	USA Navy	1947–1948	192	19	24	1 ?
Fage 1952a, b	France	Commandant Chargot	1950	74	24	18	0
Utinomi 1959	Japan	Japanese Antarct. Research	1955–1958	3	3	3	1
Stock 1965	Belgium	Antarctique 1960–1961	1960–1961	46	9	13	0
Fry and Hedgpeth 1969	New Zealand, USA	Endeavour, Eltanin Glacier	1956–1964	600	118	49	7
Arnaud 1972a, b	France	Terre Adélie, Kerguelen	1961–1970	961	121	41	0
Turpaeva 1974, 1990	Soviet Union	Akademic Koerstschatov	1965–1971	729	16	34	4
Krapp 1980	Belgium		1965–1967	37	17	16	0
Child 1987, 1994a, b, 1995a, b, c	USA, New Zealand	Eltanin, Aurora australis, Hero, Endeavour, Islas Orcadas, Vema Prof. Siedlecki 7 Coastguard	1956–1986	20,203	600 approx.	168	42
Munilla 1989, 1991, 2000 , 2001	Spain	Pescapuerta IV Hespérides (Bentart)	1986, 1994–1995	1,435	46	72	4
Bamber 1995	Canada		1970–1994	198	9	18	1
Stiboy-Risch 1992, 1993	Germany	Polarstern	1984–1985	674	81	50	2[a]
Pushkin 1993 + 10 papers	Soviet Union	Soviet Antarct. Exps.	1956–1987	5,813	616	162	25
Others							27
TOTAL	14	40+	1829–1999	38,000 approx.	2,000 approx.		251

[a] *Ammothea bicorniculata* Stiboy-Risch 1992 is probably a junior synonym for *A. allopodes*

Table 2 History of pycnogonid new species published by decades, indicating the importance of the expeditions around 1900. The Russian and USA cruises (1956–1987) published their data in the last decade. The influence of the Second World War (and the post-war) is notorious. Some species have been synonymized

Decades	New species
1835–1880	22
1881–1898	17
1900–1909	20
1910–1919	51
1920–1929	20
1930–1939	17
1940–1949	11
1950–1959	2
1960–1969	7
1970–1979	4
1980–1989	7
1990–1998	67

were recorded as endemic from Austral oceans (Table 3). Detailed data for the 251 austral species can be obtained from the author on request.

Finally, the comparison of percentages from Antarctic-living species of the main animal groups (Table 4) is relatively high for the pycnogonids: 15% versus 14.2% of the amphipods (the other most abundant group). More studies will be necessary in order to expand the knowledge of the specific biodiversity of this and other groups, mainly in the South African, Australian and Bellinghausen zones.

Ecology

Fluctuations of the abiotic parameters of the Weddell Sea (temperature, salinity, O_2, phosphates, nitrates) as an example of Antarctic environmental conditions are

Table 3 Endemicity and species richness of pycnogonid species in Antarctic and Subantarctic waters

Living in...	Total spp. number	% of spp. known worldwide (1,164 in total)	Endemics there of (no. spp.)	% of endemic spp. in respective zone	Other waters (no. spp.)
Antarctic only	120	10.3	101	84.2	19
Subantarctic only	71	6.1	60	84.5	11
Antarctic and Subantarctic	60	5.1	42	70.0	18
Total	251	21.5	203		48

Table 4 Number and percentages of recorded Antarctic species in relation to worldwide ones (main zoological groups). Sources: Arntz 1997; Brusca and Brusca 1990; Hickman et al. 1990; Wilson 1994; Corbera 1999

Zoological group	World species	Antarctic species	Percentage
Pycnogonida	1,164	180	15.5
Amphipoda	6,000	850	14.2
Polychaeta	5,500	670	12.2
Isopoda	4,000	350	8.7
Porifera	5,000	310	6.2
Crustacea	30,500	1,800	5.9
Cumacea[a]	1,200+	68	5.6
Echinodermata	6,100	300	4.9
Cnidaria	9,000	360	4.0
Mollusca	50,000	900	1.8
Pisces	23,000	300	1.3

[a] In the Southern hemisphere

very low, except for silicates (between 73 and 115 g m^{-3}) and depth (0–4,500 m; Jacques and Tréguer 1986). Apparently, the silicates are not important for the pycnogonids because they have no silicates as structural components. In respect to depth, San Vicente et al. (1997) mention that it is the only important abiotic factor of six tested in waters around Livingston Island.

Synthetic data about the bathymetric distribution of Austral species are shown in Table 5. Their main features are that two-thirds of the species have been found only on the continental shelf and upper slope (also the most sampled zones) and that the number of species decreases dramatically down to 1,000 m. However, 43.6% of the recorded Austral species have a eurybathy superior to 500 m. This fact could be explained by the stability of the environmental conditions in the water column, mainly temperature and salinity. Another cause of the conspicuous eurybathy could be the great extension of the continental glacial ice plates on the Antarctic shelves, forcing the fauna to survive at great depths.

Finally, it is necessary to carry out more coastal and very deep expeditions for a better knowledge of the biodiversity and bathymetrical distribution of the Austral pycnogonids, because the species found between 0 and 100 m and below 1,000 m are very few.

Zoogeography

The Antarctic benthos has evolved as a consequence of abiotic environmental conditions in the past and of biotic interactions (Arntz et al. 1994). The distribution of most of the benthic Antarctic fauna is circumpolar (Hedgpeth 1971; Arntz and Gallardo 1994; Clarke and Crame 1997). Circumantarctic distribution is also the most abundant with respect to pycnogonids (Fry and Hegdpeth 1969; Hegdpeth 1969a), since 55 of the 251 Austral species are circumpolar (21.9%). Only 5 are cosmopolitan.

The data in Table 3 show a high specific endemicity (81% in Austral waters, 84% and 85% in the Antarctic

Table 5 Number (*N*) of Austral species (*Sp.*) of pycnogonids related to depth and their eurybathy (*A* Antarctic, *S* Subantarctic)

	N	% of 251	Remarks
Sp. recorded exclusively between 0 and 100 m	17	6.8	4A + 13S
Sp. recorded exclusively between 0 and 1,000 m	160	63.8	Continental shelf + upper slope
Sp. recorded between 0 and 1,000 m	218	86.9	
Sp. recorded exclusively below 1,000 m	34	13.6	Very deep species
Sp. recorded below 1,000 m	88	35.2	
Sp. recorded exclusively between 1,000 and 5,000 m	28	11.2	
Sp. recorded exclusively below 5,000 m	6	2.4	2 Pantopipetta, 2 Nymphon, 1 Colossendeis, 1 Aschorynchus
Sp. with eurybathy between 500 and 1,000 m	45	18.0	
Sp. with eurybathy superior to 1,000 m	64	25.6	15 Nymphon, 13 Colossendeis, 5 Pallenopsis, 5 Ammothea + others

and Subantarctic respectively), comparable to the Isopoda (87%, Brandt 1992) or Cumacea (89%, Corbera 1999). The number of endemic species for each individual Austral region is low (10 or less) except in the Scotia Sea (26) and on the New Zealand Plateau (21). These two latter areas might be considered subcentres of speciation as a result of the separation of the Antarctic from South American and Australian continents. However, these two zones are also among those most sampled in the Southern Ocean.

Endemic genera of Antarctic waters are very few (*Dodecolopoda, Pentanymphon, Sexanymphon, Austroraptus*). Other genera (*Decolopoda, Austropallene*) which were considered by Hegdpeth (1969b) as typically Antarctic, have recently been found in Subantarctic waters; the same occurs with some species. The Subantarctic pycnogonid fauna shows origins in the Antarctic Ocean at genus level (Arnaud and Bamber 1987); this is true for example with *Colossendeis* or *Ammothea*, two genera with more than half the species in the Southern Ocean. Both, similar to other genera with abundant species, have more species in Antarctic than in Subantarctic waters. Therefore, in my opinion, the Antarctic Ocean should be considered not only as a centre of speciation (suggested by Hedgpeth 1969b) but also of geographic dispersion and evolutionary radiation, because of its great relative endemicity (101 Antarctic species versus 60 Subantarctic).

Three possible routes of migration have been proposed by Fry and Hegdpeth (1969):

1. Direct migration from the Antarctic along the sea bottom (*Colossendeis* species).
2. From Antarctica to Australasia by the Westwind Drift-Western current (*Ammothea* sibling species in both zones).
3. From Antarctica to South America via the Scotia Arc (*Achelia* sibling species in both areas).

This last route appears to be certain in the migration of actual species; 20 of the 48 Austral species recorded in other waters were detected in areas off Argentina and Brazil.

Finally, there is evidence which favours the "in situ" origin of much of the Antarctic shelf fauna during the Cretaceous or before, when the ex-Gondwana continents were still connected (Clarke and Crame 1989). Examples proposed are the gastropods (Clarke 1990), the serolid and arcturid isopods (Brandt 1992), and now the pycnogonids. This last group, like echinoderms, ascidians and gastropods, show a particular separation into closely related species, which are almost certainly the product of radiation in situ (Clarke and Crame 1989). Moreover, radiation is always connected with a high level of endemism (Arntz et al. 1994).

Acknowledgements This work was carried out under the auspices of the Spanish Council of Scientific and Technical Research (CICYT, contract no. ANT95–1011). I wish to express my gratitude to Drs. W. Arntz and A. Clarke for their suggestions and English corrections.

References

Arnaud F (1972a) Invertébrés marins. XIIème et XVème expeditions antarctiques francaises en Terre Adélie (9 Pycnogonides). Téthys 4:135–156

Arnaud F (1972b) Pycnogonides des îles Kerguelen (Sud Océan Indien): matérial nouveau et révision des spécimens du Museum National d'Histoire Naturel de Paris. Bull Mus Natl Hist Nat Paris 3 (51):801–815

Arnaud F, Bamber N (1987) The biology of pycnogonids. Adv Mar Biol 24:1–96

Arntz W (1997) Investigación antártica en biología marina: situación actual, proyectos internacionales y perspectivas. Bol R Soc Esp His Nat Secc Biol 93:13–44

Arntz W, Gallardo VA (1994) Antarctic benthos: present position and future prospects. In: Hempel C (ed) Antarctic science. Global concerns. Springer, Berlin Heidelberg New York, pp 243–277

Arntz W, Brey T, Gallardo VA (1994) Antarctic zoobenthos. Oceanogr Mar Biol Ann Rev 32:241–304

Bamber RN (1995) A new species of *Pycnogonum* Brünnich 1764 (Arthropoda, Pycnogonida) with other pycnogonid records from the Falkland and South Shetland Islands. Ophelia 40:199–205

Bouvier EL (1905) Observations préliminaires sur les Pycnogonides recueillis dans la région antarctique par la mission du " Français". Bull Mus Natl Hist Nat Paris 11:294–298

Bouvier EL (1906) Nouvelles observations sur les Pycnogonides recuellis dans les régins antarctiques au cours de la campagne dirigée par M. Jean Chargot. C R Acad Sci Paris 142:15–22

Bouvier EL (1913) Pycnogonides du "Pourquoi Pas ?" In: Calman WT (ed) Deuxiéme Expédition Antarctiques Francaise (1908–1910), vol 6. Masson, Paris, pp 1–169

Brandt A (1992) Origin of Antarctic isopods (Crustacea, Malacostraca). Mar Biol 113:415–423

Brusca RC, Brusca GJ (1990) Invertebrates. Sinauer, Sunderland, Mass.

Calman WT (1915) Pycnogonida. British Antarctic (Terra Nova) Expedition, 1910. Zoology 3 (1):1–74

Calman WT (1933) A dodecapodous pycnogonid. Nature 131:242

Child CA (1987) New and little know Pycnogonida from Antarctic and Subantarctic waters. Proc Biol Soc Wash 100:902–916

Child WA (1994a) Antarctic and Subantarctic Pycnogonida. I. The family Ammotheidae. (Biology of Antarctic Seas XXIII) Antarct Res Ser 63:1–48

Child WA (1994b) Antarctic and Subantarctic Pycnogonida. II. The family Austrodecidae. (Biology of Antarctic Seas XXIII) Antarct Res Ser 63:49–99

Child WA (1995a). Antarctic and Subantarctic Pycnogonida. III. The family Nymphonidae. (Biology of Antarctic Seas XXIV) Antarct Res Ser 69:1–68

Child WA (1995b) Antarctic and Subantarctic Pycnogonida. IV. The families Colossendeidae and Rhynchotoraxidae. (Biology of Antarctic Seas XXIV) Antarct Res Ser 69:69–112

Child WA (1995c) Antarctic and Subantarctic Pycnogonida. V. The families Pycnogonidae, Phoxichilidiidae, Endeidae, and Callipallenidae, including the genus Pallenopsis. (Biology of Antarctic Seas XXIV) Antarct Res Ser 69:113–165

Clarke A (1990) Temperature and evolution. Southern Ocean cooling and the Antarctic marine fauna. In: Kerry KR, Hempel G (eds) Antarctic ecosystems: ecological change and conservation. Springer, Berlin Heidelberg New York, pp 9–22

Clarke A, Crame JA (1989) The origin of the Southern Ocean marine fauna. In: Crame JA (ed) Origin and evolution of Antarctic biota. Geol Soc Esp Publ 47:227–246

Clarke A, Crame JA (1997) Diversity, latitude and time: patterns in the shallow sea. In: Ormond R, Gage J, Angel MV (eds) Marine biodiversity. Patterns and processes. Cambridge University Press, Cambridge, pp 122–147

Corbera J (1999) Biodiversity of Antarctic cumaceans. Abstracts of EASIZ Symposium, Bremerhaven, Germany, June 1999

Eights J (1835) Description of a new animal belonging to the Arachnides of Latreille, discovered in the sea along the shores of the New South Shetland Islands. Boston J Nat Hist 1 (2):203–206

Fage L (1952a) Pycnogonides. Missions du bâtiment polaire Commandant Chargot, récoltes faites en Terre Adélie (1950). Bull Mus Natl Hist Nat Paris 2 (24)2:180–186

Fage L (1952b) Pycnogonides de la Terre Adélie. Bull Mus Natl Hist Nat Paris 2 (24)3:263–273

Fry WG (1964) The pycnogonid fauna of the Antarctic continental shelf. Biologie Antarctique. Actual Sci Ind 1312:263–269

Fry WG, Hedgpeth JW (1969) Pycnogonida. 1. Colossendeidae, Pycnogonidae, Endeidae and Ammotheidae. (Fauna of the Ross Sea, 7) Mem N Z Oceanogr Inst 49:1–149

Gallardo VA (1987) The sublittoral macrofaunal benthos of Antarctic shelf. Environ Int 13:71–81

Gordon I (1932) Pycnogonida. Discovery Rep 6:1–138

Gordon I (1938) Pycnogonida. Sci Rep Australas Antarct Exped Zool Bot 2:1–40

Gordon I (1944) Pycnogonida. Rep Br Aust N Z Exped B 5 (1):1–72

Hedgpeth JW (1950) Pycnogonida of the United States Navy Expedition, 1947–48. Proc U S Natl Mus 100 (3260):147–160

Hedgpeth JW (1969a) Introduction to Antarctic zoogeography. Antarct Map Fol Ser Am Geogr Soc Fol 11:1–9

Hedgpeth JW (1969b) Pycnogonida. Antarct Map Fol Ser Am Geogr Soc Fol 11:26–28

Hedgpeth JW (1971) Perspectives of the benthic ecology in Antarctica. Res Antarct 1 (19):93–136

Hickman CP, Roberts LS, Hickman FM (1990) Zoología. Principios integrales. Interamericana-McGraw-Hill, Madrid

Hodgson TV (1902) Crustacea (Pycnogonida). (Report on the collections of Natural History made in the Antarctic regions during the voyage of the "Southern Cross".) British Museum, London, pp 256–258

Hodgson TV (1904) On a new Pycnogonid from the South Polar regions. Ann Mag Nat Hist 7 (14):458–462

Hodgson TV (1907) Pycnogonida. Report of the National Antarctic expedition 1901–1904. Nat Hist 3:1–72

Hodgson TV (1908) The Pycnogonida of the Scottish National Antarctic Expedition. Trans R Soc Edinb 46 (1) 6:159–188

Hodgson TV (1914) Preliminary report of the German Southpolar Expedition, 1901-1903. Zool Anz 45 (4):158–165

Hodgson TV (1915) The Pycnogonida collected by the Gauss in the Antarctic regions, 1901-1903: preliminary report. Ann Mag Nat Hist Ser 8 15 (85):141–149

Hodgson TV (1927) Die Pycnogoniden der Deutschen Südpolar-Expedition 1901–1903. Dtsch Südpolar Exped Ser II Zool 19:303–358

Hoek PPC (1881) Report on the Pycnogonida dredged by HMS Challenguer 1873–1876. Rep Sci Voyage HMS Challenguer 3 (10):1–167

Hoek PPC (1898) On four pycnogonids dredged during the cruise of the Challenguer. Tijdschr Ned Dierkd Ver (2) 5 (2–4):190–301

Jacques G, Treguer P (1986) Écosystèmes pélagiques marins. Masson, Paris, pp 130–161

Krapp F (1980) Neue Pantopodenfunde (Pycnogonida, Arthropoda) nahe der belgischen Antarktis-Station. Bull Inst R Sci Nat Belg Bruxelles Biol 52 (18):1–8

Loman JCC (1923) The Pycnogonida. Further Zool Res Swed Antarct Exped 1 (2):1–41

Möbius K (1902) Die Pantopoden der Deutschen Tiefsee-Expedition, 1898–99. Wiss Ergebn Dt Tiefsee-Exped Dampfer 'Valdivia" 1898–1899 3:177–196

Müller HG (1993) World catalogue and bibliography of the recent Pycnogonida. Laboratory for Tropical Ecosystems, Wetzlar, pp 1–388

Munilla T (1989) Dos nuevas especies de Picnogónidos antárticos. Actas del III Simposio Español deBentos Antártico, 3 al 7 de Octubre de 1989, Gredos. Comisión Interministerial de Ciencia y Tecnología, Spain, pp 74–79

Munilla T (1991) Picnogónidos capturados en la campaña 'Antártida 8811". Bol Inst Esp Oceanogr 7 (1):3–44

Munilla T (2000) A new species of *Ammothea* and other pycnogonids from around Livingston Island, South Shetland Islands, Antarctica. Antarct Sci 12 :47–51

Munilla T (2001) A new species of *Ammothea* (Pycnogonida) and other pycnogonids from Livingston Island and surrounding waters (South Shetland Islands, Antarctica). Antarct Sci 13:144–149

Pushkin AF (1974) Clavigeropallene. A new genus from Antarctic. Zool Zh 53:938–940

Pushkin AF (1975a) New species of Callipallenidae (Pantopoda) from the Subantarctic. Zool Zh 54:1402–1406

Pushkin AF (1975b) Revision of the Pycnogonids (Pantopoda) of the Pallenopsis patagonica group from Antarctic and adjacent waters. Byull Sov Antark Eksped 90:72–83

Pushkin AF (1976) New species of Endeidae (Pantopoda) from Subantarctic. Zool Zh 55:923–925

Pushkin AF (1977) New species of Austrodecidae from Antarctic. Explor Fauna Seas 21 (26):54–59

Pushkin AF (1982) On the species composition of the genus *Decolopòda* (Pantopoda). Zool Zh 59:1657–1661

Pushkin AF (1984a) A new species of the genus *Colossendeis* (Pantopoda) from the Antarctic. Zool Zh 63:1259–1261

Pushkin AF (1984b) A new species of the genus *Pycnogonum* (Pantopoda, Pycnogonida) from the Antarctic. Zool Zh 63: 1102–1104

Pushkin AF (1988) On the revision of the *Colossendeis robusta* (Pantopoda) from the Southern Ocean. Zool Zh 67:953–956

Pushkin AF (1990) A new species of the genus *Pseudopallene* (Pycnogonida, Callipallenidae) from the Antarctic. Zool Zh 69:155–157

Pushkin AF (1993) The pycnogonid fauna of the South Ocean. Biological results of the Soviet Antarctic Expeditions. [Explorations of the fauna of the seas XX (XXX)] De Acad Sci Zool Inst Semperi, Messina, pp 1–397

San Vicente C, Ramos A, Gimeno A, Sorbe JB (1997) Suprabenthic assemblages from South Shetland Islands and Bransfield Strait (Antarctica): preliminary observations on faunistical composition, bathymetric and near-bottom distribution. Polar Biol 18:415–422

Stephensen K (1947) Tanaidacea, Isopoda, Amphipoda and Pycnogonida. Sci Res Norw Anatarct Exped 1927–1928 27:1–90

Stiboy-Risch C (1992) *Ammothea bicorniculata*, eine neue art der Ammotheidae ans der Antarktis (Pantopoda, Pycnogonida). Bonn Zool Beitr 43 (2):333–338

Stiboy-Risch C (1993) Funde antarktischer und subantarktischer Pantopoden einschlieblich *Colossendeis acuta* sp.n.- sowie ein Bietrag zur Artbestimmung von *Colossendeis glacialis* Hodgson 1907 und *Colossendeis robusta* Hoek 1881. Mitt Hamb Zool Mus Inst 90:251–264

Stock JH (1965) Pycnogonides provenant de la base Antarctique Belge. Bull Inst R Sci Nat Belg 41 (31):1–2

Turpaeva E (1974) The Pycnogonids of the Scotia Sea and surrounding waters. Tr Inst Okeanol, Akad Nauk SSSR 98:277–305

Turpaeva E (1990) Pycnogonida from Southern Atlantic. Tr Inst Okeanol, Akad Nauk SSSR 126:108–126

Utinomi H (1959) Pycnogonida of the Japanese Antarctic Research Expeditions 1956-1958. (Biological results of the Japanese Antarctic Research Expeditions) Spec Publ Seto Mar Biol Lab 8:1–12

Wilson EO (1994) La Diversidad de la Vida. Drakontos, Barcelona

Polar Biol (2001) 24: 846–852
DOI 10.1007/s003000100291

E. Pisano · F. Mazzei · N. Derome · C. Ozouf-Costaz
J.-C. Hureau · G. di Prisco

Cytogenetics of the bathydraconid fish *Gymnodraco acuticeps* (Perciformes, Notothenioidei) from Terra Nova Bay, Ross Sea

Abstract In the framework of the I.C.E.FISH (International Collaborative Effort on Antarctic Fish Adaptive Evolution) project, during the 1998–1999 season at the Italian shore-based Terra Nova Bay Station, the broad fields of adaptation and evolution of the fish suborder Notothenioidei were tackled through the integration of many disciplines. As a representative contribution of I.C.E.FISH, a cytogenetic study of the bathydraconid fish *Gymnodraco acuticeps* is reported. The Bathydraconidae is a heterogeneous notothenioid taxon whose intra-family relationships are very uncertain. The conventional karyotype and the location of nuclear ribosomal genes, by means of fluorescence in situ hybridisation, contributed to species-specific characterisation and provided the basis for comparative analyses. The karyotype of *G. acuticeps* consists of 48 chromosomes (2 metacentric, 2 submetacentric and 44 acrocentric); major ribosomal cistrons correspond to a large region in a single chromosome pair. Mapping of the available karyotypic data on a molecular phylogenetic tree provided information on chromosomal diversification during the cladogenesis of the bathydraconids.

Please note: When citing this article, please use the reference to the original source in the journal Polar Biology as given on the top of this page.

E. Pisano (✉) · F. Mazzei
Department of Experimental Environmental and Applied Biology, University of Genova, Viale Benedetto XV, 5, 16132 Genova, Italy
E-mail: pisano@unige.it
Tel.: +39-010-3538052
Fax: +39-010-3538047

N. Derome · C. Ozouf-Costaz · J.-C. Hureau
National Museum of Natural History Paris, France

G. di Prisco
Institute of Protein Biochemistry and Enzymology CNR, Naples, Italy

Introduction

The I.C.E.FISH project

Within the biological project "Physiological, Biochemical and Molecular Bases of Adaptive Evolution in Antarctic Teleosts" of the Italian National Programme for Antarctic Research (PNRA), the multi-national and multi-disciplinary I.C.E.FISH (International Collaborative Effort on Antarctic Fish Adaptive Evolution) project was conducted during the 1998–1999 season in the aquarium and laboratories of the Italian shore-based Terra Nova Bay Station. Investigators were from Italy, France, the United States and New Zealand. During this international collaboration, research was focused on the ecophysiological adaptations and evolution of Antarctic fish. These themes fall within one of the main sections of the wider programme of the SCAR EASIZ (Coastal/Shelf Ecology of the Antarctic Sea Ice Zone) Research Programme, the major input from marine biological research to the SCAR IGBP programme.

I.C.E.FISH has thus proposed a multi-disciplinary, integrated approach to a research theme related to global change. The aim has been to study a homogeneous group of isolated, specialised organisms. In fact, knowledge of adaptive systems is a powerful tool in evaluating and foreseeing the impact on life of environmental changes (in particular, of rapidly developing anthropogenic changes), against which the main defence is physiological adaptation. A selected, representative I.C.E.FISH contribution is herewith reported. The complex and broad fields of adaptation and evolution of the fish suborder Notothenioidei were tackled through integration of many disciplines in the study of several biological systems, ranging in scale from the organism to the molecule. These fish are particularly important in the Antarctic waters, where they dominate the recent coastal ichthyofauna (Gon and Heemstra 1990; Balushkin 1992; Eastman and Clarke 1998). Endemism is very high within this group (95% for species and more than 85% for genera).

Cytogenetics of bathydraconids

The family Bathydraconidae of the suborder Notothenioidei includes 11 genera and 16 species living south of the Antarctic Polar Front (Eastman 1993; Skora 1995). Bathydraconids are generally described as mostly benthic and are abundant in the coldest, deepest shelf water at the highest latitudes. However, information on their biology and ecology is far from complete; data on the distribution and abundance of most of the species are also scanty, because of the characteristics of ichthyological surveys in some High Antarctic areas and of the heterogeneous sampling methods used (Eastman and Hubold 1999). It has been assumed that the Bathydraconidae is a derived family in the phylogeny of Notothenioidei (Bargelloni and Lecointre 1998). In the phylogenetic tree, established on the basis of morphological characters by Hastings (1993), the Bathydraconidae are divided into two subfamilies: Gymnodraconinae and Bathydraconinae, the latter being proposed as the sister-group of Channichthyidae. The Gymnodraconinae subfamily, previously comprising the genera *Gymnodraco* and *Psilodraco*, also includes the genus *Acanthodraco*, recently described by Skora (1995).

The intra-family relationships are extremely uncertain because most of the morphological and osteological characters are homoplasic too often to produce a reliable phylogenetic hypothesis at the specific level (Eakin 1981; Iwami 1985; Voskoboynikova 1993). Moreover, sequencing of 12S and 16S mitochondrial ribosomal DNA showed no evidence for monophyly of the taxon Bathydraconidae, and did not resolve phylogenetic problems at the specific level (Bargelloni et al. 1994; Ritchie et al. 1997).

Because of their relevance in phylogeny and evolution, cytogenetic studies have been undertaken on many Antarctic Notothenioidei, including the bathydraconids (Prirodina 1997; Pisano et al. 1998; Ozouf-Costaz et al. 1999). However, current cytogenetic information on bathydraconid fish is scarce, due to technical difficulties in sampling and in obtaining cell preparations suitable for chromosomal studies. The available data are mostly based on studies of a few specimens of each species, and are often limited to basic information (chromosome number and gross morphology). Seven bathydraconid species have been studied to date. The karyotypes of *Parachaenichthys georgianus*, *Prionodraco evansii*, *Psilodraco breviceps* have been described by Prirodina (1990); *Bathydraco marri*, *Gerlachea australis* and *Racovitzia glacialis* have been studied by Ozouf-Costaz et al. (1991), and *Cygnodraco mawsoni* by Morescalchi et al. (1996). Comparative analyses in the entire notothenioid suborder showed that bathydraconids include species with highly diversified karyotypes, the cytogenetic data being consistent with other evidence concerning the paraphyletic nature of the taxon (Pisano et al. 1998).

During the I.C.E.FISH project we focused our attention on the species *Gymnodraco acuticeps*, for which no previous cytogenetic information was available.

This bathydraconid is a circum-Antarctic fish usually found at shallow depths (Gon and Heemstra 1990). In Terra Nova Bay it is usually caught by trammel and gill nets set between 50 and 200 m, similarly to the more abundant *C. mawsoni* (Vacchi et al. 1999). It is a benthic species, characterised by a highly specialised mouth, armed with long, sharp teeth adapted for ambush predation. The conventional karyotype and the chromosomal location of the ribosomal genes, as detected through in situ DNA-DNA hybridisation, contributes to species-specific characterisation and provides elements for comparative analyses.

Materials and methods

The species under investigation

Fish were caught by means of gill nets in the vicinity of Terra Nova Bay Station at 100 m depth, and kept in aquaria with running aerated seawater at a temperature of between -2°C and 0°C.

Two female specimens of *Gymnodraco acuticeps*, caught during the I.C.E.FISH 1998–1999 project (TNB 261 and TNB 271; Muséum national d'histoire naturelle, Paris, MNHN 1999–374 and 375), were used for chromosomal studies. An additional female (TNB 41; National Antarctic Museum, Genova, MNA 1998–41), sampled in Terra Nova Bay Station during the 1997–1998 Antarctic season, was also studied.

Chromosome preparations and karyotyping

Mitotic chromosome preparations were made from head kidney after in vivo treatment according to Doussau de Bazignan and Ozouf-Costaz (1985). Chromosome spreads were made from fixed cell suspensions kept at -20°C.

The karyotype was established from conventional Giemsa-stained metaphases (Hartley and Horne 1985). Two of the three treated specimens provided mitotic chromosomes of acceptable quality for karyotyping. Conventionally, in the karyotype, the bi-armed chromosomes (metacentric, m; submetacentric, sm) were placed before the uni-armed chromosomes (acrocentric, a). Characterisation of chromosome morphology followed Levan et al. (1964).

Fluorescence in situ hybridisation

The in situ experiments were performed on the specimen TNB 261 which afforded the number and quality of metaphases suitable for hybridisation. Ribosomal DNA cistrons were located by means of fluorescence in situ hybridisation (FISH) using a fragment of 28S rDNA as probe (Ozouf-Costaz et al. 1996). The plasmid containing the rDNA fragment was nick translated with biotin-16dUTP (Boehringer), according to the manufacturer's instructions.

Slides were kept for 24–48 h at 40°C, denatured by heating at 65°C for 2 min in 70% (v/v) formamide/2XSSC, dehydrated and air-dried. The probe was denatured by heat treatment at 75°C for 10 min.

Fifteen microlitres of hybridisation buffer (50% formamide/, 10% dextran sulphate, 40 mM KH_2PO_4, 2XSSC), containing 30 ng of labelled probe, was added to each slide. Hybridisation was performed at 37°C in a moist chamber for 24 h.

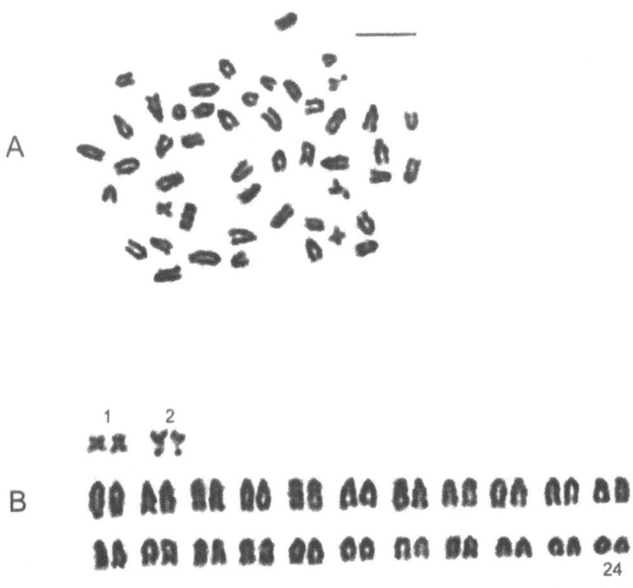

Fig. 1 Metaphase chromosomes (**a**) and karyotype (**b**) of *Gymnodraco acuticeps* (Giemsa staining; bar = 10 μm)

Post-hybridisation washings were carried out for 5 min at 45°C, twice in 50% formamide/2XSSC and 3 times in 2XSSC. The rDNA probe signal was detected with either streptavidin-FITC or streptavidin-CY3. Chromosomes were counterstained with DAPI and mounted in a standard antifade solution.

Hybridisation signals were recorded by a CCD camera (Quantix, Photometrix, Kodak) mounted on a Axioplan 2 microscope (Zeiss), using Smart Capture VP software.

Results

The karyotype of *Gymnodraco acuticeps*

Chromosomes suitable for karyotyping were obtained from specimens TNB 41 and TNB 261. Both had 48 chromosomes, which were grouped into 24 pairs; one pair (no. 1) is formed by small metacentric elements, one pair (no. 2) by submetacentric elements; all the other chromosomes are acrocentric (Fig. 1). The two homologues of pair no. 2 appeared to be homeomorphic in size in TNB 41 but heteromorphic in TNB 261. Moreover, a large region of the largest arm of the chromosomes of pair no. 2 appeared to be achromatic after Giemsa (Fig. 2a, b) and DAPI staining (not shown).

Ribosomal DNA location

After FISH, a large amount of rDNA was located in a single chromosomal domain, along the longer arm of the pair no. 2 chromosomes (Fig. 3).

The extension of the signal did not include the terminal region of the rDNA-positive arms, this being particularly evident in the larger chromosome of the pair (Fig. 5a). The position of ribosomal cistrons corresponded to Giemsa-negative regions. Moreover, as expected, the same regions were also negative to DAPI staining, indicating that these chromosomal areas are poor in AT bases. Unfortunately, the number of samples was not sufficient to perform additional and specific banding in order to better characterise rDNA-associated heterochromatin.

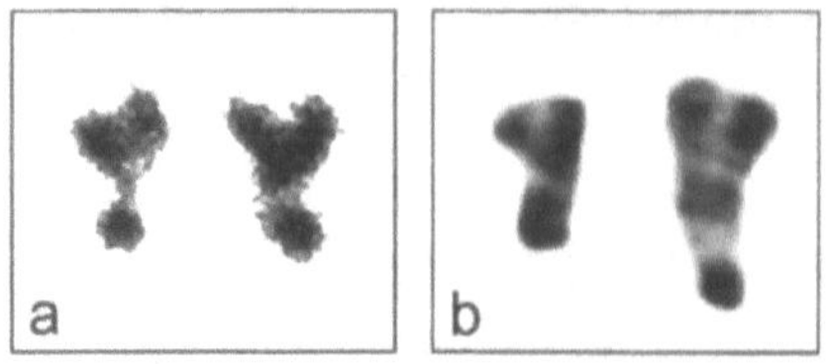

Fig. 2a, b Chromosome pair no. 2 from the karyotype of *G. acuticeps* (Giemsa staining). **a** Specimen TNB 41; **b** specimen TNB 261

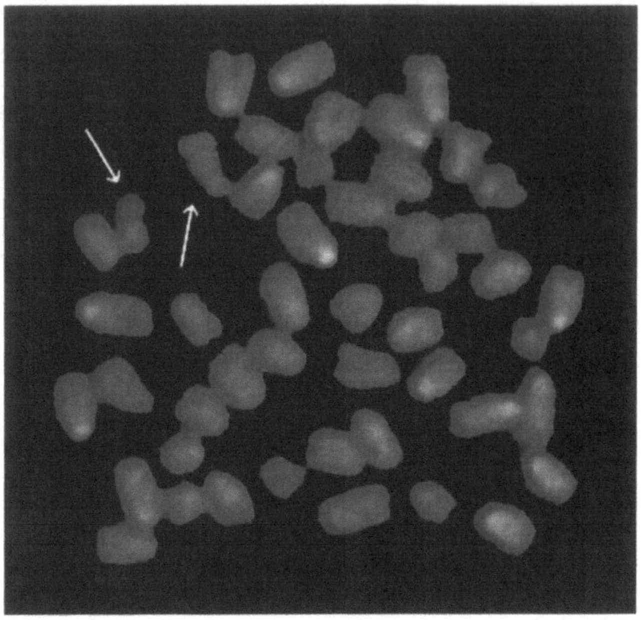

Fig. 3 Metaphase chromosomes of *G. acuticeps* after fluorescence in situ hybridisation with a 28S rDNA probe. Hybridisation signals (*arrows*) correspond to a large region of the heteromorphic chromosome pair no. 2. The probe signal is from streptavidin-CY3 fluorescence detection of biotin-labelled probe

Different amounts of rDNA per haploid set gave rise to the structural size polymorphism of chromosome pair no. 2.

Discussion

Chromosome numbers and formulae

In the hypothesis of monophyly of the suborder Notothenioids, cytogenetic evidence supports the idea that the putative ancestor of the Notothenioidei had a karyotype consisting of 48 chromosomes, all acrocentric in shape (Pisano et al. 1998). From the simple set of 48 acrocentric chromosomes, the existing karyotypes could have arisen from chromosomal rearrangements similar to those observed in other vertebrates including fish (White 1973; Sola et al. 1981; King 1993; Klinkhardt et al. 1995). These mostly include para- or pericentric inversions (leading to change in chromosomal formulae but not in number) and Robertsonian fission or fusion (leading to karyotypes highly changed in chromosome number and morphology). Moreover, additions and amplifications of heterochromatin can also produce morphological variants (Ozouf-Costaz et al. 1991; Prirodina 1997; Pisano and Ozouf-Costaz 2000).

Following the above basic assumption, the karyotype of *G. acuticeps* diverges little from the supposed ancestral notothenioid chromosomal set, i.e. the diploid number is 48 and most of the chromosomes are acrocentric (44 pairs), indicating that only two pairs of homologues are likely to be involved in extensive rearrangements. These two pairs of bi-armed chromosomes might have arisen from previous acrocentrics through pericentric inversions. Additional changes, probably involving ribosomal cistrons, have led to the strong structural heteromorphism in the submetacentric chromosomes of pair no. 2.

The eight bathydraconid species studied so far have notably different karyotypes (Table 1). *Psilodraco breviceps* has apparently retained a primitive condition of 48 acrocentric chromosomes; five species, including *G. acuticeps,* have 48 chromosomes, with various degrees of morphological changes; three species have more extensively rearranged karyotypes with reduced number of chromosomes, ranging from 36 to 20.

Although the diploid number of 48 is shared by other species, the chromosomal formula (2m + 2sm + 44a) of *G. acuticeps* appears slightly more derived than the formula described for *Psilodraco* (48a) and close to that of *G. australis* (2m + 2–4sm + 42–44a).

Derome and co-workers (personal communication) tested the phylogenetic hypotheses of Eakin (1981), Iwami (1985) and Hastings (1993) by sequencing bathydraconid mitochondrial genes, and have recently established a phylogenetic tree (Fig. 4) in which the interrelationships of the clade *Gymnodraco/Psilodraco* are not clear and thus cannot shed light on the possible origin of the formula 2m + 2sm + 44a. However, this formula is homoplasic, because it is observed elsewhere in the suborder. The other species having 2n = 48 chromosomes (*P. georgianus* and *C. mawsoni*) display at least one additional pair of rearranged elements leading to formulae of (2m + 4sm + 42a). *Parachaenichthys* and *Cygnodraco* appear clearly separated from each other in the tree; while the interrelationships of *C. mawsoni* within the family are not resolved, *P. charcoti* is clearly nested within a clade containing *Bathydraco*, *Racovitzia* and *Prionodraco*. Assuming the monophyly of the genus *Parachaenichthys,* it seems that the formula 2m + 4sm + 42a was acquired at least twice during the evolution of the bathydraconids, once in this clade with *P. charcoti*, and once or twice outside this clade. Interestingly, the clade *Bathydraco-Racovitzia* clusters species with low chromosome numbers (36, 38–39) including an important proportion of large metacentrics, which probably originate from centric fusions. *Prionodraco evansii*, which emerges as a sister-group with the clade *Bathydraco-Racovitzia/Parachaenichthys*, exhibits the lowest and most derived chromosome number (2n = 20 m).

It will be of interest in the future to study the chromosomal features of *Acanthodraco dewitti* (Skora 1995), recently found in Terra Nova Bay (Vacchi et al. 2001), and to test its phylogenetic position among the Gymnodraconinae.

Mapping the available chromosomal numbers and formulae onto the bathydraconid molecular tree shows that no clade with a homogeneous formula can be proposed, but that the formulae in the terminal taxa are significantly different. This could mean that chromosome changes occur very often in the bathydraconids, giving rise in some cases to homoplasic chromosome formulae. On the other hand, this also means that the chromosomal characters can be most informative for resolving systematic and phylogenetic problems at a low taxonomic level (genera, species), as already shown in other Notothenioids (Ozouf-Costaz et al. 1997).

Location of ribosomal genes

In order to better characterise the chromosomes of *G. acuticeps* and for further comparisons of such data within the family, the location of the major ribosomal genes was ascertained using a 28S rDNA probe for DNA-DNA hybridisation. Since the position of rDNA

Table 1 Available chromosomal data within the family Bathydraconidae

Species	Sex	2n	Formula	Reference
Bathydraco marri	M	39	3 sm + 36 a	Ozouf-Costaz et al. 1991
	F	38	4 sm + 34 a	Ozouf-Costaz et al. 1991
Cygnodraco mawsoni	F	48	2 m + 4 sm + 42 a	Morescalchi et al. 1996
Gerlachea australis	F	48	2 m + 2–4 sm + 42–44 a	Ozouf-Costaz et al. 1991
Gymnodraco acuticeps	F	48	2 m + 2 sm + 44 a	this communication
Parachaenichthys georgianus	–	48	2 m + 4 sm + 42 a	Prirodina 1990
Prionodraco evansii	–	20	20 m	Prirodina 1990
Psilodraco breviceps	–	48	48 a	Prirodina 1990
Racovitzia glacialis	F	36	4 m + 32 a	Ozouf-Costaz et al. 1991

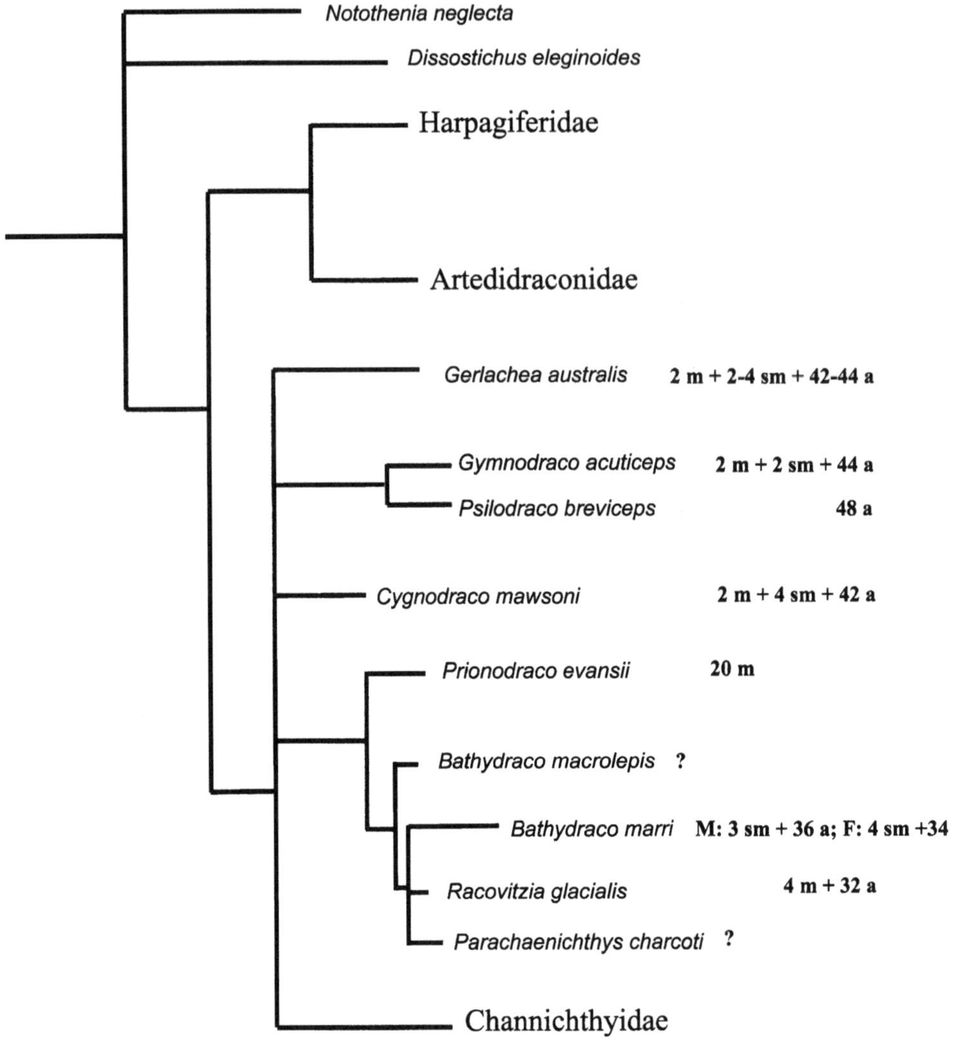

Fig. 4 A summary of bathydraconid interrelationships based on a combined parsimony analysis of cytochrome b- and d-loop sequences. Only robust nodes are shown (Derome et al., personal communication). Corresponding chromosome formulae are indicated

represents an important chromosome marker, rDNA was recently mapped in various fish species, (e.g. Pendàs et al. 1993; Foresti de Almeida Toledo et al. 1996; Phillips and Reed 1996).

In notothenioids, ribosomal cistrons were located in the chromosomes of various species of the families Bovichtidae, Nototheniidae and Channichthyidae. Comparative analyses of rDNA location in species occupying different positions within the suborder allowed some hypotheses concerning the direction of rearrangements involving ribosomal genes during karyotypic diversification of the Nototheniodei to be tested (Pisano and Ozouf-Costaz 2000).

In *G. acuticeps* the major rDNA genes are located in the longest arm of the heteromorphic chromosome pair no. 2 and different numbers of ribosomal cistrons are located in the homologous chromosomes. These findings are in accordance with the general picture of teleostean diploid species analysed to date (Galletti et al. 1995; Foresti de Almeida Toledo et al. 1996; Vinas et al. 1996) and with the majority of notothenioids (Ozouf-Costaz et al. 1996; Pisano et al. 1998). In notothenioids a large amount of rDNA is generally located in a single heterochromatic domain in a single chromosome pair. Structural differences in the amount of rDNA per haploid set were also found in all species, leading to intraspecific size polymorphism of the two homologous chromosomes bearing such a large DNA region. There are several possible explanations for the origin of the size polymorphism in this region. Genetic redundancy of rDNA regions make them more susceptible to unequal permutations, as well as to spontaneous duplications and deletions, in such a way that the amount of rDNA may be modified. Moreover, the presence of associated heterochromatin could enhance unequal crossing-over, thus facilitating heteromorphism of rDNA-bearing chromosomes.

Due to the lack of information from other bathydraconid species, the location of ribosomal DNA in *G. acuticeps* is insufficient for intrafamiliar comparative analysis. The only indirect previous data on rDNA refer to the NORs location in *C. mawsoni* detected by silver staining. In this species NORs (probably corresponding to major rDNA) were detected in the very short arm of a pair of submetacentric chromosomes (Fig. 3b in Morescalchi et al. 1996). However, these data require

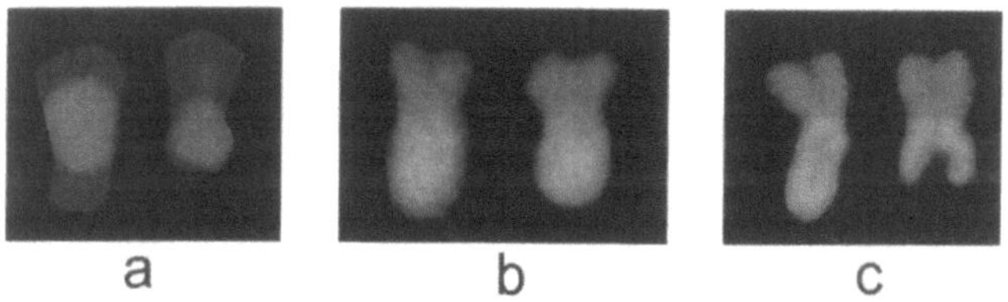

Fig. 5 rDNA bearing chromosome pair in *G. acuticeps* (**a**) and in the channichthyids *Chionodraco hamatus* (**b**) and *Champsocephalus gunnari* (**c**), after fluorescence in situ hybridisation with a 28S rDNA probe. Pink and yellow probe signals are from streptavidin-CY3 and streptavidin-FITC fluorescence detection of biotin-labeled probe, respectively

support by in situ hybridisation with the appropriate probe.

Comparative analysis of the rDNA bearing chromosomes could be attempted between *G. acuticeps* and species of the close family Channichthyidae (Fig. 5a–c). In the two channichthyid species investigated here (*Chionodraco hamatus* and *Chamsocephalus gunnari*, Fig. 5b, c) rDNA entirely covers a chromosome arm. This feature is in common with other channichthyids as well as with several nototheniid species (Pisano et al. 1997; Ozouf-Costaz et al. 1999). In all the examined metaphases of *G. acuticeps* (Fig. 5a) the hybridisation signal leaves the terminal region of the rDNA arm unlabelled, especially in the larger chromosome of the pair. This suggests several rearrangement events in this pair. Whether such a feature typically occurs in this species, and which kind of rearrangement it results from, can be ascertained by further investigations of additional specimens of both sexes and in other species.

Acknowledgements The I.C.E.FISH project was in the framework of the Italian National Programme for Antarctic Research. We are grateful to Roscoe Stanyon and Johannes Wienberg for providing facilities for FISH signal capture and analysis at the Laboratory of Genomic Diversity of the National Cancer Institute, Frederick, MD, USA.

References

Balushkin AV (1992) Classification, phylogenetic relationships, and origins of the families of the suborder Notothenioidei (Perciformes). J Ichthyol 32:90–110

Bargelloni L, Lecointre G (1998) Four years in Notothenioids systematics: a molecular perspective. In: di Prisco G, Pisano E, Clarke A (eds) Fishes of Antarctica. A biological overview. Springer, Milan, pp 259–273

Bargelloni L, Ritchie PA, Patarnello T, Battaglia B, Lambert DM, Meyer A (1994) Molecular evolution at subzero temperature: mitochondrial and nuclear phylogenies of fishes from Antarctica (suborder Notothenioidei), and the evolution of antifreeze glycopeptides. Mol Biol Evol 11:854–863

Doussau de Bazignan M, Ozouf-Costaz C (1985) Une technique rapide d'analyse chromosomique apppliquée a sept espèces de poissons antarctiques. Cybium 9:57–74

Eakin RR (1981) Osteology and relationships of the fishes of the Antarctic family Harpagiferidae (Pisces, Notothenioidei). In: Kornicker LS (ed) Antarctic Research Series 31. Biology of the Antarctic Seas IX. American Geophysical Union, Washington, pp 81–147

Eastman JT (1993) Antarctic Fish Biology: evolution in a Unique Environment. Academic Press, San Diego

Eastman JT, Clarke A (1998) A comparison of adaptive radiations of Antarctic fish with those of non-Antarctic fish. In: di Prisco G, Pisano E, Clarke A (eds) Fishes of Antarctica. A biological overview. Springer, Milan, pp 3–28

Eastman JT, Hubold G (1999) The fish fauna of the Ross Sea, Antarctica. Antarct Sci 11:293–304

Foresti de Almeida Toledo L, Stocker AJ, Foresti F, de Almeida Toledo-Filho S (1996) Fluorescence in situ hybridization with rDNA probes on chromosomes of two nucleolus organizer region phenotypes of a species of *Eigenmannia* (Pisces, Gymnotoidei, Stenopygidae). Chromosome Res 4:301–305

Galletti PM, Mestrinier CA, Monaco PJ, Rasch EM (1995) Post-zygotic modifications and intra-and inter individuals nucleolar organizing region variations in fish. Report of a case involving *Leporinus federici*. Chromosome Res 3:285–290

Gon O, Heemstra PC (eds) (1990) Fishes of the Southern Ocean. JLB Smith Institute of Ichthyology, Grahamstown, South Africa

Hartley SE, Horne MT (1985) Cytogenetic techniques in fish genetics. J Fish Biol 26:575–582

Hastings PA (1993) Relationships of the fishes of the perciform suborder Notothenioidei. In: Miller RG (ed) A history and atlas of the fishes of the Antarctic Ocean. Foresta Institute for Ocean and Mountain Studies, Carson City, pp 99–107

Iwami T (1985) Osteology and relationships of the family Channichthyidae. Mem Natl Inst Polar Res (Tokyo) Ser E 36:1–69

King M (1993) Species evolution: the role of chromosome change. Cambridge University Press, Cambridge

Klinkhardt M, Tesche M, Greven H (1995) Database of fish chromosomes. Westarp Wissenschaften, Magdeburg

Levan A, Fredga K, Sandberg AA (1964) Nomenclature for centromeric position of chromosomes. Hereditas 52:201–220

Morescalchi A, Morescalchi MA, Odierna G, Stingo V, Capriglione T (1996) Karyotype and genome size of zoarcids and notothenioids (teleostei, perciformes) from the Ross Sea: citotaxonomic implications. Polar Biol 16:559–564

Ozouf-Costaz C, Hureau J-C, Beaunier M (1991) Chromosome studies on fish of the suborder Notothenioidei collected in the Weddell Sea during EPOS 3 cruise. Cybium 15:271–289

Ozouf-Costaz C, Pisano E, Bonillo C, Williams R (1996) Location of ribosomal DNA in *Champsocephalus gunnari* (Notothenioidei, Channichthyidae) by banding and FISH techniques. Chromosome Res 4:557–561

Ozouf-Costaz C, Pisano E, Thaeron C, Hureau J-C (1997) Antarctic fish chromosome banding: significance for evolutionary studies. Cybium 21: 399–409

Ozouf-Costaz C, Pisano E, Bonillo C, Hureau JC (1999) Large scale chromosome studies of unusual fauna within a confined area: Antarctic fish suborder Notothenioidei. Cytogenet Cell Genet 85:95

Pendàs AM, Moràn P, Garcia-Vasquéz E (1993) Multi-chromosomal location of ribosomal RNA genes and heterochromatin association in brown trout. Chromosome Res 1:63–67

Phillips RB, Reed KM (1996) Application of fluorescence in situ hybridization (FISH) techniques to fish genetics: a review. Aquaculture 140:197–216

Pisano E, Ozouf-Costaz C (2000) Chromosome change and evolution of Antarctic fishes. Antarct Sci

Pisano E, Ozouf-Costaz C, Bonillo C, Williams R (1997) Cytogenetic characterization of the Antarctic icefish *Channichthys rhinoceratus* (Notothenioidei, Channichthyidae) by banding and FISH. Ninth International Congress of European Ichthyologists "Fish Biodiversity", Abstracts, p 73

Pisano E, Ozouf-Costaz C, Prirodina V (1998) Chromosome diversification in Antarctic fishes (Notothenioidei). In: di Prisco G, Pisano E, Clarke A (eds) Fishes of Antarctica. A biological overview. Springer, Milan, pp 275–286

Prirodina V (1990) Karyotypes of three species of the family Bathydraconidae from the west Antarctic with notes on its karyotypic diversity. Proc Zool Inst Leningrad 222:55–63

Prirodina VP (1997) The direction of the karyotype specialization in the suborder Notothenioidei (Pisces, Perciformes). Cybium 21:393–398

Ritchie PA, Lavoué S, Lecointre G (1997) Molecular phylogenetics and evolution of Antarctic notothenioid fishes. Comp Biochem Physiol 118A:1009–1026

Skora K (1995) *Acanthodraco dewitti* gen. et sp. n. (Pisces, Bathydraconidae) from Admiralty Bay (King George Island, South Shetland Islands, Antarctica). Arch Fish Mar Res 42:283–289

Sola L, Cataudella S, Capanna E (1981) New developments in vertebrate cytotaxonomy. III. Karyology of bony fishes: a review. Genetica 54:285–328

Vacchi M, La Mesa M, Greco S (1999) Some ecological features of the coastal fish community of Terra Nova Bay. In: Faranda F, Guglielmo L, Ianora A (eds) Ross Sea Ecology: Italian Antarctic Expeditions (1987–1995). Springer, Milan, pp 457–468

Vacchi M, La Mesa M, Ozouf-Costaz C (2001) First occurrence of *Acanthodraco dewitti* Skora, 1995 (Notothenioidei, Bathydraconidae) in the Ross Sea. Polar Biol 24: 471–472

Vinas A, Gomez C, Martinez P, Sanchez L (1996) Localization of rDNA genes in european eel (*Anguilla anguilla*) by FISH. Genome 39:1220–1223

Voskoboynikova OS (1993) Evolution of the visceral skeleton and phylogeny of Notothenioidei. J Ichthyol 33:23–47

White MJD (1973) Animal cytology and evolution, 3rd edn. Cambridge University Press, Cambridge

Polar Biol (2001) 24: 910–917
DOI 10.1007/s003000100299

Krzysztof Jazdzewski · Claude De Broyer
Magdalena Pudlarz · Dariusz Zielinski

Seasonal fluctuations of vagile benthos in the uppermost sublittoral of a maritime Antarctic fjord

Abstract Quantitative samples of benthos were taken on a stony beach in the maritime Antarctic (Admiralty Bay, King George Island, South Shetland Islands) during a complete annual cycle. The sampled habitat consisted of cobbles lying on sand and gravel in the fringe between the littoral and sublittoral zones; samples were always taken at low tide just below the water line. In this habitat, abundant macroalgal detritus was observed between stones. This stony beach appeared to be very rich in vagile fauna settled between and under stones. Macrobenthos consisted mainly of amphipods (ca. 85% of total number), gastropods (11%) and nemerteans (3%). The abundance of whole macrofauna ranged to over 50,000 ind. m^{-2} and its biomass over 600 g m^{-2} (FW). Seven species of Amphipoda and four species of Gastropoda were found. Amphipoda were dominated by *Gondogeneia antarctica* (over 70% of all amphipods) and *Paramoera edouardi* (over 20%), whereas among gastropods *Laevilitorina antarctica* prevailed (over 70%). Unexpectedly high abundance and biomass of Amphipoda were observed in the first half of winter (May/July), surpassing otherwise important summer amphipod abundance. The probable reason for this phenomenon could be high autumn abundance of decaying algae on the beach in the tidal zone, providing detritus that is probably the main food source for Amphipoda.

Please note: When citing this article, please use the reference to the original source in the journal Polar Biology as given on the top of this page.

D. Zielinski is deceased.

K. Jazdzewski (✉) · M. Pudlarz · D. Zielinski
Laboratory of Polar Biology and Oceanobiology,
Department of Invertebrate Zoology and Hydrobiology,
University of Lodz, 12/16 Banacha st., 90-237 Lodz, Poland
E-mail: kryjaz@biol.uni.lodz.pl

C. De Broyer
Departement des Invertébrés, Institut royal des Sciences
naturelles de Belgique, 29, rue Vautier, 1000 Brussels, Belgium

Introduction

Studies on Antarctic benthos have developed quickly in recent decades, allowing elaboration of several syntheses (Arnaud 1977; White 1984; Picken 1985; Gallardo 1987; Arntz et al. 1994; Knox 1994).

The shallow sublittoral zone (depth of less than 30 m) has been investigated in some detail since the establishment of permanent shore stations, and the introduction of scuba diving techniques. It has been studied recently with a renewed interest within the framework of "Global Change" research and the investigation of the potential impacts of global warming and increased UVB radiation linked to ozone seasonal depletion.

In contrast to the abundant literature on Antarctic sublittoral benthos, relatively few studies have been devoted to Antarctic very shallow coastal areas, and in particular to the usually narrow belt of littoral and uppermost sublittoral situated around the lowest water mark.

In the deep sublittoral and shelf zone, macrobenthic communities often show abundant populations, high biomasses and high species richness (e.g. Arntz et al. 1994). However, benthic communities of very shallow water usually appear much poorer in species and abundance. Most authors studying shallow sublittoral assemblages in the East Antarctic have stressed the absence or impoverishment of the uppermost sublittoral benthos (to a depth of 2–3 m) due to abrasion and disturbance by sea ice (Gruzov et al. 1967; Arnaud 1974; Kirkwood and Burton 1988; Gambi et al. 1994, 2000) and anchor-ice formation (Dayton et al. 1970; Dayton 1990). Nevertheless, Arnaud (1974) in Terre Adélie noted in sheltered rock pools and rock crevices the growth of some macroalgae and the presence of invertebrate species such as *Sterechinus neumayeri*, *Odontaster validus*, *Parborlasia corrugatus*, *Paramoera walkeri* and *Laevilitorina antarctica*.

In the West Antarctic, sea-ice impacts appear less severe, especially in summer, and allow the seasonal growth of a few macroalgae in sheltered places of the littoral/sublittoral fringe. This mostly immersed zone can host in abundance some organisms, like the patellid gastropod *Nacella concinna*, amphipods such as *Paramoera edouardi* or *Gondogeneia antarctica*, as well as gastropods of the genera *Laevilitorina* and *Laevilacunaria*, or the bivalve *Kidderia subquadratum* (Castellanos 1973; Thurston 1974a; Richardson and Whitaker 1979; Ponder 1983; Castilla and Rozbaczylo 1985; Barnes 1995).

However, relatively few year-round quantitative data on the vagile macrobenthos of the shallow sublittoral are available (Bone 1972; Bregazzi 1972a; Richardson and Whitaker 1979; Momo 1995).

The present study was undertaken to follow the year-round changes in species composition, abundance and biomass of the sublittoral fringe assemblage of Admiralty Bay.

Study area

The study was conducted at *Arctowski* station, situated at the shore of Admiralty Bay (King George Island, South Shetland Islands) (Fig. 1). Hydrological and biological characteristics of the bay are detailed in a multi-authored monograph edited by S. Rakusa-Suszczewski (1993). Surface-water temperature oscillates between −1.9°C and 3.4°C. In the littoral zone at the sampling site, the measured range in austral summer 1994/1995 was from 2.8°C to 3.8°C. Winter temperatures in 1997 (June/September) in the same site oscillated between −0.9°C and −1.7°C (J. Rudka, personal communication). Total freezing of Admiralty Bay occurs irregularly: during 20 years of operation of the *Arctowski* station (1977–1996) it was observed only 11 times (Kruszewski 1999). During the sampling year (1994), it was observed only from 13 July till 8 August. Afterwards, in August, September, October and November, large pack-ice fields entered the bay five times. These fields covered up to 70% of the bay surface, lasting for 3–8 days (Kruszewski 1999). The tidal amplitude in Admiralty Bay amounts to ca. 2 m (Catewicz and Kowalik 1984).

For this study, the beach in the immediate vicinity of the biological laboratory of *Arctowski* station was chosen. This beach is mostly composed of stones of less than 10 cm in diameter (cobbles and to a lesser extend pebbles), rounded by wave and ice action. This stony beach extends from the supralittoral level to the uppermost sublittoral, to a depth of ca. 0.5–1 m below the low-water level (Fig. 2). Below the layer of stones, there was a patchy layer of sand and gravel with relatively abundant algal detritus.

Pieces of sea ice, mostly in winter, and growlers in summer frequently cover the intertidal and shallow sublittoral zones. Through the year they cause physical abrasion, scouring, movements and displacements of the beach stones. Strong waves also contribute to these movements. At the sampling site, this very unstable habitat prevents the growth of annual macroalgae like *Adenocystis utricularis* and *Monostroma harriotti*, which can be found on sheltered rocky surfaces on neighbouring parts of the shore (Zielinski 1981, 1990).

The biodiversity of Admiralty Bay is quite well known. Owing to the efforts of Polish, Brazilian,

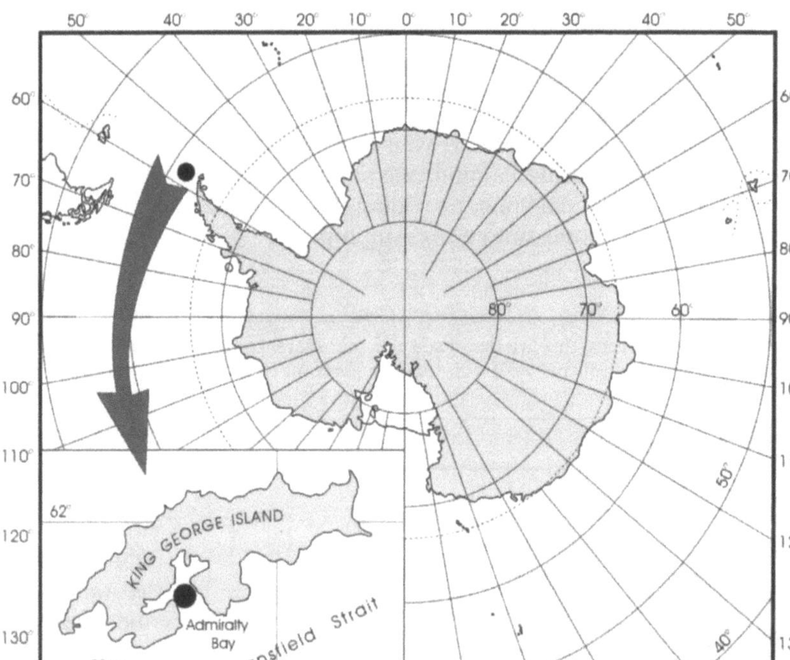

Fig. 1 Location of the study area

Fig. 2 Stony beach in the vicinity of *Arctowski* station at low tide

Belgian and German biologists, the actual list of taxa recorded from this basin amounts to nearly 500 species (Arnaud et al. 1986; Skóra and Neyelov 1992; Presler 1993; Sicinski and Janowska 1993; Jazdzewski et al. 1995; Presler and Figielska 1997).

Materials and methods

An ad hoc sampling method was designed to collect the benthos just below the water line at a depth of ca. 5–10 cm. A metal cylinder ca. 60 cm in height and 56.5 cm in diameter, with a sampling surface of 0.25 m^2, and equipped with 2 handles was used. The lower edge was cut in a saw-like manner. The cylinder was screwed into the bottom between stones to a depth of ca. 10 cm in the sediment, and all stones, underlying sand and gravel were removed by hand and hand-net to a bucket filled with seawater. All animals swimming or crawling inside the cylinder were caught with a hand-net. The sediment was sieved with a 0.5-mm mesh sieve. All macro-invertebrates were preserved in formaline and transferred into 75% ethanol. No replicates were taken.

In December 1993, the first sample was taken at low tide by the two senior authors. In 1994 and 1995, during the wintering of the 18th Polish Antarctic Expedition, the junior co-author, the late Darek Zielinski, collected the remaining samples (23) at approximately 2 week intervals, taking into account the lowest tides according to data for the neighbouring Uruguayan station *Artigas* (Servicio de Oceanographia 1992, 1993). Tide level during sampling oscillated between 18 and 63 cm, the first level being the yearly lowest (5 November 1994). Sampling dates are indicated in Figs. 3 and 4.

Preserved animals were brought to Poland and determined to species level for the main groups in the laboratory. Amphipods were identified by the authors, and gastropods were identified by Dr. Patrick M. Arnaud of Station Marine d'Endoume, Marseilles and Dr. Theresina Absher of Centro de Estudos do Mar, Universidade Federal do Parana. Alcohol wet weight was measured after blotting animals on filter paper. The weighing accuracy was 0.05 g.

Results

The sampled habitat was unexpectedly rich in animals. The two most abundant groups were determined to species level (Amphipoda: seven species, Gastropoda: four species). The other groups together, when fully determined, increased the species number to at least 20 taxa (Table 1).

Among amphipods, the pontogeneid *G. antarctica* constituted over 61% of the whole material, *Paramoera edouardi* nearly 19%, *Orchomenella* cf. *ultima* ca. 2.7% and *Cheirimedon femoratus* ca. 2%. Among the Gastropoda, *Laevilitorina antarctica* constituted over 8% of the material, followed by *Laevilacunaria bransfieldensis* (2%) and *Laevilitorina caliginosa* (0.8%). The only other animal group present in all samples were Nemertini (2.7% in the whole material).

The density of the benthos ranged from 200 ind. up to ca. 54,000 ind.·m^{-2}, with an average of about 13,000 ind. m^{-2}. The biomass (w.w.) of these assemblages ranged from 1 g m^{-2} up to 660 g m^{-2}, with an average of 180 g m^{-2}.

Figure 3 shows the yearly cycle of abundance and biomass of the benthos in the studied station. Total benthos biomass was clearly defined by the most common species *G. antarctica*. Only in May 1994 did the peaks of abundance and biomass not coincide. This was evidently caused by the change in population structure of *G. antarctica* in the second half of May when smaller animals prevailed. It can be seen that the amphipods were quite abundant in the Antarctic spring and summer (from late October till April) and very abundant in the

Fig. 3 Abundance and biomass (dotted line) of the benthos of the stony beach in the annual cycle

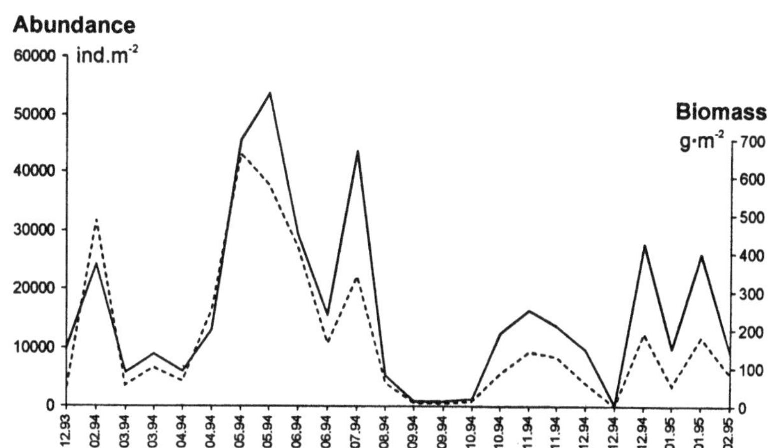

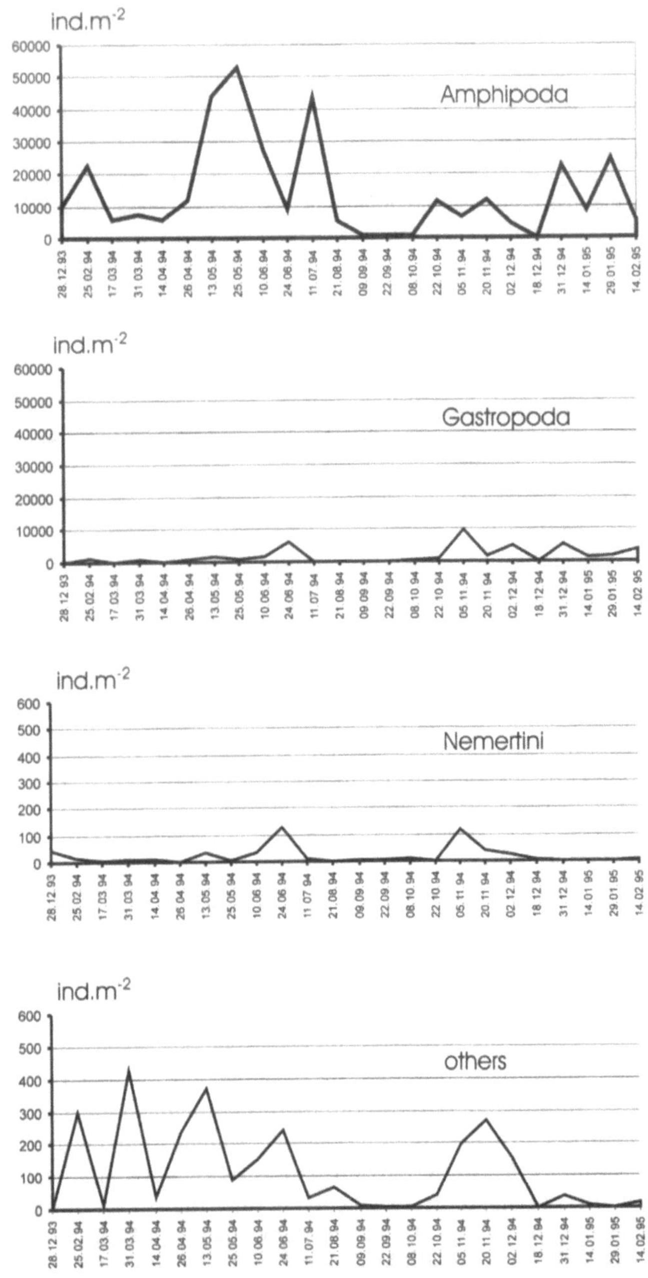

Fig. 4 Abundance of particular animal groups in the annual cycle

first half of winter (April/July) with a large drop in June (Fig. 4). This high abundance decreased sharply in late winter (August/September/beginning of October). The irregular peaks of abundance of Gastropoda occurred at the beginning of winter (June) and then in summer, from November till February. The Nemertini, as well as other groups, did not show any particular trend.

Detailed diagrams (Fig. 5) show that the large fluctuations in amphipod population were mainly due to *G. antarctica*, whereas the abundances of *Paramoera edouardi*, *C. femoratus* and *Orchomenella* cf. *ultima* were lower but more constant. In the case of Gastropoda (Fig. 5), the clearest fluctuations were observed in *Laevilitorina antarctica*. These patterns of abundance fluctuations in the habitat studied were influenced by the sampling method used.

Discussion

Nearly all species found had been previously recorded from the littoral/sublittoral zone in the maritime Antarctic. Among Amphipoda, only *Orchomenella* cf. *ultima* had not been formerly recorded from this habitat but found in somewhat deeper water (Jazdzewski et al. 1991, 1992).

The amphipod species that can be recognized as typical dwellers of this stony sublittoral fringe are primarily the pontogeneids *G. antarctica* and *Paramoera edouardi*, followed by the lysianassoids *Orchomenella* cf. *ultima* and *C. femoratus*.

Eurymera monticulosa and *Bovallia gigantea*, inhabitants of the shallow phytal zone on rocky and stony bottoms of the maritime Antarctic (Bone 1972; Thurston 1974a; Rauschert 1991), are rather poorly represented in our material, and this may be due to the unstable environment of moving stones for these relatively large-sized amphipods (*Bovallia* size ranged up to 39 mm and *Eurymera* to 26 mm in our samples).

Hippomedon kergueleni is a burrowing species, usually recorded from gravel, sandy and muddy bottoms (Thurston 1974a) but sometimes appearing also in small numbers amongst littoral stones (Factory Cove, Signy Island: Bregazzi 1972a). It is supposed to be only incidentally present in this stony habitat or linked to the sandy patches under the stones.

The two pontogeneids have epibenthic habits. *Orchomenella* cf. *ultima*, here recorded as an epibenthic dweller, is able to burrow in suitable sediments (C. De Broyer, unpublished work), and *C. femoratus* is a common burrower on sandy bottoms but can also be found on stony and rocky substrates (Bregazzi 1972a; Thurston 1974a).

The larger fluctuations observed in *G. antarctica* abundance (Fig. 5) in comparison to *Paramoera edouardi* are probably due to its higher mobility. One should cite here the precise observations of Thurston (1974b) on *G. antarctica* which "occurs frequently in littoral pools during summer months; it swims actively and thus shows a marked contrast to the cryptic habit of *Paramoera edouardi*, another abundant inhabitant of shore pools".

G. antarctica is known as an omnivorous species, primarily herbivorous on microalgae and secondarily on macroalgae, also eating small Crustacea (mainly Amphipoda and Copepoda) and detritus material (Richardson 1977; Iken 1996). It is the predominant herbivore of epontic microalgae colonizing the icefoot at Signy Island (Richardson and Whitaker 1979) or in Hope Bay, Antarctic Peninsula (Momo 1995).

C. femoratus is also an omnivorous species, with necrophagous and predatory preferences but also consuming a large proportion of micro- and macroalgal

Table 1 List of taxa found in the uppermost sublittoral of Admiralty Bay (*N* number of identified individuals; % percent in the whole material)

Taxon	*N*	%
Amphipoda		
Gondogeneia antarctica (Chevreux, 1906)	56,460	61.23
Paramoera edouardi (Schellenberg, 1929)	17,595	19.08
Orchomenella cf. *ultima* (Bellan-Santini 1972)	2,499	2.71
Cheirimedon femoratus (Pfeffer 1888)	1,804	1.95
Eurymera monticulosa Pfeffer 1888	194	0.21
Bovallia gigantea (Pfeffer 1888)	24	0.03
Hippomedon kergueleni (Miers 1875)	9	0.01
Total of Amphipoda	78,585	85.22
Gastropoda		
Laevilitorina antarctica (Smith 1902)	7,583	8.22
Laevilacunaria bransfieldensis (Preston 1916)	1,931	2.09
Laevilitorina caliginosa (Gould 1849)	754	0.82
Laevilitorina umbilicata Pfeffer 1886	172	0.19
Laevilacunaria sp. (indet.)	17	0.02
Indet.	20	0.02
Total of Gastropoda	10,477	11.36
Others		
Nemertini	2,696	2.92
Isopoda	248	0.27
Polychaeta	124	0.13
Turbellaria	32	0.04
Bivalvia	28	0.03
Oligochaeta	24	0.03
Total of other taxa	3,152	3.42
Total of all material	92,214	100

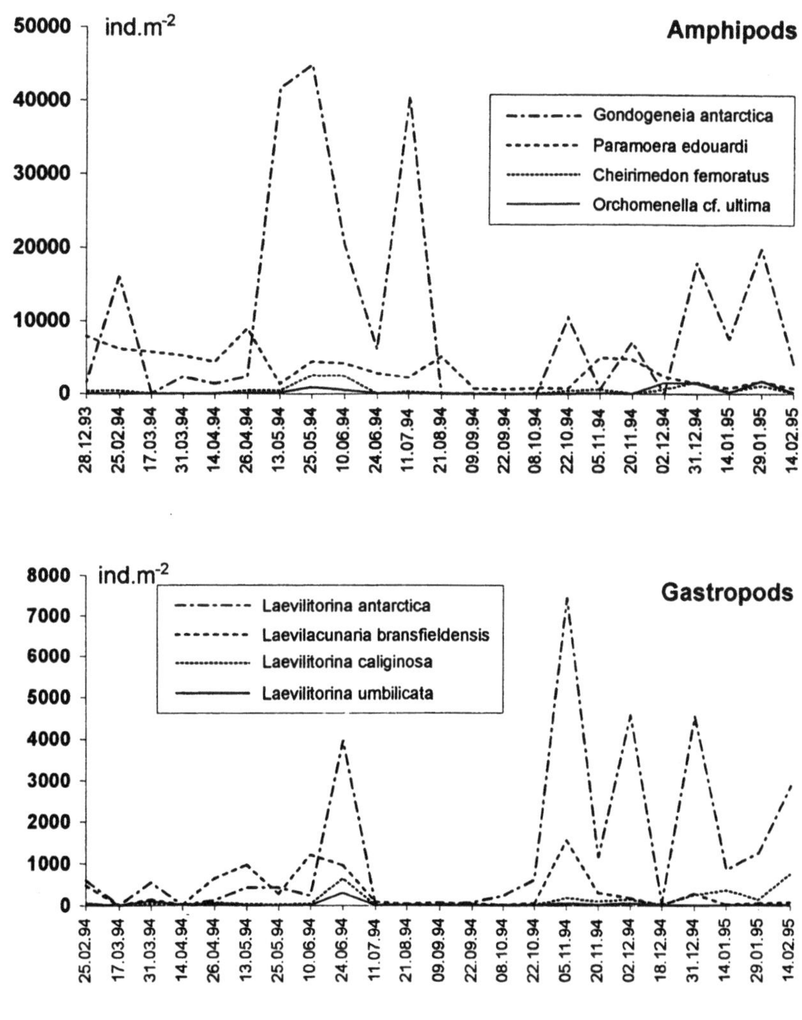

Fig. 5 Abundance of particular amphipod and gastropod species in the annual cycle

material (Bregazzi 1972a, b; Thurston 1974b). It is frequently collected in baited traps in the shallows of Admiralty Bay (Presler 1986; C. De Broyer, unpublished data).

B. gigantea is an ambush predator/omnivore eating mostly crustaceans (amphipods, copepods) but it also relies on macroalgae and diatoms for 30% of its food (Bone 1972; Iken 1996).

E. monticulosa is an omnivore consuming diatoms, filamentous algae and macroalgae such as *Desmarestia* spp. and *Adenocystis utricularis* (Scailteur and C. De Broyer, unpublished data).

The food habits of *Paramoera edouardi* and *Orchomenella* cf. *ultima* are still unknown but other *Paramoera* spp. have been characterized as omnivores/herbivores (e.g. *Paramoera walkeri*: Rakusa-Suszczewski 1972), and *Orchomenella* spp. as scavengers and detritovores (De Broyer 1983).

The dominant taxon among Gastropoda, *Laevilitorina antarctica*, was previously recorded in Admiralty Bay by Arnaud et al. (1986) as a common and abundant shallow-water species. Castilla and Rozbaczylo (1985) made similar observations in the neighbouring Robert Island (South Shetland Islands). The same species was recorded by Gambi et al. (1994) as a comparatively important component of the shallow sublittoral of Terra Nova Bay (Ross Sea), sharing the habitat with *Paramoera walkeri*, the dominant amphipod.

Laevilacunaria bransfieldensis was also recorded formerly (Arnaud et al. 1986) as a common species in Admiralty Bay, occurring at depths of 0–90 m. *Laevilitorina calignosa* is probably the most common littorinid species in the whole Antarctic and sub-Antarctic (Powell 1965; Branch et al. 1991). Gastropods of the genera *Laevilacunaria* and *Laevilitorina* are known as herbivorous animals consuming epiphytic diatoms and macroalgae (Iken 1996, 1999).

When comparing the present results with data from the shallowest station of Terra Nova Bay (Gambi et al. 1994) (hard bottom, 0.5 m, MLWL), a vicariance of the dominant amphipod species can be seen: *Paramoera walkeri* in the high Antarctic versus *G. antarctica* with *Paramoera edouardi* in the maritime Antarctic. *Paramoera walkeri*, although recorded in other West Antarctic locations (Bregazzi 1972b; Lowry and Bullock 1976; Richardson and Whitaker 1979) and, rarely, in other locations around King George Island (Rauschert 1991), has not been found so far in Admiralty Bay.

No other data from rocky or stony bottoms of Admiralty Bay are presently available for comparison. The relatively high abundance and biomass of the sublittoral fringe benthos can only be compared with quantitative data from neighbouring soft-bottom habitats of the bay (Jazdzewski et al. 1986, 1991). In the shallowest soft-bottom station (4 m), investigated by Jazdzewski et al. (1991) in the summer of 1988, in the vicinity of the present station, the total zoobenthos biomass was 87 g m^{-2}. Density of zoobenthos from this sandy bottom was about 1,900 ind.·m^{-2}, represented mainly by amphipods (ca. 480 ind m^{-2}) and polychaetes (ca. 1,000 ind. m^{-2}). The relative dominances of amphipods were: *H. kergueleni* 78%, *Cardenio paurodactylus* 6%, *Cheirimedon femoratus* 5%, and *Gondogeneia* sp., *Monoculodes scabriculosus* and *Paraphoxus rotundifrons* 2.5% each. All species, except *Gondogeneia*, are typical burrowers, which is a clear difference from the epibenthic species recorded in our study site.

Zoobenthos biomass data hitherto obtained from the shallowest sublittoral zone (less than 2 m) in the Antarctic are very limited (austral summer). In Lützow-Holm Bay (Prince Olav Coast), Nakajima et al. (1982) found on sandy bottoms biomasses of 46 g m^{-2} (only of *Adamussium colbecki* and *Sterechinus neumayeri*) for the depth 1.5–3.5 m, ranging up to 109 g m^{-2} at 5 m. The only hard-bottom data are those of Gambi et al. (1994) who measured (by indirect estimation from representative sub-samples) in Terra Nova Bay (Ross Sea), a mean biomass value of 94.4 g m^{-2} (exclusively *Laevilitorina antarctica* and *Paramoera walkeri*) at a depth of 0.5 m on bare rocky surface, colonized only by epilithic cyanophytes and diatoms. Similar values (faunal fresh weight: 96.4 g m^{-2}) were recorded at 2 m depth, where a better shelter from ice-scouring and the development of the rhodophyte *Iridaea cordata* allowed colonization by 21 different vagile animal taxa. Comparing 3 austral summer surveys, Gambi et al. (2000) reported a pronounced interannual variability in animal diversity and density in the shallowest (2–4 m) hard-bottom communities of Terra Nova Bay, which they related to variability in sea-ice development and persistence.

The unexpected peak of abundance observed during the first half of winter (May and July for the Amphipoda, June for Gastropoda and Nemertini: Fig. 3) may be due to different responses to physical factors, life-cycle (recruitment), as well as to trophic factors. The main physical factor is water temperature high enough in winter to avoid freezing of the sea fringe. The sharp decrease of benthos abundance in August/September coincides well with winter low temperatures and freezing of the bay.

We assume that the basic food source for the studied assemblage is the abundant algal detritus produced by the rich algal community of Admiralty Bay. The amount of algal biomass washed ashore by surf waves was calculated for the entire Admiralty Bay to be over 1,600 tons (FW), or 279 tons (DW), per year (Zielinski 1981, 1990). Most of these algae accumulated during spring and summer with a peak in April. Fragmentation and decomposition are quite fast (a few weeks) and one-third of this algal biomass is released as detritus and nutrients to the littoral and shallow waters (Zielinski 1981).

Another possibly important food source for at least some of these stony beach dwellers may be the epilithic, epontic and epiphytic diatoms. In addition, bacterial matter abundant on decomposing algae (Zielinski 1981) can contribute to the food of this upper sublittoral zoobenthos. On the shores of Admiralty Bay, Ligowski

(1993b, personal communication) recorded in spring and summer epilithic diatoms but these diatom assemblages disappeared in winter (May/October). Abundant epiphytic diatoms on macroalgae occur throughout the year and they are also present on stranded algae. From August till October shore-ice diatoms were also recorded by Ligowski (1993a).

Important development of diatom assemblages in spring and summer was frequently observed in the upper sublittoral zone in the East and West Antarctic (Propp 1970; Arnaud 1974; Richardson and Whitaker 1979; Sagar 1980).

Momo (1995), studying the population dynamics of *G. antarctica* in Hope Bay, showed that the total number of amphipods was positively correlated with the microalgal biomass, the peaks of which preceded, on average, by 24 days the abundance peaks of amphipods. Amphipod abundance showed maxima in March (after a peak of microalgal biomass at the seawater-bottom interface), September and November (after maximum biomass of the icefoot epontic diatoms between the end of July and end of October). Similar observations were previously made in Signy Island by Richardson and Whitaker (1979).

The present study revealed, in this preliminary phase, the occurrence of abundant populations of vagile macrobenthos, mainly amphipods, in the sublittoral fringe in winter. These specialized populations are mainly composed of herbivores/detritivores. Our results stress the ecological importance of the transitional zone between land and sea in the maritime Antarctic.

Acknowledgements Thanks are due to Professor S. Rakusa-Suszczewski, the manager of *Arctowski* station, for making the station facilities available. Information on the meteorological conditions at *Arctowski* station provided by Dr. M. Keyna is acknowledged with thanks. Thanks are also due to Camille Jamar (I.R.Sc.N.B., Brussels) for his technical help in preparing the manuscript.

References

Arnaud PM (1974) Contribution à la bionomie marine benthique des régions antarctiques et subantarctiques. Tethys 6:467–653

Arnaud PM (1977) Adaptations within the Antarctic marine benthic ecosystem. In: Llano GA (ed) Adaptations within Antarctic ecosystems. Proceedings of the Third SCAR Symposium on Antarctic Biology. Smithsonian Institution, Washington, DC, pp 135–157

Arnaud PM, Jazdzewski K, Presler P, Sicinski J (1986) Preliminary survey of benthic invertebrates collected by Polish Antarctic Expeditions in Admiralty Bay (King George Island, South Shetland Islands, Antarctica). Pol Polar Res 7:7–24

Arntz WE, Brey T, Gallardo VA (1994) Antarctic zoobenthos. Oceanogr Mar Biol Annu Rev 32:241–304

Barnes DKA (1995) Sublittoral epifaunal communities at Signy Island, Antarctica. I. The ice-foot zone. Mar Biol 121:555–563

Bone DG (1972) Aspects of the biology of the Antarctic amphipod *Bovallia gigantea* Pfeffer at Signy Island, South Orkney Islands. Br Antarct Surv Bull 27:105–122

Branch ML, Arnaud PM, Cantera J, Gianakouras D (1991) The benthic Mollusca and Brachiopoda of subantarctic Marion and Prince Edward Islands. S Afr J Antarct Res 21:45–56

Bregazzi PK (1972a) Life cycles and seasonal movements of *Cheirimedon femoratus* (Pfeffer) and *Tryphosella kergueleni* (Miers) (Crustacea: Amphipoda). Br Antarct Surv Bull 30:1–34

Bregazzi PK (1972b) Habitat selection by *Cheirimedon femoratus* (Pfeffer) and *Tryphosella kergueleni* (Miers) (Crustacea, Amphipoda). Br Antarct Surv Bull 31:21–31

Castellanos ZJA (1973) Estratificacion del complejo bentonico de invertebrados en Puerto Paraiso (Antartida). Contrib Inst Antart Argent 164:1–30

Castilla JC, Rozbaczylo N (1985) Rocky intertidal assemblages and predation on the gastropod *Nacella* (*Patinigera*) *concinna* at Robert Island, South Shetlands, Antarctica. 1. Inst Antart Chil Ser Cient 32:65–73

Catewicz Z, Kowalik Z (1994) Harmonic analysis of tides on Admiralty Bay. Oceanologia 15:97–109

Dayton PK (1990) Polar benthos. In: Smith WO Jr (ed) Polar oceanography, Part B. Chemistry, biology and geology. Academic Press, New York, pp 631–685

Dayton PK, Robillard GA, Paine RT (1970) Benthic faunal zonation as a result of anchor ice at McMurdo Sound, Antarctica. In: Laws RM (ed) Antarctic ecology. Academic Press, London, pp 244–258

De Broyer C (1983) Recherches sur la systématique et l'évolution des crustacés amphipodes gammarides antarctiques et subantarctiques. PhD Thesis, Université Catholique de Louvain

Gallardo VA (1987) The sublittoral macrofaunal benthos of the Antarctic shelf. Environ Int 13:71–81

Gambi MC, Lorenti M, Russo GF, Scipione MB (1994) Benthic associations of the shallow hard bottoms off Terra Nova Bay, Ross Sea: zonation, biomass and population structure. Antarct Sci 6:449–462

Gambi MC, Buia MC, Mazella L, Lorenti M, Scipione B (2000) Spatio-temporal variability in the structure of benthic populations in a physically controlled system of Terra Nova Bay: the shallow hard bottom. In: Farauda FM, Gugliemo L, Ianora A (eds) Ross Sea ecology. Springer, Berlin Heidelberg New York, pp 527–583

Gruzov EN, Propp MV, Puskhin AF (1967) Biological associations of coastal areas of the Davis Sea (based on the observations of divers). Sov Antarct Exped Inf Bull 6:523–533

Iken K (1996) Trophische Beziehungen zwischen Makroalgen und Herbivoren in der Potter Cove (King-George-Insel, Antarktis). Ber Polarforsch 201:1–206

Iken K (1999) Feeding ecology of the Antarctic herbivorous gastropod *Laevilacunaria antarctica* Martens. J Exp Mar Biol Ecol 236:133–148

Jazdzewski K, Jurasz W, Kittel W, Presler E, Presler P, Sicinski J (1986) Abundance and biomass estimates of the benthic fauna in Admiralty Bay, King George Island, South Shetland Islands. Polar Biol 6:5–16

Jazdzewski K, Teodorczyk W, Sicinski J, Kontek B (1991) Amphipod crustaceans as an important component of zoobenthos of the shallow Antarctic sublittoral. Hydrobiologia 223:105–117

Jazdzewski K, De Broyer C, Teodorczyk W, Konopacka A (1992) Survey and distributional patterns of the amphipod fauna of Admiralty Bay, King George Island, South Shetland Islands. Pol Polar Res 12:461–472

Jazdzewski K, Weslawski JM, De Broyer C (1995) A comparison of the amphipod faunal diversity in two polar fjords: Admiralty Bay, King George Island (Antarctic) and Hornsund, Spitsbergen (Arctic). Pol Arch Hydrobiol 42:367–384

Kirkwood JM, Burton HR (1988) Macrobenthic species assemblages in Ellis Fjord, Vestfold Hills, Antarctica. Mar Biol 97:445–457

Knox GA (1994) The biology of the Southern Ocean. Cambridge University Press, Cambridge

Kruszewski G (1999) Zlodzenie Zatoki Admiralicji w latach 1977–1996. Probl Klimatol Polarnej Gdynia 9:173–191

Ligowski R (1993a) Sea-ice algae. In: Rakusa-Suszczewski S (ed) The maritime Antarctic coastal ecosystem of Admiralty Bay. Department of Antarctic Biology of Polish Academy of Sciences, Warsaw, pp 49–51

Ligowski R (1993b) Microphytobenthos. In: Rakusa-Suszczewski (ed) The maritime Antarctic coastal ecosystem of Admiralty Bay. Department of Antarctic Biology of Polish Academy of Sciences, Warsaw, pp 53–56

Lowry JK, Bullock S (1976) Catalogue of the marine Gammaridean Amphipoda of the Southern Ocean. Bull R Soc N Z 16:1–187

Momo FR (1995) Ciclo de vida y distribucion espacial de *Gondogeneia antarctica* Chevreux (Crustacea, Amphipoda). PhD Thesis, Facultad de Ciencias Exactas y Naturales, University of Buenos Aires

Nakajima Y, Watanabe K, Naito Y (1982) Diving observations of the marine benthos at Syowa Station, Antarctica. Mem Natl Inst Polar Res Spec Issue 23:44–54

Picken GB (1985) Marine habitats – benthos. In: Bonner WN, Walton WH (eds) Key environments: Antarctica. pp 154–172

Ponder WF (1983) Rissoaform gastropods from the Antarctic and Subantarctic. Br Antarct Surv Sci Rep 108:1–37

Powell AWB (1965) Mollusca of Antarctic and Subantarctic seas. In: Van Mieghem J, Van Oye P, Schell J (eds) Biogeography and ecology in Antarctica. Junk, The Hague, pp 333–380

Presler P (1986) Necrophagous invertebrates of the Admiralty Bay of King George Island (South Shetland Islands, Antarctica). Pol Polar Res 7:25–61

Presler P (1993) Ophiuroidea. In: Rakusa-Suszczewski S (ed) The maritime Antarctic coastal ecosystem of Admiralty Bay. Department of Antarctic Biology of Polish Academy of Sciences, Warsaw, pp 117–119

Presler P, Figielska E (1997) New data on the Asteroidea of Admiralty Bay, King George Island, South Shetland Islands. Pol Polar Res 18:107–117

Propp MV (1970) The study of bottom fauna at Haswell Islands by scuba diving. In: Holdgate MW (ed) Antarctic ecology. Academic Press, London, pp 239–241

Rakusa-Suszczewski S (1972) The biology of *Paramoera walkeri* Stebbing (Amphipoda) and the Antarctic sub-fast ice community. Pol Arch Hydrobiol 19:11–36

Rakusa-Suszczewski S (ed) (1993) The maritime Antarctic coastal ecosystem of Admiralty Bay. Department of Antarctic Biology, Polish Academy of Sciences, Warsaw

Rauschert M (1991) Ergebnisse der faunistischen Arbeiten im Benthal von King George Island (Sudshetlandinseln, Antarktis). Ber Polarforsch 76:1–75

Richardson MG (1977) The ecology (including physiological aspects) of selected Antarctic marine invertebrates associated with inshore macrophytes. PhD Thesis, University of Durham

Richardson MG, Whitaker TM (1979) An Antarctic fast-ice food chain: observations on the interaction of the amphipod *Pontogeneia antarctica* Chevreux with ice-associated micro-algae. Br Antarct Surv Bull 47:107–115

Sagar PM (1980) Life cycle and growth of the Antarctic gammarid amphipod *Paramoera walkeri* (Stebbing, 1906). J R Soc N Z 10:259–270

Servicio de Oceanographia, Hidrographia y Meteorologia de la Armada (1992) Tablas de Marea. Republica Orientale del Urugyay, Montevideo

Servicio de Oceanographia, Hidrographia y Meteorologia de la Armada (1993) Tablas de Marea. Republica Orientale del Urugyay, Montevideo

Sicinski J, Janowska E (1993) Polychaetes of the shallow sublittoral of Admiralty Bay, King George Island, South Shetland Islands. Antarct Sci 5:161–167

Skora KE, Neyelov AV (1992) Fish of Admiralty Bay (King George Island, South Shetland Islands, Antarctica). Polar Biol 12:469–476

Thurston MH (1974a) The Crustacea Amphipoda of Signy Island, South Orkney Islands. Br Antarct Surv Sci Rep 71:1–133

Thurston MH (1974b) Crustacea Amphipoda from Graham Land and the Scotia Arc, collected by Operation Tabarin and the Falkland Islands Dependencies Survey, 1944–59. Br Antarct Surv Sci Rep 85:1–89

White MG (1984) Marine benthos. In: Laws RM (ed) Antarctic ecology. Academic Press, London, pp 421–461

Zielinski K (1981) Benthic macroalgae of Admiralty Bay (King George Island, South Shetland Islands) and circulation of algal matter between the water and the shore. Pol Polar Res 2:71–94

Zielinski K (1990) Bottom macroalgae of the Admiralty Bay (King George Island, South Shetlands, Antarctica). Pol Polar Res 11:95–131

Polar Biol (2001) 24: 650–656
DOI 10.1007/s003000100264

Ichiro Takeuchi · Kentaro Watanabe · Atsushi Tanimura
Mitsuo Fukuchi

Assemblages of necrophagous animals off Enderby Land, East Antarctica

Abstract A hyperbenthic baited trap was deployed at three sites in Lützow-Holm Bay and off Enderby Land, Antarctica, in the 1993/1994 summer to investigate the species composition, distribution, and abundance of necrophagous animals. The trap consisted of four chambers, with a length of 100 cm and a diameter of 38 cm. The sample at the abyssal site, 3,186 m deep, contained 74 individuals of 6 species, completely different from that at the other 2 sites on the continental shelf; it was dominated by the large amphipod *Eurythenes gryllus*, a cosmopolitan species recorded from bathyal to abyssal depths. The species composition of the continental shelf sites, 93–137 individuals of 10–12 species, was dominated by species endemic to Antarctica, such as the small amphipod *Pseudorchomene coatsi* and the isopod *Natatolana meridionalis*; in these continental shelf sites, a high degree of species diversity was observed.

Please note: When citing this article, please use the reference to the original source in the journal Polar Biology as given on the top of this page.

I. Takeuchi (✉)
Otsuchi Marine Research Center,
Ocean Research Institute,
The University of Tokyo,
2-106-1 Akahama, Otsuchi,
Iwate 028-1102, Japan

K. Watanabe · A. Tanimura · M. Fukuchi
National Institute of Polar Research,
1-9-10 Kaga, Itabashi, Tokyo 173-0003, Japan

Present address: I. Takeuchi
Department of Life Conservation Ecology,
Faculty of Agriculture, Ehime University,
3-5-7 Tarumi, Matsuyama 790-8566, Japan,
e-mail: takeuchi@agr.ehime-u.ac.jp,
Fax: +81-89-9469899

Present address: A. Tanimura
Faculty of Bioresources,
Mie University, 1515 Kamihama,
Tsu, Mie 514-0008, Japan

Introduction

The Antarctic continental slope drops steeply from about 200–600 m to the deep ocean floor of more than 3,000 m. The extensive area of the abyssal ocean floor around Antarctica makes studies on benthic and epibenthic animals difficult; they are, however, indispensable to understand ecosystem characteristics in Antarctic waters. Rapid attraction of mobile necrophagous animals to baits deployed on the abyssal ocean floor has been observed worldwide (e.g. Gage and Tyler 1991; Britton and Morton 1994). However, necrophagous animals inhabiting bathyal to abyssal depths close to the Antarctic continent are poorly known, due to technical difficulties of sampling in waters seasonally covered with sea ice, and the presence of icebergs.

Baited traps have been used to collect necrophagous invertebrates in Antarctica over the last 30 years (e.g. Hurley 1965; Arnaud 1970; Bruchhausen et al. 1979; Lipps et al. 1979; Hoshiai 1982; Stockton 1982; De Broyer 1985; Nagata 1986; Presler 1986; Slattery and Oliver 1986; De Broyer and Klages 1990; Moore 1994). Lipps et al. (1979) and Bruchhausen et al. (1979) surveyed benthic animals using fish traps at the drilling site J-9 on the Ross Ice Shelf, 400 km from the shelf, where the ice was 415 m thick. Stockton (1982), at the same site, reported that *Orchomene* dominated the mobile necrophagous animals. At McMurdo Sound, Ross Sea, *Abyssorchomene plebs* (Hurley, 1965) (as *Orchomene*) is distributed primarily in deep water (> 50 m), whereas *Orchomenella pinguides* (Walker, 1903) (as *Orchomene*) is dominant on the shallow wave-cut beach along the eastern McMurdo Sound (Slattery and Oliver 1986). Besides these ecological studies, lysianassoid amphipods as the dominant taxon of the mobile necrophagous fauna have also been taxonomically described based on the trap collections (e.g. Hurley 1965; De Broyer 1985; Nagata 1986).

However, most of these studies set the trap utilizing holes through stable fast ice adjacent to coastal areas on the continental shelf except for a few trials in the Weddell Sea during the expeditions of R/V "Polarstern" (De Broyer and Klages 1990; De Broyer et al. 1997, 1999). During the 35th JARE (Japan Antarctic Research Expedition) cruise in the 1993/1994 summer, baited trap sampling was conducted three times in Lützow-Holm Bay and off Enderby Land, as part of the SIEFS (Sea Ice Ecology and Flux Study) programme started in the 1991/1992 summer season.

The present study reveals that baited trap sampling from an icebreaker is an effective method for investigating the Antarctic benthic community in pack- or fast-ice conditions. In pack-ice areas, acoustic release is preferable to automatic release, because the recovery can be postponed until suitable weather conditions prevail, and the distance between the vessel and the acoustic release can be measured even if the trap is caught under the ice. Here, we describe the baited trap system used, and report on the assemblages of necrophagous animals collected at the three sites, two sites on and one site off the Antarctic continental shelf.

Materials and methods

Site description

Collections were made on the southern border of the Atlantic-Indian Basin on the continental shelf in the Lützow-Holm Bay and off Enderby Land, Antarctica (Fig. 1). Station L1, 3,186 m deep, is situated on the abyssal ocean floor close to the end of the continental slope off Lützow-Holm Bay, more than 10° south of the Antarctic Convergence. The Antarctic Convergence, 48–61°S, is a boundary for both phyto- and zooplankton communities of the Southern Ocean (e.g. Fukuchi 1982). Antarctic Bottom Water with a temperature of less than 0°C is restricted to the Atlantic sector and Indian Ocean-Australia sector (Ishino 1989). Thus, this station, as well as the other two stations, should be directly influenced by the Antarctic Bottom Water. Station L2, 171 m deep, is located on a rather flat area, typical of the eastern part of Lützow-Holm Bay (Moriwaki and Yoshida 1983). Fast ice (1–3 years old) covered almost all of the bay. Detailed bathymetric surveys have not been conducted on the continental shelf off Enderby Land due to the prevailing pack or fast ice. Here, the shelf extends around 150–500 m and is apparently ridged. Station E4, 353 m deep, is situated in one of the troughs between ridges.

Baited trap system

All traps and instruments were modified from the two-layered hyperbenthic baited trap, which was originally used in February 1994 during the Hakuho-Maru "KH-93-1 Cruise" to the mid-Pacific mountains (Numachi 1994). The trap is steel-framed to minimize damage by the sides of the vessel or ice during recovery (Fig. 2). Four chambers, made of PVC tube (100 cm length and 38 cm diameter), were set within the frame in two layers: the lower two were attached to the bottom of the frame, while the center position of the upper two was set at 105 cm above the bottom of the frame. Inverted entrance cones made of 1.0-mm stainless steel mesh were attached to both ends of the tubes. The entrance was 10 cm in diameter. Pacific saury, *Cololabis saira* (Brevoort) were purchased freshly frozen in Tokyo, and kept frozen at –20°C until used as bait. About 2,000 g of coarsely chopped Pacific saury was baited in each chamber. Nylon bags of 3- to 5-mm mesh were used to hold the bait in the chambers at st. L1, while bait containers made of 500-ml plastic bottles with 1.0-mm mesh, modified from Ingram and Hessler (1983), were used at sts. L2 and E4.

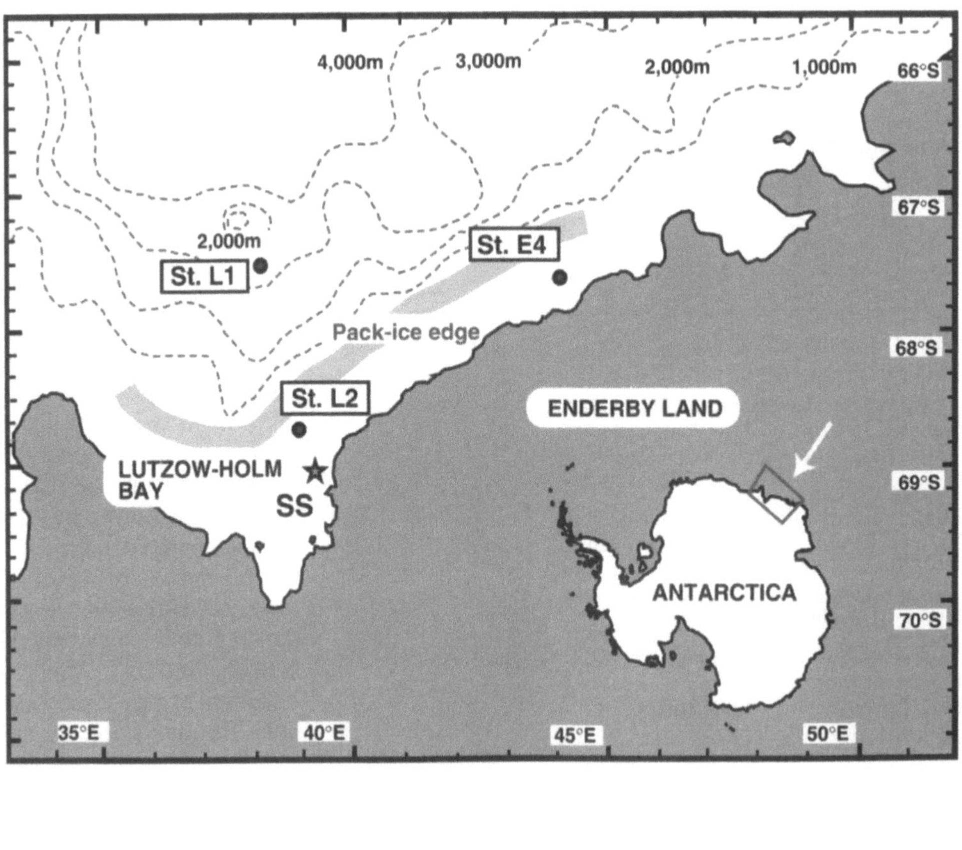

Fig. 1 Map showing the study area and sampling sites in Lützow-Holm Bay and off Enderby Land, Antarctica. Station L1 is situated on the abyssal floor at 3,168 m deep, st. L2 on the continental shelf at 171 m deep, and st. E4, 357 m deep in one of the glacial ridges on the continental shelf off Enderby Land. *SS* indicates the location of Syowa Station, Japanese Antarctic Base

Fig. 2 Recovery of the two-layered hyperbenthic baited trap system using a winch on board the ice breaker Shirase at st. L1 in December 1993

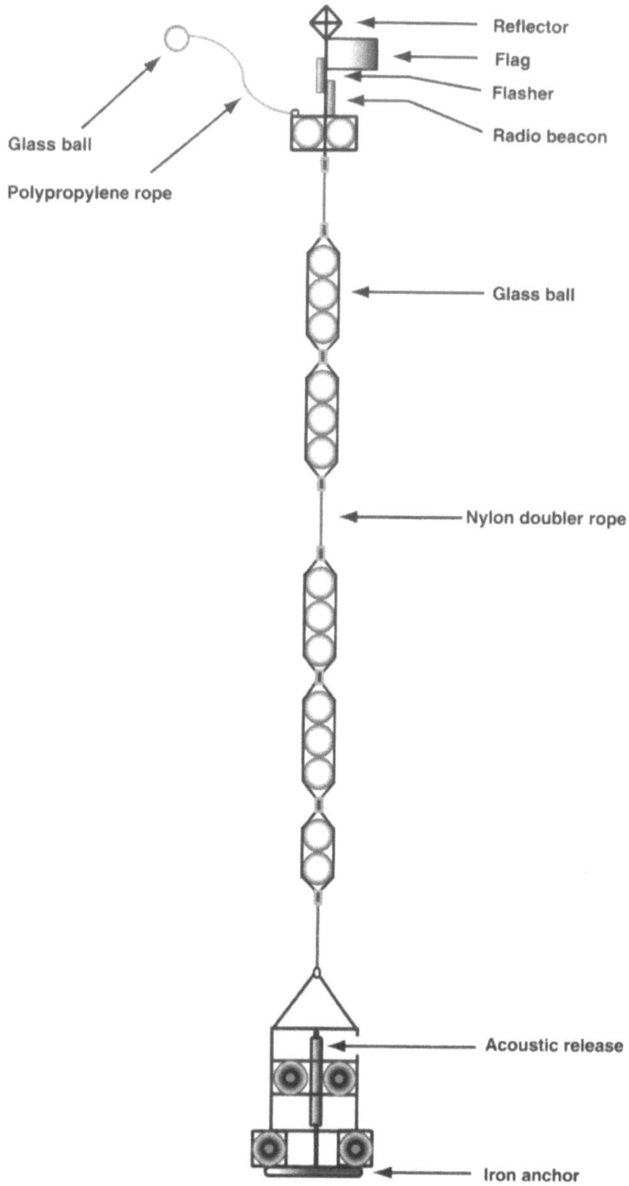

Fig. 3 Schematic view of the benthic baited trap system operated by the acoustic release method as used at st. E4 in February 1994

The methods of setting and recovering the trap were modified due to the weather and ice condition, as follows.

- Station L1. The baited trap sampling was conducted on 19 December 1993. The trap was deployed through a hole made by the icebreaker Shirase in the pack ice (Fig. 2). The vessel, together with the ice, drifted westerly about 2,000 m during the deployment of the trap. The trap was connected to the wire via a nylon rope of 500 m length with a diameter of 18 mm, to act as a shock absorber. In order to avoid kinking the wire, a 50-kg iron weight was attached to the end of it. An iron anchor of 310 kg was also attached to the bottom of the trap to avoid overturning. The trap was recovered at a speed of 0.5–1.2 m sc^{-1} using a winch on board the icebreaker, after being set for ca. 7 h on the seafloor (Table 1).
- Station L2. Sampling at this site was conducted on 7 February 1994 in stable multi-year fast ice (Table 1). The vessel did not move during the sampling. The system was similar to that used at st. L1, except that the wire was attached directly to the trap, and the iron weight on the wire was not used.
- Station E4. An acoustic release system was used at this site (Fig. 3). The acoustic release (Nichiyu Giken, Tokyo; type L) was positioned in the center of the steel frame of the trap and connected to an iron anchor of 310 kg. Flotation was provided by 15 abyssally submersible glass spheres (Benthos, North Falmouth, Mass.). The glass spheres, each housed in a plastic shroud, were inter-connected with a stainless steel chain. Nylon rope of 24 mm diameter was used to connect the floats, trap and mast. The mast, aluminum pole, 200 cm in length and 2.5 cm in diameter, was equipped with an aluminium radar reflector, red flag, radio beacon (Taiyo Electronics, Tokyo), and a flasher (Taiyo Electronics, Tokyo) to aid in recovery of the trap.

The baited trap was lowered on 15 February 1994, in a large coastal polynya containing some pancake ice floes (Table 1). However, the site was covered on 16 February with pack ice of about 1 m thickness, due to the westerly wind of more than 10 m/s which aggregated the pack ice, closing the polynya. On 18 February, when the wind speed had decreased to 5 m/s from the north, the baited trap was released from its signal under the pack ice and subsequently recovered.

Table 1 Location of the baited trap sampling during the 35th Japan Antarctic Research Expedition Cruise

Station	Latitude/longitude	Date (local time)	Bottom time (h:min)	Depth (m)	Method of recovery
L1	67.43.0–67.43.2S/ 39.09.9–39.08.1E	19 Dec. (1100–1730 hours) 1993	6:30	3186	Winch
L2	68.55.3S/38.55.3E	07 Feb. (1720 hours)– 08 (0738 hours) 1994	14:18	171	Winch
E4	67.43.15S/44.27.71E	15 Feb. (1535 hours)– 18 (1425 hours) 1994	71:10	353	Acoustic release

Sample treatment

The collected animals were preserved in neutralized 10% seawater-formalin, except for *Eurythenes gryllus* (Lichtenstein, 1822) collected from st. L1. As *E. gryllus* was collected alive, metabolic rate and swimming speed were measured in a cold room of the vessel, together with histological examinations of the gills (Matsumasa et al. 1998; Takeuchi and Watanabe 1998). After these experiments, the specimens were preserved in an ultra-low freezer (–80°C) for chemical analysis.

After returning to Japan, the specimens except for *E. gryllus* were identified to species level using a binocular microscope. The similarity index of the species assemblages between sites was obtained using Pianka's *a* index (Pianka 1973; May 1975; Kimoto and Takeda 1989). The index ranges from 0 (no common taxa in the two samples) to 1.0 (a perfect correlation of taxa in the two samples). The species diversity of each site was expressed by the Shannon-Wiener function (H') and the Pielou index of equitability (J'), both ln-based (MacArthur 1964; Kimoto and Takeda 1989). H' increases as the species diversity becomes more complex.

Results

Total abundance of necrophagous animals ranged from 71 to 137 individuals in 22 species (Table 2). All of the individuals were crustaceans, particularly Amphipoda and Isopoda, except for four individuals of two species of Antarctic endemic teleosts of the genus *Trematomus*.

The species diversity, as well as the equitability in the shallower two sites, were 1.73–1.78 by H', and 0.72–0.75 by J', both greater than those of st. L1 (Table 2). This corresponded well with the increase of S, the number of species.

The faunal composition of the abyssal site (st. L1) differed completely from that of the two shallower sites (st. L2, E4), with no overlapping species (Table 2). The fauna was dominated by the large amphipod *E. gryllus*, accounting for 64.9% of the total number of individuals. No significant difference in the level of attraction between the higher and lower chambers of the trap was observed for *E. gryllus* (χ^2-test; $P = 0.981$). Four other amphipod species and an isopod were additionally recorded.

Five species were common for both of the shallower two sites, indicating a 0.224 degree of correlation by Pianka's *a* index. Ten species were recorded from the shallower station in Lützow-Holm Bay (st. L2), with a high abundance of the lysianassoid amphipod, *Pseudorchomene coatsi* (Chilton, 1912). The number of *P. coatsi* attracted to the lower chambers was close to being significantly different to that of the higher chambers (χ^2-test; $P = 0.057$). The next two abundant species, the calanoid copepod *Xanthocalanus gracilis* (Wolfenden, 1911) and an ostracod, *Doloria levis* (Skogsberg, 1920) were collected only from the lower chambers. A detailed redescription of the copepod *X. gracilis* was provided by Ohtsuka et al. (1998). Of 12 species recorded from st. E4, the isopod *Natatolana meridionalis* (Hodgson, 1910), which was not collected from st. L2, represented nearly one-third of the fauna, followed by the leptostracan, *Nebalia* sp., and a lysianassoid amphipod, *Waldeckia* cf. *obesa* (Chevreux, 1905). *N. meridionalis* was significantly attracted to the lower chambers (χ^2-test; $P = 0.010$), while *W.* cf. *obesa* did not show any significant preference ($P = 0.780$). *P. coatsi*, the most dominant species at st. L4, was the fourth most dominant species at this site.

Discussion

Even though the three sites sampled are all probably affected by Antarctic Bottom Water, habitat segregation of the necrophagous animals by depth occurs along the Antarctic continental shelf. Two species of stegocephalid amphipods, *Parandania boecki* (Stebbing, 1888) and *Euandania gigantea* (Stebbing, 1883), collected from st. 1 have been widely recorded from bathyal to abyssal depths around the world (e.g. Barnard and Karaman 1991). *Eurythenes gryllus*, the most common species at st. L1, 3,186 m deep, off Enderby Land, is widely recorded in the Pacific and Atlantic Ocean and in both Arctic and Antarctic waters (e.g. Bowman and Manning 1972; Rauschert 1985; Barnard and Karaman 1991; De Broyer and Jazdzewski 1993). Recently, De Broyer et al. (1997) reported the collection of four specimens of *Eurythenes gryllus* in eight species of necrophagous amphipods from 2,009 m depth by baited trap; seven of the eight species were not found at shallower stations. Moreover, De Broyer et al. (1999) reported the capture of another four specimens of *Eurythenes gryllus* in a baited trap at 1,200 and 1,500 m depth.

Thus, the present study, as well as those of De Broyer et al. (1997, 1999), indicates that *Eurythenes gryllus* extends its distribution far south of the Antarctic Convergence to areas of at least 1,000 m depth near the Antarctic continental shelf. This distribution of the cosmopolitan benthic amphipod near the Antarctic continent is in contrast to the distribution of zooplankton around Antarctica. Large-scale surveys of the plankton community indicate that a sub-Antarctic community of Copepoda, which is characterized by a high species diversity and the dominance of *Neocalanus tonsus* (Brady, 1883) (as *Calanus*), is prevalent in the regions of 40°S, whereas the Antarctic community dominated by *Calanus propinquus* (Brady, 1883), *Calanoides acutus* (Giesbrecht, 1902), and *Rhincalanus gigas* (Brady, 1883) is located south of the Antarctic Convergence around 48–61°S (Fukuchi 1982).

In contrast to the abyssal site, most of the organisms recorded from the two sites on the continental shelf (Table 2) are endemic to the Southern Ocean (e.g. Barnard and Karaman 1991; De Broyer and Jazdzewski 1993). Three species of lysianassoids, *Pseudorchomene coatsi*, *Uristes murrayi* (Walker, 1903), and *Hippomedon major* (KH Barnard, 1932), have circum-Antarctic distribution. *Podoprionides incerta* (Walker, 1906) is also recorded from the eastern part of Antarctica. The eusirid, *Eusirus laticarpus* (Chevreux, 1906), was previously recorded from the western part of Antarctica; our finding is the first record from the east. In addition to

Table 2 Number of necrophagous animals collected in the baited trap during JARE-35 Cruise with the species diversity expressed by *H′*, Shannon-Wiener function. *L1*, *L2*, and *E4* indicate the sampling sites, and *U.C.* and *L.C.* the position of chambers within the trap. Data in *U.C.* (upper chambers) and *L.C.* (lower chambers) indicate mean ± standard deviation

Species	L1			L2			E4		
	U.C.	L.C.	Total (%)	U.C.	L.C.	Total (%)	U.C.	L.C.	Total (%)
Crustacea									
Amphipoda									
1. *Eurythenes gryllus* (Lichtenstein)	11.0 ± 4.2	13.0 ± 7.1	46(64.8)	0	0	0(0)	0	0	0(0)
2. *Paralicella* cf. *similis* Birstein & Vinogradov	0	4.5 ± 0.7	9(12.7)	0	0	0(0)	0	0	0(0)
3. *Parandania boecki* (Stebbing)	2.5 ± 2.1	2.5 ± 2.1	10(14.1)	0	0	0(0)	0	0	0(0)
4. *Euandania gigantea* (Stebbing)	0	0.5 ± 0.7	1(1.4)	0	0	0(0)	0	0	0(0)
5. *Waldeckia* cf. *obesa* (Chevreux)	0	0	0(0)	0	3.0 ± 2.8	6(6.5)	8.0 ± 5.7	5.0 ± 2.8	26(19.0)
6. *Waldeckia* sp.	0	0	0(0)	0	0	0(0)	0.5 ± 0.7	0	1(0.7)
7. *Uristes murrayi* Walker	0	0	0(0)	0	0.5 ± 0.7	1(1.1)	1.5 ± 0.7	3.0 ± 2.8	9(6.6)
8. *Eusirus laticarpus* Chevreux	0	0	0(0)	0	0	0(0)	1.5 ± 0.7	0.5 ± 0.7	4(2.9)
9. ? *Orchomene* sp.	1.5 ± 0.7	1.0 ± 0.0	2(5.6)	0	0	0(0)	0	0	0(0)
10. *Pseudorchomene coatsi* (Chilton)	0	0	0(0)	3.0 ± 2.8	17.0 ± 4.2	40(43.0)	0	6.0 ± 2.8	12(8.8)
11. *Hippomedon major* (K.H. Barnard)	0	0	0(0)	0	0	0(0)	0	0.5 ± 0.7	1(0.7)
12. *Podoprionides incerta* Walker	0	0	0(0)	0	5.0 ± 7.1	10(10.8)	0	0	0(0)
Isopoda									
13. Munnopsididae cf. "*Eurycope sarsi* Beddard"	0	0.5 ± 0.7	1(1.4)	0	0	0(0)	0	0	0(0)
14. *Natatolana meridionalis* (Hodgson)	0	0	0(0)	0	0	0(0)	0	14.5 ± 3.5	48(35)
15. *Gnathia* sp.	0	0	0(0)	0.5 ± 0.7	1.5 ± 2.1	4(4.3)	0	0.5 ± 0.7	1(0.7)
Mysidacea									
16. Sp.1	0	0	0(0)	0	0	0(0)	0	0.5 ± 0.7	1(0.7)
Ostracoda									
17. *Skogsbergiella* sp.	0	0	0(0)	0	0.5 ± 0.7	1(1.4)	0	0	0(0)
18. *Vargula antarctica* (Muller)	0	0	0(0)	0	0.5 ± 0.7	1(1.4)	0	3.0 ± 0.0	6(4.4)
19. *Doloria levis* Skogsberg	0	0	0(0)	0	6.0 ± 8.5	12(12.9)	0	0	0(0)
Copepoda									
20. *Xanthocalanus gracilis* Wolfenden	0	0	0(0)	0	7.5 ± 7.8	15(16.1)	0	0	0(0)
Leptostraca									
21. *Nebalia* sp.	0	0	0(0)	0	0	0(0)	0	13.5 ± 19.1	27(19.7)
Pisces									
22. *Trematomus bernacchii* Boulenger	0	0	0(0)	0.5 ± 0.7	1.0 ± 1.4	3(3.2)	0	0	0(0)
23. *Trematomus loennbergii* Regan	0	0	0(0)	0	0	0(0)	0	0.5 ± 0.7	1(0.7)
Total			74(100)			93(100)			137(100)
S; Number of species*			6			10			12
H′; Shannon-Wiener function*			1.11			1.73			1.78
J′; Pielou's index of equitability**			0.69			0.75			0.72

*$H' \simeq -\sum(p_i \cdot \ln p_i)$; p_i, ratio of *i*-th species to the total number of individuals;
**$J' \simeq -\sum(p_i \cdot \ln p_i)/\ln S$; p_i, ratio of *i*-th species to the total number of individuals

these gammarid amphipods, the isopod genus *Natatolana* is widely recorded from the east side of the Weddell Sea (Zimmermann and Brandt 1992). The genus *Trematomus*, benthic to epibenthic species, is found throughout the Antarctic and sub-Antarctic regions (Eastman 1993).

The two sites on the continental shelf with a higher species diversity show low similarity. This was caused by the difference of the dominant species, the lysianassoid, *Pseudorchomene coatsi*, on the continental shelf in Lützow-Holm Bay, and the isopod *Natatolana meridionalis* in the troughs between ridges on the shelf off Enderby Land (Table 2). As already mentioned, Slattery and Oliver (1986) found the habitat segregation of two species of lysianassoid amphipods at McMurdo Sound. De Broyer and Klages (1990) reported the dominance of *A. plebs* and *W. obesa* (Chevreux, 1905) in sampling species from the eastern Weddell Sea.

The assemblages of necrophagous animals from both stations on the continental shelf (Table 2) lacked *A. plebs*, which dominated the species compositions of lysianassoid amphipods in previous studies conducted at the South Shetland Islands and in the Ross and Weddell Seas (Hurley 1965; Arnaud et al. 1986; Slattery and Oliver 1986; De Broyer and Klages 1990). However, *A. plebs* has been recorded from both east and west Antarctica (De Broyer et al. 1999) although it was first described as *O. plebs* (Hurley, 1965) based on the specimens collected from White Islands, Ross Sea. Slattery and Oliver (1986), De Broyer and Klages (1990), and C. De Broyer (personal communication) confirmed that the *Orchomene* collected from under the Ross Ice Shelf by Stockton (1982) were composed of *A. plebs* and *A. rossi* (Walker, 1903).

Small-scale habitat segregation on the continental shelf is suggested to be caused by at least two factors: depth and topographic condition. Concentrated studies on a particular area are necessary to elucidate the mechanism of habitat segregation of necrophagous animals off Enderby Land.

Acknowledgements The authors express their sincere thanks to the captain, officers, and crew of the icebreaker "Shirase", and all members of 34th and 35th JARE for operating the baited trap sampling during the 35th JARE cruise. We thank the scientists for their cordial assistance in the identification of the benthic animals, Dr. J.K. Lowry and H. Stoddart for Amphipoda, Dr. S. Keable and Dr. G.D.F. Wilson for Isopoda, Dr. S. Hiruta for Ostracoda, Dr. S. Ohtsuka for Copepoda, Dr. T. Kikuchi for Leptostraca, and Dr. T. Iwami for Pisces. Dr. H. Marchant and Dr. C. Norman kindly reviewed early versions of the manuscript. This study forms a part of the SIEFS Programme supported by the National Institute of Polar Research and the Ministry of Education, Culture and Sciences, Japan. Nichiyu Giken Kogyo, Tokyo, cooperated in revising the design of the baited trap.

References

Arnaud PM (1970) Frequency and ecological significance of necrophagy among the benthic species of Antarctic coastal waters. In: Holdgate MW (ed) Antarctic ecology. Academic Press, London, pp 259–267

Arnaud PM, Jazdzewski K, Presler P, Sicinski J (1986) Preliminary survey of benthic invertebrates collected by Polish Antarctic Expeditions in Admiralty Bay (King George Island, South Shetland Islands, Antarctica). Pol Polar Res 7:7–24

Barnard JL, Karaman DS (1991) The families and genera of marine gammaridean Amphipoda (except marine gammaroids). Rec Aust Mus Suppl 13:1–417, 419–886

Bowman TE, Manning RB (1972) Two arctic bathyal crustaceans: the shrimp *Bythocaris cryonesus* new species and the amphipod *Eurythenes gryllus*, with *in situ* photographs from Ice Island T-3. Crustaceana 23:187–201

Britton JC, Morton B (1994) Marine carrion and scavengers. Oceanogr Mar Biol 32:369–434

Bruchhausen PM, Raymond JA, Jacobs SS, DeVries AL, Thorndike EM, DeWitt HH (1979) Fish, crustaceans, and the ocean floor under the Ross Ice Shelf. Science 203:449–501

De Broyer C (1985) Amphipodes lysianassoïdes nécrophages des îles Kerguelen (Crustacea). 1. *Orchomenella guillei* n. sp. Bull Mus Natn Hist Nat Paris 4ᵉ Sér 7:205–217

De Broyer C, Jazdzewski K (1993) Contribution to the marine biodiversity inventory. A checklist of the Amphipoda (Crustacea) of the Southern Ocean. Doc Travail L'IRScNB 73:1–154

De Broyer C, Klages M (1990) Studies on amphipod biology. In: Arntz WE, Ernst W, Hempel I (eds) The Expedition ANTARKTIS VII/4 (Epos leg 3) and VII/5 of RV "Polarstern" in 1989. Ber Polarforsch 68:113–115

De Broyer C, Rauschert M, Chapelle G (1997) Trophodynamics, biodiversity and gigantism of the amphipod Crustacea taxocoenosis. In: Arntz WE, Gutt J (eds) The Expedition ANTARKTIS XIII/3 (EASIZ I) of "Polarstern" to the eastern Weddell Sea in 1996. Ber Polarforsch 249:76–79

De Broyer C, Rauschert M, Scailteur Y (1999) Structural and ecofunctional biodiversity of benthic amphipod taxocoenoses. In: Arntz WE, Gutt J (eds) The Expedition ANTARKTIS XV/3 (EASIZ II) of RV "Polarstern" in 1998. Ber Polarforsch 301:163–174

Eastman JT (1993) Antarctic fish biology. Evolution in a unique environment. Academic Press, San Diego

Fukuchi M (1982) Biology of the Southern Ocean, primary production, and zooplankton. Sciences of the Antarctic, vol 7 (in Japanese). National Institute of Polar Research, Tokyo

Gage JD, Tyler PA (1991) Deep-sea biology. A natural history of organisms at the deep-sea floor. Cambridge University Press, Cambridge

Hoshiai T (1982) Biology along the coast, benthic animals and fishes. Sciences of the Antarctic, vol 7 (in Japanese). National Institute of Polar Research, Tokyo

Hurley DE (1965) A common but hitherto undescribed species of *Orchomenella* (Crustacea Amphipoda: Family Lysianassidae) from the Ross Sea. Trans R Soc NZ Zool 6:107–113

Ingram CL, Hessler RR (1983) Distribution and behavior of scavenging amphipods from the central North Pacific. Deep Sea Res 30:683–706

Ishino M (1989) Physical oceanography. Sciences of the Antarctic, vol 8 (in Japanese). National Institute of Polar Research, Tokyo

Kimoto S, Takeda S (1989) Introduction to community ecology (in Japanese). Kyoritsu Syuppan, Tokyo

Lipps JH, Ronan TE Jr, DeLaca TE (1979) Life below the Ross Ice Shelf, Antarctica. Science 203:447–449

MacArthur RH (1964) Environmental factors affecting bird species diversity. Am Nat 98:387–397

Matsumasa M, Kikuchi S, Takeuchi I (1998) Specialized ion-transporting epithelium around the blood vessel of the coxal gills in a deep-sea amphipod *Eurythenes gryllus*. J Crust Biol 18: 686–694

May RM (1975) Some notes on estimating the competition matrix. Ecology 56:737–741

Moore PG (1994) Observations on the behaviour of the scavenging lysianassoid *Orchomene zschaui* (Crustacea: Amphipoda) from South Georgia (South Atlantic). Mar Ecol Prog Ser 111:29–38

Moriwaki K, Yoshida Y (1983) Submarine topography of Lützow-Holm Bay, Antarctica. Mem Natl Inst Polar Res Spec Issue 28:247–258

Nagata K (1986) Amphipod crustaceans found near Syowa Station, Antarctica (1). Mem Natl Inst Polar Res Spec Issue 40:249–258

Numachi K (1994) Preliminary report of the Hakuho Maru Cruise KH 93–1. January 22–March 4, 1993. Biogeographic and population genetic study of marine organisms on the sea mounts in the northwestern Pacific. Ocean Research Institute, The University of Tokyo

Ohtsuka S, Takeuchi I, Tanimura A (1998) *Xanthocalanus gracilis* and *Tharybis magna* (Copepoda: Calanoida) rediscovered from the Antarctic Ocean with baited trap. J Nat Hist 32:785–804

Pianka ER (1973) The structure of lizard communities. Ann Rev Ecol Syst 4:53–74

Presler P (1986) Necrophagous invertebrates of the Admiralty Bay of King George Island (South Shetland Islands, Antarctica). Pol Polar Res 7:25–61

Rauschert M (1985) *Eurythenes gryllus* (Lichtenstein) (Crustacea, Amphipoda) in der marinen Fauna von King George (Südshetlandinseln, Antarktis). Milu Berlin 6:319–324

Slattery PN, Oliver JS (1986) Scavenging and other feeding habits of lysianassid amphipods (*Orchomene* spp.) from McMurdo, Antarctica. Polar Biol 6: 171–177

Stockton WL (1982) Scavenging amphipods from under the Ross Ice Shelf, Antarctica. Deep Sea Res 29:819–835

Takeuchi I, Watanabe K (1998) Respiration rate and swimming speed of the necrophagous amphipod *Eurythenes gryllus* from Antarctic deep waters. Mar Ecol Prog Ser 163:285–288

Zimmermann A, Brandt A (1992) New results of the "EPOS" leg 3 cruise to Antarctica: horizontal and vertical distribution of isopods (Crustacea) in the eastern Weddell Sea. Polar Biol 12:121–128

II.

Pelagobenthic coupling

Polar Biol (2001) 24: 473–485
DOI 10.1007/s003000100257

Josep-Maria Gili · Rafel Coma · Covadonga Orejas
Pablo J. López-González · Mikel Zabala

Are Antarctic suspension-feeding communities different from those elsewhere in the world?

Abstract This paper reviews the trophic ecology of benthic suspension feeders in Antarctic shelf communities, studied within SCAR's EASIZ Programme, in comparison with published information from other seas. Dense benthic suspension-feeder communities capture large quantities of particles and may directly regulate primary, and indirectly, secondary production in littoral food chains. Most work has been performed in temperate and tropical seas; however, little is known about suspension feeders in cold environments. Recent studies on Antarctic littoral benthic suspension feeders suggest the period of winter inactivity may last only a few weeks. This contrasts with the hypothesis that in Antarctic communities there is a prolonged period of minimal activity lasting at least 6 months during the austral winter. Results from other oceans may explain how dense benthic communities could develop under such conditions. Alternative food sources, i.e. the "fine fraction", sediment resuspension, lateral advection and efficient food assimilation may play a significant role in the development of suspension-feeder dominated, very diversified, high biomass and three-dimensionally structured communities on the Antarctic shelf.

Please note: When citing this article, please use the reference to the original source in the journal Polar Biology as given on the top of this page.

J.-M. Gili (✉) · R. Coma
Institut de Ciències del Mar (CSIC),
Plaça del Mar s/n, 08039 Barcelona, Spain
E-mail: gili@icm.csic.es
Tel.: +34-93-2216416
Fax: +34-93-2217340

C. Orejas
Alfred-Wegener-Institut für Polar- und Meeresforschung,
Columbusstrasse, 27568 Bremerhaven, Germany

P.J. López-González
Laboratorio de Biología Marina, Universidad de Sevilla,
Apartado 1095, 41080 Sevilla, Spain

M. Zabala
Departament d'Ecologia, Facultat de Biologia,
Universidad de Barcelona, Diagonal 645,
08028 Barcelona, Spain

Introduction

What are we talking about?

Primary producers, together with the products of their biological activity, generate the most abundant available food for heterotrophic organisms in the oceans: the seston. Small particles and cells predominate in such suspended communities (Wottom 1994), providing food for primary consumers and substrata for bacteria. The physical properties of seawater allow living creatures and particulate matter to remain in suspension, thereby creating a niche for a trophic strategy that does not occur on land: suspension feeding (Jørgensen 1990). Suspension feeders are common on all sea-bottom types, and are the main animal component in hard-bottom communities (Gili and Coma 1998).

Filtration rates of benthic suspension feeders are typically between 1 and 10 m^3 m^{-2} day^{-1} (Riisgård et al. 1996a) and dense assemblages of benthic suspension feeders may thus have a pronounced grazing effect (Kimmerer et al. 1994; Riisgård 1998). Large parts of Antarctic benthic communities consist of sessile suspension feeders such as sponges, cnidarians, bryozoans, ascidians and certain echinoderms (Dayton et al. 1986). Fully developed epifaunal assemblages in Antarctica may not be quite as diverse as some tropical reefs, but between 10 and 1,000 m water depth benthic biomass in the Antarctic is higher than in temperate and subtropical communities (Brey and Gerdes 1997). Antarctic benthic communities are highly structured, with a complex functional diversity and a considerable degree of patchiness in species composition at small or intermediate spatial scales (Gutt and Starmans 1998).

The ecological role of suspension feeders in Antarctica is related to the trophic ecology and environmental conditions that facilitate the processes of energy transfer between benthic and water-column systems. In this review we focus in particular on: (1) the trophic adaptations of suspension feeders that make them successful foragers worldwide, and (2) the processes responsible for the availability of food to suspension feeders, such as vertical transport, resuspension and lateral advection.

Exploitation of food resources: suspension feeding as an ecologically efficient strategy

Life in patches: a food capture strategy at individual, colony, population and community level

Suspension-feeding colonies or individuals interfere with current flow and they have evolved a multitude of adaptations designed to attract and trap particles in suspension (Gili and Coma 1998). Downstream of those structures that are exposed to the flow, such as the axis of a pennatulacean, viscosity increases and turbulence also increases slightly, generating small eddies (Vogel 1994). This hydrodynamic effect may help not only to increase the residence time of particles on the downstream side of the colony in times of heavy flow, but also to enhance settling of particles in the area adjacent to such structures. This phenomenon is of great importance when different species of suspension feeders aggregate in the same community (Fig. 1).

Fig. 1 Idealized diagram of Antarctic shelf suspension-feeder communities considering the patch theory explained in the text. The development of dense, three-dimensional communities involves a continuous process of patch spreading and aggregation during slow succession

Pliancy of colonies is one of the most widespread adaptations employed to meet the conflicting demands of minimum drag and maximum flux of capturable particles through the feeding structures (Harvell and LaBarbera 1985). As an adaptation to slow flow rates, the colonies of certain species of ascidians and bryozoans have evolved systems for circulating water through their feeding structures and refreshing the colony boundary layer (Vogel 1994). Such conditions are not a problem in active suspension feeders or in filter feeders that pump water through an internal exchange circuit

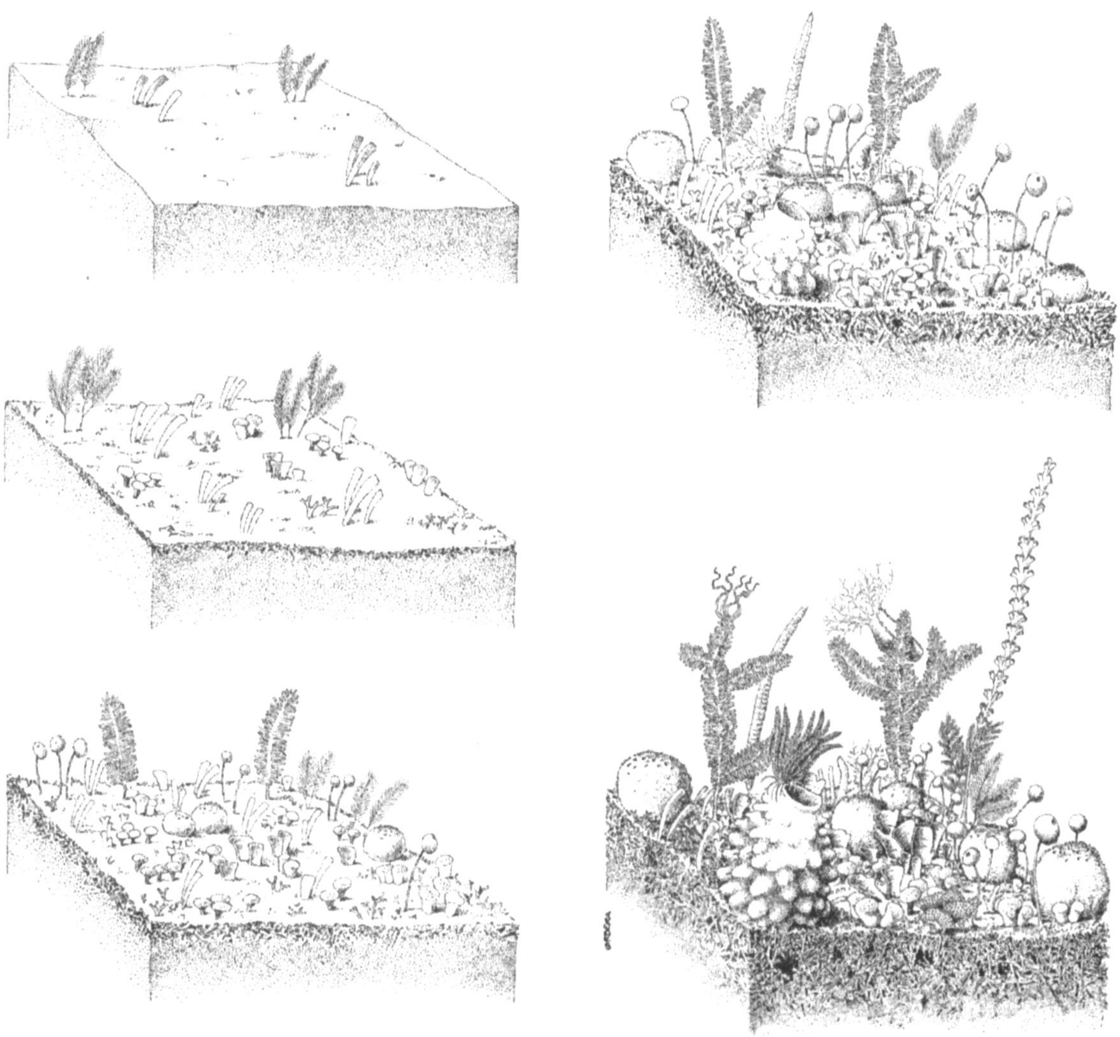

(such as sponges, ascidians, etc.), which is more efficient at moderate flow rates (Shimeta and Jumars 1991) though considerably less efficient at higher flow rates (Wildish and Miyares 1990).

The formation of monospecific patches not only lowers competition (Best and Thorpe 1986) but also improves capture rates among the colonies of the species concerned. Experiments on alcyonarians (MacFadden 1986), actinians and corals (Sebens et al. 1996) have shown that particle capture rates vary among colonies of different size within the population. This, in turn, gives rise to a spatial pattern determined by the optimum distance between colonies or individuals (Eckman and Duggins 1993). Larger colonies have been observed to be more efficient at lower rates of flow, and vice versa. Maximum and mean colony size in each patch are determined by the total food concentration and the intensity of water movement (Anthony 1997). The effect in dense aggregations of barnacles is similar (Pullen and LaBarbera 1991), with larger, asymmetrically shaped aggregates oriented towards the prevailing currents and smaller, more evenly shaped aggregates where current flow is bidirectional. Large assemblages of suspension feeders can deplete food sources under low flow conditions (Best and Thorpe 1986). However, because fast flow is the prevailing condition, enhancement of feeding among neighbours is the more significant interaction (Okamura 1990).

Benthic suspension feeders as optimal foragers

Recent studies of natural feeding show that suspension feeders can feed on a broad spectrum of prey items, which range from bacteria to zooplankton and detrital particulate organic matter (POM). Some taxa exhibit an important diet variability among species (Sorokin 1991), whereas the diet appears to be rather homogeneous in other taxa (Gili et al. 1998). The diet spectrum within most species is rather broad. Examples are octocorals (Fabricius et al. 1995; Ribes et al. 1999a), sponges (Ribes et al. 1999b), ascidians (Klumpp 1984; Ribes et al. 1998) and hydrozoans (Gili et al. 1998). Even species that feed mainly on zooplankton may prefer different prey types and sizes (Sebens and Koehl 1984) (Fig. 2). Certain suspension feeders may change their diet when different food becomes available (e.g. tropical crinoids, Rutman and Fishelson 1969; polar octocorals, Slattery et al. 1997; temperate sponges and ascidians, Ribes et al. 1998, 1999b). Thus, much of the diet variability among species of the same taxa may be due to differences in availability of resources in their environment. Few studies, however, have accounted for seasonal differences in feeding.

The exceptional ecological success of benthic suspension feeders appears to be due to two main features of prey capture mechanisms. First, there is the low cost of prey capture, which is almost nil in passive suspension feeders, while it is below 5% in active ones (Larsen and Riisgård 1996). Second, there is the heterogeneity of their diet, mainly restricted by morphological constraints (Okamura 1990), which agrees with the hypothesis that prey selection cannot be a common phenomenon in organisms that depend on flow to bring resources to their capture structures (Hughes 1980). This trophic plasticity may represent an advantage because it might attenuate the effects of seasonal environmental fluctuations in availability of different resources in the water column.

Seston ergo particles: the role of the fine fraction

The major part of marine primary production is not consumed by herbivores (Fenchel 1988) but is mainly transferred to the detrital and microbial food web (Azam et al. 1983), the latter dominating pelagic plankton communities in terms of biomass (Wottom 1994) and production (Burkill et al. 1993). The small particles and cells, which dominate water-column communities, provide surface area for bacterial colonization (Fenchel and Jørgensen 1977), which increases their quality as a feeding resource and favours their utilization (Mann 1988).

Sessile predators must be adapted to capture the most abundant food items, i.e. detrital POC and live carbon in the forms of pico- and nanoplankton. Uptake of Dissolved Organic Matter (DOM) has been reported and discussed for a wide range of invertebrates (Stephens 1982; Frost 1987), but it is generally felt that DOM represents a small source of dietary requirements for marine invertebrates (Valiela 1995).

Live carbon is assimilated to a greater degree than detrital POC but the type of live carbon will affect its assimilation by suspension feeders (Wottom 1994). Plant particles colonized by microorganisms provide a better food value (Mann 1988), besides containing a high concentration of nitrogen. Feeding rates depend on the food quality and quantity; when both are high, lower gut retention time is required to extract nutritive products, and the opposite applies especially when the quality of food is poor (Taghon 1981). Another factor that could modify feeding rates is temperature, but animals that are adapted to very cold temperatures do not show a positive relationship with temperature increase (Anderson and Dyrssen 1989).

Filter feeders can modify the seston composition by several mechanisms: sponges may break up aggregates into small particles and/or introduce other particles and

▶

Fig. 2 Prey type and prey size in different suspension-feeder groups, modified after Riedl (1966), considering the recent findings on suspension feeders' natural diet, which increased the range of quality and quantity of prey both at small size (pico- and nanoplankton, continuous prey) and at large size (zooplankton and microplankton, occasional prey). *Shaded areas* and *black crosses* refer to information that has been added to Riedl's original figure. Potential for filtering water (*right side*) of each group is also shown

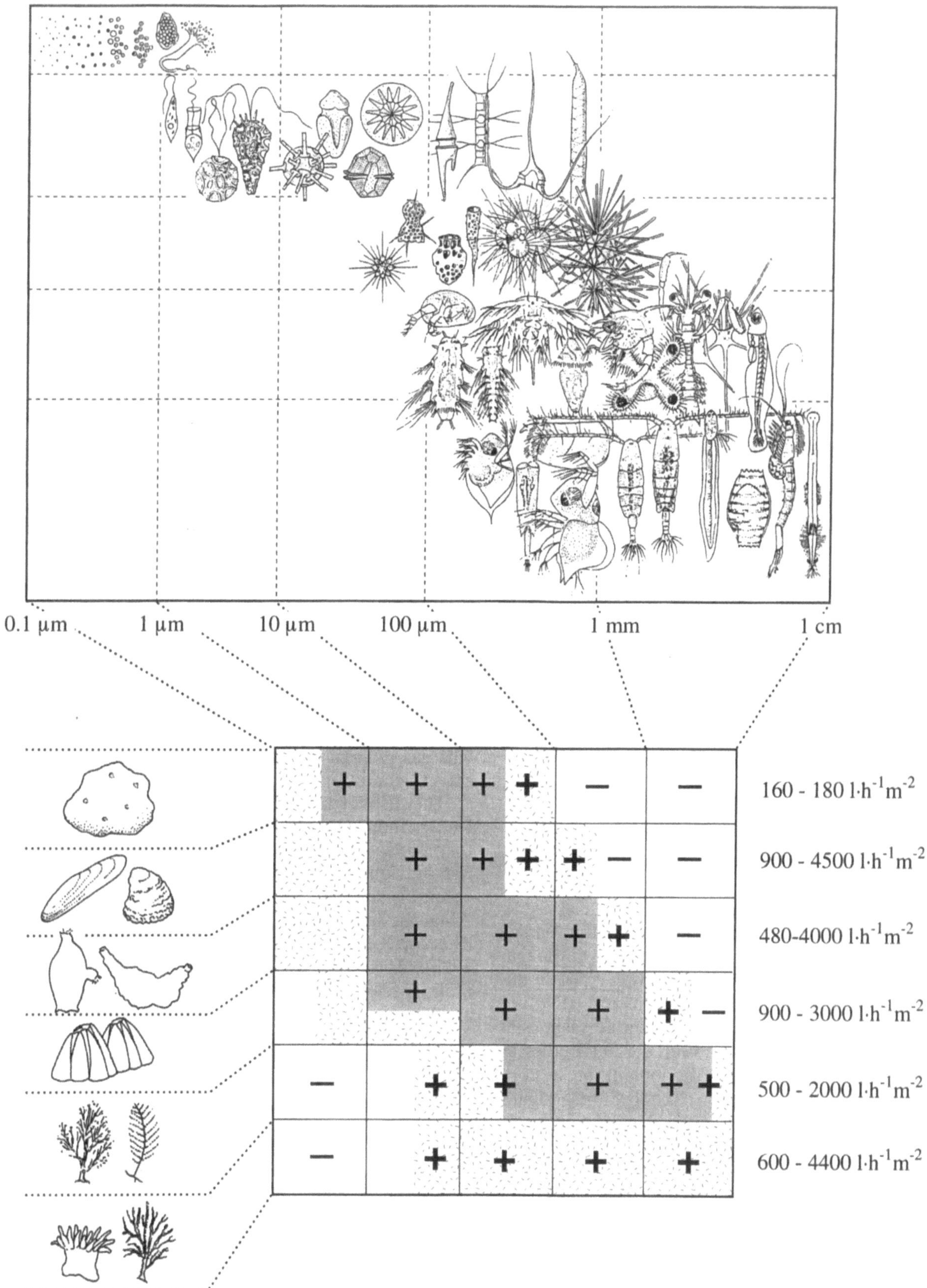

0.1 µm
1 µm
10 µm
100 µm
1 mm
1 cm
160 - 180 l·h⁻¹m⁻²
900 - 4500 l·h⁻¹m⁻²
480-4000 l·h⁻¹m⁻²
900 - 3000 l·h⁻¹m⁻²
500 - 2000 l·h⁻¹m⁻²
600 - 4400 l·h⁻¹m⁻²

substances generated by their feeding activity into the system (e.g. Witte et al. 1997); other organisms such as gorgonians decrease the velocity of the particle so they become available to other organisms. Many species aggregate small particles in the form of faecal pellets, which are an important food source for deposit feeders (e.g. Amouroux et al. 1990) or for near-bottom microbial production. The bivalve *Laternula elliptica* plays an important role in enhancing particle flux from the water column to the sea bed through biodeposition and may thus provide food for other benthic fauna, particularly in phytoplankton-impoverished near-shore waters (Ahn 1993).

Many benthic invertebrates have the capacity to feed on the fine fraction (i.e., pico-, nano-, microplankton and detrital POM) of the water column (Jørgensen et al. 1984). Studies on temperate bivalves (Fréchette et al. 1989; Riisgård 1998), tropical and temperate sponges (Pile et al. 1996) and on temperate and tropical octocorals (Ribes et al. 1999a) have shown, or suggest, a high grazing impact of benthic suspension feeders on near-bottom plankton. A recent study on the Antarctic hydroid *Oswaldella antarctica* shows a diet based on the fine fraction of seston and small zooplankton organisms (Orejas et al. 2001). The importance of the fine fraction in the diet of Antarctic suspension feeders is also demonstrated by the hydroid *Silicularia rosea* off King George Island (Gili et al. 1996), other species of sponges (Gaino et al. 1994) and brittle stars (Kellogg et al. 1982), which mainly ingest benthic diatoms and probably other detritus.

The activity of benthic suspension feeders also influences the nutrient concentration near the sea floor. High values of organic nutrients compared with their concentration in the water column are evidence for the role of benthic communities in organic matter remineralization (e.g. Davoult et al. 1990; Orejas et al. 2000).

The importance of large prey: zooplankton as an example

Capture of zooplankton prey items by benthic passive suspension feeders such as hydrozoan and anthozoan species has been demonstrated in many temperate and tropical regions (Sebens et al. 1996; Gili et al. 1998). Its contribution to the requirements of benthic suspension feeders is still a little-studied subject. Recent studies indicate the trophic significance of zooplankton in coral reef metabolism (Sorokin 1995), contributing significantly to the carbon and nutrient budget of some coral species (Dubinsky and Jokiel 1994).

Studies on the diet and feeding rate of reportedly macrophagous (mainly zooplanktivorous) suspension feeders (e.g. the hydrozoan *Campanularia everta* and the gorgonian *Paramuricea clavata*, Coma et al. 1998) have revealed that (1) the amount of zooplanktonic prey items captured could not cover the energy demand of the species, and (2) in terms of ingestion, detrital POM contributed about 50% to the diet. Mixed diets with frequent or occasional zooplankton prey capture are needed to satisfy energy needs of temperate benthic suspension feeders (Coma et al. 1995).

In the Antarctic, many zooplankton species undertake extended vertical migrations (i.e. more than 400 m), with salps, copepods or krill approaching the shelf bottom almost daily (Shulenberger et al. 1984). Moreover, several zooplankton species migrate to deep waters during austral winter to hibernate (Smith and Schnack-Schiel 1990). Recent gastrovascular cavity content analysis of Weddell Sea shelf benthic suspension feeders, such as the hydroid *Tubularia ralphii* and the anthozoans *Clavularia* cf *frankliniana* and *Anthomastus bathyproctus* (Orejas et al., in press), demonstrated significant feeding on zooplankton.

Ecological impact of suspension feeders in coastal and shelf ecosystems: the grazing impact

Benthic filter-feeding populations can have a pronounced grazing impact on the plankton in littoral systems (Kimmerer et al. 1994; Gili and Coma 1998). Without externally generated currents and mixing, however, only a thin layer of near-bottom water will be subject to "biomixing" by the action of inhalant and exhalant currents generated by the filter-feeders themselves. A number of models on the effects of filter feeders on phytoplankton concentration gradients (Nowell and Jumars 1994; Larsen and Riisgård 1996) and empirical evidence (Fréchette et al. 1989; Riisgård et al. 1996b) indicate that usually turbulence and resuspension generated by currents are the primary mixing forces, whereas biological mixing may be dominant under low current conditions.

The distinct trophic plasticity of benthic suspension feeders indicates that in littoral ecosystems their predatory impact should be higher than food selection theories previously assumed. The grazing impact of several species is summarized and compared with data on two Antarctic mollusc species, the bivalves *Adamussium colbecki* (Chiantore et al. 1998) and *L. elliptica* (Ahn 1993). According to Table 1, grazing impact of Antarctic species seems to be of the same order of magnitude as in temperate and tropical seas. Recent observations on littoral ascidians (Kowalke 1999), however, indicate filtration rates 1 order of magnitude less than those of non-Antarctic species.

Seasonality and physiological adaptations of benthic suspension feeders: a time-scale approach

The activity of benthic suspension feeders in temperate and polar seas is strongly seasonal, in contrast to most tropical environments (Harrison and Wallace 1990).

Table 1 Examples of predation impact of different benthic suspension-feeder groups on plankton communities from different ecosystems, which examined the effect of particular species (*Pot. food* potential food; *Cap. food* food captured, *GBR* Great Barrier Reef, Australia) (References: *1* Reiswig 1971; *2* Cloern 1982; *3* Jørgensen 1990; *4* Ahn 1993; *5* Kimmerer et al. 1994; *6* Coma et al. 1994; *7* Coma et al. 1995; *8* Riisgård and Larsen 1995; *9* Pile et al. 1996; *10* Gili et al. 1998; *11* Ribes et al. 1998; *12* Amouroux et al. 1990; *13* Chiantore et al. 1998; *14* Buss and Jackson 1981; *15* Linley and Koop 1986; *16* Yahel et al. 1998; *17* Ribes et al. 1999a)

Site	Level	Ecosystem	Species	Abundance	Current	Pot. Food	Cap. Food	Impact	References
Discovery Bay (Jamaica)	Species	Tropical	Sponges	5–210 cm^{-3} m^{-2}	–	Bacteria, unarmored cells, armored cells, detritus, URPOC	Bacteria, unarmored cells, armored cells, detritus, URPOC	80–1800 mg C m^{-2} day^{-1}	1
San Francisco Bay (USA)	Species	Temperate, estuarine	Bivalves	170–2700 ind. m^{-2}	–	Phytoplankton	Phytoplankton	Daily filtration of 1.2–1.8 times the South Bay volume	2
Different areas	Species	–	Bivalves	–	–	Phytoplankton	Phytoplankton	>100 m^{-3} m^{-2} day^{-1}	3
Maxwell Bay (Antarctica)	Species	Antarctic	Bivalves	36–76 ind. m^{-2}	–	Seston, Chlorophyll a	Seston, Chlorophyll a	Biodeposition of 95 mg C m^{-2} day^{-1}	4
SanFrancisco Bay (USA)	Species	Temperate	Bivalves	3500 ind. m^{-2}	–	Chorophyll, nauplii	Chorophyll, nauplii	Chorophyll: 62% daily depletion, nauplii: 8.5% daily depletion	5
Medes Islands (NW Mediterranean)	Species	Temperate	Gorgonian	32 col. m^{-2}	–	Zooplankton	Zooplankton	12.2–85.2 mg C m^{-2} day^{-1}	6
Medes Islands (NW Mediterranean)	Species	Temperate	Hydroid	21235 polyps m^{-2}	–	Zooplankton, detrital POC	Zooplankton, detrital POC	Zooplankton: 2.2 mg C m^{-2} day^{-1}; detrital POC: 1.9 mg C m^{-2} day^{-1}	7
Kertinge Nor (Denmark)	Species	Temperate, estuarine	Ascidian	20–277 ind. m^{-2}	–	Phytoplankton	Phytoplankton	Daily filtration of 0.2–1.2 times the Kertinge Nor volume	8
Gulf of Maine (USA)	Species	Temperate	Sponge	8.9% of surface, 7.6 oscula m^{-2}	–	Picoplankton, aut. eucaryotes	Picoplankton, aut. eucaryotes	29 mg C m^{-2} day^{-1}	9
Different areas	Species	Temperate, tropical, Antarctic	Hydroids	–	–	Zooplankton, detrital POC	Zooplankton, detrital POC	6–225 mg C m^{-2} day^{-1}	10
San Blas (Panamá)	Species	Tropical	Gorgonian	0.45 col. m^{-2}	1.2 cm s^{-1}	Picoplankton, microplankton, zooplankton	Microplankton, zooplankton	Microplankton: 10% daily depletion; zooplankton: 0.09–0.23 prey polyp day^{-1}	11
Banyuls-sur-mer (France)	Species	Temperate	Polychaete	Up to 3000 ind. m^{-2}	–	Picoucariotes, cyanophytes	Picoucariotes, cyanophytes	0.001–0.06 µg C $ind.^{-1}$ h^{-1}	12
Terranova Bay (Antarctica)	Species	Antarctic	Bivalve	50–60 ind. m^{-2}	–	POC, DOC, chlorophyll a	POC, DOC, chlorophyll a	Biodeposition of 21 mg C m^{-2} day^{-1}	13
Discovery Bay (Jamaica)	Community	Tropical	Bryozoans	6.2% bryozoans, 74.2% sponges	–	Bacteria, naked cells	Bacteria, naked cells	Bacteria: 89.2% depletion, naked cells: 79.4% depletion	14
One tree Island (GBR)	Community	Tropical	Reef	–	0–20 cm s^{-1}	Bacteria	Bacteria	Bacteria: 37.3%	15

Gulf of Aqaba (Red Sea)	Community	Tropical	Reef	40% live cover	–	Ultraphytoplankton, phytoplankton	Ultraphytoplankton, phytoplankton	Chorophyll a: 15–65% depletion	16
Medes Islands (NW Mediterranean)	Species	Temperate	Gorgonian	32 col. m^{-2}	1.2 cm s^{-1}	DOC, detrital POC, picoplankton, nanoplankton, microplankton	Detrital POC, nanoplankton, microplankton	Detrital POC: 2–10% daily depletion, nanoeucaryotes: 1–9% daily depl., dinoflagellates: 1–26% daily depl., diatoms: 1–22% daily depl., ciliates: 2–99% daily depl.	17

Similar seasonal activity patterns, e.g. feeding-related polyp expansion in anthozoans (Robbins and Shick 1980), have been observed in temperate (Coma et al. 2000) and cold-water (Barnes and Clarke 1995) suspension-feeder populations.

Seasonality in activity is either coupled to water temperature (Giese and Pearse 1974) or to trophic constraints (Clarke 1980). There is little direct experimental evidence for temperature control, however there is increasing evidence for trophic control of life-cycles in benthic suspension feeders inhabiting cold (Clarke 1991) and temperate (Hughes 1989; Coma et al. 2000) environments.

Life in a hydrodynamically active environment: hydrodynamics as a stochastic mechanism of food supply

Physical-biological coupling

Prey availability in the water column is dependent largely on hydrographic conditions. For benthic suspension feeders, input of suspended food is closely related to flow intensity and periodicity (Wildish and Kristmanson 1997). Therefore, water movement plays a fundamental role in determining diversity, biomass, structure and distribution of suspension-feeder communities (Fig. 3). Near-bottom currents enhanced by topographic features and internal waves create a hydrodynamical environment which favours particle sedimentation at the slope and near it (Frederiksen et al. 1992). Substratum heterogeneity and biogenic structures interfere with the flow pattern, increasing its turbulence, which enhances particle capture by benthic suspension feeders (Witte et al. 1997). These environmental conditions favour the development of dense populations of suspension feeders (Rice et al. 1990).

There are several examples of a wide distribution and dominance of benthic suspension feeders throughout continental shelves. The pennatulacean *Pennatula aculeata* represents the major component of benthic macrofauna of the Gulf of Maine (Langton et al. 1990). The ophiuroid *Ophiothrix fragilis* dominates the sessile epifaunal pebble community of the English Channel, with patches of over 200 ind. m^{-2} (Davoult et al. 1990). Banks of the coral *Lophelia pertusa* associated mainly with a rich sponge and bryozoan fauna are frequent throughout the continental shelf and slope in the north-eastern Atlantic (Jensen and Frederiksen 1992). Other examples of dense benthic suspension-feeder assemblages present in the shelf-slope region, and therefore subject to similar conditions of considerable water motion and abundant suspended material, are the high concentrations of hexactinellid sponges in the North Atlantic (Rice et al. 1990), the coral banks with associated brittle stars and crinoids in Florida (Neumann et al. 1977), and the antipatharians on top of the vertical walls of seamounts (Kaufmann et al. 1989). All these dense assemblages live in areas of considerable water movement and abundant

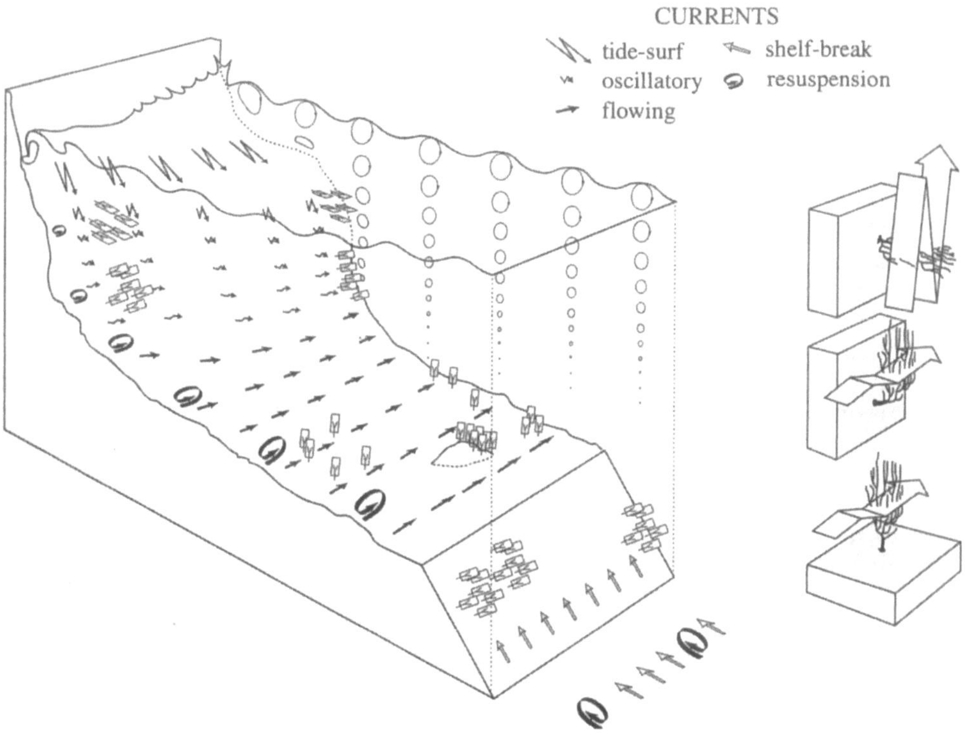

Fig. 3 Bottom-water currents along the continental shelf and slope, composed of oscillatory, longitudinal and lateral coast currents including resuspension processes (*black circular arrows* in the general view, and *white arrows* in the detailed view) (modified from Riedl 1966). Diagrams show the orientation of suspension-feeder colonies and where development of dense suspension feeder-communities is likely

suspended material. We can conclude that small- and medium-scale environmental variability, particularly hydrodynamics, is conducive to the growth of dense populations of benthic suspension feeders (e.g. Eckman and Duggins 1993; Josefson et al. 1993). Accordingly, the diverse and dense benthic suspension-feeder communities found along the eastern Weddell Sea shelf and slope (Gutt and Starmans 1998), where a structure of large sponges and bryozoans enhances the presence of other groups, appear to coincide with the hydrodynamically more active regions of this area.

The Antarctic shelf is unique regarding its depth, averaging about 500 m, with troughs to over 1,000 m (Dayton 1990). High-Antarctic benthic habitats share relatively constant physical parameters such as temperature, salinity and substrata, with few if any important barriers. The major disruptive effects on benthic suspension-feeding communities in Antarctic shallow water are abrasion by ice (Gutt et al. 1996), which replaces the wave and hydrodynamic effects in other areas, and plucking by anchor ice (Dayton 1990). In deeper waters, iceberg scouring has marked effects on the Antarctic shelf benthos (e.g. Gutt et al. 1996). Community destruction is followed by a seemingly long process of recolonization. During the second EASIZ cruise, in iceberg scour marks with early stages of recolonization, two gorgonian species, *Primnoisis antarctica* and *Ainigmaptilon antarcticus*, and bryozoans of the genus *Cellaria* appeared to be the main pioneer species (i.e. species of the first stages of recolonization after a disturbance).

The mesoscale pattern of currents in the eastern Weddell Sea influences the pattern of benthic community distribution on the shelf. Off Kapp Norvegia, diverse and dense communities are dominated by suspension feeders such as sponges and bryozoans. In contrast, in the southern part of the eastern Weddell Sea shelf area, low-biomass and low-diversity benthic communities are present, which are dominated by deposit feeders such as holothurians (Gerdes et al. 1992). Vertical particle flux off Kapp Norvegia is one of the most active in Antarctica (Bathmann et al. 1991); however, no data are available for the southern shelf off Filchner. All information suggests a picture similar to that described for the McMurdo Sound area (Dunbar et al. 1989). Off Kapp Norvegia, high water exchange, partial water mass retention due to the topographic effect of a shelf rise, and strong vertical particle flux may explain the presence of dense and diverse benthic communities dominated by suspension feeders, whereas in the southernmost part, less water renewal, additionally impoverished as it comes from underneath the Filchner Shelf, favours the dominance of deposit feeders.

Life in a boundary system: resuspension as a key process

The importance of near-bottom currents and salinity and/or temperature stratification is of crucial importance for the coupling of benthic suspension feeders to the pelagic system (e.g. Larsen and Riisgård 1996).

Generally, particles and bacteria show high abundance and a distinct small-scale vertical distribution in the benthic boundary layer (Ritzrau et al. 1997). Organic and inorganic particle fractions are sorted by resuspension (Thomsen 1999), because the organic fraction has low settling and high residence times and may form aggregates before settling or biodeposition (Thomsen and van Weering 1998). High microbial activity is associated with these particles and plays a key role in the decomposition of the particulate organic matter (Ritzrau and Thomsen 1997). Enhanced particle residence time in the water column and the formation of aggregates facilitate food availability to benthic filter feeders.

Sediment resuspension is a common phenomenon in near-shore environments surrounding Antarctica (e.g. Klöser et al. 1994), being particularly relevant during the austral winter. In shallow areas, detritus resuspended by wind-generated waves, anchor ice and currents may be a viable food source for benthic invertebrates during this season. In Maxwell Bay and Marion Cove (South Shetland Islands), the most important source of particulate matter in the water column seems to be resuspension of benthic material through a 1-year cycle (Kang et al. 1997). These authors observed two main peaks of chlorophyll, one in summer, dominated by microplankton (mainly large diatoms) and another one later and longer, during autumn and winter, dominated by pico- and nanoplankton (mainly flagellates). Obviously, organic suspended particles, including benthic diatoms resuspended by tidal currents (Ahn et al. 1997), are available practically year round for benthic suspension feeders.

Dunbar et al. (1989) found higher organic flux through the water column in the eastern compared to the western McMurdo Sound, which coincides well with the high biomass of benthic suspension feeders in the former area (Dayton et al. 1986). The high carbon accumulation in McMurdo Sound is enhanced, not only by resuspension, but also by low water temperatures, which reduce rates of organic-matter breakdown by bacteria (Hodson et al. 1981). Both the current view of water circulation and the fact that resuspension events are more common during austral winter support the hypothesis that benthic suspension feeders receive food the year round, especially in areas where they develop dense populations.

The bivalve *L. elliptica* is one of the few examples in the Antarctic where benthic suspension feeders play an important role in enhancing particle flux from the water column to the sea bed through biodeposition and, possibly, providing food for other benthic fauna, particularly in phytoplankton-impoverished near-shore waters (Ahn 1993).

Vertical flux versus horizontal transport: the role of advection

Processes related to organic decomposition in the benthic boundary layer are affected by both resuspension and advection (Jähmlich et al. 1998). Thus particles and aggregates may be transported by near-bottom currents, and recent studies in the Barents Sea (Thomsen and van Weering 1998) indicate that lateral particle fluxes are 2–3.7 times higher than vertical particle fluxes derived from sediment trap data. Vertical flux does not always supply sufficient food to benthic animals, but horizontal advection together with resuspension of material from the bottom can result in higher amounts of benthos-derived material relatively close to the substratum (Johnson 1988).

A close relationship between benthic metabolism and the pulses of vertical flux of organic matter has been observed both in shallow and deep areas (e.g. Witte et al. 1997). Near-bottom currents, however, may resuspend, spread and transport previously settled particles laterally over wide distances. These transport mechanisms enhance particle deposition on continental margins or where topography facilitates current slowdown. Not only light particles may be transported but also organisms such as foraminiferans (Brunner and Biscaye 1997).

In the Antarctic, the high variability of ice cover and its effects on hydrographic processes suggest that other mechanisms besides vertical flux play an important role controlling benthic biomass and diversity (Arntz et al. 1994). Hence, lateral advection may be more relevant than previously thought (Grebmeier and Barry 1991). Benthic organisms living under extended seasonal or even permanent ice cover rely on organic matter transported by resuspension and/or lateral advection (Fig. 4). For example, common littoral benthic diatoms such as *Navicula glaciei* have been observed in sediment traps located near the bottom (1,000 m) during winter in the Bransfield Strait (Palanques et al., in press). Internal waves associated with tidal currents may be able to advect part of the spring bloom, as well as macroalgal detritus, from the shelf areas to surrounding depths.

The role of Antarctic benthic suspension feeders: an efficient system

For Antarctic waters, a poor, flagellate-dominated pelagic community ("regenerating system" according to Eppley and Peterson 1979) is more characteristic than blooms of large phytoplankton species ("upwelling systems" according to Scharek and Nöthig 1995), which only occur in restricted areas during restricted times (Heywood and Priddle 1987). Have Antarctic suspension feeders adapted to the food resources available in a similar way to their temperate counterparts (e.g. Gili and Coma 1998)? Based on the first results obtained during the EASIZ cruises (Orejas et al. 2001), we suggest that Antarctic suspension feeders exhibit species with a high renewal rate, a high ingestion rate of microplankton and low maintenance energy requirements and, species with low ingestion, and low renewal and growth rates. The first group of species is associated with the diatom-based plankton blooms occurring in the

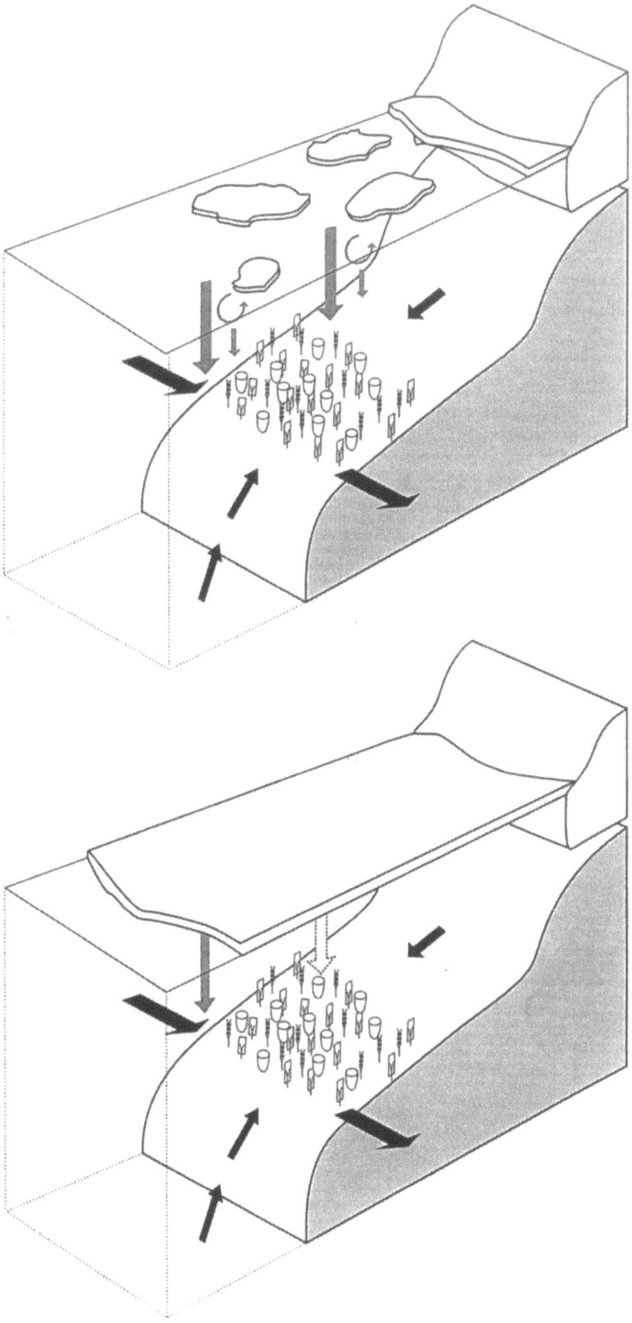

Fig. 4 Idealized diagram of food-supply processes to the Antarctic benthos in summer (*top*) and winter (*bottom*). *Grey arrows* mean microplankton and faecal pellet (summer) and pico- and nanoplankton (winter) vertical flux, *white arrow* zooplankton vertical migration (winter), and *black arrows* lateral (from the shelf, slope and shallow areas) water flow (all year)

upwelling systems. The second group is linked to the regenerating system, which is very stable throughout the whole year. Feeding strategies of sessile species belonging to the latter group will be affected significantly by sediment type and resuspension processes.

At the level of individuals or colonies, Antarctic benthic suspension feeders are efficient in energy recycling. This conclusion may be extended to the community and ecosystem level, thus leading to a new hypothesis regarding the success of these communities in the Southern Ocean: the heterogenous hydrodynamic environment, lateral particle advection and transport over long distances, continuous feeding on organic matter derived from the water column or made available by resuspension, and the ability to feed on almost all potential food lead to high recycling efficiency at the community level.

Are Antarctic benthic suspension feeders different? Conclusions

Despite low primary production, the Southern Ocean does not appear to be a food-limited system for benthic suspension feeders because of the nutritive quality of food, patch feeding (e.g. on zooplankton prey) and the role of lateral advection and resuspension. Studies to date suggest that the feeding strategy of Antarctic benthic suspension feeders does not differ from those of tropical and temperate ecosystems. Suspension feeders are able to feed on a wide range of prey items, in particular on the most abundant items, because of their capability of ingesting prey items of very different kind and size. Being remarkably efficient organisms in terms of energy transfer from the pelagial to the benthic system, benthic suspension feeders build rich but patchy communities under Antarctic conditions. As rich communities of suspension feeders may exert an important predatory impact on plankton populations and on the abundance of suspended organic matter in the water mass adjacent to the bottom, an impact at a similar order of magnitude has to be assumed for Antarctic and temperate regions.

Acknowledgements We would like to especially thank Wolf Arntz, Tom Brey, Andrew Clarke and Jean Bouillon for their constructive comments, which helped to improve the manuscript. We also thank the officers and crew of RV "Polarstern" and many colleagues for their help aboard. The programme related to Ecology of Antarctic Suspension Feeders, a cooperation between the Alfred Wegener Institute and several Spanish research institutions, started in 1998, and forms a contribution to SCAR-EASIZ (Ecology of the Antarctic Sea Ice Zone). Support for this work was provided by CICYT (Spanish Antarctic Research Programme) grants ANT97-1533-E, ANT98-1738-E and by PhD fellowships from DAAD (A/96/13073) and the European Commission (TMR-CT97-2813) to C.O. and a RED contract from the "Generalitat de Catalunya" to R.C.

References

Ahn I-Y(1993) Enhanced particle flux through the biodeposition by the Antarctic suspension-feeding bivalve *Laternula elliptica* in Marian Cove, King George Island. J Exp Mar Biol Ecol 171:75–90

Ahn I-Y, Chung H, Kang J-S, Kang S-H (1997) Diatom composition and biomass variability in nearshore waters of Maxwell Bay, Antarctica, during the 1992/1993 austral summer. Polar Biol 17:123–130

Amouroux JM, Amouroux J, Bastide J, Cahet G (1990) Interrelations in a microcosm with a suspension-feeder and a deposit-feeder. I. Experimental study. Oceanol Acta 13:61–68

Anderson L, Dyrssen D (1989) Chemical oceanography of the Arctic ocean. In: Herman Y (ed) The Arctic seas: climatology, oceanography, geology and biology. Van Nostrand Reinhold, New York, pp 93–114

Anthony KRN (1997) Prey capture by the sea anemone *Metridium senile* (L.): effects of body size, flow regime, and upstream neighbors. Biol Bull 192:73–86

Arntz WE, Brey T, Gallardo VA (1994) Antarctic zoobenthos. Oceanogr Mar Biol Annu Rev 32:241–304

Azam F, Fenchel T, Field JG, Gray JS, Meyer-Reil LA, Thingstad TF (1983) The ecological role of water-column microbes in the sea. Mar Ecol Prog Ser 10:257–263

Barnes DKA, Clarke A (1995) Seasonality of feeding activity in Antarctic suspension feeders. Polar Biol 15:335–340

Bathmann U, Fischer G, Müller PJ, Gerdes D (1991) Short-term variations in particulate matter sedimentation off Kapp Norvegia, Weddell Sea, Antarctica: relation to water mass advection, ice cover, plankton biomass and feeding activity. Polar Biol 11:185–195

Best BA, Thorpe MA (1986) Feeding-current interactions and competition for food among the bryozoan epiphytes of *Fucus serratus*. Mar Biol 93:371–375

Brey T, Gerdes D (1997) Is Antarctic benthic biomass really higher than elsewhere? Antarct Sci 9:266–267

Brunner CA, Biscaye PE (1997) Storm-driven transport of foraminifers from the shelf to the upper slope, southern Middle Atlantic Bight. Cont Shelf Res 17:491–508

Burkill PH, Mantoura RFC, Owens NJP (1993) Biogeochemical cycling in the northwestern Indian Ocean: a brief overview. Deep Sea Res II 40:643–649

Buss LW, Jackson JBC (1981) Planktonic food availability and suspension-feeder abundance: evidence of in situ depletion. J Exp Mar Biol Ecol 49:151–161

Chiantore M, Cattaneo-Vietti R, Albertelli G, Misic C, Fabiano M (1998) Role of *Adamussium colbecki* in circulation of organic matter in Terra Nova Bay (Ross Sea, Antarctica). J Mar Syst 17:411–424

Clarke A (1980) A reappraisal of the concept of metabolic cold adaptation in polar marine invertebrates. Biol J Linn Soc 14:77–92

Clarke A (1991) What is cold adaptation and how should we measure it? Am Zool 31:81–92

Cloern JE (1982) Does the benthos control phytoplankton biomass in south San Francisco Bay? Mar Ecol Prog Ser 9:191–202

Coma R, Gili JM, Zabala M, Riera T (1994) Feeding and prey capture cycles in the aposymbiotic gorgonian *Paramuricea clavata*. Mar Ecol Prog Ser 115:257–270

Coma R, Gili JM, Zabala M (1995) Trophic ecology of a marine benthic hydroid *Campanularia everta*. Mar Ecol Prog Ser 119:211–220

Coma R, Ribes M, Gili JM, Zabala M (1998) An energetic approach to the study of life history traits of two modular colonial benthic invertebrates. Mar Ecol Prog Ser 162:89–103

Coma R, Ribes M, Gili JM, Zabala M (2000) Seasonality in coastal benthic ecosystems. Trends Ecol Evol 15:448–453

Davoult D, Gounin F, Richard A (1990) Dynamique et reproduction de la population d'*Ophiothrix fragilis* (Abildgaard) du détroit du Pas-de Calais (Manche orientale). J Exp Mar Biol Ecol 138:201–216

Dayton PK (1990) Polar benthos. In: Smith WO (ed) Polar oceanography. Part B: chemistry, biology and geology. Academic Press, San Diego, pp 631–685

Dayton PK, Watson D, Palmisano A, Barry JP, Oliver JS, Rivera D (1986) Distribution patterns of benthic standing stock at McMurdo Sound, Antarctica. Polar Biol 6:207–213

Dubinsky Z, Jokiel PL (1994) Ratio of energy and nutrient fluxes regulates symbiosis between zooxanthellae and corals. Pac Sci 48:313–324

Dunbar RB, Leventer AR, Stockton WL (1989) Biogenic sedimentation in McMurdo Sound, Antarctica. Symposium on Glaciomarine Environments, INQUA XII International Congress. Mar Geol 85:155–179

Eckman JE, Duggins DO (1993) Effects of flow speed on growth of benthic suspension feeders. Biol Bull 185:28–41

Epply RW, Peterson BJ (1979) Particulate organic matter flux and planktonic new production in the deep ocean. Nature 282:667–680

Fabricius KE, Genin A, Benayahu Y (1995) Flow-dependent herbivory and growth in zooxanthellae-free soft corals. Limnol Oceanogr 40:1290–1301

Fenchel TM (1988) Marine plankton food chains. Annu Rev Ecol Syst 19:19–38

Fenchel TM, Jørgensen BB (1977) Detritus food chains of aquatic ecosystems: the role of bacteria. In: Alexander M (ed) Advances in microbial ecology. Plenum Press, London, pp 1–58

Fréchette M, Butman CA, Geyer WR (1989) The importance of boundary-layer flows in supplying phytoplankton to the benthic suspension feeder, *Mytilus edulis* L. Limnol Oceanogr 34:19–36

Frederiksen R, Jensen A, Westerberg H (1992) The distribution of the scleractinian coral *Lophelia pertusa* around the Faroe Islands and the relation to internal tidal mixing. Sarsia 77:157–171

Frost TM (1987) Porifera. In: Pandian TJ, Vernberg FJ (eds) Animal energetics. Academic Press, San Diego, pp 27–53

Gaino E, Bavestrello G, Cattaneo-Vietti R, Sarà M (1994) Scanning electron microscope evidence for diatom uptake by two Antarctic sponges. Polar Biol 14:55–58

Gerdes D, Klages M, Arntz WE, Herman RL, Galéron J, Hain S (1992) Quantitative investigation on macrobenthos communities of the southern Weddell Sea shelf based on multibox corer samples. Polar Biol 12:291–301

Giese AC, Pearse JS (1974) Reproduction of marine invertebrates: acoelomate and pseudocoelomate metazoans. Academic Press, New York

Gili JM, Coma R (1998) Benthic suspension feeders: their paramount role in littoral marine food webs. Trends Ecol Evol 13:316–321

Gili JM, Alvà V, Pagés F, Kloser H, Arntz WE (1996) Benthic diatoms as the principal food source in the sub-Antarctic marine hydroid *Silicularia rosea*. Polar Biol 16:507–512

Gili JM, Alvà V, Coma R, Orejas C, Pagès F, Ribes M, Zabala M, Arntz W, Bouillon J, Boero F, Hughes RG (1998) The impact of small benthic passive suspension feeders in shallow marine ecosystems: the hydroids as an example. Zool Verh Rijksmus Nat Hist Leiden 323:1–7

Grebmeier JM, Barry JP (1991) The influence of oceanographic processes on pelagic-benthic coupling in polar regions: a benthic perspective. J Mar Syst 2:495–518

Gutt J, Starmans A (1998) Structure and biodiversity of megabenthos in the Weddell and Lazarev Seas (Antarctic): ecological role of physical parameters and biological interactions. Polar Biol 20:229–247

Gutt J, Starmans A, Dieckmann G (1996) Impact of iceberg scouring on polar benthic habitats. Mar Ecol Prog Ser 137:311–316

Harrison PL, Wallace CC (1990) Reproduction, dispersal and recruitment of sleractinian corals. In: Dubinsky Z (ed) Ecosystems of the world: coral reef. Elsevier Science, New York, pp 133–207

Harvell CD, LaBarbera M (1985) Flexibility: a mechanism for control of local velocities in hydroid colonies. Biol Bull 168:312–320

Heywood RB, Priddle J (1987) Retention of phytoplankton by an eddy. Continental Shelf Res 7:937–955

Hodson RE, Azam F, Carlucci A, Fuhrman JA, Karl DM, Holm-Hansen O (1981) Microbial uptake of dissolved organic matter in McMurdo Sound, Antarctica. Mar Biol 61:185–209

Hughes RN (1980) Optimal foraging theory in the marine context. Oceanogr Mar Biol Annu Rev 18:423–481

Hughes RN (1989) A functional biology of clonal animals. Chapman & Hall, London

Jähmlich S, Thomsen L, Graf G (1998) Factors controlling aggregate formation in the benthic boundary layer of the Mecklenburg Bight (western Baltic Sea). J Sea Res 41:245–254

Jensen A, Frederiksen R (1992) The fauna associated with the bank-forming deepwater coral *Lophelia pertusa* (Scleractiniaria) on the Faroe shelf. Sarsia 77:53–69

Johnson AS (1988) Hydrodynamic study of the functional morphology of the benthic suspension feeder *Phoronopsis viridis* (Phoronida). Mar Biol 100:117–126
Jørgensen CB (1990) Bivalve filter feeding: hydrodynamics, bioenergetics, physiology and ecology. Olsen and Olsen, Fredensborg
Jørgensen CB, Kiørboe T, Møhlenberg F, Riisgaard HU (1984) Ciliary and mucus net filter feeding with special reference to fluid mechanical characteristics. Mar Ecol Prog Ser 15:283–292
Josefson AB, Jensen JN, Aertebjerg G (1993) The benthos community structure anomaly in the late 1970s and early 1980s – a result of a major food pulse? J Exp Mar Biol Ecol 172:31–45
Kang J-S, Kang S-H, Lee JH, Chung KH, Lee M-Y (1997) Antarctic micro- and nano-sized phytoplankton assemblages in the surface water of Maxwell Bay during the 1997 Austral summer. Korean J Polar Res 8:35–45
Kaufmann RS, Wakefield WW, Genin A (1989) Distribution of epibenthic megafauna and lebensspuren on two central North Pacific seamounts. Deep Sea Res 36:1863–1896
Kellogg D, Kellog TB, Dearborn JH, Edwards KC, Fratt DB (1982) Diatoms from brittle star stomach contents: implications for sediment reworking. Antarct J US 17:167–169
Kimmerer WJ, Gartside E, Orsi JJ (1994) Predation by an introduced clam as the likely cause of substantial declines in zooplankton of San Francisco Bay. Mar Ecol Prog Ser 113:81–93
Klöser H, Mercuri G, Laturnus F, Quartino ML, Wiencke C (1994) On the competitive balance of macroalgae at Potter Cove (King George Island, South Shetlands). Polar Biol 14:11–16
Klumpp DW (1984) Nutritional ecology of the ascidian *Pyura stolonifera*: influence of body size, food quantity and quality on filter-feeding, respiration, assimilation efficiency and energetic balance. Mar Ecol Prog Ser 19:269–284
Kowalke J (1999) Filtration in Antartic ascidians – striking a balance. J Exp Mar Biol Ecol 242:233–244
Langton RW, Langton EW, Theroux RB, Uzmann JR (1990) Distribution, behavior and abundance of sea pens, *Pennatula aculeata*, in the Gulf of Maine. Mar Biol 107:463–469
Larsen PS, Riisgård HU (1996) Biomixing generated by benthic filter-feeders: a diffusion model for near-bottom phytoplankton depletion. J Sea Res 37:81–90
Linley LAS, Koop K (1986) Significance of pelagic bacteria as a trophic resource in a coral reef lagoon, One Tree Island, Great Barrier Reef. Mar Biol 92:457–464
MacFadden CS (1986) Colony fission increases particle capture rates of a soft coral: advantages of being a small colony. J Exp Mar Biol Ecol 103:1–20
Mann KH (1988) Production and use of detritus in various freshwater, estuarine and coastal marine ecosystems. Limnol Oceanogr 33:910–930
Neumann AC, Kofoed JW, Keller GH (1977) Lithoherms in the straits of Florida. Geology 5:4–10
Nowell ARM, Jumars PA (1994) Flume experiments on food supply to the blue mussel *Mytilus edulis* L. as a function of boundary-layer flow. Limnol Oceanogr 39:1755–1768
Okamura B (1990) Behavioural plasticity in the suspension feeding of benthic animals, In: Hughes RN (ed) Behavioural mechanisms of food selection. Springer, Berlin Heidelberg New York, pp 637–660
Orejas C, Gili JM, López-González P, Arntz WE (2001) Feeding strategies and diet composition of four Antarctic cnidarian species. Polar Biol 24:620–627
Orejas C, Gili JM, Arntz WE, Ros JD, López P, Teixidó N, Filipe P (2000) Benthic suspension feeders, key players in Antarctic marine ecosystems. Contrib Sci Barcelona 1:299–311
Palanques A, Isla E, Puig P, Sánchez-Cabeza JA, Masqué P (2002) Annual evolution of settling particle fluxes in Western Bransfield Strait (Antarctica) during the FRUELA project. Deep Sea Res II 49:903–920
Pile AJ, Patterson MR, Witman JD (1996) In situ grazing on plankton < 10 μm by the boreal sponge *Mycale lingua*. Mar Ecol Prog Ser 141:95–102
Pullen J, LaBarbera M (1991) Modes of feeding in aggregations of barnacles and the shape of aggregations. Biol Bull 181:442–452
Reiswig HM (1971) Particle feeding in natural populations of three marine demosponges. Biol Bull 141:568–591
Ribes M, Coma R, Gili JM (1998) Seasonal variation of in situ feeding rates by the temperate ascidian *Halocynthia papillosa*. Mar Ecol Prog Ser 175:201–213
Ribes M, Coma R, Gili JM (1999a) Heterogeneous feeding in benthic suspension feeders: the natural diet and grazing rate of the temperate gorgonian *Paramuricea clavata* (Cnidaria: Octocorallia) over a year cycle. Mar Ecol Prog Ser 183:125–137
Ribes M, Coma R, Gili JM (1999b) Natural diet and grazing rate of the temperate sponge *Dysidea avara* (Demospongiae, Dendroceratida) throughout an annual cycle. Mar Ecol Prog Ser 176:179–190
Rice AL, Thurston MH, New AL (1990) Dense aggregations of a hexactinellid sponge, *Pheronema carpenteri* in the Porcupine Seabight (Northeast Atlantic Ocean) and possible causes. Progr Oceanogr 24:197–206
Riedl R (1966) Biologie der Meereshöhlen. Parey, Hamburg
Riisgård HU (1998) Filter feeding and plankton dynamics in a Danish fjord: a review of the importance of flow, mixing and density-driven circulation. J Environ Manage 53:195–207
Riisgård HU, Larsen PS (1995) Filter-feeding in marine macroinvertebrates: pump characteristics, modelling and energy cost. Biol Rev 70:67–106
Riisgård HU, Poulsen L, Larsen PS (1996a) Phytoplankton reduction in near-bottom water caused by filter-feeding *Nereis diversicolor* – implications for worm growth and population grazing impact. Mar Ecol Prog Ser 141:47–54
Riisgård HU, Jürgensen C, Clausen T (1996b) Filter-feeding ascidians (*Ciona intestinalis*) in a shallow cove: implications of hydrodynamics for grazing impact. J Sea Res 35:293–300
Ritzrau W, Thomsen L (1997) Spatial distribution of particle composition and microbial activity in the benthic boundary layer (BBL) of the Northeast Water Polynya. J Mar Syst 10:415–428
Ritzrau W, Thomsen L, Lara RJ, Graf G (1997) Enhanced microbial utilisation of dissolved amino acids indicates rapid modification of organic matter in the benthic boundary layer. Mar Ecol Prog Ser 156:43–50
Robbins RE, Shick JM (1980) Expansion-contraction behavior in the sea-anemone *Metridium senile*: environmental cues and energetic consequences. In: Smith DC, Tiffon Y (eds) Nutrition in the lower Metazoa. Pergamon, New York, pp 101–116
Rutman J, Fishelson L (1969) Food composition and feeding behavior of shallow-water crinoids at Eilat (Red Sea). Mar Biol 3:46–57
Scharek R, Nöthig EV (1995) Das einzellige Plankton im Ozean der Arktis und Antarktis. In: Hempel G (ed) Biologie der Polarmeere. Fischer, Jena, pp 116–127
Sebens KP, Koehl MAR (1984) Predation on zooplankton by the benthic anthozoans *Alcyonium siderium* (Alcyonacea) and *Metridium senile* (Actiniaria) in the New England subtidal. Mar Biol 81:255–271
Sebens KP, Vandersall KS, Savina LA, Graham KR (1996) Zooplankton capture by two scleractinian corals, *Madracis mirabilis* and *Monastrea cavernosa*, in a field enclosure. Mar Biol 127:303–317
Shimeta J, Jumars PA (1991) Physical mechanisms and rates of particle capture by suspension feeders. Oceanogr Mar Biol Annu Rev 29:191–257
Shulenberger E, Wormuth JH, Loeb VJ (1984) A large swarm of *Euphausia superba*: overview of patch structure and composition. J Crust Biol 4 (spec 1):75–95
Slattery M, McClintock JB, Bowser SS (1997) Deposit feeding: a novel model of nutrition in the Antarctic colonial soft coral *Germesia antarctica*. Mar Ecol Prog Ser 149:299–304
Smith SL, Schnack-Schiel SB (1990) Polar zooplankton. In: Smith WO Jr (ed) Polar oceanography. Part B: chemistry, biology, and ecology. Academic Press, San Diego, pp 251–263
Sorokin YI (1991) Biomass, metabolic rates and feeding of some common reef zoantharians and octocorals. Aust J Mar Freshwater Res 42:729–741

Sorokin YI (1995) Coral reef ecology, 2nd edn. Springer, Berlin Heidelberg New York

Stephens GC (1982) Recent progress in the study of "Die Ernährung der Wassertiere und der Stoffhaushalt der Gewässer". Am Nat 22:611–619

Taghon GL (1981) Beyond selection: optimal digestion rate as a function of food value. Am Nat 118:202–214

Thomsen L (1999) Processes in the benthic boundary layer at continental margins and their implication for benthic carbon cycle. J Sea Res 41:73–86

Thomsen L, Weering TCE van (1998) Spatial and temporal variability of particulate matter in the benthic boundary layer at the N.W. European Continental Margin (Goban Spur). Progr Oceanogr 42:61–76

Valiela I (1995) Marine ecological processes. Springer, Berlin Heidelberg New York

Vogel S (1994) Life in moving fluids: the physical biology of flow. Princeton University Press, Princeton

Wildish D, Kristmanson D (1997) Benthic suspension feeders and flow. Cambridge University Press, Cambridge

Wildish DJ, Miyares MP (1990) Filtration rate of blue mussels as a function of flow velocity: preliminary experiments. J Exp Mar Biol Ecol 142:213–220

Witte U, Brattegard T, Graf G, Spinger B (1997) Particle capture and deposition by deep-sea sponges from the Norwegian-Greenland Sea. Mar Ecol Prog Ser 154:241–252

Wottom RS (1994) The biology of particles in aquatic systems. Lewis, Boca Raton

Yahel G, Post AF, Fabricius K, Marie D, Vaulot D, Genin A (1998) Phytoplankton distribution and grazing near coral reefs. Limnol Oceanogr 43:551–563

Polar Biol (2001) 25: 41–48
DOI 10.1007/s003000100309

Irene R. Schloss · Gustavo A. Ferreyra

Primary production, light and vertical mixing in Potter Cove, a shallow bay in the maritime Antarctic

Abstract Phytoplankton photosynthesis was measured during spring–summer 1991–1992 in the inner and outer part of the shallow Potter Cove, King George Island. Strong winds characterise this area. Wind-induced turbulent mixing was quantified by means of the root-mean square expected vertical displacement depth of cells in the water column, Z_t. The light attenuation coefficient was used as a measure of the influence of the large amount of terrigenous particles usually present in the water column; 1% light penetration ranged between 30 and 9 m, and between 30 and 15 m for the inner and outer cove, respectively. Obvious differences between photosynthetic capacity [P^*_{max}; averages 2.6 and 0.6 μg C (μg chlorophyll-*a*)$^{-1}$ h^{-1}] and photosynthetic efficiency {α^*; 0.073 and 0.0018 μg C (μg chlorophyll-*a*)$^{-1}$ h^{-1} [(μmol m^{-2} s^{-1})$^{-1}$]} values were obtained for both sites during low mixing conditions (Z_t from 10 to 20 m), while no differences were found for high mixing situations ($Z_t > 20$ m). This suggests different photoacclimation of phytoplankton responses, induced by modifications of the light field, which in turn are controlled by physical forcing. Our results suggest that although in experimental work P^*_{max} can be high, wind-induced mixing and low irradiance will prevent profuse phytoplankton development in the area.

Please note: When citing this article, please use the reference to the original source in the journal Polar Biology as given on the top of this page.

I.R. Schloss (✉) · G.A. Ferreyra
Instituto Antártico Argentino,
Cerrito 1248 (C1010AAZ) Buenos Aires,
Argentina
E-mail: ischloss@dna.gov.ar
Fax: +54-11-48122039

I.R. Schloss
CONICET, Argentina

Introduction

Antarctic shallow coastal environments' (< 50 m deep) primary production is influenced by mixing and the input of suspended particles from freshwater runoff (Klöser et al. 1993, 1994) which control the light environment in the water column. Light available for photosynthesis is affected by day length, the presence of ice and the quality and, particularly in the system under study, the quantity of suspended particulate matter (Klöser et al. 1993, 1994; Schloss et al. 1997, 1998). Potter Cove is mainly characterised by strong inputs of suspended particles from meltwater of glaciers, but resuspension can be significant during periods of wind-induced destratification of the water column (Klöser et al. 1994). The greater the influence of the sediment load on the system, the less light is transmitted through the water column (Schloss et al. 1997). Water column stratification is a function of the balance between the stabilising forces (fresh-water additions, originated from melting of sea ice and glaciers, and riverine input), and destabilising forces (tides and winds). The relative weight of the stabilisation and destabilisation factors will result in different degrees of vertical mixing of the water column. The degree of mixing will, in turn, determine the time and space scales of the vertical excursions phytoplankton cells will experience (Mann and Lazier 1991). Z_t, the root-mean square expected vertical displacement of a particle in the water column allows estimation of the displacement of a particle in a turbulent flow and can be used to obtain length scales for the vertical excursions of phytoplankton as a function of time (Denman and Gargett 1983). In operational terms, it represents the vertical movement of particles as a function not only of the density field, but also of the wind speed and duration, therefore being a dynamic more than a synoptic descriptor. The depth at which a phytoplankton cell can be found in a turbulent regime is therefore well described by Z_t (Denman and Gargett 1983); the good fit of this model and the field data has been shown in Schloss (1997).

Phytoplankton photosynthesis in Antarctic waters is mainly a function of the cell's light climate (Neori and Holm-Hansen 1982), which in turn is a function of water transparency and mixing depth. The aim of the present work was to study phytoplankton primary production as a function of light and vertical displacement of cells in the water column, as well as the factors that control light and water column structure in a coastal environment characterised by the strong influence of the adjacent land.

Materials and methods

Sampling was conducted at Potter Cove (King George Island, South Shetlands, Antarctica, 62°14'S, 58°38'W), in the vicinity of Jubany Station at two stations, E1 and E2, located in the inner and outer cove, respectively (Fig. 1). Both sites present contrasting optical characteristics in the water column, with E1 being more influenced by freshwater runoff than E2. Average depth in the inner cove is around 30 m (maximum 50 m), while depth in the outer cove is around 100 m. Between October 1991 and February 1992 fieldwork was conducted on 16 dates, approximately once per week. Water samples were collected from a rubber boat with a 4.7 l Niskin bottle at five fixed depths (0, 5, 10, 20 and 30 m). Temperature was measured with an inversion thermometer, and salinity was determined with a Beckman RS9 induction salinometer. During a second sampling in spring–summer 1995–1996, water salinity and temperature were measured with an ME-ECO219 mini-CTD (conductivity-temperature-depth). Photosynthetic active radiation (PAR) was measured using a Kahlsico 268WA310 underwater radiometer during the 1991–1992 campaign, and with an LI-COR sensor added to the CTD in 1995–1996. A volume of 0.5–1 l seawater was filtered onto 0.45-mm Millipore filters, and photosynthetic pigments extracted during 24 h in cold and dark conditions with 90% acetone. Readings were made with a Hitachi Perkin Elmer UV-VIS 139 spectrophotometer and pigment concentration calculated after Strickland and Parsons (1972). Quantification of cells was done on water samples after Utermöhl (1958) on neutralised formaline (4%) - fixed samples. K_d, the light attenuation coefficient, was calculated according to Kirk (1983) as:

$$K_d = 1/z * \ln(E_0/E_z)$$

where E_0 is the surface incident irradiance and E_z is irradiance at depth z. To quantify the contribution of the main absorbing components on K_d, a second data set from the 1995–1996 season was used, since much more information was available for this than for the 1991–1992 season. K_d was related to chlorophyll-*a* (Chl-*a*), and total non-photosynthetic (suspended and resuspended) particulate materials (TPM). TPM was measured gravimetrically after filtering the samples through combusted pre-weighed Whatman GF/F filters. After filtration, filters were rinsed twice with distilled water in order to remove salts, then dried for 24 h at 60°C, and weighed. Regressions between K_d and the absorbing materials were done using Statistica for Windows version 5. The best fit for the 0–5 m data was obtained bi-linearly adjusting the data with a piecewise regression with breakpoint,

$$K_{estimated} = A_1 + C(TPM_1) * [TPM] + C(Chl-a_1) * [Chl-a], \text{ if } K_d \leq B \text{ and}$$

$$K_{estimated} = A_2 + C(TPM_2) * [TPM] + C(Chl_2) * [Chl-a], \text{ if } K_d > B \quad (1)$$

where C(TPM) and C(Chl-*a*) are the factors or coefficients for TPM and Chl-*a*, respectively. Sub-indexes 1 and 2 indicate that those coefficients correspond to values of K_d smaller or larger than the breakpoint B. The breakpoint B indicates the presence of a discontinuity in the relation between the variables. It represents the value of K_d where such discontinuous relation with the independent variables takes place. For the lower water column (5–30 m), the best fit of the data was obtained by a multiple linear regression.

Photosynthesis versus irradiance measurements (*P* vs *E*)

For the determination of gross photosynthesis as a function of light intensity, the photosynthetron method was used (Lewis and Smith 1983). Water samples collected at 5-m depth from both stations were placed in dark flasks after collection, and later 1-ml subsamples were irradiated with 12 different light intensities in the photosynthetron for 30–40 min. An activity of about 5 μCi ml^{-1} $^{14}CO_3H^-$ was added to each sample. Incubation irradiance was measured with a Li-Cor, LI-185B PAR sensor. Scintillation cocktail was added after the incubations (4 ml Aquasol), and analysed in

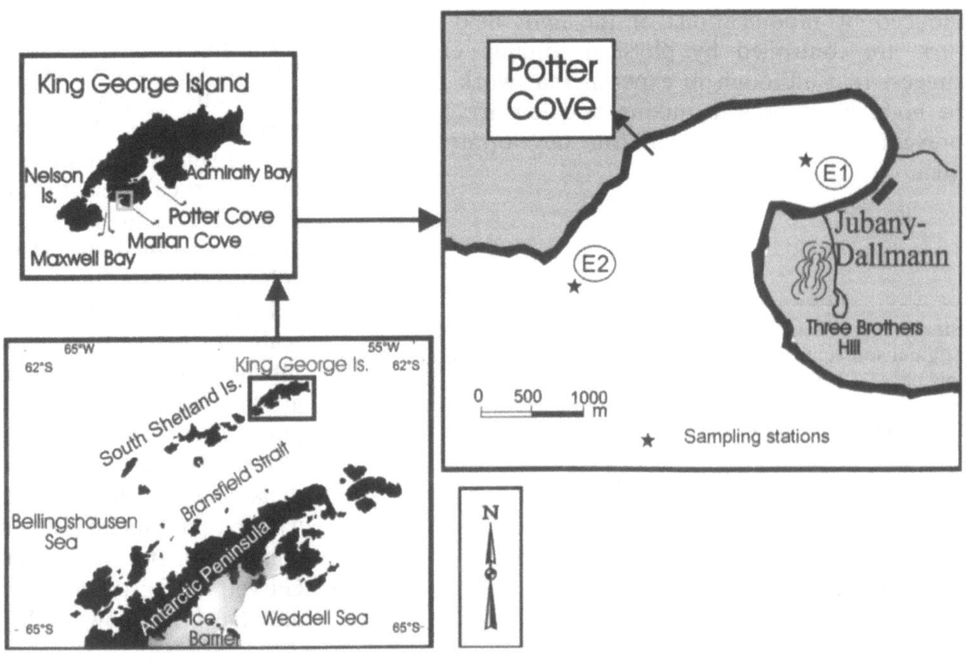

Fig. 1 Map showing the location of the sampling stations in Potter Cove, King George Island, South Shetlands

a Beckman LSDIO100C scintillation counter. The photosynthetic capacity (P^*_{max}: µg C (µg Chl-*a*)$^{-1}$ h^{-1}), photosynthetic efficiency [α^*: µg C (µg Chl-*a*)$^{-1}$ h^{-1} (µmol m^{-2} s^{-1})$^{-1}$] and the photoinhibition parameter [β^*: µg C (µg Chl-*a*)$^{-1}$ h^{-1} (µmol m^{-2} s^{-1})$^{-1}$] were calculated after Platt et al. (1980). Notation and symbols are as recommended by Sakshaug et al. (1997).

Both winds and tides can generate turbulent kinetic energy (TKE; Demers et al. 1987) that results in vertical mixing of the water column (Mann and Lazier 1991). In coastal environments such as Potter Cove, wind- more than tide-generated TKE (Schloss 1997) can disrupt water column stratification. Wind speeds were measured at Jubany's meteorological station every 3 h. The vertical transport of particles was calculated as a function of wind speed and duration, and of the water column density variation (Denman and Gargett 1983). The depths of vertical Lagrangian transport of particles, Z_t, were calculated following Denman and Gargett (1983). Equations for the calculations of Z_t can be found in the Appendix. Parameters from the *P* vs *E* curves were grouped according to the value of Z_t calculated for 12 h before the sampling moment, forming four groups.

Results

The light environment at E1 and E2

Average K_d values were 0.33 m^{-1} (range: 0.7–0.1 m^{-1}) and 0.25 m^{-1} (range 0.5–0.01 m^{-1}), for E1 and E2 respectively. Vertical profiles for temperature, salinity and the density anomaly σ_t for both stations E1 and E2, as well as light profiles and TPM in the water column during the study period, have been presented elsewhere (see Schloss et al. 1997), but a synthesis corresponding to the 1995–1996 season is shown in Fig. 2. In general, although both stations displayed the same pattern in salinity and TPM, processes in E1 were more influenced by terrigenous input than in E2, at the mouth of the cove. Figure 2 shows a typical σ_t profile for each station (Fig. 2a) as well as the contrasting TPM concentrations (Fig. 2b) and light profiles (Fig. 2c) that characterise each of them. During the 1991–1992 season, for example, light reached much deeper depths in E2 than in E1: at 5 m, where the samples for photosynthesis measurements were taken, light transmitted was only 0.8% of incident irradiance most of December and January in E1, contrasting with the 15% transmitted in E2. The different amounts of suspended particulate matter that characterises each station gave rise to different light attenuation conditions.

K_d was then analysed as a function of pigments and particulate matter in the water column. The relative importance of Chl-*a* and TPM changed with the ongoing of the season: at the beginning of the spring, after ice retreat, clear water allowed the penetration of light to the entire water column. Later, when fresh water (snow, glacier ice and land-originated creeks) containing large amounts of suspended particles entered the waters in the cove, light was absorbed in the upper few meters (see Schloss et al. 1999). Significant correlation coefficients were determined ($r=0.8$, $P<0.05$), with a breakpoint at $K_d=0.22$ for the upper water column, corresponding to the time before and after the fresh-water input. The resulting equations can then be expressed as:

$$K_{d(0-5)} = 0.109 + 0.0006 * [\mathrm{TPM}] + 0.032 * [\mathrm{Chl-a}]$$

if $K_d \leqslant 0.22$ and

$$K_{d(0-5)} = 0.181 + 0.006 * [\mathrm{TPM}] + 0.022 * [\mathrm{Chl-a}]$$

if $K_d > 0.22$ ($n=25$; explained variance 80.1%; $r=0.90$; $P<0.001$), and

$$K_{d(5-30)} = 0.0078 + [\mathrm{TPM}] * 0.013828 + [\mathrm{Chl-a}] * 0.010055$$

($n=85$; explained variance 78%; $r=0.89$; $P<0.001$). The average relative effect of each parameter resulting from this equations indicates that for the 0–5 m interval, whose $K_d<0.22$, TPM and Chl-*a* are responsible for 1.4% and 34.27% respectively, while for $K_d>0.22$, TPM and Chl-*a* contribute with 17.37% and 17.15% respectively to the value of K_d. The constants A1 and A2 (see Eq. 1) represent around 65% of K_d values, indicating another factor contributing to light attenuation. For 5–30 m, TPM represents around 47%, Chl-*a* around 43% and the constant of the regression line, only 10%.

Vertical mixing of the water column and primary production

Dominant winds blew from the western quadrant (Schloss 1997), approximately following the direction of the main axis of the cove, so that fetch was not considered in the calculations. Z_t, the vertical movement of particles as

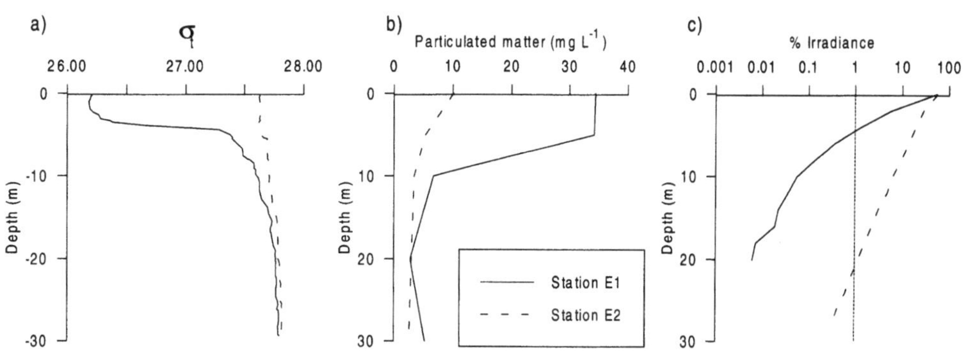

Fig. 2a–c Schematic characterisation of the water column during summer, after the massive riverine water discharge into the cove. **a** Density anomaly σ_t profiles. **b** Vertical distribution of total particulate matter concentration in the water column. **c** Percentage of surface irradiance penetrating the water column. The *straight line* indicates 1% irradiance

a function of density field and the wind speed and duration, has been calculated in each of the sampling stations for the average stratification conditions that characterised each site and for the wind intensities measured during the 1991–1992 season (Fig. 3). The most frequent wind intensities were between 8 and 12 m s^{-1} (see Fig. 3). If, as an example, a 24-h period is considered (which is adequate for polar phytoplankton that undergo less than one duplication per day, Smith and Sakshaug 1990; Sakshaug and Slagstad 1991), at 12 m s^{-1}, it can be clearly seen that wind-induced turbulent mixing may transport cells to deeper depths in E2 (around 50 m) than in E1 (around 37 m). Due to greater density stratification at E1 as compared to E2, vertical displacement depths can be expected to be smaller. However, it has to be considered that mixing depths are usually large enough to carry cells almost down to the bottom in E1 (with maximum depths close to 50 m and averages around 30 m; Fig. 3). Z_t values were then specifically computed for those dates when photosynthesis was measured, in this case considering the winds in a 12-h period before the sampling moment. When only these selected dates were considered, Z_t values ranged between 0.3 and 10.5 m (average 4.5 m) in E1, while in E2, Z_t values were between 0.1 and 21.5 m (average 6.5 m). This information will be discussed below in the context of the photosynthetic parameters calculated.

Phytoplankton samples for the photosynthesis measurements were dominated by species from the genus *Thalassiosira* all through the spring–summer season, in both E1 and E2. The most abundant species was *T. antarctica*. Maximum cell concentrations were observed in E2, on 26 November, at 20 m (5.6×10^4 cells l^{-1}), in an assemblage where *Thalassiosira* spp. co-dominated with *Corethron criophilum*. This was especially so when turbulent mixing was more intense. Another species related to higher Z_t values was *Chaetoceros socialis*, which was mostly present during the high turbulence mixing events in E2.

The parameters characterising the *P* vs *E* curves are given in Schloss et al. (1998). A selection of them is presented in Table 1. Parameters calculated for the *P* vs *E* curve showed a wide range of values, with P^*_{max} values ranging from 0.19 to 4.58 μg C (μg Chl-*a*)$^{-1}$ h^{-1}, α^* varying between 0.007 and 0.158 μg C (μg Chl-*a*)$^{-1}$ h^{-1} [(μmol m^{-2} s^{-1})$^{-1}$], without significant differences between stations, if the pooled data from stations E1 and E2 were considered (Schloss et al. 1998). Photoinhibition was hardly observed at the light intensities used in the experiments (maximum corresponding to those irradiance values observed in the field at 5-m depth), although in some cases photoinhibition was indeed observed [average = 0.0012 μg C (μg Chl-*a*)$^{-1}$ h^{-1}(μmol m^{-2} s^{-1})$^{-1}$].

According to the Z_t value calculated for each sampling date, the photosynthesis measurements were divided into four groups, for E1 and E2 and for the dates when the calculated Z_t was higher or lower than the average Z_t. These "low" and "high" average Z_t values were between 10 and 30 m for E1, and 16 and 60 m for E2, respectively. Photosynthetic parameters that corresponded to dates when Z_t was in the range of the average Z_t ± standard error of Z_t were not included in this analysis; final $n = 16$ (Table 1). Four average *P* vs *E* curves were then estimated (one for each group) and the resulting curves presented in Fig. 4. Average K_d values are also presented in the figure. The average P^*_{max} and α^* from the data in each group are indicated. Clear differences in both parameters P^*_{max} and α^* can be seen between E1 and E2 for low Z_t (Fig. 4). The highest production value was measured in E1 for those dates when vertical turbulent mixing was low (Fig. 3a), when the highest α^* value was also computed. Considering these four situations, photoinhibition was only detected in E2, being higher when low Z_t values were observed.

Discussion

Polar primary producers frequently are light limited (Smith and Sakshaug 1990). Phytoplankton development in Potter Cove is constrained by light limitation due to the seasonal ice cover – especially if the ice is also covered by snow (Palmisano et al. 1986) – at the beginning of the growing season. When ice has melted or has been carried away, heavy fresh-water inflow containing large amounts of organic and inorganic particles causes turbidity so that frequently the lower limit of the euphotic zone (even considering 0.1% of surface irradiance) can be found at very shallow depths (5–10 m, Schloss et al. 1997).

Unlike other Antarctic coastal or oceanic environments where water transparency is highly dependent on pigment concentrations (Mitchell et al. 1991; Tilzer et al.

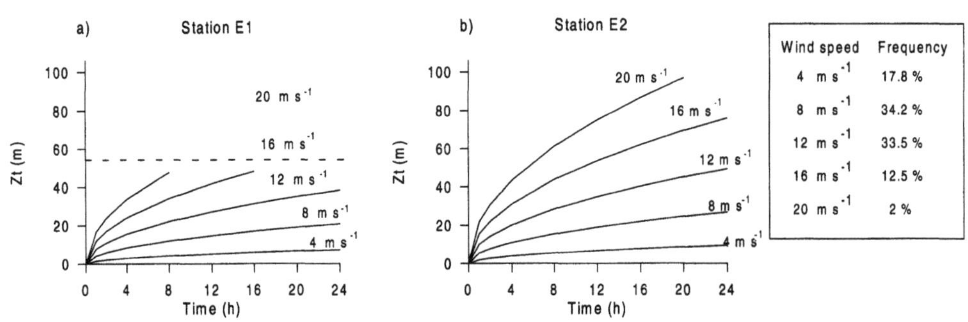

Fig. 3a, b Calculated Z_t, the root-mean square expected vertical displacement of a cell in the water column, for the average density situation in each of the sampling stations and for different wind intensities as a function of time that wind is blowing. Frequencies of the observations for the different wind speeds are indicated below. **a** Station E1; **b** Station E2

Table 1 Parameters of the photosynthesis vs irradiance curves resulting from production measured experimentally with the photosynthetron: the photosynthetic capacity [P^*_{max}: µg C (µg Chl-a)$^{-1}$ h^{-1}], the photosynthetic efficiency [α^*: µg C (µg Chl-a)$^{-1}$ h^{-1} (µmol m^{-2} s^{-1})$^{-1}$], the photoinhibition parameter [β^*: µg C (µg Chl-a)$^{-1}$ h^{-1} (µmol m^{-2} s^{-1})$^{-1}$] and E_k, which indicates the irradiance at which the rate of photon absorption by PSII and the rate of electron transport from water to CO_2 are balanced. Z_t values (in m), the root-mean square expected vertical displacement of a cell in the water column, have been classified according to their values into category 1 (low Z_t) or 2 (high Z_t). Only dates when Z_t was higher or lower than the average ± the standard deviations were considered

Date	Station	P^*_{max}	α^*	β^*	E_k	Z_t	Z_t category
18 November 1991	E1	1.454	0.015		96.93	45.90	2
18 November 1991	E2	1.536	0.030		51.20	45.47	2
26 November 1991	E2	1.350	0.057		23.68	27.81	2
29 November 1991	E2	0.435	0.023		18.91	12.88	1
4 December 1991	E1	4.577	0.034		134.62	11.23	1
4 December 1991	E2	0.608	0.026	0.0002	23.40	18.64	1
8 December 1991	E1	1.491	0.037	0.0002	40.31	99.78	2
14 December 1991	E1	0.920	0.007		12.60	30.09	2
14 December 1991	E2	2.700	0.020		135.00	70.43	2
20 December 1991	E2	1.045	0.033		32.14	67.04	2
6 January 1992	E1	2.396	0.059		40.61	7.00	1
17 January 1992	E1	2.337	0.158		14.83	7.63	1
17 January 1992	E2	1.236	0.021	0.0029	58.88	10.44	1
4 February 1992	E1	0.937	0.042		22.31	17.51	1
4 February 1992	E2	0.501	0.011	0.0002	47.65	19.82	1
14 February 1992	E2	0.185	0.008		24.67	21.18	1

1994; Basterretxea and Arístegui 1999) in Potter Cove the particulate fraction may contribute to up to 47% light absorption (in the lower water column, from 5–30 m). Particle load and Chl-a are highly variable both in time and space in Potter Cove. This is well represented in the relative weight of Chl-a and TPM for low and high K_d values. During winter and early spring in surface (0–5 m) waters K_d is controlled by Chl-a, which contributes to 34% of K_d. With the ongoing season, particle load becomes more important as riverine discharge increases, in relation with summer temperature increase. The partition made by the breakpoint reflects this temporal change. The relative weight in light absorption of one factor or another will also depend on the distance from the sampling site to the particle source, i.e. the riverine discharge area. It is noteworthy that for the upper water column (0–5 m) the constant A (see Eq. 1) value in both regressions is very high, representing around 65% of K_d value. This may be due to several reasons such as the absorption produced by dissolved organic carbon (DOC), and by the fact that inorganic particles alter light penetration by scattering, which increases the average path length of the photons (Kirk 1983). Therefore, the integrated effect of TPM on the

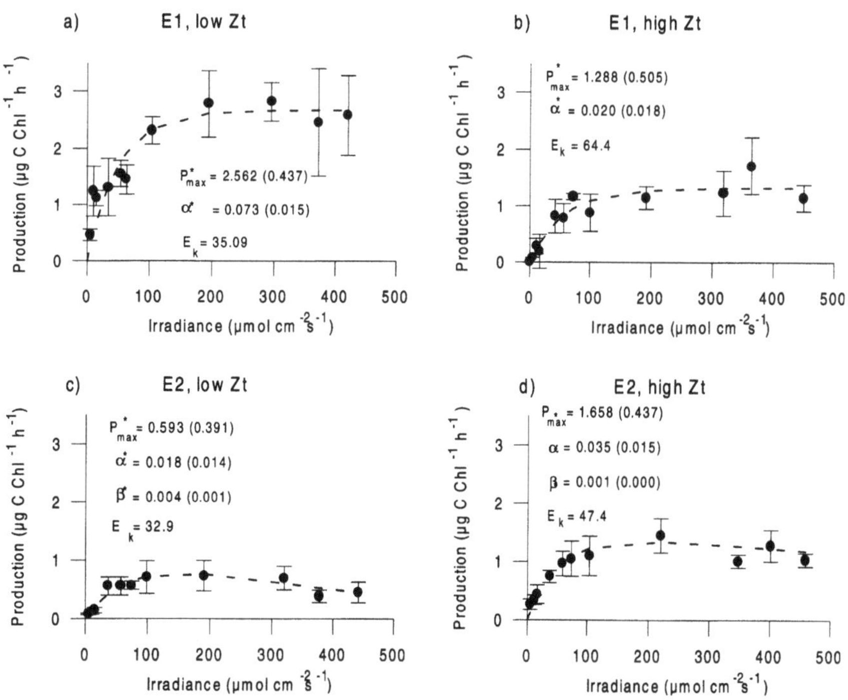

Fig. 4a–d Representation of the photosynthesis vs irradiance characteristics for each of the sampling stations, grouped according to the calculated expected vertical displacement of the cells in the water column, Z_t (i.e. if they were higher or lower than the average Z_t). The photosynthetic capacity [P^*_{max}: µg C (µg Chl-a)$^{-1}$ h^{-1}], photosynthetic efficiency [α^*: µg C (µg Chl-a)$^{-1}$ h^{-1} (µmol m^{-2} s^{-1})$^{-1}$], photoinhibition parameter [β^*: µg C (µg Chl-a)$^{-1}$ h^{-1} (µmol m^{-2} s^{-1})$^{-1}$], and the light saturation parameter (E_k in µmol m^{-2} s^{-1}) correspond to the P vs I curve calculated for the averaged points of each group. For E1/low Z_t, K_d = 0.49 (SD 0.30; **a**); for E1/high Z_t, K_d = 0.17 (SD 0.06; **b**); for E2/low Z_t, K_d = 0.27 (SD 0.24; **c**); and for E2/high Z_t, K_d = 0.12 (SD 0.02; **d**)

system will be higher than the factor indicated in the regressions, since a significant proportion would be explained by scattering, as shown for both clear and turbid waters by Phillips and Kirk (1984). This interpretation can be reinforced by the fact that low DOC concentrations were measured in Potter Cove (Abele et al. 1999). For the dates when DOC was measured (Abele et al. 1999) the constant A in the regressions was smaller, although the fraction of K_d explained by DOC was less than 1% (data not shown). The constant value A for the lower water column (5–30 m) was responsible for around 10% of the K_d value, indicating that the error source, from one or other origin, is less important in the lower than in the upper water column.

The depth of vertical turbulent wind-induced mixing is biased by the fact that sampling was only possible when wind intensities allowed the fieldwork (in Potter Cove: below 10 m s^{-1}). However, in the computation of Z_t average wind intensities corresponding to different temporal extensions can be included (Denman and Gargett 1983), therefore making it a dynamic and more adequate measure of particle displacements in relation to water column structure than just the upper mixed layer depth. The mixed layer is usually defined on the density gradient (as in Moline and Prézelin 1997), which is, in turn, also included in the calculations of Z_t (see Appendix). Moreover, it has been observed in Potter Cove that when TPM concentration is high, low mixing conditions lead to a low Z_t–high K_d situation ($P < 0.01$), resulting in the light environment being controlled by means of turbulent mixing. Low Z_t values have a different meaning in each station in relation to the particle load. In E1, with an important particle load present in the water, low mixing means that the materials entering the system are not mixed and a dark layer shades the water column below the halocline that is also associated with the fresh-water runoff. The opposite is also true, in that high mixing dilutes the particulate material, augmenting the availability of PAR to the phytoplankton cells. The same effect, although attenuated, could be observed in E2. However, it has to be considered that the two sampling stations E1 and E2 are only 4 km apart, so that the high wind-induced vertical mixing situation is associated with the mixing of the whole system, so that the difference between the stations is also a priori reduced.

Species composition did not present a clear temporal pattern, except for the fact that in E2 under high turbulence conditions species like *Corethron criophilum* and *Chaetoceros socialis* appear in higher concentrations than in the rest of the samples. The overall abundance of *Thalassiosira* spp during the studied season shaded the difference among samples. Therefore, the variation in species composition was considered not to affect the interpretation of the photoacclimation results.

When grouping the P vs E curve parameters according to the Z_t (and hence, K_d) values, different situations could be identified in the study area. Clear differences resulted from the combined effect of the two factors, the sampling station and the depth of Z_t. The α^* values calculated for low mixing situations indicate that cells are responding to light conditions in the system (α^* in E1 > α^* in E2) with less light available in E1 than in E2. When mixing is high in E2 (Z_t reaching up to 100 m), deep mixing will again condition the cells to adapt to lower light conditions. This could explain the higher α^* calculated for E2 than for E1. Photoinhibition was only observed in E2, in relation to a smaller amount of materials in the water column than in E1. During the low mixing situation, E_b, the photoinhibition light intensity, was 370 μmol cm^{-2} s^{-1}, an irradiance value often reached at the sampling depth. With high mixing E_b was 1,576 μmol cm^{-2} s^{-1}. This high intensity has rarely been measured at a 5-m depth. Therefore, it can be considered that almost no photoinhibition will affect the cells undergoing deep mixing in E2. In general, the β values calculated are low, when comparing them with similar studies (Figueiras et al. 1994).

E_k indicates the irradiance at which the rate of photon absorption by PSII and the rate of electron transport from water to CO_2 are balanced. Increased irradiances result in higher values of this parameter (Sakshaug and Slagstad 1991). For this reason, E_k is considered a good indicator of photoacclimation of cells (Sakshaug et al. 1997). In open waters, an inverse relation between E_k and mixing depth should be expected. Our results for Potter Cove, however, show an opposite pattern, with lower E_k values corresponding to low mixing scenarios for both E1 and E2 (35.09 and 32.9 μmol photons m^{-2} s^{-1} respectively), than for high mixing (64.4 and 47.4 μmol photons m^{-2} s^{-1} respectively). This is so because optical conditions in the water column are mainly controlled by shadowing of particles from near the surface, more than by vertical excursion of the cells. However, under high mixing cells are transported up and down in a more rapid way, so that the photosynthetic antenna will be adapted to the slightly higher light intensities (as shown by E_k = 64.4 μmol photons m^{-2} s^{-1}), with values close to the average irradiance measured in a mixed water column, as reported by Figueiras et al. (1994). Several coastal environments have been described as presenting hydrographic characteristics like Potter Cove (Klöser et al. 1994 and references therein). Consequently, the observed behaviour of photosynthetic parameters may be a common feature in such areas of Antarctica.

The maximum photosynthetic rates were achieved in E1, meaning that the inner Potter Cove could be a very productive environment. Moreover, all the P^*_{max} values calculated in the present study are in the range described for Antarctic coastal waters (Sakshaug and Holm-Hansen 1986; Tilzer et al. 1986; see Smith et al. 1996 for a review), and in particular agree with information available for the South Shetlands (Hapter et al. 1983). However, and although in Z_t the winds considered have been averaged over 12 h before the sampling moment, the structure of the water column under the calm weather that allowed the sampling does not represent the most common situation observed in Potter Cove. In fact,

Z_t values around 10 m were only calculated for 5% of the water column samplings (Schloss 1997). The interaction of the effects of the light environment and the mixing conditions due to wind influence on phytoplankton productivity is well documented (Holm-Hansen and Mitchell 1991; Sakshaug et al. 1991; Brandini and Rebello 1994). This is, however, the first time that the wind influence is quantified in relation to P vs E curve parameters in Antarctic shallow coastal waters, and indirectly related to the amount of particles that characterise environments like Potter Cove, heavily influenced by the land inputs. Although these combined effects of wind and light conditions in Potter Cove are usually bad for phytoplankton development, as also stated by the relation between the critical depth and the turbulent mixing depth calculated in Schloss (1997), some temporal windows with favourable light conditions may remain open. One could occur after the end of the sea-ice melt and before the particle input coming with the riverine flow started. At this point, it has to be considered that year-to-year variation in ice conditions could affect the temporal span for phytoplankton development. Light conditions could also be favourable at the end of the summer, when much less fresh water enters the sea. It is during these periods when phytoplankton could grow and attain significant biomass concentrations, whenever these temporal windows last for at least 10–14 days, which is the lag observed experimentally for the detection of relevant biomass accumulation (Schloss 1997). At those times, other factors like cell grazing, sedimentation, horizontal advection among others, have to be considered. Additionally, the relative contribution of different components of the system to light absorption at a given wavelength is in proportion to their absorption coefficients at that wavelength (Kirk 1983). In a more general way, irradiance spectral distribution has to be determined in order to get a better understanding of the role of the different components of K_d in light attenuation, which in turn will determine the light environment for primary producers. These types of studies are currently going on at Potter Cove.

For the above reasons, the Potter Cove coastal ecosystem can be defined as a low phytoplankton biomass accumulation system. Other sources of organic matter are supposed to fuel the profuse benthic fauna in this and other similar areas (Ahn et al. 1993; Sahade et al. 1997; Mercuri et al. 1998). In particular, microphytobenthic resuspension as well as the products of macroalgal degradation are considered as the most probable food source for the benthic filter- and suspension-feeders there present. The consequences for the whole ecosystem have been discussed elsewhere (Schloss et al. 1997, 1999).

Acknowledgements We would like to thank Tony Curtosi, Guillermo Mercuri, Heinz Klöser and the divers of the Prefectura Naval Argentina for their assistance in the field, and for the logistic support in Jubany-Dallmann. Doris Abele provided the results of the DOC analysis. The present work is part of the co-operation between the Instituto Antártico Argentino and the Alfred Wegener Institute.

Appendix

To calculate vertical displacement of cells Z_t (Denman and Gargett 1983), which is a function of time and the wind-generated turbulent kinetic energy (TKE) the equation used was:

$$Z_t^2 = 2K_z t,$$

where t = time (in s) and K_z = turbulent vertical diffusion coefficient ($m^2\ s^{-1}$; Mann and Lazier 1991):

$$K_z = 0.25N^{-2}\varepsilon_V$$

where N = Brunt-Väisäla frequency, which is defined as:

$$N = \frac{g}{\bar{\rho}}\left(\frac{\delta\rho}{\delta z}\right)^{\frac{1}{2}}$$

with $\rho_{(water)}$ = average water density (1.0277×10^3 kg m^{-3} in our study); $\delta\rho$ = density difference between two water masses (in kg m^{-3}) and δz = depth difference among those water masses (in m).

ϵ_V = Rate of dissipation of the TKE due to the winds:

$$\varepsilon_V = \frac{\rho_{(air)}}{\rho_{(water)}}\frac{C_{10}}{H}K\,U_{10}^3$$

where C_{10} = drag coefficient for the winds at 10 m height = 1.3×10^{-3}, H = depth of the water column in m, K_w = correction factor for the winds at surface = 4×10^{-2} (Pingree et al. 1978), U_{10} = wind speed at 10 m above sea level.

References

Abele D, Ferreyra GA, Schloss IR (1999) H_2O_2 accumulation from photochemical production and atmospheric wet deposition in Antarctic off-shore waters of Potter Cove, King George Island, South Shetland Islands. Antarct Sci 11:131–139

Ahn IY, Kang JS, Kang S-H (1993) Primary food sources for shallow benthic fauna in Marian Cove, King George Island, during an austral summer. Korean J Polar Res 4:67–72

Basterretxea G, Arístegui J (1999) Phytoplankton biomass and production during late austral spring (1991) and summer (1993) in the Bransfield Strait. Polar Biol 21:11–22

Brandini FP, Rebello J (1994) Wind field effect on hydrography and chlorophyll dynamics in the coastal pelagial of Admiralty Bay, King George Island, Antarctica. Antarct Sci 6:433–442

Demers S, Therriault J-C, Bourget E, Bah A (1987) Resuspension in the shallow sublittoral zone or a macrotidal estuarine environment: wind influence. Limnol Oceanogr 32:327–339

Denman KL, Gargett AE (1983) Time and space scales of vertical mixing and advection of phytoplankton in the upper ocean. Limnol Oceanogr 28:801–815

Figueiras FG, Pérez FF, Pazos Y, Ríos AF (1994) Light and productivity of Antarctic phytoplankton during austral summer in an ice edge region in the Weddell-Scotia Sea. J Plankton Res 16:233–253

Hapter R, Wozniak B, Dobrowolski K (1983) Primary production in Ezcurra Inlet during the Antarctic summer of 1977/78. Oceanologia 15:175–184

Holm-Hansen O, Mitchell BG (1991) Spatial and temporal distribution of phytoplankton and primary production in the western Bransfield Strait region. Deep Sea Res 38:961–980

Kirk JTO (1983) Light and photosynthesis in aquatic ecosystems. Cambridge University Press, Cambridge

Klöser H, Ferreyra G, Schloss IR, Mercuri G, Laturnus F, Curtosi A (1993) Seasonal variation of algal growth conditions in sheltered Antarctic bays: the example of Potter Cove (King George Island, South Shetlands). J Mar Syst 4:289–301

Klöser H, Ferreyra G, Schloss IR, Mercuri G, Laturnus F, Curtosi A (1994) Hydrography of Potter Cove, a small fjord-like inlet on King George Island, South Shetlands. Estuar Coast Shelf Sci 38:523–537

Lewis MR, Smith JC (1983) A small volume, short incubation-time method for measurement of photosynthesis as a function of incident irradiance. Mar Ecol Prog Ser 13:99–102

Mann KH, Lazier JRN (1991) Dynamics of marine ecosystems. Biological-physical interactions in the oceans. Blackwell, Oxford, 466 pp

Mercuri G, Iken K, Ledesma B, Dubois R (1998) On the distribution patterns and density of the Antarctic infaunal bivalve *Laternula elliptica* in Potter Cove, King George Island, Antarctica. In: Wiencke C, Ferreyra GA, Arntz W, Rinaldi C (eds) The Potter Cove coastal ecosystem, Antarctica. Rep Polar Res 299:137–143

Mitchell BG, Brody E, Holm-Hansen O, McClain C, Bishop J (1991) Light limitation of phytoplankton biomass and macronutrient utilization in the Southern Ocean. Limnol Oceanogr 36:1662–1677

Moline M, Prézelin B (1997) High-resolution time-series data for 1991/1992 primary production and related parameters at a Palmer LTER coastal site: implications for modelling carbon fixation in the Southern Ocean. Polar Biol 17:39–53

Neori A, Holm-Hansen O (1982) Effect of temperature on rate of photosynthesis in Antarctic phytoplankton. Polar Biol 1:33–38

Palmisano AC, Soo Hoo JB, Moe RL, Sullivan CW (1986) Sea ice microbial communities. VII. Changes in under-ice spectral irradiance during the development of Antarctica sea ice microbial communities. Mar Ecol Prog Ser 35:165–173

Phillips DM, Kirk JTO (1984) Study of the spectral variation of absorption and scattering in some Australian Coastal waters. Aust J Mar Freshw Res 35:635–644

Pingree RD, Holligan PM, Mardell GT (1978) The effects of vertical stability on phytoplankton distributions in the summer on the northwest European Shelf. Deep Sea Res 25:1011–1028

Platt T, Gallegos CL, Harrison WG (1980) Photoinhibition of photosynthesis in natural assemblages of marine phytoplankton. J Mar Res 38:687–701

Sahade R, Tatián M, Kowalke J, Kuehne S, Esnal G (1997) Benthic faunal associations on soft bottom at Potter Cove, King George Island, Antarctica. Polar Biol 19:85–91

Sakshaug E, Holm-Hansen O (1986) Photoadaptation in Antarctic phytoplankton: variation in growth rate, chemical composition and PvsI curves. J Plankton Res 8:459–473

Sakshaug E, Slagstad D (1991) Light and productivity of phytoplankton in polar marine ecosystems: a physiological view. In: Sakshaug E, Hopkins CEE, Øritsland NA (eds) Proceedings of the Pro-Mare Symposium on Polar Marine Ecology, Trondheim, 12–16 May 1990. Polar Res 10:69–85

Sakshaug E, Slagstad D, Holm-Hansen O (1991) Factors controlling the development of phytoplankton blooms in the Antarctic ocean: a mathematical model. Mar Chem 35:259–271

Sakshaug E, Bricaud A, Dandonneau Y, Falkowski PW, Kiefer DA, Legendre L, Morel A, Parslow J, Takahashi M (1997) Parameters of photosynthesis: definitions, theory and interpretation of results. J Plankton Res 19:1637–1670

Schloss IR (1997) Escalas temporo-espaciales de variabilidad del fitoplancton costero antártico. PhD Thesis, University of Buenos Aires, Buenos Aires

Schloss IR, Ferreyra GA, Curtosi A, Klöser H, Mercuri G, Pinola E (1997) Factors governing phytoplankton and particulate matter variation in Potter Cove, King George Island (Antarctica). In: Battaglia B, Valencia J, Walton DWH (eds) Antarctic communities. Cambridge University Press, Cambridge, pp 135–141

Schloss IR, Ferreyra GA, Curtosi A (1998) Primary production and conditions for phytoplankton growth in Potter Cove, King George Island. In: Wiencke C, Ferreyra GA, Arntz W, Rinaldi C (eds) The Potter Cove coastal ecosystem, Antarctica. Rep Polar Res 299:67–73

Schloss IR, Ferreyra GA, Mercuri G, Kowalke J (1999) Particle dynamics in an Antarctic coastal shallow environment: a sediment trap study. In: Arntz W, Ríos C (eds) Magellan-Antarctic: ecosystems that drifted apart. Sci Mar 63:99–111

Smith RC, Dierssen HM, Vernet M (1996) Phytoplankton biomass and productivity in the western Antarctic Peninsula region. In: Ross RM, Hofmann EE, Quetin LB (eds) Foundations for ecological research west of the Antarctic Peninsula. Antarct Res Ser 70:333–356

Smith WO Jr, Sakshaug E (1990) Polar phytoplankton. In: Smith WO Jr (ed) Polar oceanography, part B: Chemistry, biology and geology. Academic Press, New York, pp 477–525

Strickland JDH, Parsons DR (1972) A practical handbook of seawater analysis. J Fish Res Bd Can Bull 167:1–310

Tilzer MM, Elbrächter M, Gieskes WW, Beese B (1986) Light-temperature interactions in the control of photosynthesis in Antarctic phytoplankton. Polar Biol 5:105–111

Tilzer MM, Gieskes WW, Heusel R, Fenton N (1994) The impact of phytoplankton on spectral water transparency in the Southern Ocean: implications for primary productivity. Polar Biol 14:127–136

Utermöhl H (1958) Zur Vervollkommung der quantitativen Phytoplankton–Methodik. Mitt Int Ver Theor Angew Limnol 9:1–38

Polar Biol (2001) 24: 875–882
DOI 10.1007/s003000100286

P. Povero · M. Chiantore · C. Misic · G. Budillon
R. Cattaneo-Vietti

Land forcing controls pelagic-benthic coupling in Adelie Cove (Terra Nova Bay, Ross Sea)

Abstract Benthic-pelagic coupling processes were studied in Adelie Cove, a small 70-m deep V-shaped bay, located on the coast of Terra Nova Bay (Ross Sea). This cove is set apart from the open sea by a 12- to 15-m deep sill. Samplings were carried out over 20 days in summer 1997/1998 in the water column, in the sediment and in material collected from the sea-bottom interface by means of a sediment trap. Benthic communities were studied mainly using ROV video tapes. The complex and peculiar interactions between physical, chemical and biological processes, both in the water column and in the sediments, were tightly linked to a katabatic wind pattern, which is the main constraint in water mass circulation. A clear evolution in organic matter distribution and composition was observed. These processes led to a high build-up of materials in the sediments in the deepest part of the cove, which favoured deposit-feeding species and largely contributed to the organic enrichment of the sediments.

Introduction

During the Antarctic summer, a high amount of biogenic material originating from intense phytoplankton blooming reaches the sea floor, sustaining the high biomass of benthic invertebrates living on the continental shelf (Dayton 1990; Matsuda et al. 1990; Fabiano et al. 2000a). In order to evaluate such pelagic-benthic coupling processes, qualitative and quantitative changes of particulate matter in the water column must be investigated. While extensive data about particle fluxes have been reported for the open sea, little information is available about fluxes occurring in shallow waters (Bathmann et al. 1991; Handa et al. 1992; Nedwell et al. 1993; Fabiano et al. 1997; Pusceddu et al. 1999). This is the area where particularly rich benthic communities can be found, both in terms of biomass and diversity, and where much more information is available about benthic community structure, dynamics, and the role of key species in employing and transferring organic matter between the water column and the sediment (Knox 1990; Ahn 1993; Chiantore et al. 1998; Cattaneo-Vietti et al. 1999). Here, the annual particle flux is by several orders of magnitude larger than in the off-shore pelagic system. Also, it is particularly affected by terrestrial input, wind, resuspension, and coastal morphology (Berkman et al. 1986; Bathmann et al. 1991; Fabiano et al. 1997).

The aim of this study is to evaluate short-term changes in the composition and fate of sinking material and to highlight relations linking pelagic and benthic compartments in a confined area [Adelie Cove, Terra Nova Bay, an area proposed as part of the future Terra Nova Bay ASPA (Antarctic Specially Protected Area)] which is under terrestrial influences, such as strong katabatic events and the presence of an Adelie penguin rookery.

Please note: When citing this article, please use the reference to the original source in the journal Polar Biology as given on the top of this page.

P. Povero (✉) · M. Chiantore · C. Misic · R. Cattaneo-Vietti
Dipartimento per lo Studio del Territorio e delle sue Risorse (DIP.TE.RIS.), Università degli Studi di Genova, C.so Europa 26, 16132 Genoa, Italy
E-mail: povero@unige.it
Fax: +39-10-3538140

G. Budillon
I.M.O., Istituto Universitario Navale di Napoli, Naples, Italy

Materials and methods

Study site

Adelie Cove, a small 70-m deep V-shaped bay, located on the coast of Terra Nova Bay (Ross Sea), hosts an Adelie penguin rookery (Fig. 1A). It is set apart from the open sea by a 12- to 15-m deep sill, which forms a barrier to in- and outflows (Fig. 1B). This area is strongly affected by katabatic wind events, flowing down a glacial valley towards the open sea.

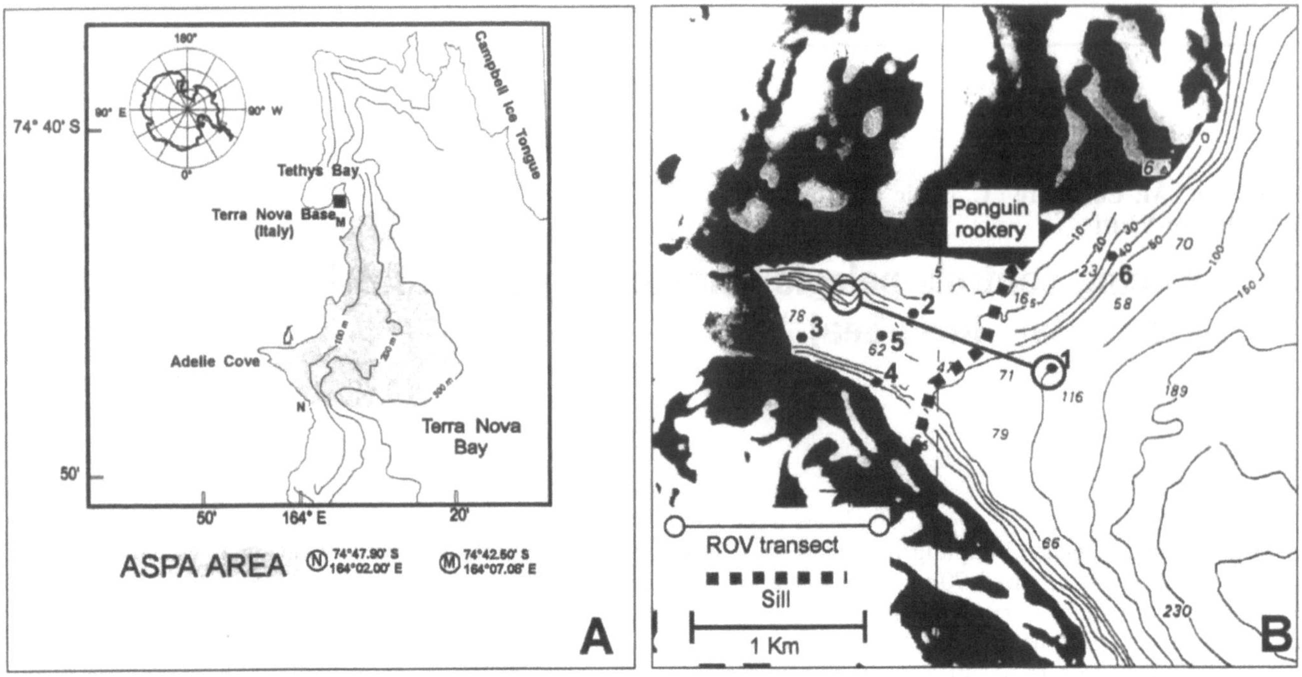

Fig. 1 **A** Study area and **B** sampling stations and ROV transect

Sample collection and analyses

In order to evaluate the physical, chemical, and biological structure of the bay in the summer season, samplings were carried out in one station (St. 5) located in the middle of the cove (53 m depth) from 31 January to 14 February 1998, during the 13th Italian Antarctic Expedition.

Water temperature and salinity data were recorded using a multiparametric Meerestechnik CTD probe. Water samples were collected at the surface and in the bottom layer (53 m depth), using Niskin bottles (5 l) on an ME rosette sampler. Five additional stations were studied for water column parameters at the beginning and at the end of the study period: three were inside the cove (Sts. 2, 3 and 4) at depths ranging between 30 and 70 m, and two (Sts. 1 and 6) were placed outside (depth ranging between 80 and 100 m; Fig. 1B).

Settling material was collected by means of sediment traps placed near the bottom at the centre of the Cove (St. 5). The cylindrical traps (inside diameter: 12 cm; height: 60 cm) were deployed 3 times at a 50-m depth. The short deployment period was designed to reduce contamination by zooplankton and by changes of organic matter composition due to respiration and dissolution of living material (Bloesch and Burns 1980). No preservatives were used in the traps to avoid significant and undesirable side-effects (Gundersen and Wassmann 1990).

Chemical analyses for nutrient concentration (silicates, phosphates, nitrates) were carried out according to Hansen and Grasshoff (1983) using a Technicon II AutoAnalyzer, after prefiltering on 0.45-μm filters.

Niskin and trap samples were pre-filtered on a 200-μm mesh net, and filtered through Whatman GF/F filters in order to determine particulate organic carbon and nitrogen (POC, PON), lipid (LIP), carbohydrate (CHO), and protein (PRT) concentrations, and on Nuclepore membrane filters (0.4 μm) in order to determine total suspended matter (TSM) and total matter (TM) from sediment traps. All analyses were carried out according to Fabiano et al. (1997, 2000b).

Particulate fluxes were calculated according to Baines and Pace (1994). Bottom sediments were collected by means of Van Veen grab in St. 5 at the beginning and end of the sampling period. The surface sediment layer (0–2 cm) was used for the analyses. Total organic matter (OM), chlorophyll *a* and phaeopigments (Chl-*a* and Phaeo), CHO, and PRT and LIP contents were assessed according to Fabiano et al. (2000b). The sum of carbon equivalents of proteins, carbohydrates, and lipids (C-OM) was used as labile fraction (i.e. easily digestible) estimate of total organic matter according to Fabiano et al. (1996).

Megalobenthic species distribution and density were evaluated along a visual transect covering an area of about 700 m^2 through the sill (Fig. 1B), using ROV (ROV ROMEO; IAN-CNR, Genoa, Italy) video images. The estimation of the observed area was calculated using two parallel beams a distance of 40 cm apart, the average speed of the vehicle and the time elapsed.

Wind direction and speed, air humidity, and temperature were measured at the Meteorological Terra Nova Bay Station.

Results

Katabatic events during the sampling period can be identified by the simultaneous occurrence of high wind speed, offshore wind direction (i.e. about 270°), lowest temperature and humidity values (Fig. 2). Such events affected temperature values and water mass dynamics, triggering rapid changes in their structure.

While until 2 February, a strong thermal stratification was observed both inside and outside the sill (Fig. 3), some days later (10 February 1998), after a week of strong katabatic wind (Fig. 2), the water column was nearly isothermal (T ≅ −1.2°C), with salinity increasing from the surface to the bottom (Fig. 3), thus favouring the vertical stability observed until 14 February (Fig. 3). Nutrients and particulate organic matter (POC, PON, PRT, CHO, LIP) distribution in the water column showed a quick temporal development (Fig. 4, Table 1).

In the surface layer of Adelie Cove (St. 5) from 31 January to 2 February, nutrients decreased (nitrate

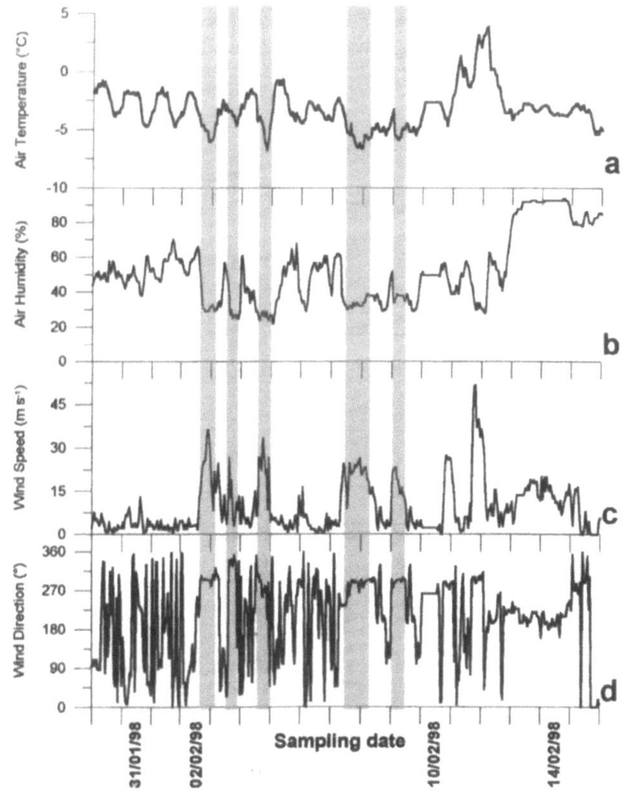

Fig. 2a–d Meteorological data in Terra Nova Bay, during summer 1998. Katabatic events are evidenced in *grey*. On *x*-axis the sampling dates are reported

concentration dropped from 28.2 to 17.2 μM, phosphates from 1.9 to 0.9 μM, silicates from 63.0 to 43.3 μm), while particulate organic matter concentration, except the lipids fraction, increased (POC rose from 120.6 to 231.0 μg C l^{-1}, PON from 23.4 to 40.7 μg N l^{-1}, PRT from 143.3 to 280.8 μg l^{-1}, CHO from 67.4 to 108.6 μg l^{-1}, LIP from 134.5 to 50.1 μg l^{-1}) Conversely, in the deep layer, very similar nutrient concentrations were observed (nitrate concentrations ranging between 29.0 and 29.6 μM, phosphates between 1.9 and 2.2 μM, and silicates between 65.9 and 69.1 μM). The organic matter showed a trend similar to the surface layer (POC raised from 73.7 to 130.8 μg C l^{-1}, PON from 12.2 to 21.1 μg N l^{-1}, PRT from 75.8 to 88.2 μg l^{-1}, CHO from 27.4 to 54.8 μg l^{-1}, LIP decrease from 74.1 to 18.3 μg l^{-1}).

At the end of the katabatic events (10 February), inside the cove (Fig. 4) a more homogeneous distribution, throughout the whole water column, both of nutrients (nitrates from 12 to 16 μM, phosphates from 0.8 to 1.4 μM, silicates from 48 to 55 μM) and of particulate organic matter was observed, reaching higher values (POC average 300 μgC l^{-1}, PON 50 μg N l^{-1}, PRT 330 μg l^{-1}, CHO 110 μg l^{-1}, LIP up to 85 μg l^{-1}). In the

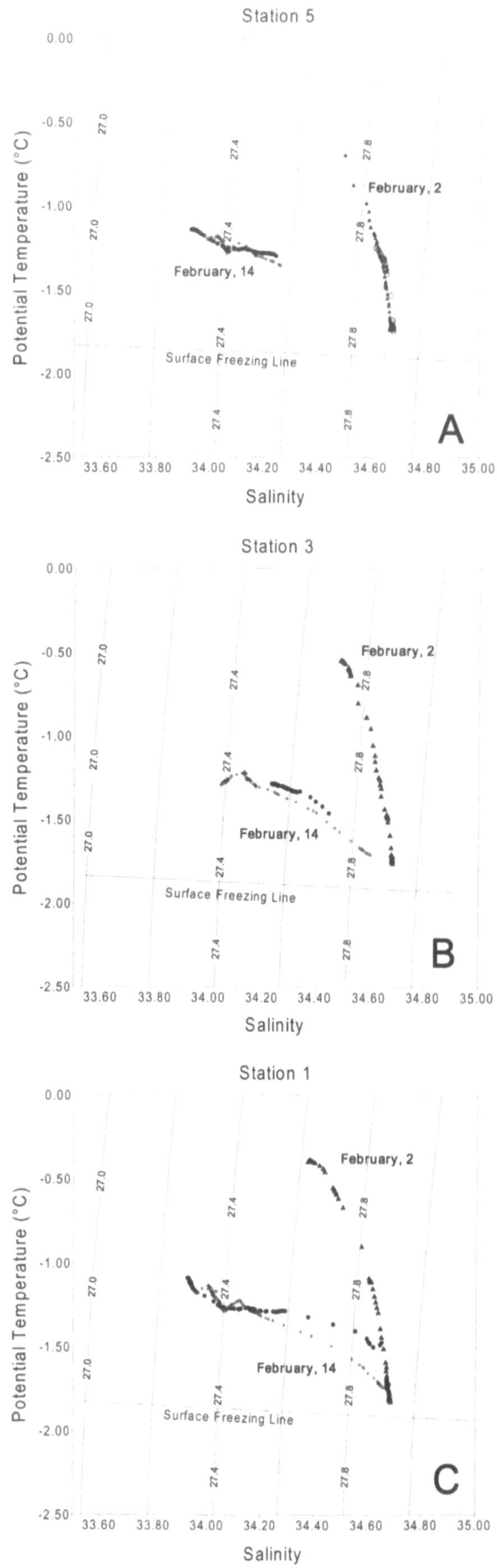

▶

Fig. 3 Temperature/salinity diagrams of stations inside Adelie Cove (**A** St. 5 and **B** St. 3), and offshore (**C** St. 1)

Fig. 4 Temporal evolution inside Adelie Cove (St. 5) in the surface and deep layer of: nitrates (*N* μM), phosphates (*P* μM), and silicates (*Si* μM) (**A**); particulate organic carbon and nitrogen (*POC* μg C l^{-1}, *PON* μg N l^{-1}) (**B**); and proteins (*PRT* μg l^{-1}), carbohydrates (*CHO* μg l^{-1}) and lipids (*LIP* μg l^{-1}) (**C**)

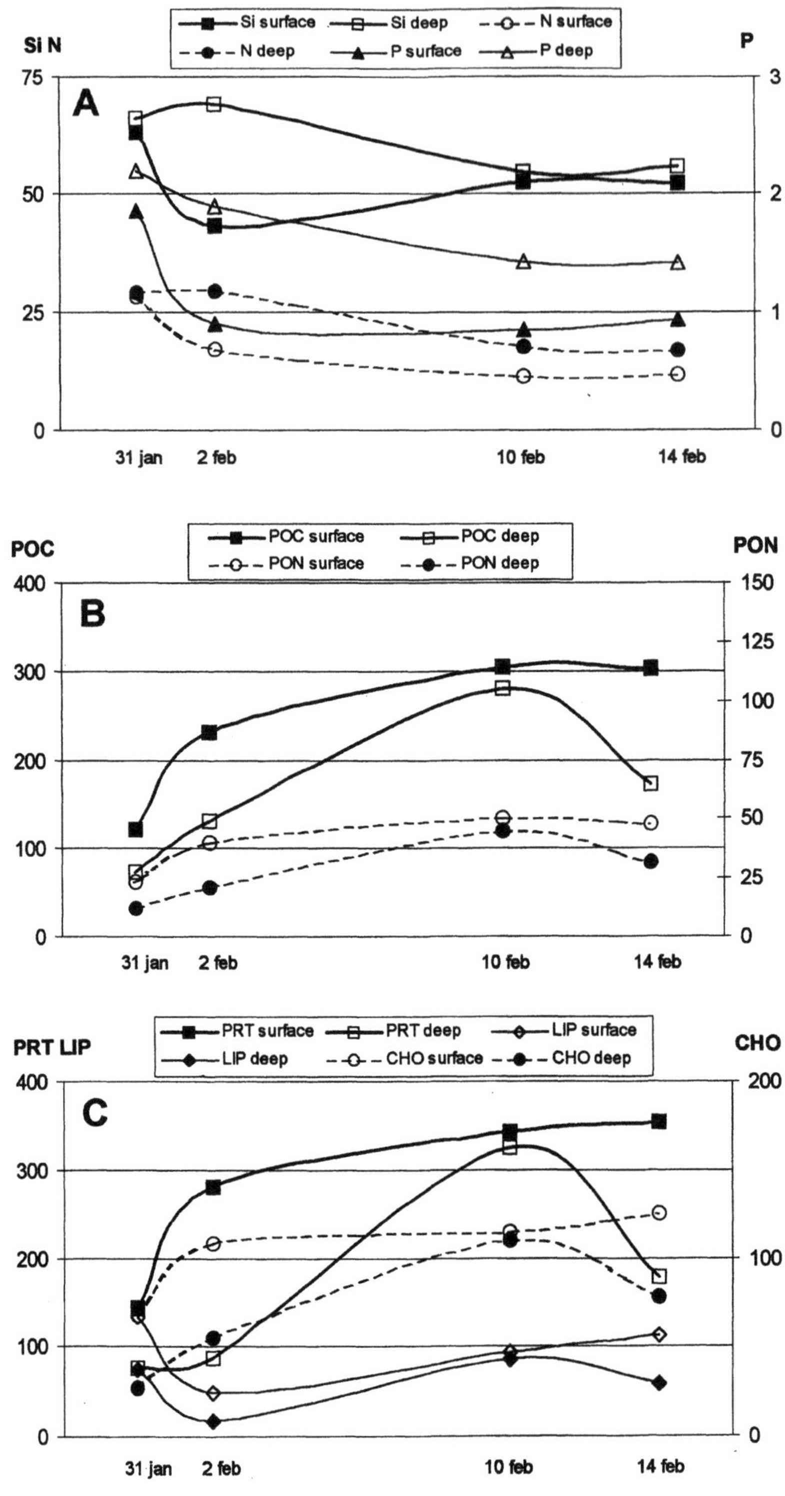

deep layer 4 days later, particulate organic matter decreased (POC dropped to 172.5 μg C l^{-1}, PON to 31.9 μg N l^{-1}, PRT to 178.7 μg l^{-1}, CHO to 78.4 μg l^{-1} and LIP to 58 μg l^{-1}), while the change in nutrient concentrations was negligible (nitrates 16.6 μM, phosphates 1.4 μM and silicates 55.9 μM).

To evaluate the role of the Adelie penguin rookery located along the northern side of the cove, data from St. 6, located out of the sill, and those from the Cove were compared, taking into account nutrient concentrations and POM quantitative and qualitative composition (Table 1). Although samplings had not been performed at the same time, POC and nutrient values outside the sill were on average higher than inside (in St. 6 POC > 500 μg C l^{-1} and nitrates > 22 μM), while lower labile fraction values (C-POM:POC < 50%) were recorded.

Table 1 Temporal evolution inside Adelie Cove (*St. 2, 3* and *4*) and offshore (*St. 1* and *6*) of: nitrates (*N* μM), phosphates (*P* μM), silicates (*Si* μM), particulate organic carbon and nitrogen (*POC* μg C l^{-1}, *PON* μg N l^{-1}), proteins (*PRT* μg l^{-1}), carbohydrates (*CHO* μg l^{-1}) and lipids (*LIP* μg l^{-1})

Station	Date	Layer	N	P	Si	POC	PON	PRT	CHO	LIP
1	2 February 1998	Surface	14.47	0.86	31.56	453.3	69.6	490.0	150.5	136.1
1	2 February 1998	Deep	26.72	1.85	53.26	104.5	14.6	81.2	41.3	37.6
1	10 February 1998	Surface	12.25	0.99	53.71	252.8	49.7	329.5	111.5	103.1
1	10 February 1998	Deep	25.99	1.85	57.35	112.3	32.6	122.9	44.2	50.6
1	14 February 1998	Surface	11.57	0.62	51.44	271.4	50.3	335.1	172.7	116.1
1	14 February 1998	Deep	28.37	1.97	72.74	129.4	38.5	95.0	56.1	51.4
2	2 February 1998	Surface	15.98	1.21	41.05	218.8	35.2	341.4	107.5	85.0
2	2 February 1998	Deep	28.40	2.03	67.27	198.0	27.0	180.0	63.0	50.0
2	14 February 1998	Surface	11.22	0.33	49.55	340.2	59.3	190.0	189.4	123.9
2	14 February 1998	Deep	–	–	–	285.0	47.0	180.0	133.2	90.0
3	2 February 1998	Surface	16.87	1.02	45.57	318.3	41.0	317.4	116.8	41.8
3	2 February 1998	Deep	29.99	2.20	68.00	150.0	–	120.0	40.0	–
3	14 February 1998	Surface	11.26	0.93	50.67	319.7	46.8	243.7	99.0	81.5
3	14 February 1998	Deep	27.00	2.20	71.68	120.0	15.2	95.0	49.4	48.5
4	2 February 1998	Surface	18.29	1.19	46.48	250.0	28.6	249.6	101.7	33.0
4	2 February 1998	Deep	29.99	2.34	71.34	118.4	19.9	76.2	38.4	19.2
4	14 February 1998	Surface	11.26	0.80	4255	383.5	63.2	406.8	147.4	124.7
4	14 February 1998	Deep	17.88	1.56	54.49	141.6	23.5	–	57.7	51.0
6	5 February 1998	Surface	20.50	1.28	44.26	542.3	94.6	257.8	63.4	50.2
6	5 February 1998	Deep	22.82	1.80	51.54	284.9	43.7	180.9	68.4	52.9

Particle matter flux values and biochemical composition of the matter trapped inside the cove from 31 January 1998 to 14 February 1998 are reported in Table 2. Total mass flux values, in accordance with all the measured parameters, were comparable at the beginning and end of the study period. Conversely, immediately after katabatic wind forcing, all flux data showed a marked decrease (>40%).

The Adelie Cove bottom is characterised by muddy sand; 60% of its surface is covered by a layer of benthic diatoms (Cattaneo-Vietti et al. 2000). Here, high concentrations of organic matter were measured (OM ranging between 13.7 and 16.6 mg g^{-1}DW). Moreover, the high trophic value of this organic material is confined by high protein (more than 3 mg g^{-1}DW) and chlorophyll-*a* concentrations (ranging between 3.8 and 9.2 μg g^{-1}DW) (Table 3).

The structure of benthic communities inside and outside the sill is illustrated in Fig. 5. In the deeper part of Adelie Cove, at about 70 m depth, the benthic community is characterised by the molluscs *Laternula elliptica* (up to 45 ind. m^{-2}) and *Yoldia eightsi* (up to 70–80 ind. m^{-2}), sabellid polychaetes, and the large spiny polychaete *Flabelligera mundata*. At this depth, the Antarctic scallop *Adamussium colbecki*, which is very common at comparable depth in other areas of Terra Nova Bay, has a very scattered distribution. Conversely, between 50 and 40 m depth, *A. colbecki* reaches densities of about 35 ind. m^{-2}. Also, several echinoderms are present: *Ophionotus victoriae*, *Diplasterias brucei*, and irregular echinoids of the genus *Abatus*. On the rare and sparse cobbles, gorgonians are frequent. Outside the sill, sediments are coarser, with cobbles covered by coralline algae and alcyonarians; here the echinoid *Sterechinus neumayeri* (40 ind. m^{-2}) grazing on coralline algae, as well as several filter feeding holothuroids, sponges, and gorgonians are dominant.

Discussion

In Adelie Cove (Terra Nova Bay), over a short time scale, relevant physical (water column stability, vertical and lateral transport), chemical (nutrient evolution) and biological (primary production and degradation) changes can be observed.

Water mass circulation within the cove is regulated by two mechanisms. The first one is linked to the general southward coastal flow from Terra Nova Bay (Stocchino and Manzella 1992). The second one is due to strong

Table 2 Biochemical characterisation of sinking material in Adelie Cove. Total matter (*TM* g m^{-2} day^{-1}), particulate organic carbon and nitrogen (*POC* mg C m^{-2} day^{-1}, *PON* mg N m^{-2} day^{-1}), proteins (*PRT* mg m^{-2} day^{-1}), carbohydrates (*CHO* mg m^{-2} day^{-1}) lipids (*LIP* mg m^{-2} day^{-1}) and chlorophyll *a* (*Chl a* mg m^{-2} day^{-1}), C-particulate organic matter (*C-POM* mg C m^{-2} day^{-1}), C:N, PRT:CHO and C-POM: POC ratios

Trap deployment date	TM	POC	PON	PRT	CHO	LIP	Chl *a*	C-POM	C:N	PRT:CHO	C-POM:POC
31.01–5.02.1998	2.1	168.1	25.0	144.0	101.3	41.8	0.67	142.5	6.7	1.4	84.8
05.02–10.02.1998	1.4	116.2	20.1	117.6	52.9	36.7	0.32	106.3	5.7	2.2	91.5
10.02–14.02.1998	2.1	178.0	26.1	192.4	37.6	43.6	0.79	141.9	6.8	5.1	79.8

Table 3 Sediment data. Organic matter (*OM* in mg g^{-1}), proteins (*PRT* μg g^{-1} DW), carbohydrates (*CHO* μg g^{-1}) and lipids (*LIP* μg g^{-1} DW), chlorophyll *a* (*Chl a* μg g^{-1} DW) and phaeopigments (*Phaeo* μg g^{-1} DW)

Sampling date	Layer	OM	PRT	CHO	LIP	Chl *a*	Phaeo
01.02.1998	0–2	16.6	3207.3	1332.5	513.0	3.8	17.5
20.02.1998	0–2	13.7	3448.6	1227.0	697.2	9.2	25.0

off-shore winds which, periodically, bring about a major water renewal: surface water characterised by low salinity enters the bay, suggesting that renewal is confined to surface circulation. However, occasionally, concurrently with strong katabatic events, a coastal upwelling can be assumed to form along the sill. This is likely to favour the entrapment of dense waters below 50 m depth inside the cove. Such waters are not involved in a frequent turn-over, as indicated in Fig. 3, where bottom waters of St. 3 closely resemble those of the bottom layer of St. 1 and show little change over time.

During our 20 days of investigations, we observed a time sequence in the quality and quantity of chemical particulate matter composition, pointing to the key role played by katabatic events. Indeed, the homogeneity of all parameters is clearly observed after the wind stress (Fig. 5).

The comparison (Table 1) between nutrient concentration, quantity and quality of particulate organic matter in stations close to the Adelie penguin rookery, but located on opposite sides of the sill (St. 2 and St. 6), suggests that the influence of the colony within the Adelie Cove marine system is irrelevant, but, owing to wind action, it has a more significant impact on the outside coastal area.

Particle fluxes (Table 2) were comparable to data from adjacent non-confined areas (Fabiano et al. 1997; Pusceddu et al. 1999). In the first period of sampling (31 January 1998–5 February 1998), without any wind stress, flux data indicate the presence of abundant and quite refractory material, linked to the breakdown of a previous *Phaeocystis* bloom, immediately before diatom blooming, as observed in many polar coastal areas in spring time (Stoecker et al. 1995; Smith et al. 1996). This

Fig. 5 Distribution and percentage frequency of megalobenthic species inside Adelie Cove (*A–D*) and offshore (*E, F*) according to ROV tape analysis

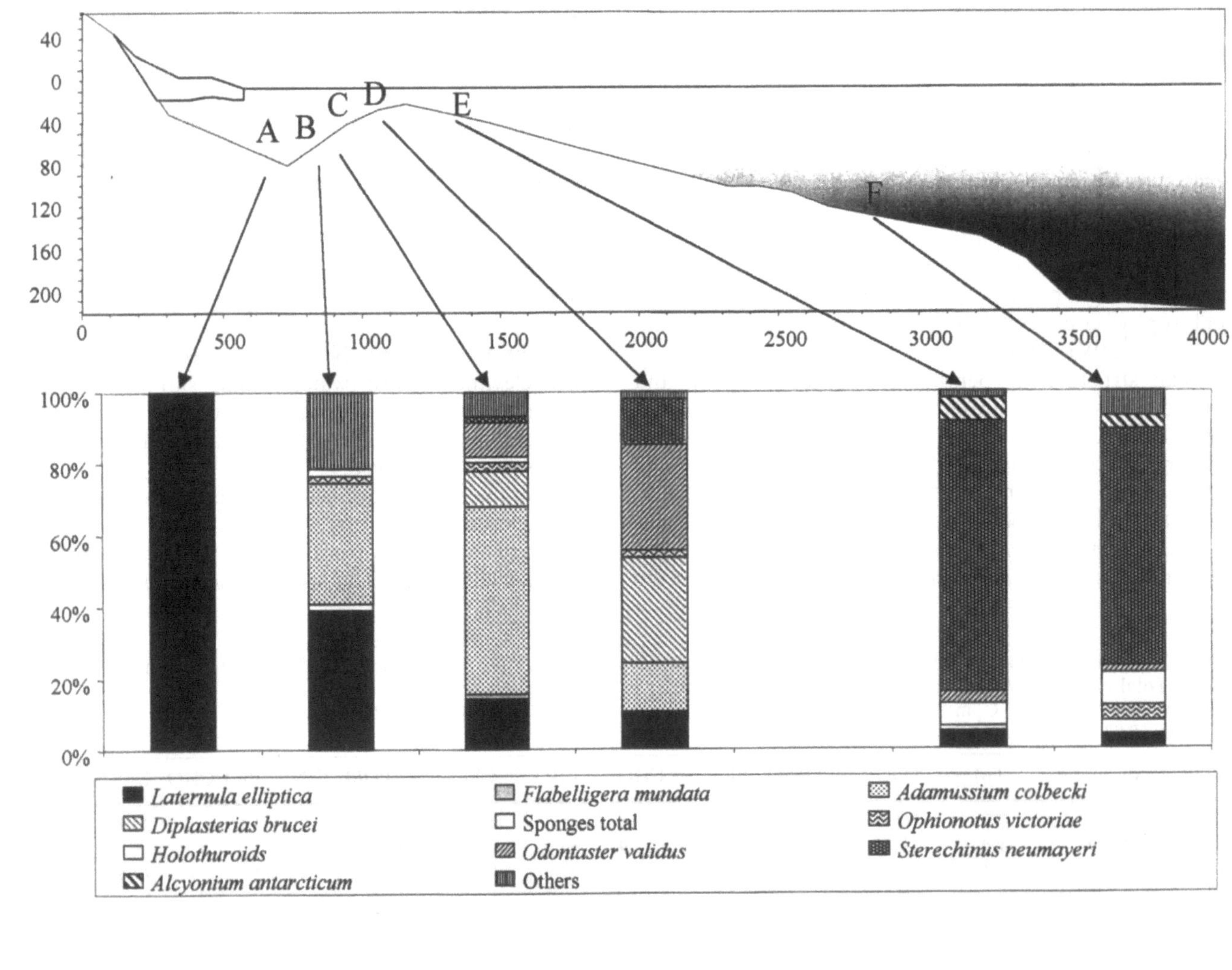

is confirmed by the ratio of 19′-hexanoyloxyfucoxanthin vs. fucoxanthin, which are characteristic pigments of prymnesiophytes and diatoms, respectively (19′HF: Fucox = 0.70; V. Saggiomo, personal communication).

From 5 to 10 February in the central area of Adelie Cove, katabatic events caused the mixing of the water column from the surface to a 50-m depth, and favoured the output of saltier and colder water and advection from the outside of less saline surface water masses (Fig. 3a–c), concurrent with a remarkable decrease (> 40%) in sinking matter (Table 2). This event relies both on lower primary production in the upper layer of the water column and on wind-induced lateral transport. The wind simply causes a mechanical forcing without any land-derived inorganic matter transport, since no remarkable inorganic fraction increase was observed (POC:TM values did not change). Moreover, the high quality of sinking material in this period is confirmed by an evident decrease in the C:N ratio (C:N = 5.7) and an increase in the PRT:CHO ratio (PRT:CHO = 2.2), related to algal material produced with a high labile fraction (> 90%, with the prevalence of proteins) in the preceding period, due to a change in phytoplankton surface community (from a *Phaeocystis*-diatom mixed crop to a diatom dominated crop, as indicated by the 19′HF:Fucox = 0.05 ratio).

In the last sampling period, renewed vertical stability is achieved, favouring an increase in particle flux. Actually, vertical stability favours the increase in particulate matter production and accumulation in the water column, as previously observed by Povero et al. (2000) in the Ross Sea area. Despite vertical stability, which should prevent sinking, the quantitative increase in zooplankton community (C. Ossola, personal communication) and the subsequent increase in grazing activity supply the deep layer of fast sinking, free algae-detritus, densely packed in faecal pellets (Fabiano et al. 1997), all of which is largely found in the sediment traps. The high PRT:CHO ratio values found in this period suggest the presence of nitrogen-enriched detritus produced by microbial populations (Fabiano et al. 1996).

The presence of a sill favours entrapment not only of denser waters, but also of carbon (Table 3). Values regarding organic matter content of sediments at 50 m depth (St. 5) are 3–4 times higher than those reported for other off-shore areas at the same depth (Fabiano et al. 2000b). Moreover, in Adelie Cove, even higher values are reported at 70 m depth (Fabiano et al. 2000b). This favours the abundance of some detritus-feeding species, such as *Yoldia eightsi* and *Flabelligera mundata* and of large-sized infaunal suspension-feeding species, such as *Laternula elliptica*. The high organic matter build-up in the sediments is mainly related to two different processes: a high local production of faecal material from large benthic organisms (Ahn 1993; Chiantore et al. 1998, 2000) coupled with a low hydrolytic enzymatic activity (unpublished data from this area) in the sediments, as previously observed in different areas of the Ross Sea (Fabiano and Danovaro 1998). Moving up the sill, the benthic community structure is more similar to the one observed in other shallow waters located within the ASPA area. On the seaward side of the sill, the very different community structure is explained by bottom features, mainly consisting of large pebbles.

In conclusion, the physical, chemical, and biological processes occurring in the water column, affecting the flux and consequently the structure of the sediments and of the benthic communities at Adelie Cove, are strongly influenced by katabatic winds. These pulsing events are one of the main forces in the Antarctic littoral ecosystem.

Acknowledgements This work was supported by Progetto Nazionale Ricerche in Antartide (P.N.R.A.) funded by the Italian Government. The authors are indebted to the scientific staff of the project who contributed with several suggestions and comments and made available all meteorological data. We also thank anonymous referees for valuable suggestions and criticism.

References

Ahn IY (1993) Enhanced particle flux through the biodeposition by the Antarctic suspension-feeding bivalve *Laternula elliptica* in Marian Cove, King George Island. J Exp Mar Biol Ecol 171:75–90

Baines SB, Pace ML (1994) Relationships between suspended particulate matter and sinking flux along a trophic gradient and implication for the fate of planktonic primary production. Can J Fish Aquat Sci 51:25–36

Bathmann U, Fisher G, Muller PJ, Gerdes D (1991) Short-term variations in particulate matter sedimentation off Kapp Norvegia, Weddell Sea, Antarctica: relation to water mass advection, ice covered, plankton biomass and feeding activity. Polar Biol 11:185–195

Berkman PA, Marks DS, Shreve GP (1986) Winter sediment resuspension in McMurdo Sound, Antarctica and its ecological implications. Polar Biol 6:1–3

Bloesch J, Burns NM (1980) A critical review of sediment trap technique. Schweiz Z Hydrol 42:15–20

Cattaneo-Vietti R, Chiantore M, Misic C, Povero P, Fabiano M (1999) The role of pelagic-benthic coupling at Terra Nova Bay (Ross Sea) and inside the Strait of Magellan. Sci Mar 63: 113–121

Cattaneo-Vietti R, Chiantore M, Gambi MC, Albertelli G, Cormaci M, Di Geronimo I (2000) Spatial and vertical distribution of benthic littoral communities in Terra Nova Bay. In: Faranda FM, Guglielmo L, Ianora A (eds) Ross Sea ecology: Italian Antarctic Expeditions (1986–1995). Springer, Berlin Heidelberg New York, pp 503–514

Chiantore M, Cattaneo-Vietti R, Albertelli G, Misic C, Fabiano M (1998) Role of filtering and biodeposition by *Adamussium colbecki* in circulation of organic matter at Terra Nova Bay (Ross Sea, Antarctica). J Mar Syst 17:411–424

Chiantore M, Cattaneo-Vietti R, Povero P, Albertelli G (2000) The population structure and ecology of the Antarctic scallop *Adamussium colbecki* in Terra Nova Bay. In: Faranda FM, Guglielmo L, Ianora A (eds) Ross Sea ecology: Italian Antarctic Expeditions (1986–1995). Springer, Berlin Heidelberg New York, pp 563–573

Dayton PK (1990) Polar benthos. In: Smith WO (ed) Polar oceanography, part B. Academic Press, London, pp 637–676

Fabiano M, Danovaro R (1998) Enzymatic activity, bacterial distribution and organic matter composition in sediments of the Ross Sea (Antarctica). Appl Environ Microbiol 64:3838–3845

Fabiano M, Povero P, Danovaro R (1996) Particulate organic matter composition in Terra Nova Bay (Ross Sea, Antarctica) during summer. Antarct Sci 8:7–13

Fabiano M, Chiantore M, Povero P, Cattaneo-Vietti R, Pusceddu A, Misic C, Albertelli G (1997) Short-term variations in particulate matter flux in terra Nova Bay, Ross Sea. Antarct Sci 9:143–149

Fabiano M, Povero P, Misic C (2000a) Spatial and temporal distribution of particulate organic matter in the Ross Sea. In: Faranda FM, Guglielmo L, Ianora A (eds) Ross Sea ecology: Italian Antarctic Expeditions (1986–1995). Springer, Berlin Heidelberg New York, pp 135–149

Fabiano M, Danovaro R, Chiantore M, Pusceddu A (2000b) Bacteria, protozoa and organic matter composition in sediments of Terra Nova Bay (Ross Sea). In: Faranda FM, Guglielmo L, Ianora A (eds) Ross Sea ecology: Italian Antarctic Expeditions (1986–1995). Springer, Berlin Heidelberg New York, pp 503–514

Gundersen K, Wassmann P (1990) Use of chloroform in sediment traps: caution advised. Mar Ecol Prog Ser 64:187–195

Handa N, Nakatsuka T, Fukuchi M, Hattori H, Hoshiai T (1992) Vertical fluxes and ecological significance of organic materials during the phytoplankton bloom during austral summer in Breid Bay, Antarctica. Mar Biol 112:469–478

Hansen HP, Grasshoff K (1983) Automated chemical analysis. In: Grasshoff K, Ehrhardt M, Kremling K (eds) Methods of seawater analysis, 2nd edn. Verlag Chemie, Weinheim, pp 347–379

Knox GA (1990) Primary production and consumption in McMurdo Sound, Antarctica. In: Kerry KR, Hempel G (eds) Antarctic ecosystems, ecological change and conservation. Springer, Berlin Heidelberg New York, , pp 115–128

Matsuda O, Ishikawa S, Kawaguchi K (1990) Seasonal variation of particulate organic matter under the Antarctic fast ice and its importance to benthic life. In: Kerry KR, Hempel G (eds) Antarctic ecosystems, ecological change and conservation. Springer, Berlin Heidelberg New York, pp 143–148

Nedwell DB, Walker TR, Ellis-Evans JC, Clarke A (1993) Measurement of seasonal rates and annual budget of organic carbon fluxes in an Antarctic coastal environment at Signy Island, South Orkney Islands, suggest a broad balance between production and decomposition. Appl Environ Microbiol 59:3898–3995

Povero P, Fabiano M, Catalano G (2000) Particulate organic matter and nutrient utilization in the mixed layer of the Ross Sea. In: Faranda FM, Guglielmo L, Ianora A (eds) Ross Sea ecology: Italian Antarctic Expeditions (1986–1995). Springer, Berlin Heidelberg New York, pp 121–134

Pusceddu A, Cattaneo-Vietti R, Albertelli G, Fabiano M (1999) Origin, biochemical composition and vertical flux of particulate organic matter under pack ice in Terra Nova Bay (Ross Sea, Antarctica) during late summer 1995. Polar Biol 22:124–132

Smith WO, Nelson DM, Di Tullio GR, Leventer R (1996) Temporal and spatial patterns in the Ross Sea: phytoplankton biomass, elemental composition, productivity and growth rates. J Geophys Res 101:18455–18465

Stocchino C, Manzella GMR (1992) La circolazione delle correnti a Baia Terra Nova (Mare di Ross – Antartide). In: Gallardo VA, Ferretti O, Moyano HI (eds) Oceanografia in Antartide. ENEA Progetto Antartide, Concepcion, pp 39–40

Stoecker DK, Putt M, Moisan T (1995) Nano and microplankton dynamics during the spring *Phaeocystis* sp. bloom in McMurdo Sound, Antarctica. J Mar Biol Assoc UK 75:815–832

Polar Biol (2001) 24: 670–676
DOI 10.1007/s003000100267

Enrique Isla · Albert Palanques
Victor Alvà · P. Puig · J. Guillén

Fluxes and composition of settling particles during summer in an Antarctic shallow bay of Livingston Island, South Shetlands

Abstract A moored experiment using a sediment trap was conducted at Johnson's Dock, Livingston Island from 11 December 1997 to 24 February 1998, as part of the EASIZ Programme activities carried out at the Juan Carlos I Spanish Antarctic base. Total mass vertical fluxes ranged from 23,235 mg m^{-2} day^{-1} to 89,073 mg m^{-2} day^{-1} during the experiment, with a mean value of 42,857 mg m^{-2} day^{-1}. Lithogenic components were the major contributors to the settling particulate flux. Organic components accounted for a low fraction of the settling particulate matter, showing an inverse relation to total mass flux. Nevertheless, the fluxes of organic components at Johnson's Dock are as high as in the open sea. The increases in chlorophyll *a* in water were related to increases in the organic carbon content, which dominated over inorganic carbon during the whole experiment. Calcium carbonate particles settle without being significantly altered in the water column and are dissolved in the upper centimetres of the bottom sediments, once they are buried. The settling material consisted of fine particles, with coarse clasts transported by icebergs. Antarctic shallow environments receive important sediment fluxes from the erosion and transport action of ice.

Introduction

Particle fluxes in polar oceans show a high seasonal variability linked to the sea-ice dynamics and its relation to living organisms and glacier-melting events. Maximum activity takes place mainly during the polar spring and summer when the ice sheets break and melt and the ocean surface becomes ice free (Huntley et al. 1991; Wefer and Fischer 1991).

Most of the research carried out in Antarctic areas has focused on offshore coastal and oceanic environments, and very little has been done in the small bays and water inlets along the enormous coastal line of the Antarctic continent and its surrounding islands.

These inshore coastal waters have very peculiar oceanographic and climatic conditions (Klöser et al. 1994; Kowalewski and Wielbinska 1983; Pruszak 1980) and are different environments to offshore oceanic waters. In contrast to Arctic coastal systems, Antarctic shallow environments show weak estuarine circulation, little riverine input and low sedimentation rates (Klöser et al. 1994). These conditions, however, should not be taken as a rule for the whole Antarctic. The singular relations between many factors such as biological productivity, sediment inputs and sea ice give unique characteristics to the small bays. Settling particle flux behaviour in Antarctic shallow nearshore environments is poorly known. Most of the continental shelves in the Bransfield Strait are mainly covered by coarse sediment (sand and gravel) (Yoon et al. 1992). Accumulation of fine sediment is restricted to fjords and bays (Griffith and Anderson 1989), so coastal settling sediment fluxes can only be studied in these confined environments. The existing data are mainly qualitative (Griffith and Anderson 1989) and the few experimental measurements do not include identification of the major components of the fluxes (Cripps and Clarke 1998).

This paper studies the settling particle fluxes and their composition, in order to quantify and characterise the particle inputs entering the Antarctic coastal ecosystem.

Johnson's Dock, a semi-enclosed and well-controlled small bay, was chosen to study these fluxes. Johnson's Dock is in Livingston Island's South Bay (South Shetland Islands), beside the Spanish Antarctic Base (Fig. 1). This dock is approximately 750 m long and 550 m wide.

Please note: When citing this article, please use the reference to the original source in the journal Polar Biology as given on the top of this page.

E. Isla (✉) · A. Palanques · V. Alvà · P. Puig · J. Guillén
Institut de Ciències del Mar (CSIC),
Passeig Joan de Borbò s/n, Barcelona 08039, Spain
E-mail: isla@icm.csic.es
Fax: +34-93-2217340

It has several depressions of more than 20 m depth that are confined by the till deposits of the frontal morraine of the Johnson's glacier, and it is connected to Livingston Island South Bay through a small mouth about 75 m wide and 40 cm deep.

Materials and methods

As part of the EASIZ Programme activities carried out at the Juan Carlos I Spanish Antarctic base, a mooring line equipped with a Technicap pps 4 sediment trap installed 4.5 m above the bottom, from 11 December 1997 to 24 February 1998, was deployed at 60°23.2′W and 62°39.5′S in Johnson's Dock at 19.5 m water depth (Fig. 1). The sediment trap used in this study has a carousel with 12 rotary collectors, and the sample-collecting intervals were preselected at 6–7 days. The upper part of the internal-collecting hull is cylindrical and has an inner diameter of 25 cm. At the mooring site, water samples were collected at 10 m depth and a sediment core was taken using a grab.

The sample tubes were filled with 5% formalin solution to avoid organic matter degradation in the sediment trap. The collected samples were processed according to Heussner et al. (1990). Zooplankton organisms that entered the trap, also called "swimmers", were removed by hand picking and stored for further analysis. Sample dry weight was determined using three subsamples filtered on 47-mm-diameter and 0.45-μm-pore preweighed Millipore cellulose filters. Total mass flux was calculated from the sample dry weight, the collecting trap area and the time-sampling interval.

Organic carbon, calcium carbonate and nitrogen of the settling particles and bottom sediment were analysed in duplicate using a LECO CN2000 analyser. Two subsamples were used to determine the total carbon and the nitrogen content, and another two subsamples were digested with HCl 6 M in a LECO CC100, measuring the inorganic carbon delivered by this reaction in the analyser. The difference between total and inorganic carbon was considered as the organic carbon content. Biogenic silica was analysed using a wet alkaline extraction with sodium carbonate, following the method described by Mortlock and Froelich (1989).

The lithogenic component was estimated as the difference between the total mass and the sum of the biogenic components: organic matter (twice the percentage of organic carbon), calcium carbonate (inorganic carbon percentage multiplied by 8.3331) and opal.

Faecal pellets were counted under a dissection microscope from the subsample used for major components analysis. Faecal pellets flux was calculated from the faecal pellets number, the collecting trap area and the time-sampling interval.

Clasts coarser than 1 mm were separated by sieving and also counted from the subsample used for major components analysis.

The chlorophyll *a* content in water samples taken from 10 m depth was analysed fluorometrically (Strickland and Parsons 1972). Samples (100 ml) were filtered through 25-mm GF/F glass fibre filters and immediately frozen at −70°C. The filters were then left

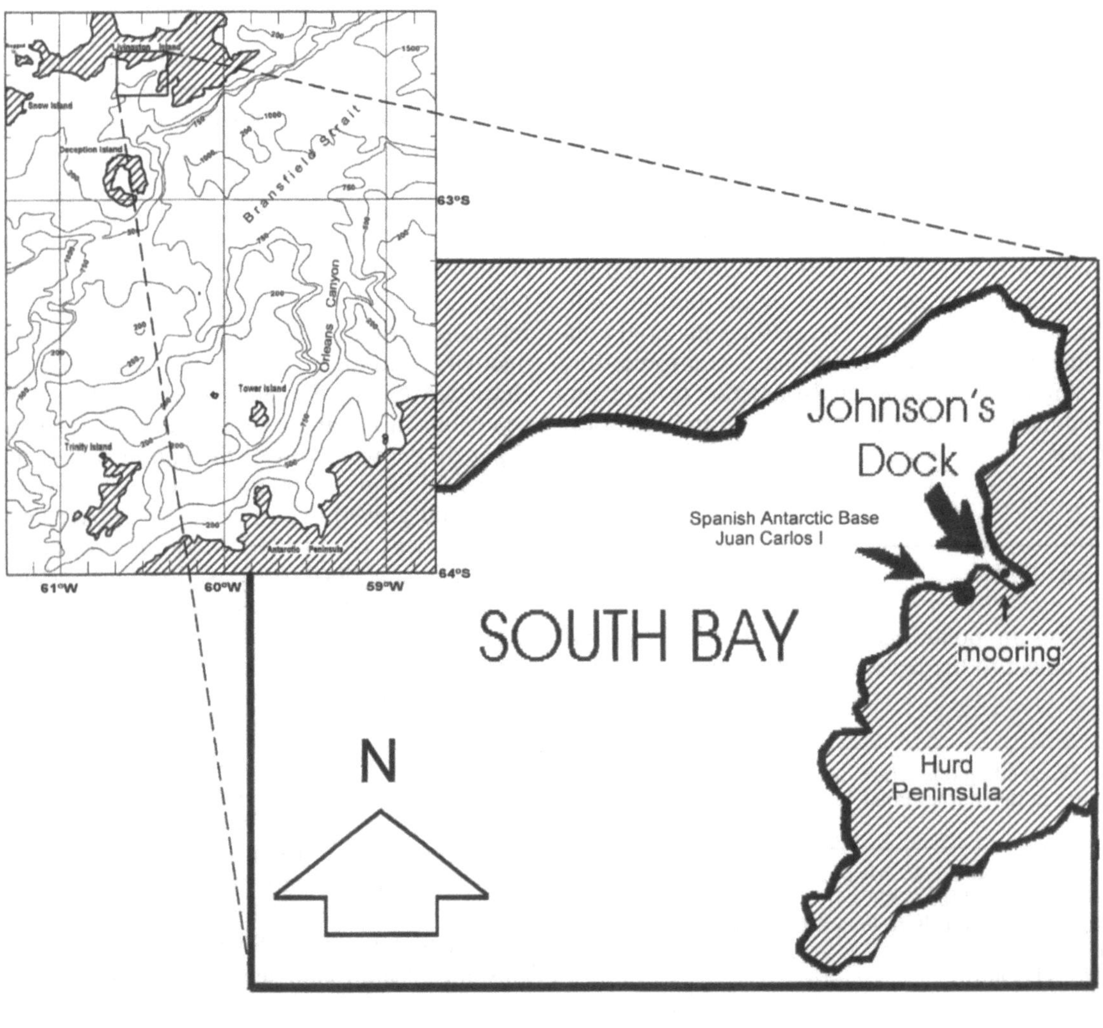

Fig. 1 Map showing the location of the sediment trap site at Johnson's Dock, Livingston Island, South Shetland Islands, Antarctica

for 12 h in 90% acetone at 4°C in the dark for pigment extraction. The fluorescence of the extract (before and after acidification) was measured using a Turner Designs fluorometer.

Results

Settling particulate matter

The mean total mass flux collected by the sediment trap and the mean content and fluxes of the major components (organic and inorganic carbon, calcium carbonate, organic matter, biogenic opal, total nitrogen, faecal pellets and aluminosilicates) are listed in Table 1. All the samples were a mixture of fine particles with poorly classified clastic material.

Time series of total mass flux (TMF) show a relatively constant value throughout the study period, although four peaks can be recognised (Fig. 2). The maximum value, 89,073 mg m^{-2} day^{-1}, was registered in late January and the minimum value, 23,235 mg m^{-2} day^{-1}, in mid-late December (Fig. 2). The mean total mass flux during the study period was about 43,000 mg m^{-2} day^{-1}.

The organic carbon (OC) content ranged from 0.3 to 0.8% (Fig. 3a). Maximum OC values were in late and mid-February, coinciding with low TMF values, whereas minimum values were in mid-late January when TMF was high. In fact, taking into account all the samples, there is a clear inverse relation between OC content and TMF.

OC flux values showed similar trends to those of the TMF except in mid-January and early February (Fig. 3a). OC fluxes ranged between a minimum of 125 mg m^{-2} day^{-1}, during the 1st week of February and a maximum of 315 mg m^{-2} day^{-1}, during the 4th week of January.

Calcium carbonate content in the trapped sediment was very low, ranging between 0 and 1.23% (Fig. 3b). The mean $CaCO_3$ content was only 0.72%, showing the maximum contents in mid-December and mid-February, whereas in late January it was practically zero [when the inorganic carbon (IC) values were below detection level (0.003%)]. The calcium carbonate flux showed a different pattern to TMF, with maximum values of more than 400 mg m^{-2} day^{-1} in mid-December and early January and minimum values (0 mg m^{-2} day^{-1}) in late January, just when TMF was maximum (Fig. 3b).

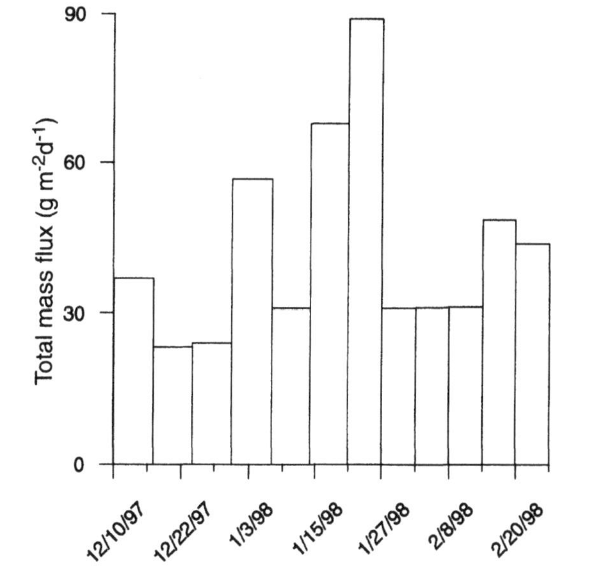

Fig. 2 Time series of the total mass flux at Johnson's Dock, Livingston Island during the study period

The OC/N relation showed low values, ranging between 6 and 8, except in late January when it increased to about 20. The OC/IC ratio was also relatively constant, ranging between 3.3 and 8, except in mid- and late January when it increased sharply to 23.2. Both OC/N and OC/IC increases were produced by a decrease in N and IC contents (Fig. 4a, b).

The opal content ranged from 1.75 to 3.08%. The maximum opal content corresponded to the 2nd week of January, coinciding with a peak of the OC content. Minimum opal content occurred in mid-late January and also coincided with minimum OC values and maximum TMF. The opal flux was influenced by the TMF and shows trends opposite to those of the opal content (Fig. 3c). The opal flux ranged from 524 mg m^{-2} day^{-1}in late December to 1,562 mg m^{-2} day^{-1} in late February. Mean opal flux was 962 mg m^{-2} day^{-1}.

The lithogenic components were the major fraction of the settling particulate matter, ranging from 94.75 to 97.54%. Maximum lithogenic values were registered in mid- and late January (Fig. 3d). The lithogenic flux clearly controlled the total mass flux and ranged from 86,881 mg m^{-2} day^{-1} in late January to 22,074 mg m^{-2}

Table 1 Maximum, minimum and mean values of total mass flux, major components, faecal pellets and clasts (content expressed in units) content and fluxes at Johnson's Dock, Livingston Island

	Max	Content (%)			Flux (mg/m^2 per day)	
		Min	Mean	Max	Min	Mean
Total flux				89,073	23,235	42,857
Organic carbon	0.78	0.32	0.52	315	125	206
Inorganic carbon	0.15	0	0.09	54	0	31
Calcium carbonate	1.23	0	0.72	452	0	258
Organic matter	1.56	0.65	1.05	630	250	412
Opal	3.08	1.75	2.39	1,563	524	962
Lithogenics	97.54	94.75	95.84	86,881	22,074	41,225
Total nitrogen	0.12	0.02	0.07	39	15	28
Faecal pellets	101,000	2,160	22,295	288,571	7,200	65,463
Clasts (units)	804	11	250	9,143	5	2,181

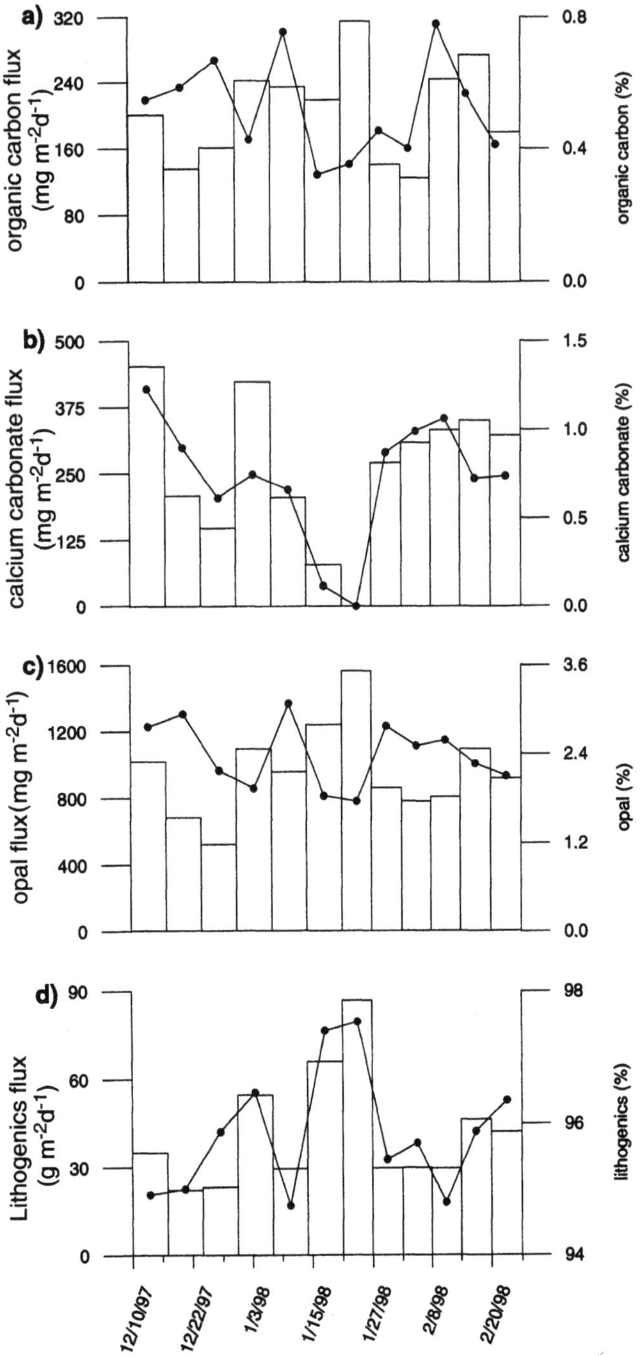

Fig. 3 Time series of organic carbon (**a**), calcium carbonate (**b**), opal (**c**) and lithogenics (**d**) fluxes (*histograms*) and contents (*filled circles*) during the study period

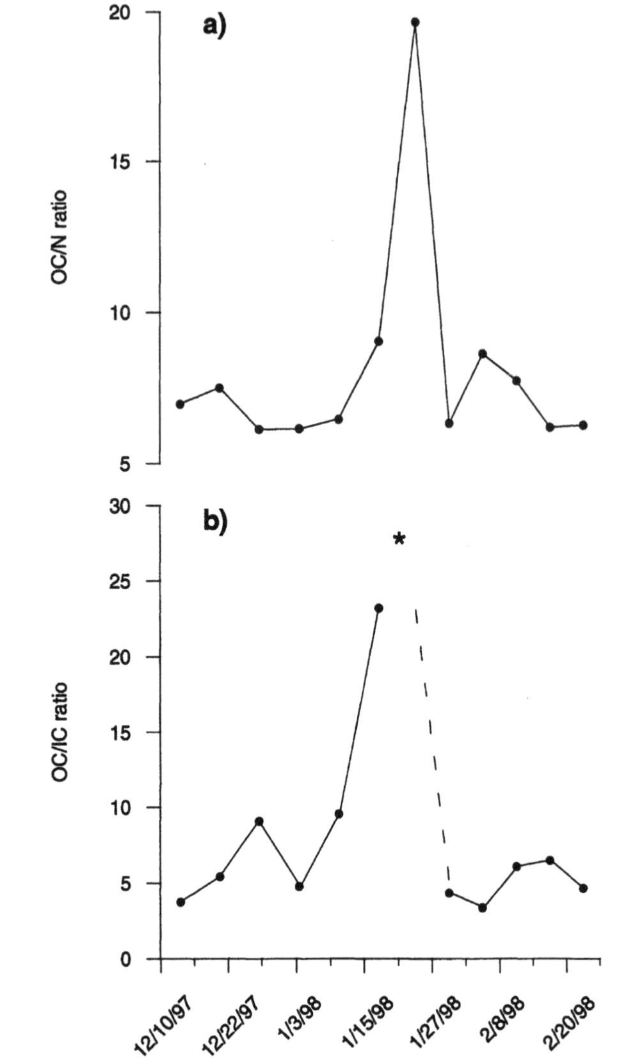

Fig. 4 Temporal variation of the organic carbon (*OC*) to total nitrogen (*N*) ratio (**a**) and organic carbon to inorganic carbon (*IC*) (**b**) ratio. [* missing point because inorganic carbon content in that sample was below detection level (0.003%)]

day^{-1} in late December, with an average value of 41,225 mg m^{-2} day^{-1}.

The faecal pellets (FP) flux showed a decreasing trend with time from 100,000 to 10,000 fp m^{-2} day^{-1}, but with a maximum peak of 288,571 fp m^{-2} day^{-1} in early January and a minimum peak of 7,200 fp m^{-2} day^{-1} in mid-January, which corresponded to the highest TMF period (Fig. 5a). However, FP content did not show a clear relation to any major component.

The sediment trap also collected detritic clasts. The number and flux of the clasts coarser than 1 mm showed an increasing trend with time, from 10 to more than 800, and from 5 mg m^{-2} day^{-1} to 9,143 mg m^{-2} day^{-1}, respectively. Minimum values were recorded at the beginning of the experiment and maximum peaks in early and late February (Fig. 5b).

Water samples

There were relatively constant chlorophyll *a* content values during the sampling period (Fig. 6). These values ranged between 0.3 and 0.5 mg m^{-3} day^{-1}, with two peaks of 0.9 and 1 mg m^{-3} day^{-1}, which corresponded to mid-January and mid-February, respectively. These two peaks occurred when organic carbon content was high and the OC/N ratio had relatively low values.

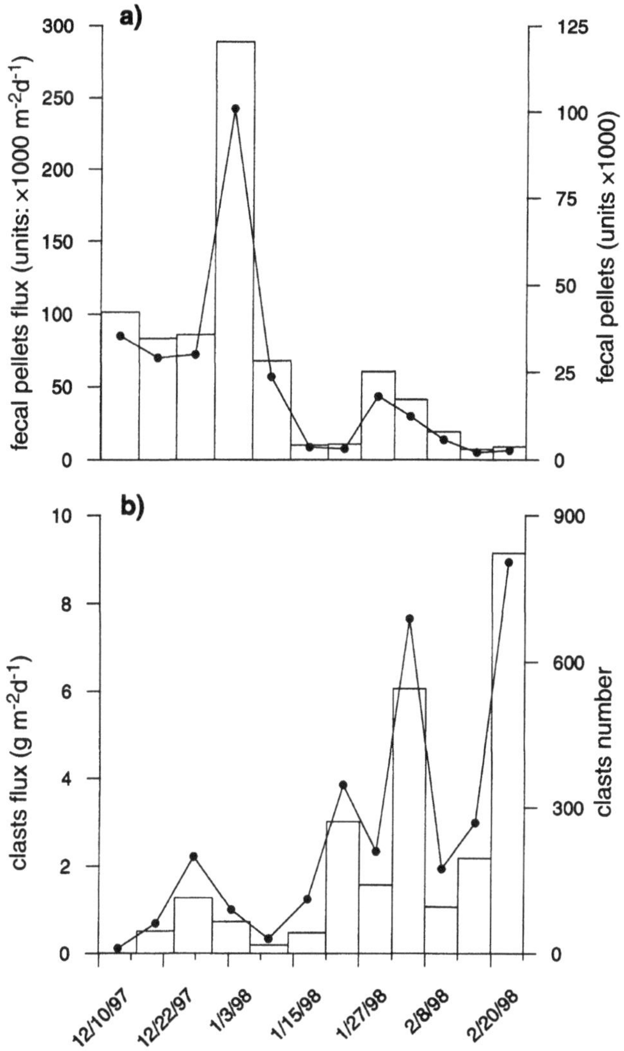

Fig. 5 Time series of clasts (**a**) and faecal pellets (**b**) fluxes (*histogram*) and contents (*filled circles*) during the study period

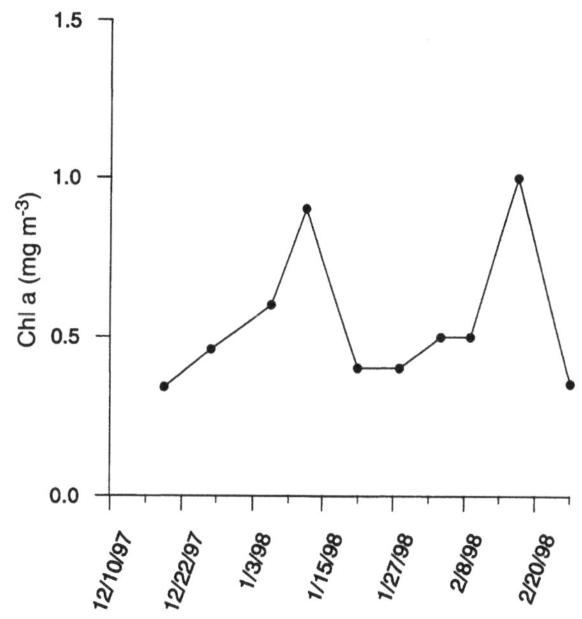

Fig. 6 Temporal variation in chlorophyll *a* at 10 m depth during the study period

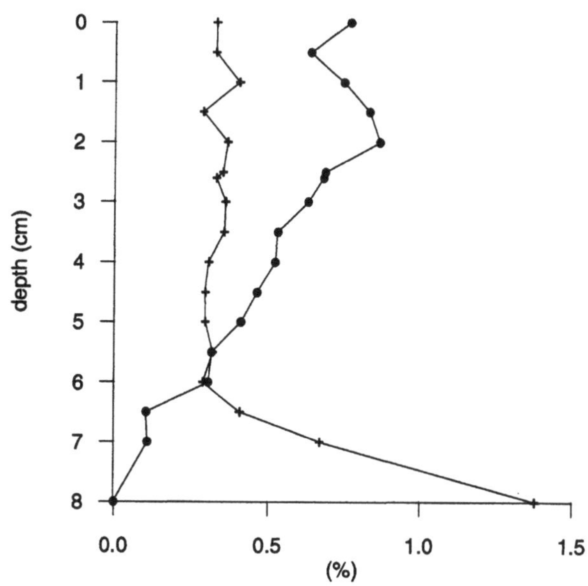

Fig. 7 Vertical profile of organic carbon (*filled circles*) and calcium carbonate (*crosses*) content in the bottom sediment at the mooring site

Bottom sediment

The organic carbon content in the bottom sediment varied between 0.28 and 1.38%, with an average value of 0.42% (Fig. 7). The calcium carbonate content ranged from 0 to 0.86%, with a mean value of 0.51% (Fig. 7). The OC content remained quite constant in the upper centimetres of the sediment record, and increased at the same depth where the carbonate content decreased to zero.

Discussion and conclusions

Johnson's Dock is a shallow closed system that receives icebergs and water from the Johnson's glacier. The interchange with the open sea is restricted and takes place only via its narrow and shallow mouth. The settling total mass fluxes of this bay are of the same order of magnitude as those measured in other Antarctic coastal environments (Cripps and Clarke 1998). The TMF collected during the summer period is higher than 20,000 mg m^{-2} day^{-1}, with a very high lithogenic content, and shows an inverse relation to the organic components content. In Johnson's Dock, TMF increased by a factor of 2 in mid-December and mid-February, and by a factor of 3–4 in early and late January. These TMF increments were linked to increases of the lithogenic content and to decreases of organic carbon and biogenic opal contents. However, the fluxes of the organic components increased with the TMF, suggesting that the increase in the lithogenic content caused the decrease in the organic components content by dilution.

Lithogenic inputs come mainly from erosive action and melting of the glaciers. Clasts coarser than 1 mm must have been transported by icebergs and settled into the trap during ice melting. The clasts flux shows an increasing trend with time, indicating that ice melting increases during summer, but the clasts flux trend is not directly related to the TMF trends. This is because the flux of fine lithogenic components is more than 5 times higher than that of > 1-mm lithogenic clasts, so fine lithogenic flux is dominant. The fine lithogenic material has a different dynamics to coarse lithogenic material, because fine particles can remain suspended in the water column along density gradients (Domack et al. 1994), and their sinking is controlled by other hydrographic processes.

Regarding the organic components, their fluxes were similar to, or significantly higher than those measured in deeper regions of the Bransfield Strait (DeMaster et al. 1987; Karl et al. 1991; Palanques et al. 2002; Wefer et al. 1988), even though their relative contents were much lower than those from offshore. This is not because there are low organic inputs, but because they are diluted by the high lithogenic inputs of this bay.

The chlorophyll *a* in the surface water is related to primary productivity and biogenic component inputs. The temporal evolution of chlorophyll *a* in the bay does not show a direct relation to the content and fluxes of organic components and faecal pellets during the study period (Fig. 6). However, the peaks of chlorophyll *a* occurring in mid-January and late February correspond to increases in organic carbon content. Both chlorophyll *a* peaks coincide with low OC/N values, suggesting that settling organic matter was fresher during these periods of time. The peak value of faecal pellets indicates that a bloom of zooplankton could have happened in early January before the mid-January chlorophyll *a* peak. The February peak could have some relation to the biomass bloom that tends to happen in the open water of the Bransfield Strait during this month (Holm-Hansen and Mitchell 1991). The higher OC/N values correspond to the highest lithogenic fluxes, which probably indicates that organic matter during these events could have been recycled and/or degraded.

Settling OC was dominant over IC and ranged from 77 to 100% of the total carbon. Organic carbon and calcium carbonate content in the surface sediment of the study site showed similar values to those of the settling particles (Fig. 7). This means that the particles in the water column reach the sediment without being significantly altered. The residence time of the particles in shallow environments is probably too short to allow the transformations that happen in Antarctic deeper environments, where carbonate in sediments is lower than in the settling particulate matter (Palanques et al. 2002). In Johnson's Dock, the carbonate content in the sediment column decreases with depth, indicating that dissolution is gradual from the surface to about 8 cm depth, where the $CaCO_3$ content is zero (Fig.7). This indicates that the shells of carbonated organisms settle and accumulate in the bottom sediment where they are dissolved after being buried.

The results obtained from this experiment reveal that even though there are no large river systems in the Bransfield Strait, high sediment fluxes occurred in the shallow environments mainly due to sediment supplies from the action of ice. The lithogenic fine-sediment fraction is the dominant contributor to these fluxes, but it is important to note that the input of coarse clasts from iceberg melting is also very relevant in the study area. However, not only the processes associated with ice melting seem to control the temporal variability of the settling fluxes of particles, and hydrodynamic processes should also be studied in these shallow environments to understand the mechanisms that control their particle fluxes.

Acknowledgements This research was supported by the project MAR96-1781-CO2-01, funded by the "Comisión Interministerial de Ciencia y Tecnología" and by the "Plan Nacional de Investigación en la Antártida" as a Spanish contribution to the EASIZ-SCAR programme. It was also benefited by fellowship 92766 from the Consejo Nacional de Ciencia y Tecnología (México).

References

Cripps GC, Clarke A (1998) Seasonal variation in the biochemical composition of the particulate material collected by sediment traps at Signy Island, Antarctica. Polar Biol 20:414–423

DeMaster DJ, Nelson TM, Nittrouer CA, Harden SL (1987) Biogenic silica and organic carbon accumulation in modern Bransfield Strait sediments. Antarct J US 12:108–110

Domack EW, Foss DJP, Syvitski JPM, McClennen CE (1994) Transport of suspended particulate matter in an Antarctic fjord. Mar Geol 121:161–170

Griffith TW, Anderson JB (1989) Climatic control of sedimentation in bays and fjords of the northern Antarctic Peninsula. Mar Geol 85:181–204

Heussner S, Ratti C, Carbonne J (1990) The PPS 3 time-series sediment trap and the trap sample processing techniques used during the ECOMARGE experiment. Cont Shelf Res 10: 943–958

Holm-Hansen O, Mitchell BG (1991) Spatial and temporal distribution of phytoplankton and primary production in the western Bransfield Strait region. Deep Sea Res 38:961–980

Huntley M, Karl DM, Niiler P, Holm-Hansen O (1991) Research on Antarctic Coastal Ecosystems Rates (RACER): an interdisciplinary field experiment. Deep Sea Res 38:911–941

Karl DM, Tilbrook BD, Tien G (1991) Seasonal coupling of organic matter production and particle flux in the western Bransfield Strait, Antarctica. Deep Sea Res 38:1097–1126

Klöser H, Ferreyra G, Schloss I, Mercuri G, Laturnus F, Curtosi A (1994) Hydrography of Potter Cove, a small fjord like inlet on King George Island (South Shetlands). Estuarine Coastal Shelf Sci 38:523–537

Kowalewski J, Wielbinska D (1983) Characteristics of variation of meteorological elements in Ezcurra inlet during the Polish Academy of Sciences' second Antarctic expedition from 20 December 1977 to 16 March 1978. Oceanologia 15:7–19

Mortlock RA, Froelich PN (1989) A simple method for the rapid determination of biogenic opal in pelagic marine sediments. Deep Sea Res 36:1415–1426

Palanques A, Isla E, Puig P, Sanchez-Cabeza JA, Masqué P (2002) Annual evolution of settling particle fluxes in Western Bransfield Street (Antarctice) during the FRUELA project. Deep Sea Res II 49:903–920

Pruszak Z (1980) Currents circulation in the waters of Admiralty Bay (region of Arctowski station on King George Island). Pol Polar Res 1:55–74

Strickland JDH, Parsons TR (1972) A practical handbook of seawater analysis, 2nd edn. Fish Res Board Can Bull 167:1–326

Wefer G, Fischer G (1991) Annual primary production and export flux in the Southern Ocean from sediment trap data. Mar Chem 35:597–613

Wefer G, Fischer G, Fütterer D, Gersonde R (1988) Seasonal particle flux in the Bransfield Strait, Antarctica. Deep Sea Res 35:891–898

Yoon HI, Han MW, Park BK, Han SJ, Oh JK (1992) Distribution, provenance, and dispersal pattern of clay minerals in surface sediment, Bransfield Strait, Antarctica. Geo-Mar Lett 12: 223–227

Polar Biol (2001) 24: 620–627
DOI 10.1007/s003000100272

Covadonga Orejas · Josep Maria Gili
Pablo J. López-González · Wolf E Arntz

Feeding strategies and diet composition of four Antarctic cnidarian species

Abstract The diet of four species of Antarctic cnidarians, two hydroids and two anthozoans, was investigated. One hydroid, *Tubularia ralphii*, and one anthozoan, *Anthomastus bathyproctus*, seem to basically consume zooplankton whereas the other hydroid species, *Oswaldella antarctica*, has a diet mainly based on the fine fraction of seston. The last cnidarian investigated, the stoloniferan alcyonacean, *Clavularia* cf. *frankliniana*, feeds mainly on resuspended material. The wide range of diets of these Antarctic cnidarians indicates opportunistic behaviour by feeding on different sources and taking advantage of the available food sources. Data related to high C/N ratios in the sediment and high NO_2 concentrations in areas with dense communities of benthic suspension feeders, along with data on capture rates, lead us to hypothesize that these organisms play an important role in the recycling processes of organic matter in Antarctic benthic ecosystems.

Please note: When citing this article, please use the reference to the original source in the journal Polar Biology as given on the top of this page.

C. Orejas (✉) · W.E. Arntz
Alfred Wegener Institut für Polar- und Meeresforschung,
Columbusstrasse, 27568 Bremerhaven,
Germany
E-mail: corejas@awi-bremerhaven.de
Tel.: +49-471-4831304/1327
Fax: +49-471-4831149

J.M. Gili
Institut de Ciències del Mar (CSIC),
Plaça del Mar s/n, 08039 Barcelona,
Spain

P.J. López-González
Laboratorio de Biología Marina,
Universidad de Sevilla,
Apartado 1095, 4080 Sevilla,
Spain

Introduction

Cnidarians are one of the most conspicuous taxonomic groups on the Antarctic continental shelf, being part of the rich benthic communities which have been described from several locations around Antarctica (e.g. Arntz et al. 1994; Gutt and Starmans 1998; Starmans et al. 1999). Like other suspension feeders, cnidarians depend on the available food suspended in the water column and on the near-bottom currents for its renewal.

The Southern Ocean is principally an oligotrophic environment dominated by flagellates and other small organisms of the pelagic retention system, with short overlying outbursts of microalgae (bloom system) in spring and summer (Scharek and Nöthig 1995) which are, however, essential for larger zooplankton. For quite some time, these blooms were suspected of determining the feeding strategies of the benthic suspension feeders in Antarctica and the composition of their diet. These organisms were thought to feed heavily during a few weeks in summer and to be able to sustain long periods of starvation during most of the year. This would resemble overwintering as resting stages or the hibernation periods that have been described for other latitudes (Gili and Hughes 1995; Garrabou 1999). However, Barnes and Clarke (1995) demonstrated recently that certain species of Antarctic shallow-water benthic suspension feeders are able to feed throughout the year with only short periods of starvation. Phytoplankton (mainly microplankton) in these areas is also dominant in the water column and the sediment only during short periods of the year, but pico- and nanoplankton (between 0.2 and 20 μm) dominate the water column and the sediment during most of the year, and the suspension feeders take advantage of them (Cripps and Clarke 1998). From these findings, the question arose as to whether this situation might apply to deeper shelf communities as well.

In the Southern Ocean, vertical fluxes (Nöthig and von Bodungen 1989) are not the only source of food for the benthic suspension feeders, but zooplankton, seston, elements of the microbial food web and resuspension processes could also contribute to the potential available food for this group. Although cnidarians have traditionally been considered mainly carnivorous (Hyman 1940), more recent studies have found evidence of herbivory (Elyakova et al. 1981; Fabricius et al. 1995a, b), which is supported by a study on an Antarctic hydroid, *Silicularia rosea*, which consumes mainly diatoms (Gili et al. 1996b). However, gorgonian diet varies from dissolved organic matter via plankton to benthic organisms, and is very well coupled with the dynamics of their potential prey (Ribes et al. 1998, 1999).

Considering the various potential food sources for benthic suspension feeders, the goal of the present study was to determine the diet of four species of Antarctic cnidarians: *Oswaldella antarctica* (Hydrozoa), *Tubularia ralphii* (Hydrozoa), *Clavularia* cf. *frankliniana* (Anthozoa) and *Anthomastus bathyproctus* (Anthozoa). We also consider the relation between their feeding strategies, the Antarctic environment and the factors which regulate the availability of food at the sea floor. We present new insights into the significance of plankton production, resuspension processes and nekton migration for the feeding strategies of these organisms, and on the role of the group in the trophic food web of the Southern Ocean.

Materials and methods

Four benthic cnidarian species were studied (Fig. 1). *O. antarctica* (Jäderholm 1904) is a benthic hydroid found on hard substrates in a depth range between 50 and 200 m, approximately, forming arborescent colonies and presenting very small polyps (200–375 μm diameter). *T. ralphii* Bale 1884 is a common Antarctic hydroid, which forms creeping colonies on stones and is mainly distributed in shallow areas (Stepanjants 1979). *T. ralphii* presents large hydranths with long aboral tentacles, and the hydrocaulus is usually more than 2 cm above the substratum. The stoloniferous alcyonacean, *C.* cf. *frankliniana* Roule, 1902 was one of the most common species collected in shallow waters during the second EASIZ (Ecology of the Antarctic Sea Ice Zone) cruise. This species forms encrusting colonies covering, in some cases, great parts of the stones. *A. bathyproctus* Bayer, 1993 is a common alcyonacean in the waters of the Antarctic Peninsula area. Their colonies live mainly on sandy substrates. They were caught between 400 and 450 m depth; however, off the South Shetland Islands, specimens have been found at 1,153 m depth (Bayer 1993).

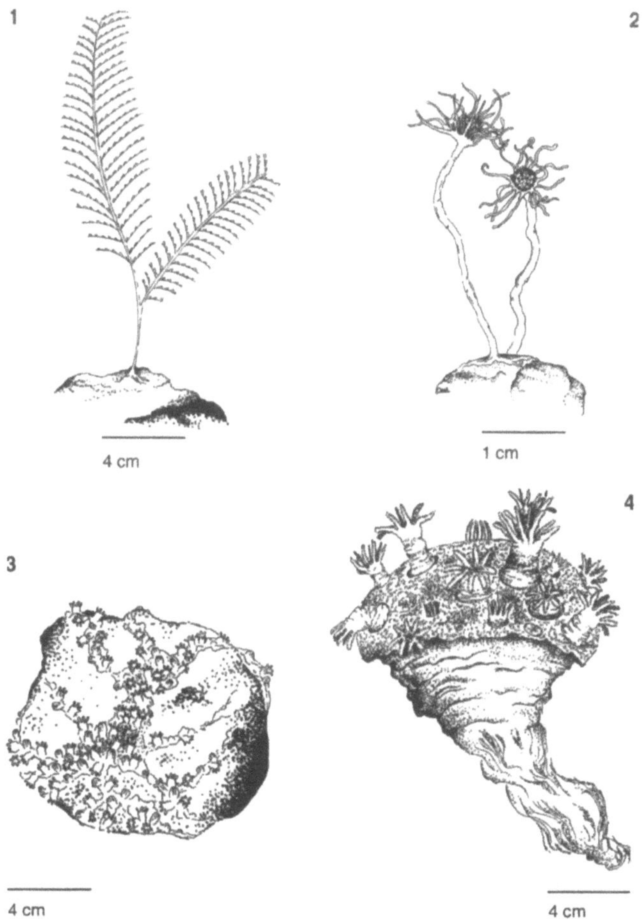

Fig. 1 Studied species: *1 Oswaldella antarctica, 2 Tubularia ralphii, 3 Clavularia* cf. *frankliniana, 4 Anthomastus bathyproctus*

Sampling, experimental design and analysis of gastrovascular contents

Sampling on board RV "Polarstern" was carried out during the second EASIZ cruise between January and March 1998 at two sites: for *O. antarctica*, *T. ralphii* and *C.* cf. *frankliniana* in the eastern Weddell Sea, and for *A. bathyproctus* off the Antarctic Peninsula (Table 1). Sampling was conducted with an Agassiz trawl, a bottom trawl and a large TV grab of 1.82 m^2 sampled area.

In order to determine the trophic role of the seston (also of its fine fraction) for suspension feeders, data are derived from analyses of cnidarian gastrovascular contents and from feeding experiments. Polyps of all four species were dissected to analyse the gastrovascular contents under a binocular and a light microscope. The number of dissected polyps was between 110 and 148, depending on the species, because of their varying abundance in the samples and in the studied area.

Feeding experiments were done only with the hydrozoan species, *O. antarctica*. A modified prototype of incubator developed by the Instituto de Ciencias del Mar (Ribes et al. 1998) was used to study the diet of benthic suspension feeders. The incubator consists of a chamber in which the colony is placed, and an empty control chamber. Each closed chamber was connected to a pump that

Table 1 Sampling stations for the four studied species (*KG* King George Island; *KN* Kapp Norvegia; *N/KN* north of Kapp Norvegia; *AGT* Agassiz trawl; *BT* bottom trawl; *TV grab* bottom grab 1.82 m^2 with video camera)

	Area	Station	Gear	Latitude (S)	Longitude (W)	Depth (m)
Oswaldella antarctica	N/KN	71	BT	70°49.3′	10°28.6′	281
	KN	210	TVgrab	71°07.2′	11°28.4′	67
Tubularia ralphii	KN	210	TVgrab	71°07.2′	11°28.4′	67
Clavularia cf. *frankliniana*	KN	210	TVgrab	71°07.2′	11°28.4′	67
Anthomastus bathyproctus	KG	303	AGT	62°16.5′	58°43.1′	430
	KG	304	AGT	62°16.4′	58°43.9′	447

recirculated seawater at a renewal rate of 3.3 l min^{-1}. Water was collected as close to the sea floor as possible by means of a CTD equipped with a rosette. Temperature and salinity were measured simultaneously using a CTD. The chambers and all necessary material and instrumentation were placed in a cool-room container equipped for operation at Southern Ocean ambient temperature (-1°C to +0.5°C). Colonies were kept in seawater within the cool room until the incubation experiments. Animals were placed in the experimental chamber and were allowed to fully expand before the experiment started. After the acclimation time (which was from 3–4 h for some colonies to 1 day for others), the incubation chambers were closed and initial water samples of 2,000 ml were taken from both chambers. After 6 h another set of water samples was collected. Capture rates were calculated using the decreasing concentration levels of bacteria, flagellates, ciliates, and phytoplankton in the water samples, considering their growth rates during the experiment. To quantify heterotrophic bacteria, water samples were stored for flow cytometry by standard methods (Gasol and del Giorgio 2000). Quantification and measurement of picoplankton was done using epifluorescence microscopy. To quantify, measure and identify phytoplankton and ciliates, water samples were stored in acid Lugol. Nutrients (NO_2, NO_3, Si, PO_4) for the water samples were also measured using a Technikon Autoanalyzer II System von Bran and Lübbe and following standard methods. Carbon and nitrogen were measured using a Leco CN 2000 analyser.

In the sampling areas, a visual check of benthic fauna was made just when the gear arrived on board, to find possible relations between nutrient concentrations, C/N ratio and densities of benthic suspension feeders. At all sampling sites, a sample of 2 cm^3 of surface sediments was collected and frozen in liquid nitrogen, and carbon and nitrogen content were measured using a Leco CN 2000 analyser.

To determine the abundance of zooplankton near the bottom, complementary data collected during the second EASIZ cruise by means of a epibenthic sledge were revised and compared with the zooplankton abundance and distribution in the water column (Brandt et al. 1999; Schiel, personal communication).

Statistical and numerical treatment

Results are noted as means ± standard deviation, and the differences between the samples are tested using ANOVA model 2 at different significance levels. The Wilcoxon test was used to test possible differences of prey net growth rate, POC (Particulate Organic Carbon), DOC (Dissolved Organic Carbon) and nutrient values (NO_2, NO_3, Si, PO_4) in the experimental and control chambers. Depletion rates of the plankton were calculated, assuming exponential growth and clearance of prey as described in Ribes et al. (1998).

Results

Gastrovascular contents

O. antarctica contained three different types of prey number: diatoms (94%), invertebrate eggs (4%) and invertebrate larvae (2%) (Fig. 2). Sizes of the prey were from 20 µm (smallest) to 235 µm (largest); however, 81% of prey belonged to the size range 20–42 µm (Fig. 3). Twenty-five of the dissected polyps (23%) contained some prey in their gastrovascular cavities whereas 85 were empty (77%) (Fig. 4). Statistically, significant differences were found in prey size among different polyps (ANOVA, $F=5.35$, $N=110$; $P<0.05$).

One hundred and four (71%) polyps of *T. ralphii* contained prey and 43 (29%) were empty (Fig. 4). The diet was composed mainly of copepods (85%), followed by invertebrate eggs (7%), organic matter (7%) and copepod nauplii (~1%) (Fig. 2). Prey number and prey size showed significant differences between polyps (ANOVA, $F=74.876$, $N=147$; $P<0.0001$; ANOVA, $F=11.165$, $N=147$; $P<0.005$, respectively). The size range of prey items was very wide (45–1,145 µm) (Fig. 3).

Of a total of 148 polyps of *C.* cf. *frankliniana* (belonging to different colonies), 36 (24%) were empty and 112 (76%) showed prey items in the gastrovascular cavity (Fig. 4) The diet was diatoms (81%), protozoans (8%), foraminiferans (5%), nematodes (4%) and invertebrate larvae (2%) (Fig. 2). Size range of prey varied between 48 and 1,920 µm, with 92% of prey being in the range 48–235 µm (Fig. 3). Prey number varied significantly between polyps of different diameter (ANOVA, $F=5.928$, $N=148$; $P<0.05$), but not between different prey size.

The 8 studied colonies of *A. bathyproctus* had a total of 136 polyps. Salps belonging to the species *Salpa*

Fig. 2 Percentage (by number) of full and empty gastrovascular cavities for the four studied species

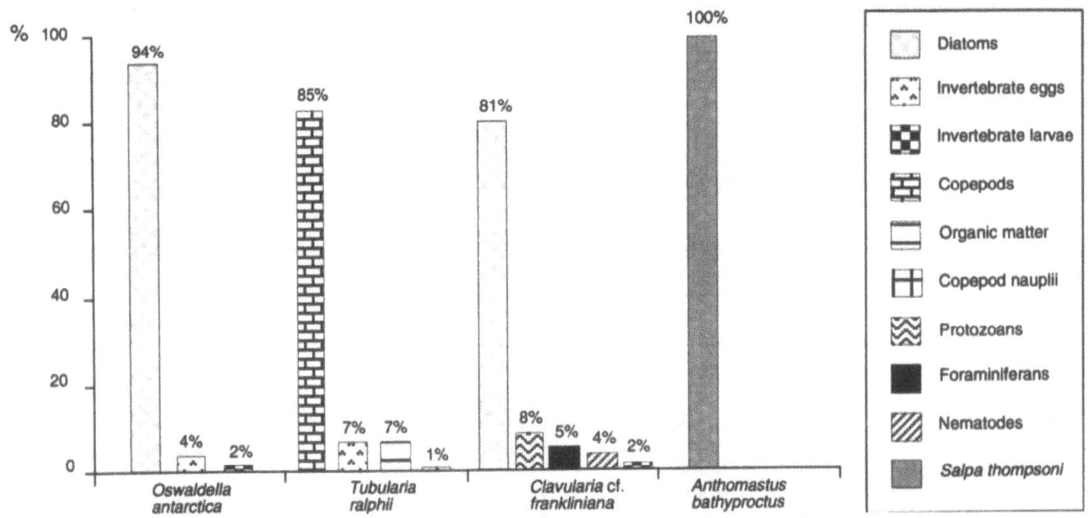

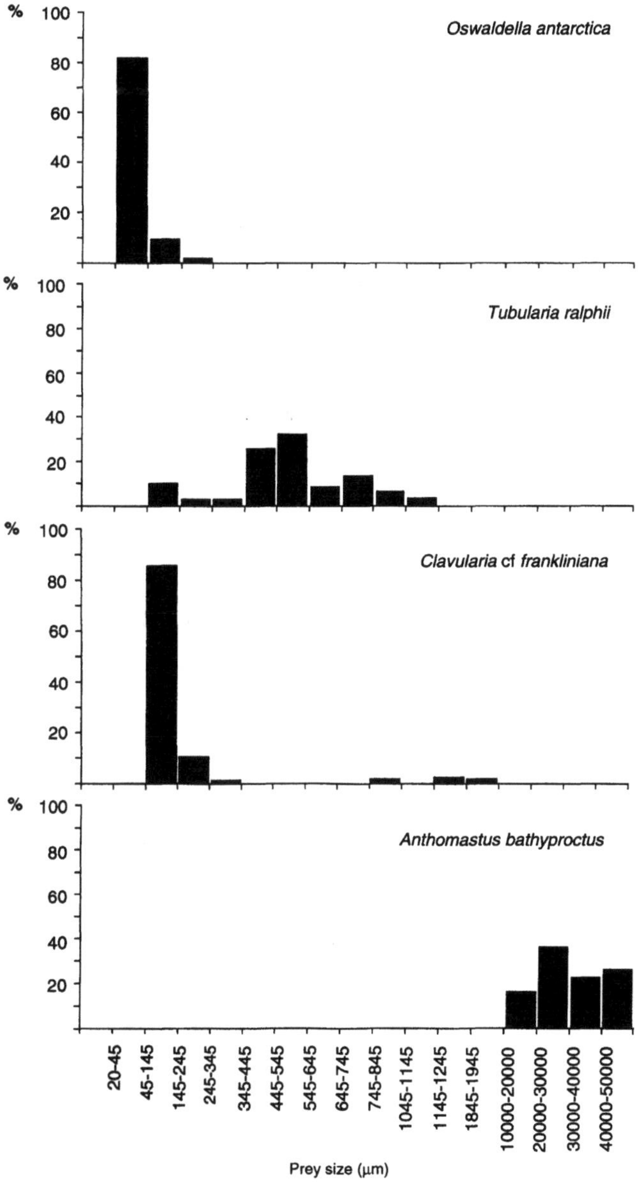

Fig. 3 Percentage (by number) of prey items of different size range found in the dissected polyps for the four studied species

thompsoni were found in 113 (83%) of them whereas 23 (17%) were empty (Fig. 4). The size of the salps was between 10 and 50 mm (Fig. 3). Many polyps contained other items (mainly diatoms of the genus *Corethron*, and also centric diatoms and fecal pellets of small size). It seems, however, that these items came from prey gut contents, which was confirmed by dissecting the prey guts. No statistically significant differences were found between number and size of prey items and polyp diameter, or between number of prey items and capitulum diameter. However, there were significant differences between number of prey items and height of the capitulum (ANOVA, $F=10.649$, $N=8$; $P<0.01$), and between prey size and height of the capitulum (ANOVA, $F=6.087$, $N=8$; $P<0.05$). The number of polyps/capitulum in relation to height of the capitulum and also with capitulum diameter showed significant differences (ANOVA, $F=61.898$, $N=8$; $P<0.001$; ANOVA, $F=139.033$, $N=8$; $P<0.001$, respectively).

Feeding experiments

Net prey growth rates were calculated for the potential prey items of *O. antarctica* working at the level of taxonomic group: bacteria, nanoflagellates, dinoflagellates, ciliates and diatoms; POC and DOC values were also calculated. Sizes of the prey items were also taken into account (excluding in this case values of bacteria). Statistically significant differences ($P<0.01$) were found for nanoflagellates (Wilcoxon test). Results are shown in Fig. 5. However, a pattern could also be observed for the other groups: dinoflagellates showed a clear decrease in the experimental chamber, as did *Chaetoceros* species and other phytoplankton groups. Different types of diatoms were lumped because of the low number of specimens found in the samples.

Nutrients and C/N ratio

The concentration of nutrients NO_2, NO_3 and PO_4 varied between 0.02 and 0.20 μmol/l, 15.94 and 40.35 μmol/l and 0.77 and 1.49 μmol/l, respectively. C/N values in the sediment varied between 6 and 293, showing a clear pattern in relation to the distribution of benthic suspension-feeder communities: areas which showed a high presence of benthic suspension feeders presented the highest values (Fig. 6). Conversely, the areas with higher nitrogen values correspond to areas with a low density of benthic suspension feeders. The same pattern was observed for nutrient values, with

Fig. 4 Percentage (by number) of prey items in the dissected polyps for each studied species. *Black colour* shows coelenteron with prey items; *white colour* without prey items

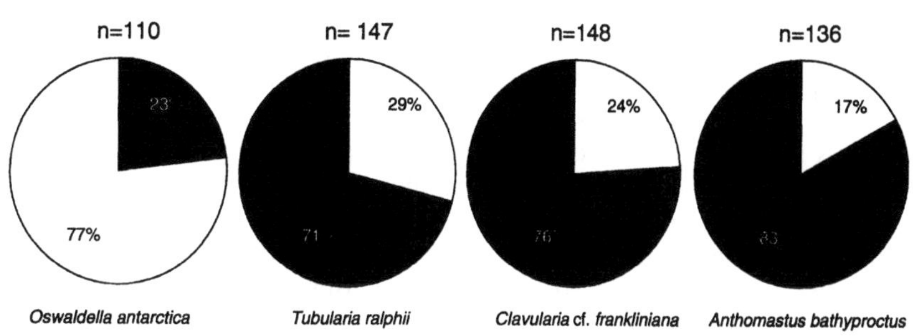

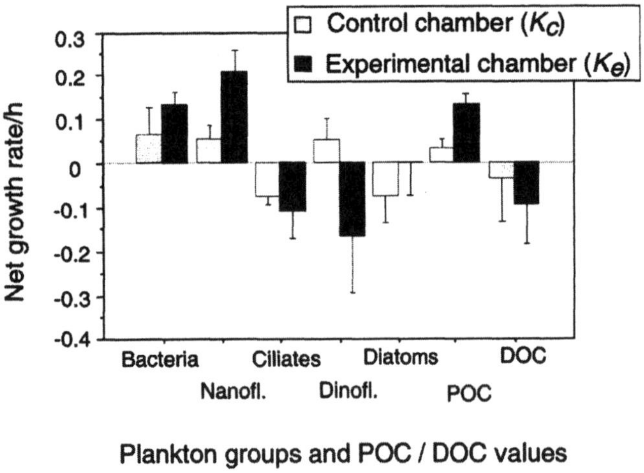

Fig. 5 Net growth rate in control (*Kc*) and experimental (*Ke*) chambers of the different groups of plankton and POC/DOC

higher values in areas where benthic suspension feeder communities were abundant.

Zooplankton abundance

Data of zooplankton densities (focusing on copepod data because of their abundance) were extracted from preliminary results of the second EASIZ cruise. Near-bottom copepod abundances, obtained by epibenthic sledge, were registered by Brandt et al. (1999) on the Weddell Sea shelf, reaching summer values between 232 and 702 individuals/1,000 m^2 at 400 m depth. Copepod abundance (calanoids) was also measured by Schiel (personal communication) in the upper 200 m of the water column in the same area. The values ranged between 1,000 and 4,500 ind./1,000 m^3 during the Antarctic summer.

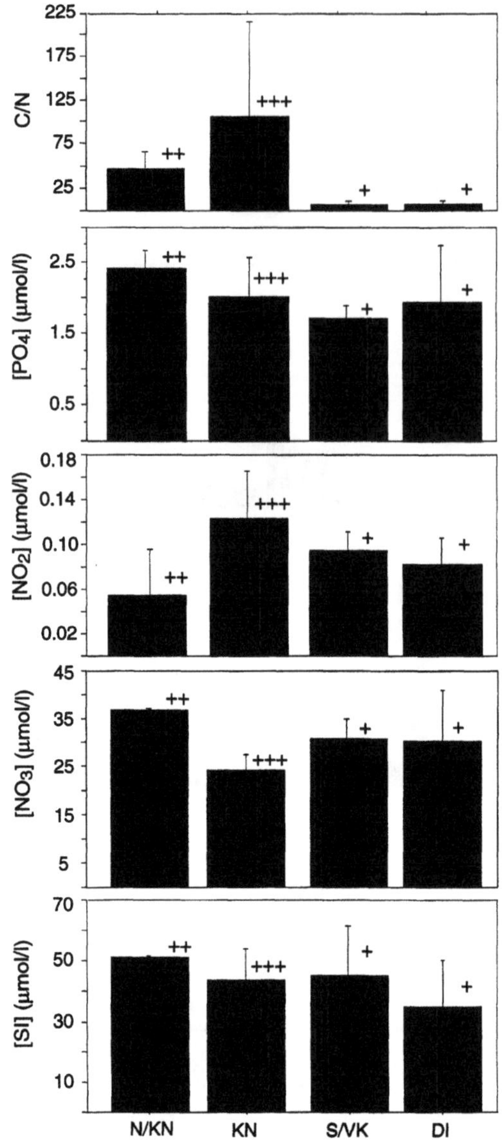

Fig. 6 Values of C/N rate and nutrients for the areas north of Kapp Norvegia (*N/KN*), Kapp Norvegia (*KN*), south of Vestkapp (*S/VK*) and Drescher Inlet (*DI*) (+ + + high dominance of benthic suspension feeders; + + regular abundance of benthic suspension feeders; + occasional presence of benthic suspension feeders)

Discussion

Anthozoans and hydrozoans are groups which have received great attention in tropical and temperate waters, and substantial data on their natural diet are available (anthozoans, e.g. Coma et al. 1994; Ribes et al. 1998; hydrozoans, e.g. Gili et al. 1998). Scarce information exists, however, for Antarctic and Arctic species (Gili et al. 1996b; Slattery et al. 1997). Benthic zooplanktivorous species, which feed on substratum-associated organisms or larvae, also consume a few holoplanktonic animals such as calanoid copepods (Sebens and Koehl 1984). We name these species "macrophagous" but introduce a new food item, the fine fraction of seston, calling the species which feed on this fraction, "microphagous". The four studied species reflect this whole spectrum of prey items (Fig. 2).

It has been hypothesized that benthic organisms just feed during favourable periods, remaining in hibernation or in resting stages during cold seasons and under unfavourable environmental conditions (for review see Gili and Hughes 1995). This is true for the "macrophytoplankton system" (Smetacek et al. 1990; Scharek and Nöthig 1995) which occurs in summer, and most zooplanktonic (hence macrophagous) species profit from these summer blooms to grow and reproduce. However, the "flagellate system" (Smetacek et al. 1990; Scharek and Nöthig 1995), consisting mainly of nanoplankton and members of the microbial food web, is also active during winter (Clarke 1988; Ahn 1993). This system might be capable of sustaining communities of benthic suspension feeders during the whole year, as has been observed by Barnes and Clarke (1995) at Signy Island.

Cnidarian diet: macrophagous and microphagous suspension feeders

Macrophagous species: Tubularia ralphii, Clavularia cf. frankliniana, Anthomastus bathyproctus

T. ralphii showed a diet highly dependent on the water column; 85% of their diet were copepods, some of which have been identified as pelagic harpacticoids, indicating a dependence on the planktonic community. The lack of relation between polyp size and prey size suggests an opportunistic strategy contrasting with that observed in a congeneric species from the North Atlantic, *T. larynx* (Gili et al. 1996a).

The feeding strategy of the stoloniferan *C.* cf. *frankliniana* could be linked with resuspension processes at the sea floor. Its polyps are located very close to the substrate and they capture prey moving around their tentacle crown. This kind of feeding is quite frequent in massive and encrusting anthozoans such as alcyonaceans and actiniarians (Sebens and Koehl 1984). Sediment might be continually resuspended by the effect of currents, but also by various animals such as bivalves (e.g. Stockton 1984; Alexander and DeLaca 1987), echinoderms and demersal fish.

The diet of *A. bathyproctus* is probably related to the patchiness of plankton. They might profit from episodes of high availability of food, as occurs during the salps' downward migration. The occurrence of gelatinous zooplankton close to the sea floor has been shown in some cases (Gili et al., 2001). Vertical migration of Antarctic zooplankton to the sea bed has been documented for salps (Weibe et al. 1979; Casareto and Nemoto 1986) and krill (Gutt and Siegel 1994). In gorgonians (Coma et al. 1994), plankton patches can cause events of high prey capture rates. The presence of large, although occasional, prey in the cavities of the studied species may be quite significant for their energetic balance (Coma et al. 1998). In the cavities of the *A. bathyproctus* specimens, we also found diatoms (mainly *Corethron* spp.) which seem, however, to come from prey gut contents, as has been described for other prey items in other cases (Sebens and Koehl 1984).

Two possible strategies might be applied by *A. bathyproctus*: (1) they might feed on zooplankton, i.e. prey that occasionally passes by, in an opportunistic way, but feed on the fine fraction of seston too, using a "sieving" mechanism; (2) they might feed episodically only on large prey items. This latter possibility does not appear to be realistic because they would depend completely on Antarctic summer production to survive. A more realistic scenario is that they use, in different proportions, both resources. In any case, *A. bathyproctus* is neither a deposit nor a sediment feeder. Its anatomical characteristics show the animal to be capable of doubling its size by way of hydrostatic mechanisms. Thus it can take better advantage of the water column, the food composition of which can vary greatly from the bottom to some centimetres above it (Schloss et al. 1999).

Microphagous species: Oswaldella antarctica

Little information exists on the role of types of prey other than zooplankton in hydroid feeding (Gili and Hughes 1995). Several studies reported that dinoflagellates, diatoms, tintinnids and bacteria are sparse in hydroid gut contents (Gili et al. 1996a). However, nanoplankton, mainly of plant origin, is the most abundant food source in the Southern Ocean, and Coma et al. (1995) demonstrated that detritus from plant matter or particles of unspecified origin represent more than 90% of prey captured in a Mediterranean hydroid. Diatoms have been also reported to be an important food source for the Antarctic hydroid *Silicularia rosea* (Gili et al. 1996b).

Recently it has been observed that the fine fraction of seston (pico- and nanoplankton) may represent an important part of the diet of benthic suspension feeders (Pile et al. 1996; Ribes et al. 1999). The studies by Bröckel (1981) showed that nanoplankton cells are also a very important component of the planktonic community in nearshore Antarctic waters. The case of *O. antarctica* is the first where the importance of this fraction has been shown for hydroids. Our feeding experiments show that components from the fine fraction of seston contribute to the diet of the hydroid (Fig. 5). This is not surprising taking into account that particles smaller than 2 μm and between 2 and 20 μm diameter dominate the pelagic planktonic community in terms of biomass and production (Platt et al. 1983). It has also been demonstrated that suspension feeders can remove very effectively particles of these sizes, and that active filtration is not the only method by which suspension feeders efficiently extract small particles from the water (Rubinstein and Koehl 1977; Okamura 1990). The importance of small prey in the diet of hydroids indicates that other mechanisms of capture may also occur, e.g. inertial impact or gravitational deposition, and creation of currents (Koehl 1977a, b; LaBarbera 1984; Harris 1990).

A wide variety of diets in benthic cnidarians has recently been shown also for two gorgonian species: *Plexaura flexuosa* (Ribes et al. 1998) in the Caribbean, and *Paramuricea clavata* (Ribes et al. 1999) in the Mediterranean Sea, with prey items comprising bacteria to zooplankton. Apparently, feeding strategies and diets in this group are not very different in Antarctica compared with other latitudes. An example is the capture rates of *T. ralphii*, which are in the same order of magnitude as for related species in temperate waters (Gili et al. 1998).

Role of the suspension-feeder communities in the Antarctic benthic communities

Particle composition in the water column in Antarctica may be very different during the year (Clarke and Leakey 1996). In the Southern Ocean, a strong vertical flux of organic matter occurs during a short time, but due to the low decomposition rate of the sediments, they keep their nutritional value longer (Bathmann et al. 1991). The data of Palanques et al. (2002) show high values of nitrogen even in winter. Benthic communities may respond to sedimentation events, increasing their activity and production (e.g. Grémare et al. 1997).

The high values of C/N ratio observed (Fig. 6) in places where the benthic suspension feeders are very abundant might reflect their feeding activity consuming the sediment organic component (POC or small organisms). Benthic cnidarians might be able to use sediment material due to resuspension processes. A similar set of sediment C/N values has been observed in the Bransfield Street (Palanques et al., 2002) and the authors hypothesize that this organic sediment decomposition could be related to the activity of benthic organisms, and not to the different organic composition of material sedimenting from the water column. Parallel to these near-bottom feeding processes, high nutrient values such as the observed nitrite values may be expected (Fig. 6). Below the depth limit of photosynthetic activity, nitrite cycling is controlled by microbial nitrification, as well as denitrification in the water layer close to the bottom, but benthic filter feeders can increase nitrite cycling by ammonium from their excretion processes. The role of nitrite as an obligatory intermediate in nitrification processes and the high values found during our experimental work point to the possible role of benthic suspension feeders in the recycling and remineralization of the organic matter and their role as very active organisms in the dynamics of benthic boundary layers.

In summary, cnidarians exploit both patch food and seasonally available food. This information on the natural diet of Antarctic benthic cnidarians supports the hypothesis that they are adapted to perform a wide spectrum of prey capture, which enhances their survival in a stochastic environment. Furthermore, it indicates that the activity and production of Antarctic benthos does not stop during winter. The role of benthic suspension feeders in such communities is still far from being clear, but they must play a principal role both in the benthic pelagic coupling processes and in organic matter recycling.

Acknowledgements Financial support for this study was provided by a DAAD fellowship (A/96/13073), by a European Commission fellowship from the Programme "Training and Mobility of Researchers" (TMR-CT97-2813) and by CICYT (Spanish Antarctic Research Programme) grants ANT97-1533-E and ANT98-1738-E. We also thank Ms. Patricia Pinto, Ms. Nuria Teixidó and Dr. Joandomènec Ross for their help on board. We are grateful to Dr. Jens Kowalke and three anonymous referees for their criticism and comments, which improved the manuscript. Many colleagues who helped on board and the readiness to help by the crew of RV "Polarstern" are also gratefully acknowledged.

References

Ahn I-Y (1993) Enhanced particle flux through the biodeposition by the Antarctic suspension-feeding bivalve *Laternula elliptica* in Marian Cove, King George Island. J Exp Mar Biol Ecol 171:75–90

Alexander SP, DeLaca TE (1987) Feeding adaptations of the foraminiferan *Cibides refulgens* living epizoically and parasitically on the Antarctic scallop *Adamussium colbecki*. Biol Bull Woods Hole Mass 173:136–159

Arntz WE, Brey T, Gallardo VA (1994) Antarctic zoobenthos. Oceanogr Mar Biol Annu Rev 32:241–304

Barnes DKA, Clarke A (1995) Seasonality of feeding activity in Antarctic suspension feeders. Polar Biol 15:335–340

Bathmann U, Fischer G, Müller PJ, Gerdes D (1991) Short-term variations in particulate matter sedimentation off Kapp Norvegia, Weddell Sea, Antarctica: relation to water mass advection, ice cover, plankton biomass and feeding activity. Polar Biol 11:185–195

Bayer F (1993) Taxonomic status of the octocoral genus *Bathyalcyon* (Alcyoniidae: Anthomastinae), with descriptions of a new subspecies from the Gulf of Mexico and a new species of *Anthomastus* from Antarctic waters (Plates 1–9). Precious Coral Octocoral Res 1:3–13

Brandt A, Hilbig B, Linse K, Wegener G (1999) Biodiversity and zoogeography of Mollusca, Polychaeta, and Crustacea Peracarida. In: Arntz WE, Gutt J (eds) The Expedition Antarktis XV/3 (EASIZ II) of RV "Polarstern" in 1998. Ber Polarforsch 301:135–149

Bröckel K (1981) The importance of nanoplankton within the pelagic Antarctic ecosystems. In: Rheinheimer G, Fluegel H, Lenz J, Zeitzschel (eds) Lower organisms and their role in the food web, vol 5. European Marine Biology Symposium, Kiel. pp 61–64

Casareto BE, Nemoto T (1986) Salps of the Southern Ocean (Australian sector) during the 1983–84 summer, with special reference to the species *Salpa thompsoni*, Foxton 1961. Mem Natl Inst Polar Res Ser 40:221–239

Clarke A (1988) Seasonality in the Antarctic marine environment. Comp Biochem Physiol 90B:461–473

Clarke A, Leakey RJG (1996) The seasonal cycle of phytoplankton, macronutrients, and the microbial community in a nearshore Antarctic marine ecosystem. Limnol Oceanogr 41:1281–1294

Coma R, Gili JM, Zabala M, Riera T (1994) Feeding and prey capture cycles in the aposymbiotic gorgonian *Paramuricea clavata*. Mar Ecol Prog Ser 115:257–270

Coma R, Gili JM, Zabala M (1995) Trophic ecology of the benthic marine hydroid *Campanularia everta*. Mar Ecol Prog Ser 119:211–220

Coma R, Ribes M, Gili JM, Zabala M (1998) An energetic approach to the study of life-history traits of two modular colonial benthic invertebrates. Mar Ecol Prog Ser 162:89–103

Cripps GC, Clarke A (1998) Seasonal variation in the biochemical composition of particulate material collected by sediment traps at Signy Island, Antarctica. Polar Biol 20:414–423

Elyakova LA, Shevchenko NM, Avaeva SM (1981) A comparative study of carbohydrase activities in marine invertebrates. Comp Biochem Physiol 69B:905–908

Fabricius KE, Genin A, Benayahu Y (1995a) Flow-dependent herbivory and growth in zooxanthellae-free soft corals. Limnol Oceanogr 40:1290–1301

Fabricius KE, Benayahu Y, Genin A (1995b) Herbivory in asymbiotic soft corals. Science 268:90–92

Garrabou J (1999) Life-history traits of *Alcyonum acaule* and *Parazoanthus axinellae* (Cnidaria, Anthozoa), with emphasis on growth. Mar Ecol Prog Ser 178:193–204

Gasol JM, Giorgio P del (2000) Using flow cytometry for counting natural planktonic bacteria and understanding the structure of planktonic bacterial communities. Sci Mar 64:197–224

Gili JM, Hughes RG (1995) Ecology of benthic hydroids. Oceanogr Mar Biol Annu Rev 33:351–422

Gili JM, Hughes RG, Alvà V (1996a) A quantitative study of feeding by the hydroid *Tubularia larynx* Ellis and Solander, 1876. Sci Mar 60:43–54

Gili JM, Alvà V, Pagès F, Klöser H, Arntz WE (1996b) Benthic diatoms as the major food source in the sub-Antarctic marine hydroid *Silicularia rosea*. Polar Biol 16:507–512

Gili JM, Alvà V, Coma R, Orejas C, Pagès F, Ribes M, Zabala M, Arntz WE, Bouillon J, Boero F, Hughes R (1998) The impact of small benthic passive suspension feeders in shallow marine ecosystems: the hydroids as an example. Zool Verh Rijksmus Nat Hist Leiden 323:99–105

Gili JM, Coma R, Orejas C, López-González P, Zabala M (2001) Are Antarctic suspension-feeding communities different from those elsewhere in the world? Polar Biol 24:473–485

Grémare A, Amouroux JM, Charles F, Dinet A, Riaux-Gobin C, Baudart J, Medernach L, Bodiou JY, Vetion G, Colomines JC, Albert P (1997) Temporal changes in the biochemical composition and nutritional value of the particulate organic matter available to surface deposit-feeders: a two year study. Mar Ecol Prog Ser 150:195–206

Gutt J, Siegel V (1994) Observations on benthopelagic aggregations of krill (*Euphausia superba*) on the deeper shelf of the southeastern Weddell Sea. Deep Sea Res 41:169–178

Gutt J, Starmans A (1998) Megabenthic structure and biodiversity in the Weddell and Lazarev Seas (Antarctic): ecological role of physical parameters and biological interactions. Polar Biol 20:229–247

Harris VA (1990) Sessile animals on the sea shore. Chapman & Hall, London

Hyman LH (1940) The invertebrates: Protozoa through Ctenophora. McGraw-Hill, New York

Koehl MAR (1977a) Effects of sea anemones on the flow forces they encounter. J Exp Biol 69:87–105

Koehl MAR (1977b) Water flow and the morphology of zoanthid colonies. Proceedings 3rd International Coral Reef Symposium, vol I

LaBarbera M (1984) Feeding currents and particle capture mechanisms in suspension feeding animals. Am Zool 24:71–84

Nöthig E, Bodungen B von (1989) Occurrence and vertical flux of faecal pellets of probably protozoan origin in the southeastern Weddell Sea (Antarctica). Mar Ecol Prog Ser 56:281–289

Okamura B (1990) Particle size, flow velocity, and suspension feeding by the erect bryozoans *Bugula neritica* and *B. stolonifera*. Mar Biol 105:33–38

Palanques A, Isla E, Puig P, Sánchez-Cabrera JA, Masqué P (2002) Annual evolution of settling particle fluxes in Western Bransfield Strait (Antarctice) during the FRUELA project. Deep Sea Res II 49:903–920

Pile AJ, Patterson MR, Witman JD (1996) *In situ* grazing on plankton < 10 μm by the boreal sponge *Mycale lingua*. Mar Ecol Prog Ser 141:95–102

Platt T, Subba Rao DV, Irwin B (1983) Photosynthesis of picoplankton in the oligotrophic ocean. Nature 301:702–704

Ribes M, Coma R, Gili JM (1998) Heterotrophic feeding by gorgonian corals with symbiotic zooxanthellae. Limnol Oceanogr 43:1170–1179

Ribes M, Coma R, Gili JM (1999) Heterogeneous feeding in benthic suspension feeders: natural diet and grazing rate of the temperate gorgonian *Paramuricea clavata* (Cnidaria: Octocorallia) over a year cycle. Mar Ecol Prog Ser 183:125–137

Rubinstein DI, Koehl MAR (1977) The mechanisms of filter feeding: some theoretical considerations. Am Nat 111:981–994

Scharek R, Nöthig E (1995) Das einzellige Plankton im Ozean der Arktis und Antarktis. In: Hempel I, Hempel G (eds) Biologie der Polarmeere. Fischer, Jena, pp 116–127

Schloss I, Ferreyra G, Mercuri G, Kowalke J (1999) Particle flux in an Antarctic shallow coastal environment: a sediment trap study. In: Arntz WE, Rios C (eds) Magellan-Antarctic: ecosystems that drifted apart. Sci Mar 63 [Suppl 1]:99–111

Sebens KP, Koehl MAR (1984) Predation on zooplankton by the benthic anthozoans *Alcyonum siderium* (Alcyonacea) and *Metridium senile* (Actiniaria) in the New England subtidal. Mar Biol 81:255–271

Slattery M, McClintock JB, Bowser SS (1997) Deposit feeding: a novel mode of nutrition in the Antarctic colonial soft coral *Germesia antarctica*. Mar Ecol Prog Ser 149:299–304

Smetacek V, Scharek R, Nöthig EM (1990) Seasonal variation in the pelagial and its relationship to the life history cycle of krill In: Kerry KR, Hempel G (eds) Antarctic ecosystems. Ecological change and conservation. Springer, Berlin Heidelberg New York, pp 103–114

Starmans A, Gutt J, Arntz WE (1999) Mega-epibenthic communities in Arctic and Antarctic shelf areas. Mar Biol 135:269–280

Stepanjants SD (1979) Hydroids of the Antarctic and Subantarctic waters. Biological results of the Soviet Antarctic Expeditions, 6. Issled Fauny Morei 20:1–200

Stockton WL (1984) The biology and ecology of the epifaunal scallop *Adamussium colbecki* on the west side of McMurdo Sound, Antarctica. Mar Biol 78:171–178

Weibe PH, Madin LP, Haury LR, Harbison GR, Philbin LM (1979) Diel vertical migration by *Salpa aspera* and its potential for large scale particulate organic matter transport to the deep sea. Mar Biol 53:249–255

Polar Biol (2001) 24: 901–909
DOI 10.1007/s003000100297

Joachim Plötz · Horst Bornemann · Rainer Knust
Alexander Schröder · Marthan Bester

Foraging behaviour of Weddell seals, and its ecological implications

Abstract Time-depth recorder data of eight adult Weddell seals (*Leptonychotes weddellii*) provided simultaneous dive records over 8 days in the Drescher Inlet, eastern Weddell Sea coast, in February 1998. The seals primarily foraged within two depth layers, these being from the sea surface to 160 m where temperature and salinity varied considerably, and near the bottom from 340 to 450 m where temperature was lowest and salinity highest. While both pelagic and benthic diving occurred during daylight, the seals foraged almost exclusively in the upper water column at night. Trawling during daytime confirmed that *Pleuragramma antarcticum* were by far the most abundant fish both in the pelagial and close to the bottom. Pelagic night hauls at 110–170 m depth showed highly variable biomass of *P. antarcticum*. The temporal changes in the local abundance of *P. antarcticum*, particularly in the pelagial, may explain the trends in the seals' pelagic and benthic foraging activities. This study describes the jaw movements of a hunting seal, which are presumably indicative of feeding events. Trophic links from the Weddell seal to fish, zooplankton and krill, *Euphausia superba*, are discussed.

Introduction

Research projects that focus on the ecology of the Antarctic sea-ice zone require investigations on the trophic links between top predators and their prey. The Weddell seal is adapted to exploit coastal shelf waters, which are largely covered by fast ice for most of the year. Complementary fishery and hydrographic data from locations where Weddell seals are diving and foraging are sparse. *Pleuragramma antarcticum* is the dominant pelagic fish in the Ross Sea (e.g. DeWitt 1970) and Weddell Sea (Hubold 1985) and, consequently, the main prey for Weddell seals in the Ross Sea (e.g. Castellini et al. 1984; Burns et al. 1998) and Weddell Sea (Plötz 1986). The present paper attempts to relate the diving behaviour of the Weddell seal both to the vertical distribution of its principal prey and to local hydrographic processes in a fast-ice environment that is characterised and strongly influenced by the seasonal ice break-up, particularly during late summer.

Please note: When citing this article, please use the reference to the original source in the journal Polar Biology as given on the top of this page.

J. Plötz (✉) · H. Bornemann · R. Knust · A. Schröder
Alfred-Wegener-Institut für Polar- und Meeresforschung,
Postfach 120161, 27515 Bremerhaven, Germany
E-mail: jploetz@awi-bremerhaven.de

M. Bester
Mammal Research Institute, Department of Zoology &
Entomology, University of Pretoria, Pretoria 0002, South Africa

Materials and methods

During the EASIZ (II) Weddell Sea cruise of RV *Polarstern*, a joint seal-fisheries project was carried out from 26 January to 25 February 1998 at Drescher Inlet (72°52′S, 19°26′W), a 25-km-long and 1- to 2-km-wide crack in the Riiser Larsen Ice Shelf. The sea bed under the ice shelf extends over 100 km to the nearest grounding line of Dronning Maud Land (Schenke et al. 1998). According to bathymetric surveys of RV *Polarstern* (Graffe and Niederjasper 1997), the depth of the seafloor inside the inlet ranges from 430 m in the inner section to 380 m over a central 6-km-wide bank, and to 520 m at the inlet mouth. The depth outside the inlet gradually increases, reaching the 600-m isobath about 2 km distant from the inlet mouth.

Recording and processing of seal data

The seal studies were conducted from a camp on the ice shelf. Twenty-five Weddell seals were equipped with time-depth recorders (TDRs) (Driesen and Kern, Bad Bramstedt, Germany). Strong gales and sea swell induced ice break-up, which was most intensive from 13 to 16 February. Over a few days, the fast ice progressively retreated into the inlet so that the recapture of 11 instrumented seals was impossible. For this study, the data sets of eight TDRs were selected as they provided simultaneous records which allowed a direct comparison of the diving and haul-out patterns of eight seals (three males, five females) over a period of 8 days (9–17

February). Immobilisation and deployment techniques used in this study followed Bornemann et al. (1998). The TDRs (8 bit, 130 kb) measured hydrostatic pressure (0–80 bar) at 8-s intervals, which allowed continuous data records for a maximum of 12 days. The logger housing, a titanium tube, was 12×1.8 cm. An additional reed-contact-magnet logger was deployed on 5 of the 25 seals fitted with TDRs to measure jaw action and dive depths simultaneously. The glass-encapsulated reed-contact switch and its shielded extension cable (both 2 mm in diameter) were coated with polyurethane resin for stabilisation and protection against seawater. The cable was screw-sealed where it led into the logger housing. The reed-contact and magnet (10×10×2 mm) were glued onto the hair-covered parts of the upper and lower flews, respectively. The reed-contact was disengaged (switched) by jaw movement (mouth opening), and the number of jaw action switches totalled over the 8-s measuring intervals. The synchronous link between the dive- and jaw-data information was achieved using the computer program Loglink (Jensen Software Systems, Germany). The independent DOS-application was programmed to analyse the logger-generated binary files. The minimum depth that was considered to be a dive was set at 6.25 m, i.e. twice the resolution of the pressure transducer. During analysis, each dive was subdivided into a descent, bottom and ascent phase, followed by its successive surface interval. Timing of the start and end of all phases was based on user-defined thresholds and slope values for the rate of depth changes per unit time. The median maximum dive depth (used in Figs. 2, 4, 5) was calculated from all 8-s sampling points recorded in the bottom phase of a dive. The bottom phase was user-selected and defined as the period spent at maximum depth between the end of the descent and the start of the ascent.

A Seacat SBE19 CTD profiler was deployed through cracks in the ice (Dieckmann and Thomas 1999) to measure hydrographic parameters within the water column of Drescher Inlet.

Fish trawling

Trawling was done from RV *Polarstern*. For the pelagic trawls, a bentho-pelagic net (25×16 m opening, cod-end mesh size 10 mm) was used; hauls were carried out for ~40 min (38–50 min) at depth, controlled by a Scanmar net sensor system. Demersal trawling was done with a 140-ft standard otter trawl (cod-end inlet mesh size 20 mm). The haul duration was ~15 min (10–17 min) on the ground due to the voluminous by-catch, mainly sponges. Prior to and after trawling, the hydrographic profile of the water column was recorded by CTD. Previous studies on Weddell seals in the Drescher Inlet indicated that pelagic foraging was most common around 150 m (Plötz et al. 1997). Pelagic trawling (this study) was carried out in this depth layer to identify potential food items of seals. In total, 12 hauls were taken in the inlet. These were six pelagic tows (110–170 m) during two consecutive nights (26–28 Jan.), two pelagic (30–70 m, 150–250 m) and three bottom hauls (~ 400 m) during daytime (3 and 22 Feb.), and one bottom haul (~ 400 m) at night (25 Feb.). To measure the vertical distribution pattern of fish during daytime, 3 of the total of 12 trawls were done around noon (1000, 1200, 1400 hours local time) at different depths (close to surface, 30–70 m; pycnocline, 150–250 m; and on the bottom, ~400 m depth). The same depth-stratified sampling procedure was planned for the night-time but due to a broken net this sampling could not be conducted. Fish were identified to species level and individual total length (TL) and wet weight (WW) measured. Numbers and biomass were standardised to 1 h trawling time. Fish guts were examined for prey items. In addition, fish specimens vomited by a seal onto the ice (11 Feb.) were collected and analysed in conjunction with the trawl catches.

Results

Diurnal activity and haul-out

All individual haul-out and diving records for the eight seals, measured over 8 days, were pooled to illustrate the proportion of time the seals spent on the ice and at sea (Fig. 1). At 1200 hours local time, for example, 58% and 42% of seals were hauled-out on ice and at sea, respectively. Of the total of 1,536 h recorded, time spent on the ice was 31% compared to 69% spent at sea. Overall, Fig. 1 shows the seals' preference for hauling-out during daylight and being at sea during darkness. Nocturnal periods at sea generally extended to the early morning hours, and from 0700 hours onwards the seals apparently had a tendency to haul-out onto the ice, with a distinct peak at 1300–1500 hours local time. Nevertheless, even during that peak haul-out not all the seals were on the ice every day nor were they always active during the night, and it was not uncommon to observe a few animals, including instrumented ones, resting on the ice at night-time.

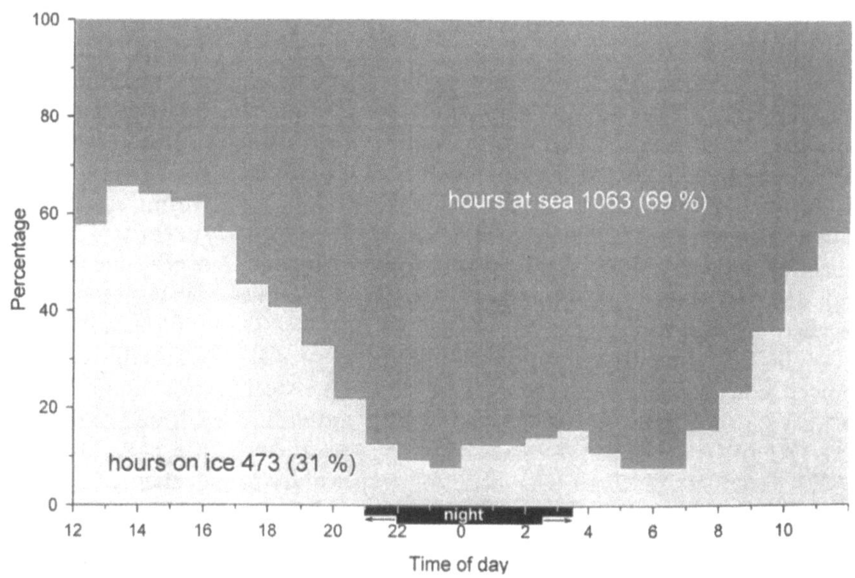

Fig. 1 Diurnal activity and haul-out patterns for Weddell seals ($n = 8$) combined for the period 9–17 Feb. 1998. Indented ends of the *black bar* indicate the increase in the duration of night time during the 8-day period

Fig. 2 Frequency distributions of dive depths of Weddell seals ($n = 8$) recorded from 9 to 17 Feb. 1998. Original identification number (e.g. Seal 1) and total number of dives for each seal are indicated on the individual histograms

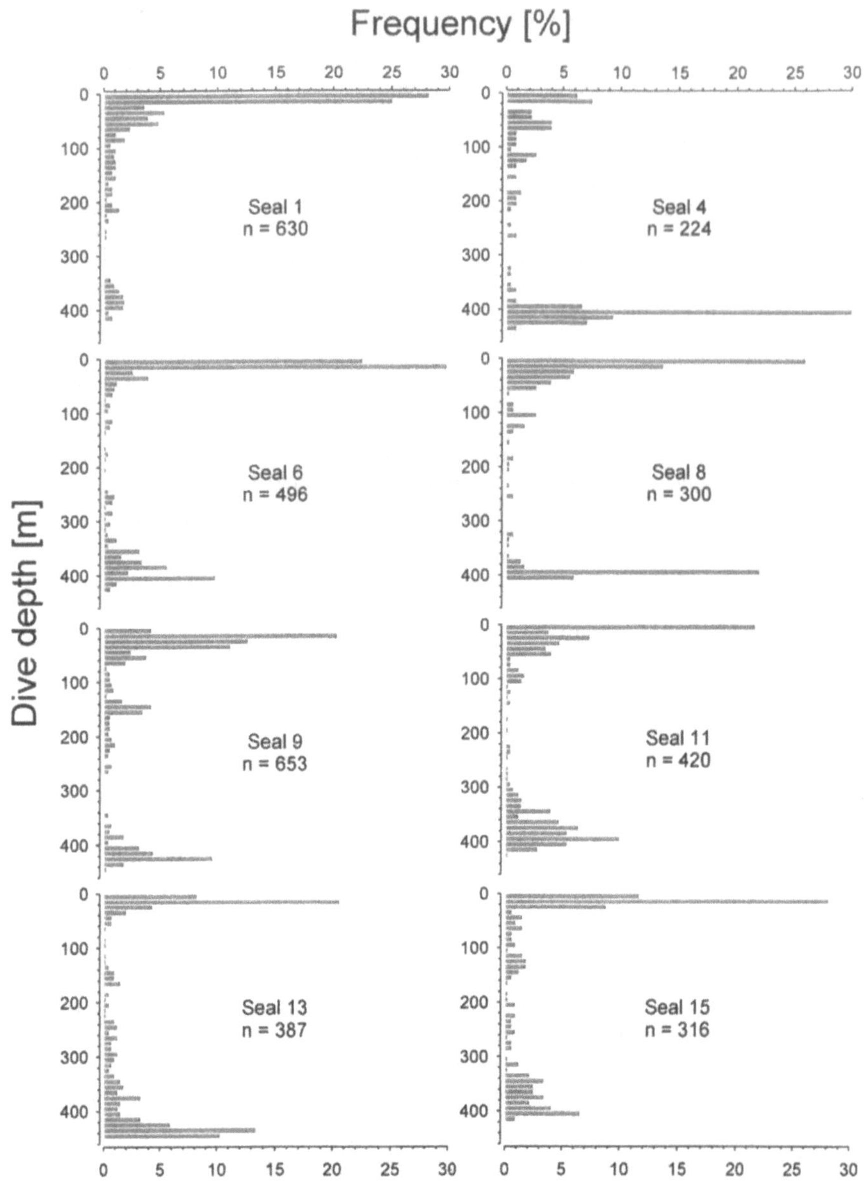

General diving pattern

The frequencies of maximum dive depths were summed for each seal (Fig. 2). The range of benthic dives from 380 to 450 m reflects the irregular bottom topography inside the Drescher Inlet. Dives to 410 m (Seal 8, female), 420 m (Seals 1, male and 15, female), 430 m (Seals 6, female and 11, male), 440 m (Seal 4, female) and 450 m (Seals 9, male and 13, female) were the deepest recorded.

A broad comparison amongst the dive profiles showed considerable variation in dive depths between (ANOVA on ranks: $P < 0.001$) and within individual seals (ANOVA on ranks: P-values between < 0.001 and 0.042), except seal 8 ($P = 0.124$). Series of deeper dives to the seafloor which were alternated with abrupt shifts to series of shallower dives were evident in all the seals studied. Numbers of dives recorded for each seal varied greatly, with Seal 4 ($n = 224$) diving the least and Seal 9 ($n = 653$) being the most active diver (Fig. 2). Seal 9 performed a few very long dives ($n = 10$); the maximum duration was 67 min and the remaining 9 dives lasted 47–61 min. These dives were single events occurring within a 5-day period (11–16 Feb.), and all were consistent to 150 m (± 20 m) and most of them ($n = 7$) were night-time dives. Seal 15 showed a progressive change in depths (Fig. 3). Deepest dives were made from 0800–1000 hours in the morning (15 Feb.) and then they gradually decreased in depth. At night, foraging was most common around 50 m. After a haul-out period of 2 h on the ice, the animal continued pelagic foraging in the early morning hours, and then dived increasingly deeper. Such sporadic events of a progressive change in depths were evident in all seals but differed in appearance and duration. They commonly lasted 2, 3 or up to 7 h and dive depth over time did not necessarily show a

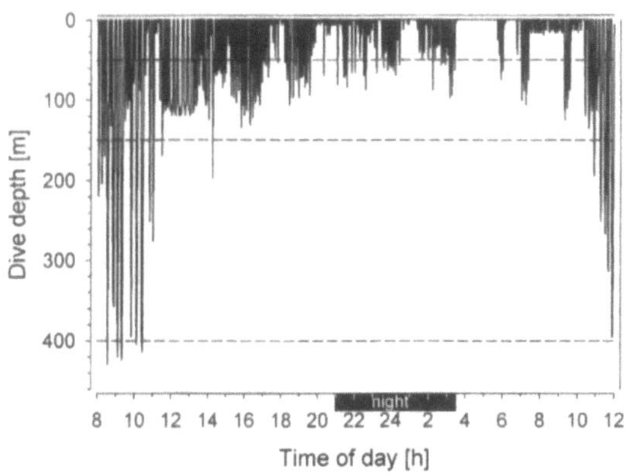

Fig. 3 Diurnal dive pattern of a Weddell seal (Seal 15) during 15/16 Feb. 1998

regular dome-shaped distribution, such as over the maximum record of 28 h presented here (Fig. 3). Most of such dive sequences included a diurnal component, with progressively shallower dives during dusk and progressively deeper dives towards dawn.

The frequency distribution of maximum dive depths for all seals combined is shown in Fig. 4. Highest frequencies were recorded to 70 m (55% of dives). A moderate increase of pelagic diving appeared from 80 to 160 m (9% of dives). The slight increase in this 80- to 160-m depth range is more or less pronounced in all individuals as indicated by additional peaks at these particular depths (Fig. 2). The mid-water layer from 170 to 330 m (6% of dives) was mainly used by the seals to commute so as to reach the seabed, where there was an activity peak at around 400 m. Overall, the seals primarily foraged within two depth layers, these being from the sea surface to 160 m (64% of dives) and from 340 to 450 m (30% of dives).

The distribution pattern of dive depths is related to CTD profiles (Fig. 4), which show the variation in the water temperature and salinity regime of the Drescher Inlet before and after an intensive ice break-up. A characteristic hydrographic feature is the stable thermo(pycno)cline between 130 and 230 m corresponding to the temperature maximum of –1.05°C at around 150 m. The fast ice was up to 2 m and the underlying platelet-ice layer up to 5 m thick. Before ice break-up, the surface layer was cooled to a depth of about 40 m, forming a weak thermocline in between, while a remnant of warmer water was still trapped between 50 and 150 m. After ice break-up, the upper water column was mixed down to the stable pycnocline, resulting in a warming up of the entire water body above 150 m, and consequent melting of masses of brash and platelet ice, as is reflected by a distinct temperature increase (Ta) and decrease in salinity (Sa). The 150-m depth range of large variation in temperature (Tb and Ta) and salinity (Sb and Sa) coincides with the seals' most intensive pelagic foraging activity, as identified from the frequency of dives to these depths. This situation changed in the mid-water layer between 170 and 330 m where the decrease in foraging activity is consistent with the gradual temperature decrease. A further distinct increase in foraging activity occurred from 340 m depth down to the seafloor where temperature was lowest (Tb and Ta) and salinity highest (Sb and Sa) with little variation.

Figure 5 shows that both dive frequency and dive depth varied with time of day. Of the total of 3,426 dives recorded, 34.9% occurred during the short (max. 6.5 h) night period. According to the seals' diurnal activity (see Fig. 1), dive frequency began to increase at least 2 h before dusk and remained high for about 4 h into the morning crepuscular period. The diurnal pattern in dive depth indicates a preference for foraging deeper during daylight and shallower during darkness (Mann-Whitney

Fig. 4 Frequency distribution of the depths of all dives ($n = 3426$) for Weddell seals ($n = 8$) during 9–17 Feb. 1998 combined with temperature and salinity profiles recorded on the day before (*Tb* and *Sb*) and 3 days after (*Ta* and *Sa*) an intensive ice break-up from 13 to 16 Feb. 1998

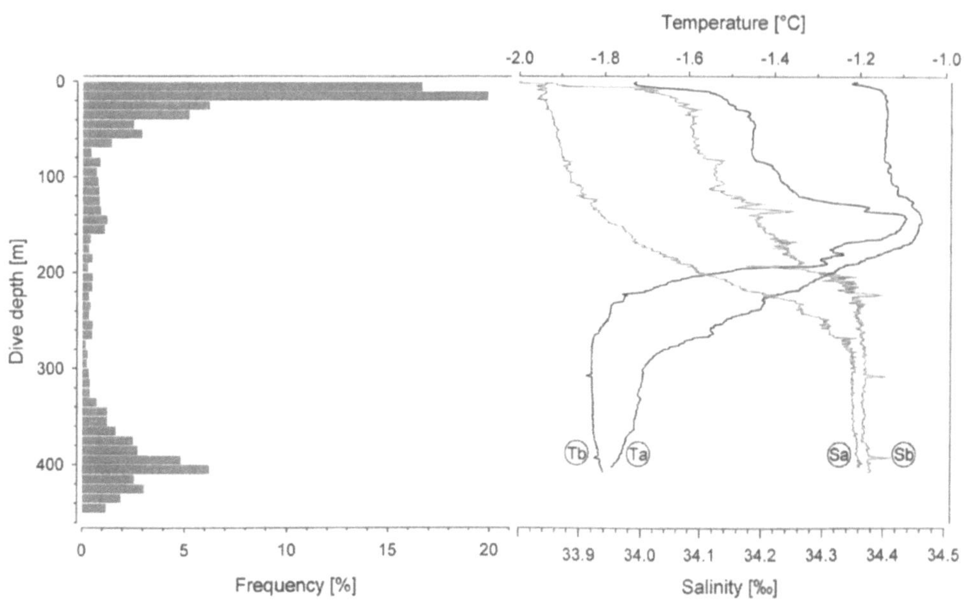

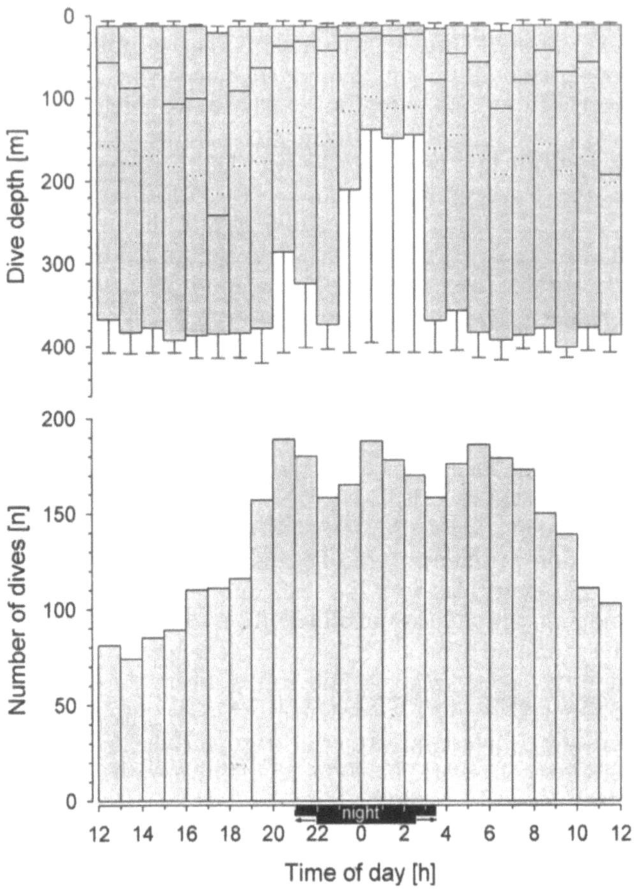

Fig. 5 Box plot (*above*) and frequency diagram of the diurnal variation in depth and number of all dives ($n = 3426$) made by Weddell seals ($n = 8$) during 9–17 Feb. 1998. The *whisker caps* define the 5th and 95th percentiles. Median *solid lines*, mean *stippled lines*

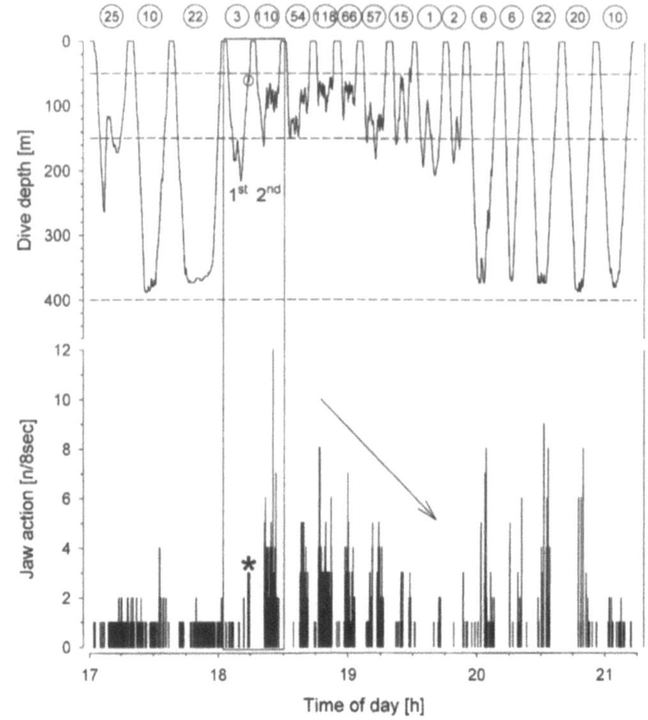

Fig. 6 Synchronous records of diving and jaw activity of a Weddell seal (Seal 1). The sequence of selected dives was measured on 9 Feb. 1998. Jaw action signals (JASs) were accumulated over 8-s intervals. A single JAS means that the seal opened its mouth once within 8 s. Therefore, 2/3/4 JASs in a single vertical stroke translates into 2/3/4 mouth openings within 8 s. Numbers (in *circles*) correspond to the number of JASs measured during the bottom phase of each dive. Explanation for 1st and 2nd (in *frame*) see text; *asterisk* indicates 9 JASs within 24 s coinciding with a short interruption (at 60 m, *unfilled circle*) in the ascent phase of the 1st dive. *Arrow* indicates decrease of JASs

rank sum test, $P < 0.001$). For daytime dives, the hourly variation in the medians of the depths (Fig. 5, solid lines) is significantly higher compared to those carried out at night (Mann-Whitney rank sum test, $P = 0.042$). This shows that the seals concentrated their diving activities during daylight hours over a wider depth range by foraging either in the water column above the pycnocline, or considerably deeper down to the seafloor as is also demonstrated in Fig. 4. Night-time dives between 2300 and 0300 hours, in particular, were the shallowest and almost exclusively to less than 220 m, most of them concentrated well above the pycnocline. There was an abrupt change to considerably deeper dives just after sunrise, deep dives continuing throughout the daylight hours.

Feeding dives

Two of five seals instrumented with reed-contact-magnet loggers were not recaptured. The remaining three reed-contact-magnet units provided partial records of up to 6 days, which only in one case (Seal 1) overlapped with the 8-day period of the present study. Figure 6 illustrates one of many scenarios of diving behaviour and concomitant jaw movements. This first demonstration of jaw action was also chosen as it constitutes a representative sample of diving characteristics observed in virtually all individuals studied. Dive duration in this sequence of 17 dives ranged from 9.6 to 22.0 min for the 7 benthic dives and 7.3 to 15.8 min for the 10 pelagic dives. The post-dive intervals at the sea surface were consistently brief (range 1.7–2.9 min). All dives showed a similar rapid descent, followed by a sustained activity period of 0.9–12.4 min bottom time before a rapid ascent. The movements from or to the sea surface sometimes coincided with a short interruption (spike) at various depths. During the bottom time, the seal exhibited up and down movements. This vertical zigzag pattern was most intensive in the bout of nine pelagic dives between 1800 and 2000 hours, with the shallowest two dives being made in mid-period.

During the first three dives, the last two of which reached the seafloor, the seal opened its mouth mostly once per 8-s measuring interval. A significant increase in jaw action signals (JASs) occurred when the seal switched from benthic to pelagic dives. These signals were most intensive within the 50–150 m depth layer, where a remnant of warmer water was trapped before the ice break-up (Fig. 4). The slight increase in signals (9 JASs within 24 s) in the first pelagic dive (see frame in Fig. 6) coincided with a short interruption (spike) in the ascent at a depth of 60 m. At that particular depth the seal apparently encountered prey in high density because the subsequent (second) dive showed a dramatic increase of up to 12 JASs per 8-s interval, coincident with an intensive zigzag dive pattern, also in the 4 following pelagic dives, covering a range of as much as 60 m. After the second dive, the frequency of signals declined more or less gradually, followed by a subsequent increase in jaw action which coincided with the seal's switching to a series of deep dives about 1 h before sunset.

Natural regurgitation

Haul-out sites of seals often had food remnants on the ice, a grey mixture of partially digested flesh and hard-part remains of prey. On one occasion (11 Feb.; 0930 hours) we observed an apparently gorged Weddell seal which, upon leaving the water, vomited 230 specimens of *P. antarcticum* onto the ice. Their length-frequency distribution is presented in Fig. 7. With the exception of small specimens (3–11 cm) caught by net in the pycnocline, all size classes were represented in the seal's diet. The total wet weight of the regurgitated meal was 7.6 kg, which amounted to an average weight of 33 g per fish. Apart from a few tooth-marks found on only 13 specimens, all fishes were undamaged, apparently caught recently, and swallowed whole. This also shows that the canines and post-canines of the Weddell seal (if used at all) serve to catch and hold but not chew prey items up to 21 cm in length.

Pelagic and bottom trawling

Eleven fish species were caught in eight pelagic trawls (Fig. 8). *P. antarcticum* were by far the most abundant fish (87%), and predominated in biomass (~75%). *Anotopterus pharao*, a pelagic fish predator, occurred in small numbers (1%) but due to its large size (up to 95 cm TL and 490 g WW) this fish contributed about 20% of the total biomass. Abundance and biomass of all other fish species were much lower. Apart from single specimens of adult *Chionodraco myersi*, *C. hamatus*, *Dacodraco hunteri* and *Pagothenia borchgrevinki*, the catches comprised juveniles and larval stages only (Fig. 8). Pelagic trawls also contained considerable amounts of krill, *Euphausia superba*, unidentified gelatinous zooplankton and small numbers of pelagic amphipods, mainly *Themisto gaudichaudii*. The krill biomass in the first night of pelagic trawling (26/27 Jan.: 52.5 kg/h ± 0.01 SE) was 4 times that of the second night

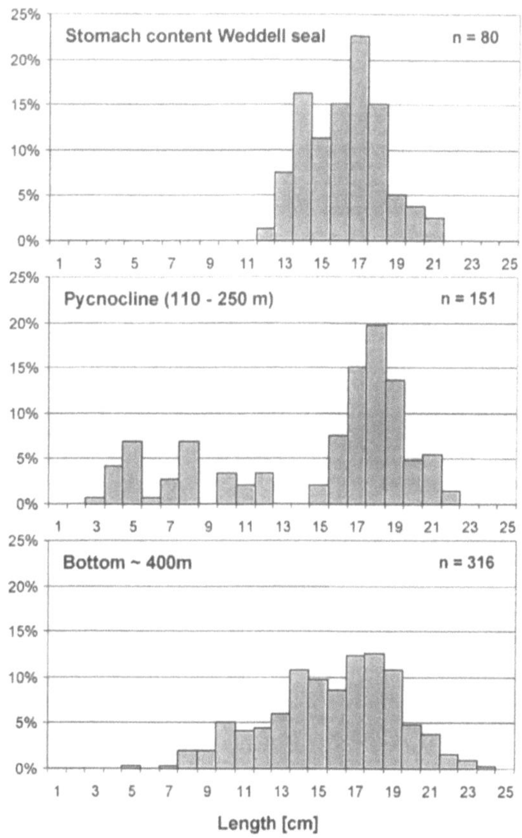

Fig. 7 Length-frequency distribution of *Pleuragramma antarcticum* caught by a Weddell seal (*top*) by pelagic trawling in the pycnocline (*middle*) and by bottom trawling (*bottom*)

(27/28 Jan.: 12.4 kg/h ± 0.005 SE). The amount of gelatinous zooplankton was about 9 kg/h in both nights.

Demersal catches showed a high diversity in the composition of species (Table 1). A total of 1,179 fish specimens, belonging to 26 species and unidentified Liparididae and Zoarcidae, were caught in 4 bottom trawls. Channichthyids dominated in biomass, followed by nototheniids, which was the most abundant group.

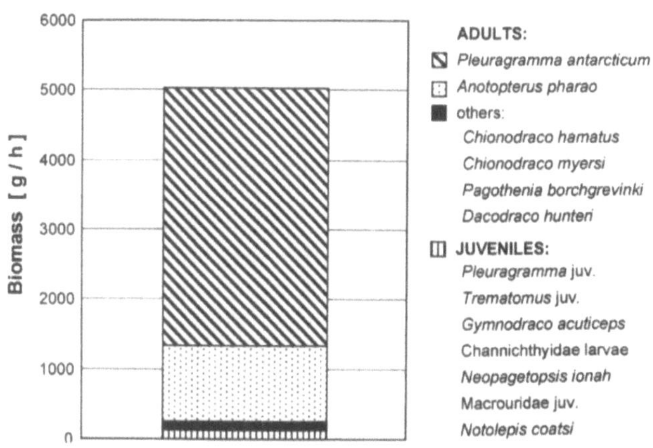

Fig. 8 Biomass and species composition of the pelagic fish community in Drescher Inlet. Average biomass (wet weight) per hour trawling summarised over all pelagic hauls ($n = 7$)

Table 1 Catch composition of demersal trawls in Drescher Inlet

Species	% of total biomass	Average number/h
Channichthyidae	59.8	343
Chionodraco myersi	22.6	121
Chionodraco hamatus	15.7	70
Cryodraco antartcicus	19.0	96
Dacodraco hunteri	2.1	47
Pagetopsis maculatus	0.3	8
Chaenodraco wilsoni	< 0.1	1
Nototheniidae	30.0	856
Pleuragramma antarcticum	14.0	553
Aethotaxis mitopteryx	0.9	5
Trematomus scotti	1.6	163
Trematomus eulepidotus	5.2	53
Trematomus lepidorhinus	3.2	49
Trematomus hansoni	2.7	9
Trematomus loennbergii	1.8	11
Trematomus nicolai	0.5	2
Trematomus pennellii	0.1	7
Trematomus spp. juv.	≪0.1	3
Bathydraconidae	2.9	85
Gerlachea australis	2.2	77
Other Bathydraconidae	0.7	7
Artedidraconidae	0.4	53
Dolloidraco longedorsalis	0.2	25
Artedidraco loennbergi	< 0.1	9
Other Artedidraconidae	0.2	19
Bathyrajidae	6.6	4
Bathyraya maccaini	6.6	4
Liparididae	0.1	15
Zoarcidae	< 0.1	2

Among channichthyid species, the most common fish caught were *C. myersi*. The nototheniid *P. antarcticum*, however, was by far the most abundant fish in both the demersal and pelagic catches.

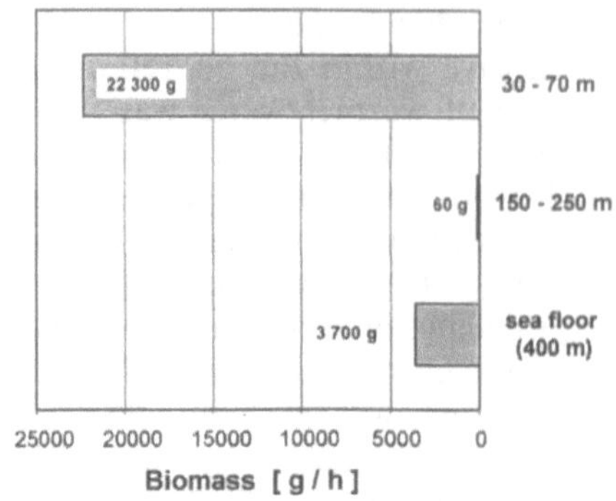

Fig. 9 Vertical distribution of *Pleuragramma antarcticum* in Drescher Inlet during daytime (between 1000 and 1400 hours). Biomass expressed as wet weight per hour trawling

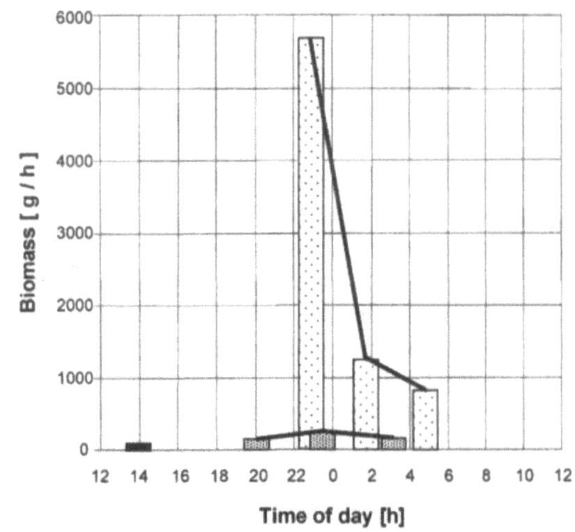

Fig. 10 Pelagic biomass of *Pleuragramma antarcticum* in the pycnocline of Drescher Inlet. Data derived from 6 hauls (110–170 m) over 2 consecutive nights (26 and 27 Jan. *stippled bars*) and 1 haul (150–250 m) during daytime (22 Feb. *black bar*). The depth of the trawls was adjusted according to the position of the pycnocline as determined by CTD profiles

Vertical distribution of *Pleuragramma antarcticum*

During daytime, the biomass of *P. antarcticum* was highest above the pycnocline (30–70 m), lowest in the pycnocline (150–250 m), and high at the bottom (~400 m) (Fig. 9). A comparable picture of the vertical distribution of *P. antarcticum* cannot be shown for the night-time due to the damaged gear. Nevertheless, the results of 6 pelagic trawls (110–170 m) on 2 consecutive nights indicated that the biomass of *P. antarcticum* was highly variable in this particular depth layer (Fig. 10). During the first night (26/27 Jan.), the highest value in biomass occurred at around midnight and then gradually decreased towards dawn. This finding indicates nightly vertical migration of *P. antarcticum* into this layer of warmer water which corresponds to the temperature maximum of −1.05°C at around 150 m depth (see Fig. 4). The following night (27/28 Jan.), only single specimens of *P. antarcticum* were caught in the same depth layer, and no trend of vertical migration was detectable. The length-frequency distribution of *P. antarcticum* caught between 110 and 250 m depth was bimodal (3–12 cm and 15–22 cm TL). At the bottom, sizes of *P. antarcticum* ranged between 5 and 24 cm TL with an emphasis on larger (14–21 cm) adult individuals (Fig. 7).

Discussion

Seals that use the same area on the same days conceivably are feeding under similar conditions. They should, furthermore, behave in a comparable manner with respect to their pattern of diving to certain water depths where hunting may be most efficient. The present study shows that during a period of 8 days, all eight Weddell seals confined their foraging activities within the Drescher Inlet, as evidenced by their benthic dives. The diving behaviour within and amongst individuals was, however, highly variable, as has been documented for yearlings (Burns et al. 1997) and adults studied in McMurdo Sound (Kooyman 1967; Castellini et al. 1991;

Testa 1994; Schreer and Testa 1996). Burns et al. (1997) concluded that average diving behaviour is not sufficient for comparing or understanding foraging strategies of Weddell seals, a statement with which we fully agree. Although the Weddell seal is one of the best-studied pinnipeds, the large variations in diving patterns make interpretation difficult.

Correlation of CTD profiles with the distribution pattern in dive depths of the seals (Fig. 4) indicates a close relationship between diving behaviour and the hydrography of the Drescher Inlet. In general, pelagic foraging was most intensive within the layer of warmer water above the pycnocline. By contrast, the 160-m mid-water layer between 170 and 330 m, where the temperature gradually decreased (Fig. 4), appears to be prey-deficient, as the seals mostly did not forage there but commuted through this zone in order to forage near the bottom. The abrupt shift from a series of benthic to pelagic dives observed in all Weddell seals studied is of particular interest. We believe that in many such cases the seals located pelagic prey (see Fig. 6), which also demonstrates that a hunting seal descending from the surface would not switch to benthic foraging as long as prey is readily available in the upper water column.

Trawling in the Drescher Inlet suggested that fish, primarily *P. antarcticum*, apparently moved into the upper water column at night and, consequently, night-time dives of Weddell seals were the shallowest. Based on this we assume that the warmer water layer above the pycnocline attracts and aggregates prey, enabling effective hunting on *Pleuragramma* at night. Foraging in the pelagial also occurred during daylight hours, but over a wider depth range, in addition to a high frequency of deeper benthic dives. Correspondingly, trawling during daytime confirmed large numbers of fingerlings and adults of *P. antarcticum* in the pelagial, and of adults near the bottom. This leads us to conclude that the seals' considerable variations in pelagic and benthic foraging activities corresponded with dives to feed primarily on *Pleuragramma*, in order to optimise food intake during the day and particularly during the short (6 h) night period.

Testa (1994) proposed a continued reliance on *P. antarcticum* as the principal prey of Weddell seals in the Ross Sea region, though other fish (and cephalopods) cannot be excluded, as observed in the present study (Table 1), despite the fact that *P. antarcticum* was the most abundant species in the trawl catches, and the exclusive prey item in the regurgitated meal of a Weddell seal (Fig. 7). The predominantly intact appearance of the 230 specimens in our sample is surprising because the flesh of *P. antarcticum* is very soft. From this it appears that most fishes (of up to 21 cm in length) were sucked in by the Weddell seal in one gulp, as reported for a captive crabeater seal sucking in small fish over distances of up to 50 cm (Klages and Cockcroft 1990).

The frequency in jaw movements (see Fig. 6) cannot provide the number of prey items actually ingested by the seal, although we expect that the majority of JASs have to do with feeding events or at least with prey-catch attempts. We surmise that a high frequency of jaw movements of a hunting seal principally indicates high prey density. The 4-h sequence of 17 consecutive dives, characterised by rapid descents and ascents from/to the surface, implies that the seal foraged at a particular site. The bout of nine pelagic dives lasted nearly 2 h during which the abrupt increase in JASs (second dive in Fig. 6) gradually decreased and phased out in the last two pelagic dives. As each of these two dives, as well as the first dive (see frame in Fig. 6), showed a single, pronounced up and down excursion during the bottom time that coincided with a few JASs only, we suggest that such dives are more typical for searching rather than for pursuing and catching prey. Within the same pelagic diving bout, however, there are five dives that showed an intensive vertical zigzag pattern during the relatively long bottom time, coincident with a high frequency of JASs. Moreover, the intensive zigzags occurred exactly within the 50–150 m depth layer where a remnant of warmer water was trapped before the ice break-up. Since zigzag dive patterns have been referred to as prey-pursuit movements (PPMs) in crabeater seals and in a Ross seal (Bengtson and Stewart 1992, 1997), our findings suggest that PPMs indeed reflect hunting behaviour during foraging dives also in Weddell seals.

The progressive change in dive depths was observed in all seals, as illustrated in Figs. 3 and 6. Although sporadic, such distinct patterns were exhibited most in crepuscular- and night-dives, suggesting that the Weddell seals' foraging behaviour is finely tuned to vertically migrating prey. Similar patterns were observed in the diving behaviour of crabeater seals (Bengtson and Stewart 1992), where the distinct diel pattern of dive timing and depth were related to prey-capture strategy of the seals, and also to possible predator avoidance behaviour by their principal prey, Antarctic krill *E. superba*. Virtually nothing is known about the vertical migratory behaviour of *P. antarcticum*, although this pelagic schooling fish appears to pursue the swarms of krill on which it feeds. For example, *Pleuragramma* have been reported as by-catch in hauls of Antarctic krill (Rembizewski et al. 1978). Juveniles feed on small zooplankton such as calanoid copepods and euphausiid larvae (Hubold and Hagen 1997), while adult *Pleuragramma* additionally ingest a higher proportion of larger euphausiids 10–40 mm long (Wöhrmann et al. 1997). Moreover, a video system-equipped vehicle observed *E. superba* patchily distributed and most plentiful underneath vast ice sheets, while vertical net catches confirmed that krill were scarce in the underlying water column (Marschall 1988). Laboratory observations (Stretch et al. 1988) additionally showed that melting ice releases algae into the water column, which induces area-intensive foraging behaviour in krill. Night trawls in the upper water column of the Drescher Inlet revealed considerable amounts of *E. superba*. At midnight all the trawled fish samples had food in their guts. Most abundant prey items were copepods and *E. superba*

(Olaso 1999). It seems, therefore, that *Pleuragramma* are more active at night and that their movements in shallower depths are linked to pelagic feeding. It is, moreover, not surprising that the guts of the many specimens of *Pleuragramma* vomited by a Weddell seal onto the ice primarily contained Antarctic krill.

Conclusions

The fish fauna of the Drescher Inlet was dominated by *P. antarcticum*, which was by far the most abundant fish both in the upper water layer and near the bottom during day and night. It is reasonable to assume that the abundance of *Pleuragramma* in the upper water layer had been caused by ice break-up in that late summer season, which was reflected in a distinct salinity decrease and temperature increase above the pycnocline. Melting brash and platelet ice released masses of algae and other organic particles into the surface layers where intermediate and upper level trophic interactions would have been maximal at that time. The continuous replenishment of biomass, formerly bound to the fast and platelet ice, constituted locally attractive feeding spots which may have induced intensive foraging behaviour in zooplankton, krill and fish aggregating underneath melting sea ice. This, in turn, may have affected the foraging behaviour of Weddell seals, which responded to changes in the vertical patchy distribution and abundance of their principal prey, *P. antarcticum.*

Acknowledgements Thanks are due to Gerhard Dieckmann and David Thomas for the measurement and interpretation of CTD-profiles taken in the Drescher Inlet, Dieter Gerdes for providing the CTD measurements on board RV *Polarstern*, and Wolf Arntz for co-ordinating the EASIZ programme.

References

Bengtson JL, Stewart BS (1992) Diving and haulout behavior of crabeater seals in the Weddell Sea, Antarctica. Polar Biol 12:635–644

Bengtson JL, Stewart BS (1997) Diving patterns of a Ross seal near the eastern coast of the Antarctic Peninsula. Polar Biol 18: 214–218

Bornemann H, Mohr E, Plötz J, Krause G (1998) The tide as zeitgeber for Weddell seals. Polar Biol 20:396–403

Burns JM, Schreer JF, Castellini MA (1997) Physiological effects on dive patterns and foraging strategies in yearling Weddell seals (*Leptonychotes weddellii*). Can J Zool 75:1796–1810

Burns JM, Trumble S, Castellini M, Testa W (1998) The diet of Weddell seals in McMurdo Sound as determined from scat collections and stable isotope analysis. Polar Biol 19: 272–282

Castellini M, Davis R, Davis M, Horning M (1984) Antarctic marine life under the McMurdo ice shelf at White Island: a link between nutrient influx and seal population. Polar Biol 23: 229–231

Castellini M, Davis R, Kooyman G (1991) Annual cycles of diving behavior and ecology of the Weddell seal. Bull Scripps Inst Oceanogr 28:1–54

DeWitt HH (1970) The character of the midwater fish fauna of the Ross Sea, Antarctica. In: Holdgate MW (ed) Antarctic ecology, vol 1. Academic Press, London, pp 305–314

Dieckmann G, Thomas D (1999) Organic matter production within, and export from Antarctic sea ice platelet layer. In: Arntz W, Gutt J (eds) Expedition Antarktis XV/3 (EAZIS II) of RV Polarstern 1998. Rep Polar Res 301:118–123

Graffe D, Niederjasper F (1997) Meeresbodenkartierung mit dem Hydrosweep-System. In: Jokat W, Oerter H (eds) Expedition Antarktis XII of RV Polarstern 1995. Rep Polar Res: 219: 37–44

Hubold G (1985) Stomach contents of the Antarctic silver fish *Pleuragramma antarcticum* from the southern and eastern Weddell Sea (Antarctica). Polar Biol 5:43–48

Hubold G, Hagen W (1997) Seasonality of feeding and lipid content in juvenile *Pleuragramma antarcticum* from the southern Weddell Sea. In: Battaglia B, Valencia J, Walton D (eds) Antarctic communities: species, structure and survival. University Press, Cambridge, pp 277–283

Klages NTW, Cockcroft VG (1990) Feeding behaviour of a captive crabeater seal. Polar Biol 10:403–404

Kooyman GL (1967) An analysis of some behavioral and physiological characteristics related to diving in the Weddell seal. In: Llano GA, Schmitt WL (eds) Antarctic Research Series (Biology of Antarctic Seas III). Am Geophys Union 11:227–261

Marschall HP (1988) The overwintering strategy of Antarctic krill under the pack-ice of the Weddell Sea. Polar Biol 9:129–135

Olaso I (1999) The pelagic fish food web. In: Arntz W, Gutt J (eds) Expedition Antarktis XV/3 (EAZIS II) of RV Polarstern 1998. Rep Polar Res 301:110–118

Plötz J (1986) Summer diet of Weddell seals (*Leptonychotes weddellii*) in the eastern and southern Weddell Sea, Antarctica. Polar Biol 6:97–102

Plötz J, Bornemann H, Gleitz M, Günther S (1997) Biologische Untersuchungen im Drescher Inlet. In: Jokat W, Oerter H (eds) Expedition Antarktis-XII/3 of RV Polarstern 1995. Rep Polar Res 219:61–66

Rembizewski JM, Krzeptowski M, Linkowski TB (1978) Fishes as by-catch in fisheries of krill *Euphausia superba*. Pol Arch Hydrobiol 25:677–695

Schenke HW, Diijkstra S, Niederjasper F, Schöne T (1998) The new bathymetric charts of the Weddell Sea: AWI BCWS. Am Geophys Union, Antarct Res Ser 75:371–380

Schreer JF, Testa JW (1996) Classification of Weddell seal diving behavior. Mar Mammal Sci 12:227–250

Stretch JJ, Hamner PP, Hamner WM, Michel WC, Cook J, Sullivan CW (1988) Foraging behavior of Antarctic krill *Euphausia superba* on sea ice microalgae. Mar Ecol Prog Ser 44:131–139

Testa JW (1994) Over-winter movements and diving behavior of female Weddell seals (*Leptonychotes weddellii*) in the southwestern Ross Sea, Antarctica. Can J Zool 72:1700–1710

Wöhrmann A, Hagen W, Kunzmann A (1997) Adaptations of the Antarctic silverfish *Pleuragramma antarcticum* (Pisces:Nototheniidae) to pelagic life in high-Antarctic waters. Mar Ecol Prog Ser 151:205–218

III.

Ice biota and ice impact on benthic communities

Polar Biol (2001) 24: 553–564
DOI 10.1007/s003000100262

Julian Gutt

On the direct impact of ice on marine benthic communities, a review

Abstract Ice has a significant impact on the polar and sub-polar benthos, but relationships between corresponding physical and biological processes are not yet sufficiently understood. Sea ice contributes to a vertical zonation in shallow waters, which also experience other important disturbances. Due to the length of the non-glaciated coastline, sea ice is of greater relevance in the north than in the south. Scouring by icebergs and ridged sea ice causes an increased diversity when different recolonisation stages coexist. Frequently scoured areas do not recover, especially in the Antarctic, due to slow growth rates of the fauna. Iceberg grounding in the Arctic is mainly restricted to the western Eurasian and northeastern American shelf, including Greenland. Around Antarctica, scouring is more evenly distributed. Glacier termini prevent sessile animals from settling in their proximity where only few motile species occur.

Introduction

When biologists decipher the functioning of shallow-water ecosystems in cold regions, ice is commonly mentioned as a highly relevant factor (Stephenson and Stephenson 1972; White 1973, 1984; Curtis 1975; Gallardo 1987; Clarke 1996; Conlan et al. 1998; Barnes 1999; Gutt 2000). Depending on water depth and climatic/hydrographic conditions, the direct disturbance by ice can range from being locally most important, to negligible. Corresponding studies are not only of interest to scientists engaged in fundamental research, but also because of applied aspects: the understanding of ecological processes can support the non-exploitive utilisation of nature, and coastal management requires information on the resilience of the system.

This review summarises current knowledge on consequences for the fauna and flora resulting from direct disturbance by different types of ice. In contrast to the nekton or plankton, for the benthos such impact is of particular relevance, because most organisms can neither escape before, reinvade after an event nor drift back into their usual habitat. The impact of ice generally fulfils White and Pickett's (1985) definition of disturbance: "...discrete event in time that disrupts ecosystem, community, or population structure and changes resources, substrate availability, or the physical environment". If its ecological effect is quantitatively relevant, it seems to be necessary to register the "disturbance regime descriptors" such as spatial distribution, frequency, predictability, size of the area affected, intensity and severity.

The World Meteorological Organization (WMO) classified ice into two major categories, sea and glacier ice. Sea ice undergoes conspicuous seasonality; first-year ice has an average thickness of 0.5–2 m. Multi-year or ridged sea ice rarely reaches depths >40 m. Anchor ice is attached to the sediment in shallow waters. Glacier ice, either as part of a glacier or of a continental ice cap, disintegrates at the grounding line or floats before tabular icebergs calve. Due to their drift being mainly determined by the current (Sanderson 1988), they can run aground up to 600 m depth.

In order to correlate quantitatively ice dynamics with benthic structures, it is essential to know some basic biological characteristics, e.g. population structure, number of species in a given area, growth rates at least for the key species, their substrate preferences and interactions. It is assumed that the early life-history of many polar invertebrates is highly relevant, but there is

Please note: When citing this article, please use the reference to the original source in the journal Polar Biology as given on the top of this page.

J. Gutt
Alfred Wegener Institute for Polar and Marine Research,
Columbusstrasse, 27568 Bremerhaven,
Germany
E-mail: jgutt@awi-bremerhaven.de
Tel.: +49-471-4831333
Fax: +49-471-4831149

still little known about it. The Arctic benthos is characterised by a well-developed depth zonation (Ellis and Wilce 1961; Lewis 1972; Stephenson and Stephenson 1972; Golikov and Averincev 1977; Schmid and Piepenburg 1993; Piepenburg and Schmid 1996) where echinoderms are among the most common motile organisms encountered. The infauna is comparatively rich in species and abundance but locally epifauna can be dominant (Starmans et al. 1999). Species assemblages can represent an impoverished temperate Atlantic fauna or a mix of Atlantic, Arctic and Pacific species. Due to the geography of the continents, a latitudinal shift of assemblages and a gene flow were possible, which is relevant especially during periods of climatic change. Growth rates are similar to those measured in species from adjacent boreal regions.

Antarctic benthic invertebrates at >30 m depth generally have a circumpolar distribution; however, high-resolution studies revealed a pronounced patchiness (Gutt and Starmans 1998). Where sessile species cover the seafloor totally, they provide three-dimensional microhabitats for a rich associated fauna (Gutt and Schickan 1998). Due to the geographical isolation for approximately the last 20 million years (Lawver et al. 1992), a high percentage of species are endemic. It seems as if growth rates, especially those of species that help structure the environment, are several factors lower than those of related temperate species. However, there are also a few exceptionally fast-growing species (Dayton 1989). Planktonic larvae seem to be common in some Arctic areas (Clough et al. 1997); however, it is still impossible to quantify their role in community resilience (for further information, see Pearse et al. 1991; Clarke 1992).

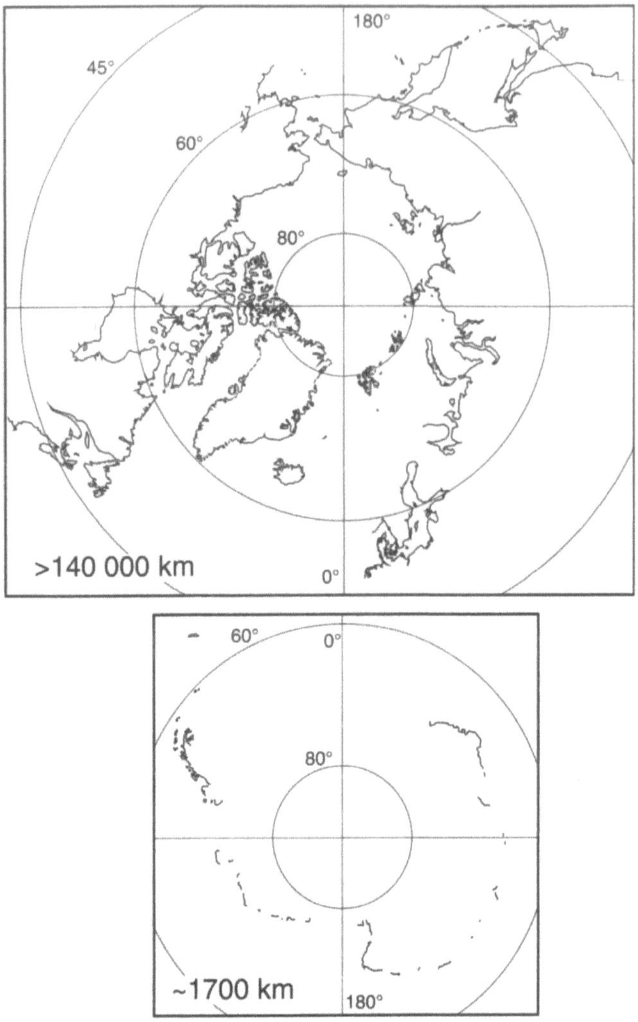

Fig. 1 Non-glaciated coasts where sea ice affects the benthos in the intertidal and upper sublittoral. Arctic ice occurrence mainly after Polar Regions Atlas, Central Intelligence Atlas, National Foreign Assessment Centre, CIA; Antarctic map simplified after Choubert and Faure-Muret (1976)

Sea ice and icefoot

Occurrence of sea ice

In the northern hemisphere, the length of the coastline generally affected is >140,000 km, based on values given by Cooper et al. (1983) and Conlan et al. (1998) (Fig. 1). It covers a gradient ranging from year-round presence to rare occurrence only in extremely cold winters. Regionally, the ice cover is well documented, especially where shipping or petroleum development is of relevance. However, most ice-charts only show a ≥20% chance of encountering ice with >3/10 coverage on a rough grid. Less attention is paid to its southern limits, with a low probability of presence, where ecological studies can provide valuable information on the resilience of benthic systems which are also subject to anthropogenic damage. Around Antarctica, sea ice is found everywhere, while it is rare at southern South America. Compared to the northern hemisphere, its effect on the benthos is restricted to the non-glaciated 5–8% of the coastline (Kristensen 1983), where a true intertidal exists (Fig. 1).

Occurrence and definitions of the icefoot

According to the WMO an icefoot is "A narrow fringe of ice attached to the coast, unmoved by tides and remaining after the fast ice has moved away". It is formed in the upper intertidal, mainly when the tidal amplitude is greater than the maximum thickness of the sea ice (Ellis and Wilce 1961). It can also develop in association with glaciers, ice shelves and anchor ice (Joyce 1950; Whitaker 1977; Hunter and Powell 1998).

Impact of sea ice and icefoot

Along polar and subpolar shores, benthic life in the infralittoral fringe and uppermost metres of the littoral is regularly exposed to ice. But additional environmental factors are essential to this part of the ecosystem, e.g.

wave action and the resultant reworking of the sediment, desiccation, dewatering, bioturbation, predation and food or nutrient supply, as well as extreme changes in temperature, light, and salinity due to ice formation, evaporation, precipitation, or river runoff (Anderson 1983; Dayton 1990; Carey 1991). Most of these shape an environment which can be considered as "harsh" for organisms. However, a limited number of species is found densely populating mudflats and rocky shores. According to Ellis and Wilce (1961), this is only true for hard substrata where the thickness of sea ice is less than the tidal amplitude, which coincides with areas where an icefoot can develop, e.g. in the eastern Arctic and subarctic North America. Under such conditions, the most conspicuous organisms are periwinkles in the shallowest fringe, which react to environmental events and can hide in crevices. Below these, barnacles followed by blue mussels are dominant, which can survive temperatures colder than –10°C. Fucoid macroalgae in the barnacle fringe are mainly restricted to crevices. Based on studies carried out in Greenland, Petersen (1977) concluded that the icefoot has both a positive and negative effect. However, the "protective effect ... should not be overexaggerated ... densest populations are found ... where the ice foot is not well developed ... and where ice floes have the best chance of scraping the rocks". However, life in crevices is most common at the high water mark where the icefoot best protects the coast. The use of this microniche is not a peculiarity of polar waters, as it can also be explained, for example, by humidity or shade. Consequently, there was no correlation found between biological zonations in areas with and without an icefoot (Petersen 1962). Lee (1973) argued that ice scouring is unimportant for the lack of plants in intertidals because denuded substrata can be colonised by the green algae *Ulothrix*, *Urospora* or *Enteromorpha* within < 15 days.

Bergeron and Bourget (1986) assumed the physical impact of ice to be an overwhelming factor responsible for the bareness of Arctic, subarctic and boreal shores and they compared it with predation pressure in tropical assemblages. Ice floes and the icefoot have a similar effect on sediments and, thus, an indirect effect on the fauna and flora. As long as the ice is attached to rocks, it protects the shore against erosion and waves. When it disintegrates, a lot of hard or soft material can be rafted away (Everson and Cohn 1979; Nielsen 1988). At a rocky shore in Newfoundland, Bolton (1983) observed "very little effect" of ice on macroalgae in the littoral fringe, but the physical impact reduced the abundance and species number mainly in the mid and lower intertidal, where Petersen (1977) found highest densities, especially of barnacles. However, at Bolton's study site, barnacles and *Enteromorpha umbilicalis* were more abundant immediately after an ice impact than before. These investigations obviously indicate that populations of both species show a low susceptibility. The fact that these results or interpretations are contradictory reflects both the lack of a correlation between quantitative faunistic and environmental data, and general differences in various biological systems.

Kiirikki (1996) successfully used a statistical approach when he demonstrated the relevance of each of three environmental factors in the zonation of eight common macroalgae found in the Baltic Sea between 0.2 and 2.0 m depth. His model showed that the shallowest (upper and lower) limits were best explained by wave action, the intermediate ones mainly by ice and the lower limits by the water clarity. Synergistic effects were not taken into account; these can be better detected, for example, by the BIO-ENV statistics developed at the Plymouth Marine Laboratory.

Sandy or muddy shores are rare in the entire Canadian Arctic Archipelago and are mainly populated by mobile sublittoral animals together with a few algae (Ellis and Wilce 1961). In the Alaskan Beaufort Sea, Feder and Schamel (1976) even found "no macrofaunal marine invertebrates ... along the narrow intertidal zone". Data from the Siberian Arctic indicate similar conditions (Golikov and Averincev 1977). In more southern mudflats, macrobenthic species have apparently developed two strategies to persist in areas subject to periodical ice impact (for reviews and concepts, see Connell and Slatyer 1977; Boesch and Rosenberg 1981). The stock size or biomass of sensitive species, e.g. the amphipod, *Corophium volutator*, or macroalgae, varies considerably due to a fluctuation between high mortality and very successful recruitment or the ability of species to avoid direct physical impact. The bivalve, *Macoma baltica*, for instance, is most abundant in protected areas, and the polychaete, *Heteromastus filiformis*, burrows deep (Gordon and Desplanque 1983). A third strategy is the capacity to survive being frozen in ice and to withstand the mechanical impact. Examples are the hard substrata favouring bivalves *Mytilus edulis*, *Crassostrea virginica* and *Mya arenaria* (Medcof and Thomas 1974), the periwinkle *Littorina saxatilis* (Petersen 1977), and the barnacles, *Balanus balanoides* and *Chthamalus dalli* (Madsen 1940; Southward and Southward 1967). However, *Mytilus edulis* banks can also be subject to massive erosion by ice floes. Generally, the pronounced seasonality in the growth of micro- and macroorganisms in northern, temperate, muddy intertidals is mainly controlled by wave action, dewatering, rain and temperature (Anderson 1983). Ice has either a stabilising or disturbing effect only during its formation and disintegration in winter.

For the Arctic subtidal, there have been few studies published on the relationship between ice scouring and diversity. At a site in eastern Newfoundland, Keats et al. (1985) clearly demonstrated that in years where there was no ice formation, only < 1% of the light penetrated through the canopy of the macroalga *Alaria esculenta*. In years when the *Alaria* canopy was removed by ice-scouring and light was not a limiting factor, additional annual species were most abundant and, thus, plant diversity was higher compared to years without disturbance when mainly perennials were able to co-exist.

However, the canopy can redevelop fast if the ice disturbance occurred early on in that year. At a more southerly site, which was fully exposed to wave action but where ice was rare, more perennial species sensitive to ice were present. These contributed to a latitudinal increase in diversity. Such a situation was also found by O'Clair (1981) in the Bering Sea where he compared "frequently", "rarely" and "never scoured" sites of which the last were the most diverse. Similar results were obtained by Heine (1989) in the same area. These examples demonstrate that within an ecological time scale of only a few years, one can either confirm or reject Huston's (1979) "intermediate disturbance hypothesis".

The generally well-documented gradient of decreasing benthic devastation with increasing depth can also be superimposed by physical pecularities along the coast of the Beaufort and part of the Chukchi Sea, as well as the Canadian Archipelago and Bering Sea between 10 and 40 m (Barnes et al. 1984; Weeks et al. 1984; Dayton 1990; Carey 1991). The landfast ice zone is an area usually 15 km wide but extending up to 60 km offshore between the coast and a grounded wall of ice, generated where rotating pack ice grinds against the landfast ice (Newbury 1983) in combination with wind stress, currents and seabed topography (Reimnitz and Kempema 1984). Here, ice scouring is rare due to the sea-ice cover for three-quarters of the year. Besides the role that landfast ice and resulting seasonal and interannual variations play for life in the water column, different effects on the seafloor and its fauna can be identified. Also, in shallow waters (< 6 m) the microrelief of the sediment is shaped by ice; however, a nearshore pumping effect by tides can lead to a fast smoothening. Macroepibenthic animals that seem to play a key role in a shallow biological system under landfast ice are mobile crustaceans, such as gammarids, amphipods, mysids and isopods (Newbury 1983). The intensity of mechanical disturbance is highest where the sea ice is ridged, mainly between 15 and 23 m depth. As a consequence, species diversity is assumed to be higher at these intermediate depths compared with areas beyond the shelf break (> 70 m). This is due to the relatively low overall benthic biomass found in the shallower areas, the fact that abundances are much lower within rather than outside single scours, and sediment heterogeneity caused by the ice (Carey 1991).

Deeper scours (30–45 m) are assumed not to be recent due to their considerable overlap and that such "bottom irregularities can remain visible for hundreds or ... even thousands of years" (Carey 1991). The lack of evidence of biological patterns caused by ice scouring between 8 and 54 m depth in the adjacent Chukchi and Bering Seas (Grebmeier and McRoy 1989; Feder et al. 1994) was attributed to the recurring of winter polynyas and consequently a low probability of ridged sea ice; however, this finding remains in contrast to other results already mentioned.

Excellent results of how keels of ridged sea ice shape the benthic structure in the central high Canadian Arctic at depths of between 3 and 15 m were provided by Conlan et al. (1998). The undisturbed assemblage dominated by predators and suspension feeders (large infaunal bivalves, *Mya truncata*, *Serripes groenlandicus* and a mix of *Macoma calcarea*, *Macoma moesta*, and *Macoma loveni*, the sea urchin, *Strongylocentrotus pallidus*, and the kelp, *Agarum cribrosum*) is completely transformed to an assemblage with a higher proportion of scavengers and deposit feeders (small spionid and errant polychaetes, lysianassid amphipods, cumaceans, buccinid gastropods, nemerteans, ostracods and pennate diatoms, e.g. *Parlibellus* sp.). The less diverse fauna found in the scours at various depths (3–15 m) resembled that in shallow (3 m), inshore waters where kelp indicated that it had not been recently affected. However, side-scan images showed that the abundance of scours in these shallow areas is, over a longer period of time, relatively high. Multivariate analysis did not show a higher variability at ice-disturbed sites, as could be expected according to Warwick and Clarke (1993).

A secondary effect of scouring by multi-year ice, observed in the Canadian Arctic Archipelago, are the so-called "black pools". These depressions in the sediment are filled with hypoxic, sulphide-rich water, which persists from its formation by the release of brine from the sea ice until the first strong wind. Not only can they cause the mortality of the infauna and sessile epifauna, but they also act as a lethal trap for motile animals (Kvitek et al. 1998).

In the Southern Ocean, less attention has been paid to the ecological functioning of the benthos in the intertidal and upper sublittoral. In contrast to the Arctic zonation, studies along the Antarctic coast have shown a gradient from a very scant life to richer communities at greater depths. The Antarctic species, which similar to the northern barnacles, periwinkles and blue mussels, can survive in the intertidal despite the threat of ice scour in great numbers are the limpets, *Nacella concinna* and *Patinigera polaris*. Neushul (1965), Dayton et al. (1970), Stephenson and Stephenson (1972), Etcheverry (1983), Zamorano (1983), Heywood and Whitaker (1984) and Sahade et al. (1998) all described gradients of assemblages within and below the intertidal. Most authors assumed that ice is the major factor shaping the fauna and flora in shallow water. However, Gambi et al. (1994) pointed out that the light regime and biological interactions are also important; Westermeier et al. (1992) also mentioned wave action in this context.

The attempt of Klöser et al. (1994) to compare the exposure to environmental factors with macroalgal assemblages off King George Island, revealed that the species composition can vary considerably within a small area, and their interpretation is in contrast to some more generally accepted patterns. *Desmarestia anceps* and *D. menziesii* are supposed to be not adapted "...to conditions disturbed by turbulences and ice", whereas *Himantothallus grandifolius* is. *Desmarestia* was only abundant "...where the probability of disturbance is low" and is thought to be dominant over *H. grandifolius*.

It is most likely that the discrepancies between these results and those summarised by Dayton (1990) are due to the use of non-defined terms: "extremely", "severely" or "exposed" (to what?).

In McMurdo Sound, ice scouring apparently determines the upper limit of algal distribution (Miller and Pearse 1991), where *Iridea cordata* is the shallowest species (3–10 m), followed by *Phyllophora antarctica* (6–18 m). The basal crusts of both species allow them to continue as perennials without upright blades and, thus, they are not generally susceptible to ice scouring. In a study at Signy Island, Barnes (1995a) found that frequent disturbance by ice at < 15 m depth restricted community development (both macrofauna and macroflora) to a high degree. However, intermediate disturbance at 25 m depth prevented the assemblage being dominated by only a few species. A continual change in the benthic assemblages of the upper subtidal with an increase in diversity, number of sessile species, abundance and biomass with depth is generally attributed to a reduced disturbance by sea ice, small pieces of icebergs (growlers) or anchor ice. Barnes (1995b) identified substratum type, salinity, wave action, current, substratum profile and length of the period of ice cover, and not the mechanical effects of sea ice, as the main factors influencing the development of sublittoral communities. However, deviations from a general zonation scheme can apparently develop where sheltered microniches, such as small depressions or the bases of large rocks, exist (Sato et al. 1992).

Anchor ice

Definition/formation

The definition of the WMO is: "submerged ice attached or anchored to the bottom, irrespective of the nature of its formation". As it is situated below the intertidal, it differs from an icefoot; physical conditions under which it is formed are not yet completely understood. Littlepage (1965) described a situation in the Antarctic where cold air caused a convection current in the water until freezing point was reached, with its lower limit at 33 m characterised by a halocline (Dayton et al. 1969). As a consequence, masses of ice platelets, 12 cm in diameter and 0.2 cm thick, adhered to the bottom and formed the anchor ice (Joyce 1950; Martin 1981). Based on its origin from platelet ice, which has also been observed at 250 m depth (Dieckmann et al. 1986), it is thought that it may also form in deeper water, e.g. under ice shelves (Pearse 1962). Reimnitz et al. (1987) described its formation in the Arctic during storms, which lead to a continuous cover of anchor ice from the shore down to 2 m depth, a patchy occurrence down to 4.5 m and its possible presence down to 20 m depth. It has also been shown to originate from fresh water along a 30-km-long stretch of coast, where it was formed by run-off water forced up through the seafloor since it was unable to penetrate deeper into the sediment due to permafrost (Sadler and Serson 1981). Problems of mapping anchor ice sufficiently are related to the method by which it is best detected; observations by divers are rare in autumn and winter when it seems to be most abundant. In the Antarctic, it is known to be common in McMurdo Sound and at a few other places in the east Antarctic, but it is rare or weakly described for the Peninsula area and South Orkney Islands. In the Arctic, anchor ice seems to be scattered everywhere, even occurring in the Black, Baltic and Norvegian Seas.

Impact

The only comprehensive ecological studies on the effect of anchor ice on benthic assemblages are those of Dayton et al. (1970) and Dayton (1989). Apparently, fouling of the epifauna by this kind of ice and the rise of large pieces from the bottom or its absence significantly structure the benthos in McMurdo Sound (Antarctica) between depths of 0 and 33 m. A piece of ice the size of 2 m^2 can carry 25 kg of sediment, including animals such as the asteroid, *Odontaster validus* and the nemertean, *Parborlasia corrugatus*, the echinoid, *Sterechinus neumayeri*, the isopod, *Glyptonotus antarcticus*, pycnogonids and the fish, *Trematomus bernacchii*, *T. centronotus*, and *T. nicolai*, as well as macroalgae. As a consequence, anchor ice is apparently the main factor that determines the upper limit of a unique and generally rich benthic assemblage, containing slow-growing sponges and an associated diverse fauna. One site with ideal conditions for the formation of anchor ice was completely devoid of this assemblage above 33 m, while at a site where anchor ice was "greatly reduced", sponges extended into shallower water up to 23 m depth (Dayton et al. 1969). The 15- to 33-m zone was characterised by coelenterates such as alcyonarians, actinians, stoloniferans and hydrozoans, as well as ascidians, which are either resistant to anchor ice formation or have very fast recruitment, or both (Dayton et al. 1970). In the shallowest waters (0–10 m), the effect of anchor ice is superimposed by that of ice-flows.

Intermediate-term oscillations, and not only its long-term probability of occurrence, mediate the effects of anchor ice. During the 1960s, Dayton (1989) observed extensive formation of anchor ice (blocks 1–2 m in diameter and up to 0.7 m high) between 0 and 30 m depth. This was reduced during the 1970s and, during the same period, a massive recruitment of *Homaxinella balfourensis* led to 80% seafloor cover by this normally rare sponge. Even though a study of 30 marked specimens revealed a mortality of 87%, the population quadrupled between 1975 and 1977. In 1984, anchor ice formed again and covered 5–30% of the seafloor. Consequently, 60% of the subsurface of the sea ice had clumps of uplifted anchor ice containing *Homaxinella balfourensis*, which were almost eliminated from the bottom. Ninety percent of the few remaining individuals were fouled by ice. Again in 1988, very little anchor ice was observed and *Homaxinella balfourensis* once more started to recruit in some places. With the "explosion" of the population in the mid 1970s, its predators, the asteroids *O. validus*, *O. meridionalis* and young *Perknaster fuscus antarcticus*, also became more abundant in the 15- to 30-m zone and larger individuals were observed, probably due to faster growth. These fluctuations may reflect a local upwelling of deep cold water related to ENSO events in 1957–1959 and 1982–1983, which led to the variation in the presence of anchor ice.

The foraminiferan fauna at an anchor ice-affected site also in McMurdo Sound was poor in species and abundance, resulting in a low diversity. Juveniles with a supposed seasonal recruitment were most common (Bernhard 1987), which was explained by the input of a higher organic level from inshore melt pools and the reduced activity of detritivores in the anchor-ice area. Thus, ice disturbance locally contributes to the patchy pattern of foraminiferan assemblages. The macrofauna at this site was dominated by the scallop *Adamussium colbecki*, whose mobility might be the reason for its high abundance in recently disturbed areas. In the same region, Lenihan and Oliver (1995) found mainly motile and opportunistic polychaetes such as *Capitella* spp, *Ophryotrocha claperedii*, and *Gyptis* spp, in areas with a high impact. Anchor ice affects the benthos also at the Haswell Islands (66°32′S, 093°00′E), mainly between 7 and 10 m depth. The benthos here comprises the hydroid, *Tubularia ralphii*, the echinoid, *S. neumayeri*, the asteroid, *Odontaster validus*, alcyonarians and *Lithothamnion*, of which *Tubularia ralphii* is apparently killed by anchor ice while others can survive. As a consequence, there is a pronounced seasonal variability in the most susceptible species (Gruzov 1977).

Kirkwood and Burton (1988) concluded that the absence of anchor ice in Ellis Fjord at least partly explained the high abundance of epifauna and its diversity found within the fjord in contrast to nearby inshore waters, where anchor ice was "extremely abundant". It is still unclear whether anchor ice was responsible for the anoxic sediments and poor deposit-feeding fauna in one of the three assemblages found by Everitt et al. (1980) in the adjacent Davis Bay. At Arthur Harbor, off Anvers Island, anchor ice has little effect on the biota (Shabica 1972) due to its unconsolidated nature and because it was only found down to a depth of less than 3 m. Here the limpet, *Patinigera polaris*, and the alga, *Iridea* sp., were abundant, and they are known to be able to survive being frozen in ice (Hargens and Shabica 1973; Miller and Pearse 1991).

Another phenomenon that has attracted the attention of scientists is the occurrence of benthic animals on the surface of the McMurdo ice shelf (for literature see Kellogg et al. 1990). It seems clear that these animals move from the subsurface to the surface of the 30- to 40-m-thick ice shelf due to ablation and bottom freezing. Several hypotheses about their origin have been discussed in the past. The only relevance to this present review is whether benthic animals are uplifted by anchor ice in ice-shelf covered areas, as proposed by Pearse (1962) and Dayton et al. (1970). If this is true, the composition of invertebrates found on the surface of the ice shelf would represent the faunal composition under the ice shelf and, thus, provide information about benthic structure and ecological function in areas that are virtually inaccessible to research activities. However, the most recent ^{14}C data of Kellogg et al. (1990) indicate that this mechanism is probably uncommon, although it cannot be excluded for shallow water.

A similar phenomenon was described for Victoria Strait in the Canadian Arctic. Here, the existence of sea ice laden with sediment and benthic invertebrates was explained by the suspension freezing and anchor ice, which uplifted from a water depth of 25–30 m (Reimnitz et al. 1992; Weeks 1994). The organisms found in the ice were 7 species of bivalves, 4 gastropods, 12 ostracods and 1 foraminiferan; there were also sea urchins, as well as algae. This had already been described in early Arctic expeditions, e.g. the Fram expedition north of the New Siberian Islands, northwest of Svalbard and in the Baltic (Zubov 1943).

Icebergs

Definitions, origin and occurrence

The WMO defined "Iceberg: a massive piece of ice ... > 5 m above sea-level ...". They have an average size of > 300 m^2, which includes ice islands; however, these are separately defined to be 30–50 m thick and from a few thousand square metres to 500 km^2 or larger. Small pieces of icebergs are called bergy bits (< 5 m freeboard, 100–300 m^2) or growlers (< 1 m above sea surface, ~20 m^2).

Along the Antarctic coast, tabular icebergs calve from the ice shelves, which make up 46% of the coastal length mainly in the Ross and Weddell Seas, and into which 62% of the land-originated ice discharges (Kristensen 1983). Smaller icebergs originate from an additional 46% of the coastline where there are glaciers, ice streams and grounded ice walls. A total "population" of > 300,000 icebergs (> 10 m wide) were calculated for the entire Antarctic in 1981–1984 by Orheim (1985). If they drift around the continent in the East Wind Drift and do not reach warmer northerly waters (Hamley and Budd 1986), they have an average half-life expectancy of 2–4 years (Orheim 1985), which results in a relatively homogenous dispersion of icebergs throughout the coastal waters. In the Arctic, icebergs mainly originate from true glaciers, most of them in Greenland, where 40,000 icebergs calve annually from 100 glaciers (Sanderson 1988). The majority of icebergs come from Disco Bay but also from the northwestern Canadian Archipelago, Svalbard, Franz Joseph Land, Novaya and Severnaya Zemlya and Southern Alaska. Although most analyses do not provide data suitable for a global comparison, Dayton (1990) concluded that icebergs are less abundant and smaller in the Arctic than in the Antarctic and that they are less homogenously distributed in the Arctic (see also Charlesworth 1957).

The most important parameter concerning the size of the icebergs in the context of this review is their draught. Deepest plough marks in both polar areas have been reported from approximately 600 m depth (Marienfeld 1992; Dowdeswell et al. 1993; Vogt et al. 1994; Viehoff and Li 1995; Harris and O'Brien 1996). Figure 2 shows maps of those shelf areas that generally experience iceberg scouring. Although the abundance of scours can be recorded using fine-scale topographical measurements (e.g. Charlesworth 1957; Just 1970; Fossum et al. 1982; Syvitski et al. 1983; Pereira et al. 1988; Sanderson 1988; Lien et al. 1989; Rearic et al. 1990; Woodworth-Lynas et al. 1991; Niessen and Musatov 1997), little is known about the frequency of iceberg scouring which is suited for a polar comparison, especially over larger areas.

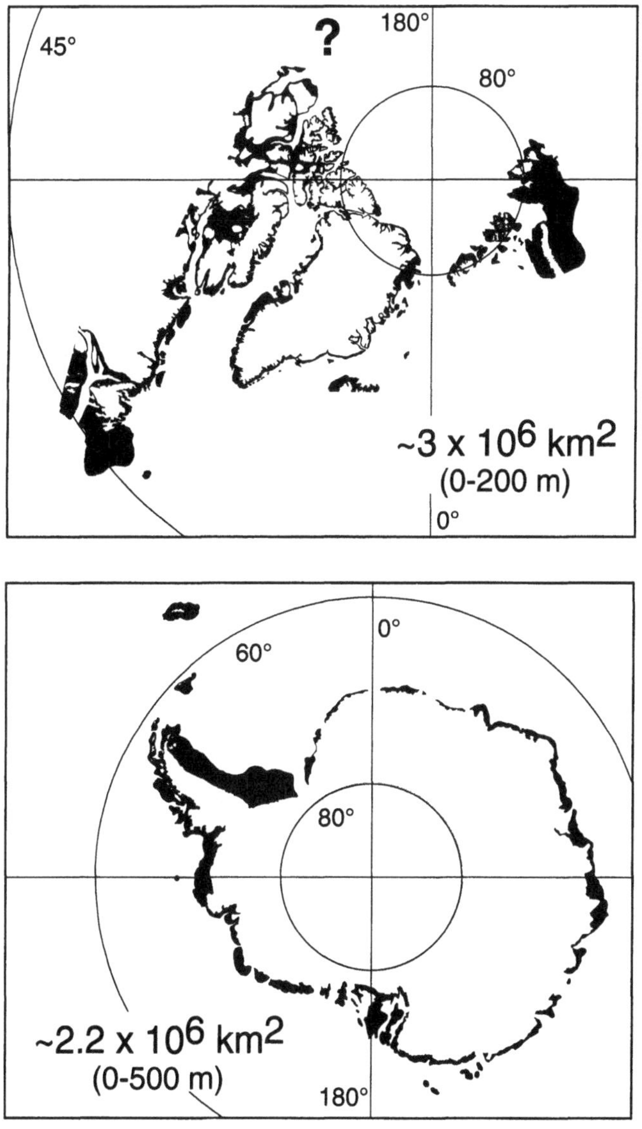

Fig. 2 Shelf areas which generally experience iceberg scouring. Southernmost records in the northern hemisphere close to Bermuda and Azores, northernmost records in the southern hemisphere at Tierra del Fuego. Size of areas estimated according to Everson (1977) and Gierloff-Emden (1982)

For the Arctic, there are many records that ice disturbs sediments at depths > 15 m. However, there is, so far, only one publication on the ecological effects within this depth (Gutt et al. 1996). One major result was that scours were observed by underwater video down to 70 m, which leaves the question of whether these were caused by ridged sea ice, pieces of an ice shelf of uncertain origin or true icebergs. However, the answer is irrelevant as far as the effects on the benthos are concerned. Based on the assumption that the scours comprising 9.5% of the seafloor were no older than 5 years, a period of 53 years was calculated in which each square metre was disturbed once on average.

A similar calculation for the Antarctic was made for the entire depth range of the shelf (> 500 m), revealing one disturbance per square metre every 230 years, which has had to be changed to 340 years because of a recent age determination of the pioneer species, the bryozoan *Cellaria incula* (Gutt and Starmans 2001). Data also indicated that iceberg scouring contributes significantly to a high between-habitat diversity (Gutt 2000). Peck and Bullough (1993) came to similar conclusions on the general relevance of physical disturbance for a subtidal assemblage at Signy Island at ~9 m depth, which experiences ice scouring by growlers or bergy bits approximately every 50–75 years. The meiofauna was less affected than the macrofauna which was also demonstrated for the Weddell sea by H. Lee et al. (unpublised work). Three different recolonisation mechanisms were identified: locomotion, e.g. by isopods and amphipods, advection of adult animals by storms, e.g. the small bivalve, *Mysella charcoti*, and larval recolonisation by most other species (Peck et al. 1999). Complete recovery of the benthos at this site is not reached before 65 years because the assemblage is dominated by the very slow-growing bivalve, *Yoldia eightsi*. The main components of rocky intertidal assemblages near Halifax (Nova Scotia, Canada) consisting of barnacles, mussels and algae, recovered to a large extent from sea-ice disturbance after a short period of 2–3 years (McCook and Chapman 1991; Minchinton et al. 1997), apparently because they are better adapted to stress. Such data were used for a global comparison of the effects of natural catastrophes on large ecosystems (Gutt and Starmans 2001). Also, demersal fish in the Weddell Sea (Antarctica) make use of specific microhabitats partly shaped by icebergs (Gutt and Ekau 1996; Brenner et al. 2001). These results were only made possible by the combined use of imaging methods and fishing gear.

An initial stage in the recolonisation of scour marks at Arthur Harbour (Anvers Island) was characterised by the immigration of motile animals, among which grazers on macroalgae attracted larger scavengers and carnivores (Richardson and Hedgpeth 1977). Kauffman

(1974) even showed for the same area, a "jelly-like bacterial growth" on a ridge half a year after an ice impact and the meiofauna was described as being comparable but not identical to the undisturbed situation. Scours of bergy bits in McMurdo Sound between 25 and 40 m depth were colonised by significantly less infauna. A few crustaceans, e.g. the ostracod *Philomedes* sp., the cumacean, *Eudorella splendida*, an unidentified isopod and the motile polychaetes *Tharyx* sp. and *Polygordius* sp. were more abundant than in the surrounding habitat (Lenihan and Oliver 1995).

Glaciers

Occurrence

In the Antarctic, glaciers with a tidewater cliff make up 13% of the entire coastline (Cooper et al. 1983) and are scattered all around the continent. In the northern hemisphere, the most productive glaciers are found in Franz Joseph Land, Severnaya and Novaya Zemlya, Svalbard, Greenland and the northwest Canadian Arctic.

Impact

A comprehensive study by Wlodarska et al. (1996) in the vicinity of two glaciers in West Svalbard and Franz Josef Land showed that the benthos was dominated by deposit feeders, indicating a terrigenous input. The differences in community structure, however, showed a clear gradient with depth rather than with distance from the glacier termini. Thus, any direct effects of the glaciers were either unimportant, masked by other factors or undetectable because the distances of 300–2,000 m from the cliff were not long enough. Korsun and Hald (1998) investigated the impact of three glaciers in Novaya Zemlya. Along transects from the termini into open water, which were ≤ 20 km long, they found an increase in water transparency, a change in the sediment to coarser particles and higher organic content. They also recorded an increase in the number of foraminiferan species and their abundances.

In the Antarctic, at Arthur Harbour (Anvers Island), Hyland et al. (1994) observed a substantial change of the macroinfauna at a shallow site, which was initially situated < 60 m from the cliff of a glacier (Richardson and Hedgpeth 1977). Within 18 years, the front retreated by 250 m while the species number doubled and the abundance increased by a factor of 5.5. A striking conclusion from this change was that environmental instability very close to the glacier causes a relatively poor benthos, dominated by the polychaete *Tharyx cincinnatus*. Also, the presence of H_2S was attributed to the calving of this glacier and indicated the inefficient utilisation of food. A similar effect caused by keeled sea ice in the Arctic is called "black pools" (Kvitek et al. 1998, see above) or hypothesised as a consequence of Antarctic anchor ice (Everitt et al. 1980).

Close to the grounding line around the floating Mackay Glacier Tongue in McMurdo Sound, Antarctica, Powell et al. (1996) and Dawber and Powell (1997) recorded a benthic fauna with large, extremely slow-growing hexactinellid sponges. Species composition did not differ greatly from that at a reference site 8 km from the glacier. In some areas, there was a reduction in benthic abundance but not in the number of species, leading to an increase in diversity. There was only one local grounding point within 250 m of the ice margin where the density and diversity of the community gradually decreased. Here, hexactinellid sponges were totally absent. However, the demosponge, *Homaxinella*, which is apparently able to react quickly to short-term environmental events in terms of recruitment and population growth (Dayton 1989), was abundant. The poor fauna at this pinning point was thought to be due to the retreat of the glacier by 3 km in 30 years or by 10 km in approximately 72 years before the study.

Discussion

From the references cited, it can be concluded that ice is of relevance to the shelf-inhabiting benthos in large areas of both polar regions. This then raises the question for many scientists about just what is the role of this environmental parameter. Before attempting to answer this question, it should be considered that such a physical effect is part of a complex ecological system. In other words, on a general level the question can be answered simply: this parameter shaped the biological structure because the fauna and flora developed under the given conditions to the stage we see it at today. If the environment had been significantly different, the biological system would have developed differently too. Scientifically more useful are specific questions which, if answered, would contribute to a broader understanding of ecological relationships. In this context the intermediate disturbance hypothesis (Huston 1979), or alternatively, the stability time hypothesis (Grassle and Sanders 1973) or many others can be tested under defined conditions, e.g. concerning temporal and spatial scales. Corresponding quantitative biological and physical results would reduce the risk of misinterpretations, which mainly arise due to anecdotal and, thus, non-reproducible findings.

Although it is still early days for quantitative ecology associated with the effects of ice, it is possible to draw a few conclusions. Benthic assemblages in the polar and subpolar intertidal and the upper metres of the subtidal are exposed to a very high degree of variability in physical parameters and mechanical damage. As a consequence, benthic life is poor either in terms of diversity, number of species or even regionally in the levels of abundance. Only those animals and algae which can cope in this specific situation are found in such shallow waters. Just which of the environmental parameters are most significant in shaping the benthic structure in this fringe depends on the local conditions and particular characteristics of the species involved. Modern statistical tools and a suitable sampling design can help to decipher such relatively complex relationships. Environmental stability increases with increasing depth. As a consequence, species in the lower subtidal, especially in the Antarctic, have adapted to these conditions by developing a pronounced stenoecy and extremely slow growth. However, the assemblages also commonly experience physical disturbance by highly ridged sea ice or icebergs. It is only beyond a certain depth or latitude where ice is rare, that corresponding consequences can be considered as alien to the ecosystem.

The local devastation of benthic life by ice scouring and the subsequent recolonisation creates a patchy pattern on the seafloor. The biological development within such patches may or may not be predictable. In the first case, the mosaic-cycle (Reise 1991) and in the second case, the patch dynamics (White and Pickett 1985) concepts can be applied. The higher the variability of succession with time and between stages of similar age, the larger the increase in between-habitat diversity. The results of Conlan et al. (1998) indicate, due to the low variability within the disturbed sites, a predictable development. The locally varying pioneer species found on the deeper Antarctic shelf (Gutt et al. 1996) suggest rather unpredictable processes. Generalisations about the temporal change in within-habitat diversity during recolonisation are difficult. If only a few species are very successful at an early stage, the dominance pattern will be very uneven. At a late stage, the frequency of the disturbance and the intensity of competetive displacement may determine whether a high diversity can develop and thus, the stability time hypothesis must be rejected for ecologically effective time scales. It also remains unclear if a long-term change in the bathymetry and sediment composition (see Syvitski et al. 1983; Vorren et al. 1983; Lien 1986) has an influence on the benthos. Belderson et al. (1973), studying the Rockhall Bank in the northeast Atlantic, speculated that coarser sediment on the berms, thrown up during scouring, favours the growth of the coral, *Lophelia prolifera*, which acts as a special habitat for a variety of accompanying species. However, Conlan et al. (1998) did not find the expected differences in the smaller fraction of the fauna investigated between the scour berms and troughs.

It is also conceivable that ice dynamics affect the origin of benthic invertebrates. Ice surges happened mainly during glacial periods in the past million years but they are also assumed to have taken place 25–30 million years ago. These surges could interrupt the geneflow among meta-populations and act as a main "player" in a vicariance event (Clarke and Crame 1997). Another possibility is that devastated areas provide a habitat for animals that cannot survive in climax communities due to competition pressure. However, there is no clear evidence yet that the intermediate disturbance hypothesis, and not only the stability time hypothesis, can be applied on an evolutionary time scale.

Comparing both polar areas, it is possible to conclude that in shallow waters ice is most important in the Arctic, due to the great length of its coastline. Disturbances of the deeper benthos by true icebergs seem to be very patchy in the Arctic but are a common event everywhere on the Antarctic shelf. Glaciers have a direct effect only in their close vicinity. From this it can also be taken that the ice-shelf edge has no great direct effect. Indirect consequences (Dayton and Oliver 1977; Barry and Dayton 1988), however, should not be ignored.

Ice scouring is among those natural catastrophes that have a significant effect on large ecosystems. This is mainly due to slow growth, and consequently, slow recolonisation in combination with an intermediate level of disturbance. It affects approximately 15% of the entire world's shelf areas (see Fig. 2).

References

Anderson FE (1983) The northern muddy intertidal: seasonal factors controlling erosion and deposition – a review. Can J Fish Aquat Sci 40 [Suppl 1]:143–159

Barnes DKA (1995a) Sublittoral epifaunal communities at Signy Island, Antarctica. I. The ice-foot zone. Mar Biol 121:555–563

Barnes DKA (1995b) Sublittoral epifaunal communities at Signy Island, Antarctica. II. Below the ice-foot zone. Mar Biol 121:565–572

Barnes DKA (1999) The influence of ice on polar nearshore benthos. J Mar Biol Assoc UK 79:401–407

Barnes PW, Rearic DM, Reimnitz E (1984) Ice gouging characteristics and processes. In: Barnes PW, Schell DM, Reimnitz E (eds) The Alaskan Beaufort Sea: ecosystems and environments. Academic Press, Orlando, pp 185–212

Barry JP, Dayton PK (1988) Current patterns in McMurdo Sound, Antarctica, and their relationship to local biotic communities. Polar Biol 8:367–376

Belderson RH, Kenyon, Wilson JB (1973) Iceberg plough marks in the Northeast Atlantic. Paleogeogr Paleoclimatol Paleoecol 13:215–224

Bergeron P, Bourget E (1986) Shore topography and spatial partitioning of crevice refuges by sessile epibenthos in an ice disturbed environment. Mar Ecol Prog Ser 28:129–145

Bernhard JM (1987) Foraminiferal biotopes in Explorers Cove, McMurdo Sound, Antarctica. J Foraminiferal Res 17:286–297

Boesch DF, Rosenberg R (1981) Response to stress in marine benthic communities. In: Barrett GW, Rosenberg R (eds) Stress effects on natural ecosystems. Wiley, London, pp 179–200

Bolton JJ (1983) Effects of short-term ice scouring on a Newfoundland rocky shore community. Astarte 12:39–43

Brenner M, Buck BH, Cordes S, Dietrich L, Jacob U, Mintenbeck K, Schröder A, Brey T, Knust R, Arntz W (2001) The role of iceberg scours in niche seperation within the Antarctic fish genus *Trematomus*. Polar Biol (in press)

Carey AG Jr (1991) Ecology of North American Arctic continental shelf benthos: a review. Cont Shelf Res 11:865–883

Charlesworth JK (1957) Drift ice. In: Charlesworth JK (ed) The Quarternary era. Arnold, London, pp 177–208

Choubert G, Faure-Muret A (1976) Geological world atlas. Sheet 17. Unesco and Commission for the Geological Map of the World, Paris

Clarke A (1992) Reproduction in the cold: Thorson revisited. Invertebr Reprod Dev 22:175–184

Clarke A (1996) Marine benthic populations in Antarctica: patterns and processes. Antarct Res Ser 70:373–388

Clarke A, Crame JA (1997) Diversity, latitude and time: patterns in the shallow sea. In: Ormond RFG, Gage JD, Angel MV (eds) Marine biodiversity. Cambridge University Press, Cambridge, pp 122–147
Clough LM, Ambrose WG Jr, Ashjian CJ, Piepenburg D, Renaud PESSL (1997) Meroplankton abundance in the Northeast Water Polynya: insights from oceanographic parameters and benthic abundance patterns. J Mar Syst 10:343–357
Conlan KE, Lenihan HS, Kvitek RG, Oliver JS (1998) Ice scour disturbance to benthic communities in the Canadian High Arctic. Mar Ecol Prog Ser 166:1–16
Connell JH, Slatyer RO (1977) Mechanisms of succession in natural communities and their role in community stability and organisation. Am Nat 111:1119–1144
Cooper APR, Drewry DJ, Jordan SR (1983) Geophysical database and Antarctic coastline compilation. In: Drewry DJ (ed) Antarctic: glaciological and geophysical folio. Scott Polar Research Institute, Cambridge
Curtis M (1975) The marine benthos of arctic and sub-arctic continental shelves. Polar Rec 17:595–626
Dawber M, Powell RD (1997) Epifaunal distribution at Antarctic marine-ending glaciers: influences of ice dynamics and sedimentation. In: Ricci CA (ed) The Antarctic region: geological evolution and processes. Terra Antarctica Publication, Siena, pp 875–884
Dayton PK (1989) Interdecadal variation in an Antarctic sponge and its predators from oceanographic climate shifts. Science 245:1484–1486
Dayton PK (1990) Polar benthos. In: Smith WO Jr (ed) Polar oceanography, Part B Chemistry, biology, and geology. Academic Press, San Diego, pp 631–685
Dayton PK, Oliver JS (1977) Antarctic soft-bottom benthos in oligotrophic and eutrophic environments. Science 197:55–58
Dayton PK, Robilliard GA, DeVries AL (1969) Anchor ice formation in McMurdo Sound, Antarctica, and its biological effects. Science 163:273–274
Dayton PK, Robillard GA, Paine RT (1970) Benthic faunal zonation as a result of anchor ice at McMurdo Sound, Antarctica. In: Holdgate MW (ed) Antarctic ecology, vol 1. Academic Press, London, pp 244–258
Dieckmann G, Rohardt G, Hellmer H, Kipfstuhl J (1986) The occurrence of ice platelets at 250 m depth near Filchner Ice Shelf and its significance for sea-ice biology. Deep Sea Res 33:141–148
Dowdeswell JA, Villinger H, Whittington RJ, Marienfeld P (1993) Icebergs scouring in Scoresby Sound and on the East Greenland continental shelf. Mar Geol 111:37–53
Ellis DV, Wilce RT (1961) Arctic and subarctic examples of intertidal zonation. Arctic 14:224–235
Etcheverry H (1983) Algas bentónicas de la Antártica chilena. Ser Cient INACH 30:97–124
Everitt DA, Poore GCB, Pickard J (1980) Marine benthos from Davis station, eastern Antarctica. Aust J Mar Freshwater Res 31:829–836
Everson EB, Cohn BP (1979) The ice-foot complex: its morphology, formation and role in sediment transport and shoreline protection. Z Geomorphol 23:58–75
Everson I (1977) The living resources of the Southern Ocean. FAO, UNDP, Rome
Feder HM, Schamel D (1976) Shallow-water benthic fauna of Prudhoe Bay. In: Hood DW, Burrell DC (eds) Assessment of the Arctic marine environment: selected topics. Occasional publication no. 4. Institute of Marine Science, University of Alaska, Fairbanks, pp 329–359
Feder HM, Naidu AS, Jewett SC, Hameedi JM, Johnson WR, Whitledge TE (1994) The northeastern Chukchi Sea: benthos-environmental interactions. Mar Ecol Prog Ser 111:171–190
Fossum BA, Maisey GH, Torsen HO (1982) Marine geophysical research in the Weddell Sea during the Norwegian research expedition of 1976–77. Publ Int Union Geol Sci ser B 4:397–404
Gallardo VA (1987) The sublittoral macrofaunal benthos of the Antarctic shelf. Environ Int 13:71–81
Gambi MC, Lorenti M, Russo GF, Scipione MB (1994) Benthic associations of the shallow hard bottoms off Terra Nova Bay, Ross Sea: zonation, biomass and population structure. Antarct Sci 6:449–462
Gierloff-Emden HG (1980) Geographie des Meeres: Ozeane und Küsten. de Gruyter, Berlin
Golikov AN, Averincev VG (1977) Distribution of benthic and ice biocoenoses in the high latitudes of the polar basin and their part in the biological structure of the world ocean. In: Dunbar ML (ed) Polar oceans. Arctic Institute of North America, Calgary, pp 331–364
Gordon DC Jr, Desplanque C (1983) Dynamic and environmental effects of ice in the Cumberland Basin of the Bay of Fundy. Can J Fish Aquat Sci 40:1331–1342
Grassle JF, Sanders HL (1973) Life histories and the role of disturbance. Deep Sea Res 20:643–659
Grebmeier JM, McRoy CP (1989) Pelagic-benthic coupling on the shelf of the northern Bering and Chukchi Seas. II. Benthic community structure. Mar Ecol Prog Ser 51:253–268
Gruzov EN (1977) Seasonal alternations in coastal communities in the Davis Sea. In: Llano GA (ed) Adaptation within Antarctic ecosystems. Gulf, Houston, pp 263–278
Gutt J (2000) Some "driving forces" structuring communities of the sublittoral Antarctic macrobenthos. Antarct Sci 12:297–313
Gutt J, Ekau W (1996) Habitat partitioning of dominant high Antarctic demersal fish in the Weddell Sea and Lazarev Sea. J Exp Mar Biol Ecol 206:25–37
Gutt J, Schickan T (1998) Epibiotic relationships in the Antarctic benthos. Antarct Sci 10:398–405
Gutt J, Starmans A (1998) Structure and biodiversity of megabenthos in the Weddell and Lazarev Seas (Antarctic): ecological role of physical parameters and biological interactions. Polar Biol 20:229–247
Gutt J, Starmans A (2001) Quantification of iceberg impact and benthic recolonisation patterns in the Weddell Sea (Antarctica) Polar Biol (in press)
Gutt J, Starmans A, Dieckmann G (1996) Impact of iceberg scouring on polar benthic habitats. Mar Ecol Prog Ser 137:311–316
Hamley TC, Budd WF (1986) Antarctic iceberg distribution and dissolution. J Glaciol 32:242–251
Hargens AR, Shabica SV (1973) Protection against lethal freezing temperatures by mucus in an Antarctic limpet. Cryobiology 10:331–337
Harris PT, O'Brien PE (1996) Geomorphology and sedimentology of the continental shelf adjacent to MacRobertson Land, East Antarctica: a scalped shelf. Geo-Mar Lett 16:287–296
Heine JN (1989) Effects of ice scour on the structure of sublittoral marine algal assemblages of St. Lawrence and St. Matthew Islands, Alaska. Mar Ecol Prog Ser 52:253–260
Heywood RB, Whitaker TM (1984) The Antarctic marine flora. In: Laws RM (ed) Antarctic ecology, vol 2. Academic Press, London, pp 373–419
Hunter LE, Powell RD (1998) Ice foot development at temperate tidewater margins in Alaska. Geophys Res Lett 25:1923–1926
Huston M (1979) A general hypothesis of species diversity. Am Nat 113:81–101
Hyland J, Laur D, Jones J, Shrake J, Cadian D, Harris L (1994) Effects of an oil spill on the soft-bottom macrofauna of Arthur Harbour, Antarctica compared with long-term natural change. Antarct Sci 6:37–44
Joyce JRF (1950) Notes on ice-foot development, Neny Fjord Graham Land, Antarctica. J Geol 58:646–649
Just J (1970) Marine biological investigations of Jørgen Brønlund Fjord, North Greenland. Medd Gron 184:1–42
Kauffman TA (1974) Seasonality and disturbance in benthic communities, Arthur Harbor, Antarctic Peninsula. Antarct J US 9:307–310
Keats DW, South GR, Steele DH (1985) Algal biomass and diversity in the upper subtidal at a pack-ice disturbed site in eastern Newfoundland. Mar Ecol Prog Ser 25:151–158

Kellogg TB, Kellogg DE, Stuiver M (1990) Late quaternary history of the southwestern Ross Sea: evidence from debris bands on the McMurdo Ice Shelf, Antarctica. Antarct Res Ser 50:25–56
Kiirikki M (1996) Mechanisms affecting macroalgal zonation in the northern Baltic Sea. Eur J Phycol 31:225–232
Kirkwood JM, Burton HR (1988) Macrobenthic species assemblages in Ellis Fjord, Vestfold Hills, Antarctica. Mar Biol 97:445–457
Klöser H, Mercuri G, Laturnus F, Quartino ML, Wiencke C (1994) On the competitive balance of macroalgae at Potter Cove (King George Island, South Shetlands). Polar Biol 14:11–16
Korsun S, Hald M (1998) Modern benthic foraminifera off Novaya Zemlya tidewater glaciers, Russian Arctic. Arct Alp Res 30:61–77
Kristensen M (1983) Iceberg calving and deterioration in Antarctica. Prog Phys Geogr 7:313–328
Kvitek RG, Conlan KE, Iampietro PJ (1998) Black pools of death: hypoxic, brine-filled ice gouge depressions become lethal traps for benthic organisms in a shallow Arctic embayment. Mar Ecol Prog Ser 162:1–10
Lawver LA, Gahagan LM, Coffin MF (1992) The development of paleoseaways around Antarctica. Antarct Res Ser 56:7–30
Lee RKS (1973) General ecology of the Canadian Arctic benthic marine algae. Arctic 26:32–43
Lenihan HS, Oliver JS (1995) Anthropogenic and natural disturbances to marine benthic communities in Antarctica. Ecol Appl 5:311–326
Lewis JR (1972) The ecology of rocky shores. The English University Press, London
Lien R (1986) An iceberg scouring model; a remedy for survey planning, data interpretation and technical evaluation. Proceedings of the Fifth International Offshore Mechanics and Arctic Engineering. pp 119–133
Lien R, Solheim A, Elverhøi A, Rokoengen K (1989) Iceberg scouring and sea bed morphology on the eastern Weddell Sea shelf, Antarctica. Polar Res 7:43–57
Littlepage JL (1965) Oceanographic investigations in McMurdo Sound, Antarctica. In: Llano GA (ed) Biology of the Antarctic Seas. II. Antarct Res Ser 5:1–31
Madsen H (1940) A study of the littoral fauna of Northwest Greenland. Medd Grøn 124:1–24
Marienfeld P (1992) Recent sedimentary processes in Scoresby Sund, East Greenland. Boreas 21:169–186
Martin S (1981) Frazil ice in rivers and oceans. Annu Rev Fluid Mech 13:379–397
McCook LJ, Chapman ARO (1991) Community succession following massive ice-scour on an exposed rocky shore: effects of *Fucus* canopy algae and of mussels during late succession. J Exp Mar Biol Ecol 154:137–169
Medcof JC, Thomas MLH (1974) Surfacing on ice of frozen-in marine bottom materials. J Fish Res Board Can 31:1195–1200
Miller KA, Pearse JS (1991) Ecological studies of seaweeds in McMurdo Sound, Antarctica. Am Zool 31:35–48
Minchinton TE, Scheibling RE, Hunt HL (1997) Recovery of an intertidal assemblage following a rare occurrence of scouring by sea ice in Nova Scotia, Canada. Bot Mar 40:139–148
Neushul M (1965) Diving observations of sub-tidal Antarctic marine vegetation. Bot Mar 8:234–243
Newbury TK (1983) Under landfast ice. Arctic 36:328–340
Nielsen N (1988) Observations of sea ice influence on the littoral sediment exchange, North Zealand, Denmark. Geogr Tidsskr 88:61–67
Niessen F, Musatov E (1997) Marine sediment echosounding using PARASOUND. Ber Polarforsch 226:118–128
O'Clair CE (1981) Disturbance and diversity in a boreal marine community: the role of intertidal scouring by sea ice. In: Hood DW, Calder JA (eds) The eastern Bering Sea shelf: oceanography and resources 2.University of Washington Press, Seattle, pp 1105–1130
Orheim O (1985) Iceberg discharge and the mass balance of Antarctica. In: Glaciers, ice sheeets, and sea level: effect of a CO_2-induced climatic change. Report of a workshop held in Seattle, Washington, September 13–15, 1984. United States Department of Energy, Washington, DC, pp 210–215
Pearse JS (1962) Letters. Sci Am 207:12
Pearse JS, McClintock JB, Bosch I (1991) Reproduction of Antarctic benthic marine invertebrates: tempos, modes, and timing. Am Zool 31:65–80
Peck LS, Bullough LW (1993) Growth and population structure in the infaunal bivalve *Yoldia eightsi* in relation to iceberg activity at Signy Island, Antarctica. Mar Biol 117:235–241
Peck LS, Brockington S, Vanhove S, Beghyn M (1999) Community recovery following catastrophic iceberg impacts in a soft-sediment shallow-water site at Signy Island, Antarctica. Mar Ecol Prog Ser 186:1–8
Pereira CPG, Woodworth-Lynas CMT, Barrie JV (1988) Iceberg scour investigations and sedimentology of the southeast Baffin Island continental shelf. Arctic 41:221–230
Petersen GH (1962) The distribution of *Balanus balanoides* (L.) and *Littorina saxatilis*, Olivi, var. *groenlandica*, Mencke in Northern West Greenland. Medd Grøn 159:1–40
Petersen GH (1977) Biological effects of sea-ice and icebergs in Greenland. In: Dunbar ML (ed) Polar oceans. Arctic Institute of North America, Calgary, pp 319–329
Piepenburg D, Schmid MK (1996) Distribution, abundance, biomass, and mineralization potential of the epibenthic megafauna of the Northeast Greenland shelf. Polar Biol 125:321–332
Powell RD, Dawber M, McInnes JN, Pyne AR (1996) Observation of the grounding-line area at a floating glacier terminus. Ann Glacol 22:217–223
Rearic DM, Barnes PW, Reimnitz E (1990) Bulldozing and resuspension of shallow-shelf sediment by ice keels: implications for Arctic sediment transport trajectories. Mar Geol 91:133–147
Reimnitz E, Kempema EW (1984) Pack ice interaction with Stamukhi Shoal Beaufort Sea, Alaska. In: Barnes PW, Schell DM, Reimnitz E (eds) The Alaskan Beaufort Sea: ecosystems and environments. Academic Press, Orlando, pp 159–183
Reimnitz E, Kempema EW, Barnes PW (1987) Anchor ice, seabed freezing, and sediment dynamics in shallow Arctic seas. J Geophys Res 92:14671–14678
Reimnitz E, Marincovich L Jr, McCormick M, Briggs WM (1992) Suspension freezing of bottom sediment and biota in the Northwest Passage and implications for Arctic Ocean sedimentation. Can J Earth Sci 29:693–703
Reise K (1991) Mosaic cycles in the marine benthos. In: Remmert H (ed) The mosaic-cycle concept of ecosystems. Springer, Berlin Heidelberg New York, pp 61–82
Richardson MD, Hedgpeth JW (1977) Antarctic soft-bottom, macrobenthic community adaptation to a cold, stable, highly productive, glacially affected environment. In: Llano GA (ed) Adaptation within Antarctic ecosystems. Gulf, Houston, pp 181–196
Sadler HE, Serson HV (1981) Fresh water anchor ice along an Arctic beach. Arctic 34:62–63
Sahade R, Tatián M, Kowalke J, Kühne S, Esnal GB (1998) Benthic faunal associations on soft substrates at Potter Cove, King George Island, Antarctica. Polar Biol 19:85–91
Sanderson TJO (1988) Ice mechanics. Risks to offshore structures. Graham & Trotman, London
Sato T, Sakurai H, Takasaki A, Watanabe K, Hirano Y (1992) Underwater observation of Antarctic fishes and invertebrates with a note on collection and transportation techniques for research and exhibition in the aquarium. Ser Cient INACH 42:95–103
Schmid MK, Piepenburg D (1993) The benthos zonation of the Disko Fjord, West Greenland. Medd Grøn Biosci 37:3–21
Shabica SV (1972) Tidal zone ecology at Palmer Station. Antarct J US 7:184–185
Southward AJ, Southward EC (1967) On the biology of an intertidal chthamalid (Crustacea, Cirripedia) from the Chukchi Sea. Arctic 20:8–20
Starmans A, Gutt J, Arntz WE (1999) Mega-epibenthic communities in Arctic and Antarctic shelf areas. Mar Biol 135:269–280

Stephenson TA, Stephenson A (1972) Life between tidemarks on rocky shores. Freeman, San Francisco
Syvitski JPM, Fader GB, Josenhans HW, MacLean B, Piper DJW (1983) Seabed investigations of the Canadian East coast and Arctic using *Pisces*. IV. Geosci Can 10:59–68
Viehoff T, Li A (1995) Iceberg observations and estimation of submarine ridges in the western Weddell Sea. Int J Remote Sensing 16:3391–3408
Vogt PR, Crane K, Sundvor E (1994) Deep Pleistocene iceberg plowmarks on the Yermak Plateau: sidescan and 3.5 kHz evidence for thick calving ice fronts and a possible marine ice sheet in the Arctic Ocean. Geology 22:403–406
Vorren TO, Hald M, Edvardsen M, Lind-Hansen O-W (1983) Glacigenic sediments and sedimentary environments on continental shelves: general principles with a case study from the Norwegian shelf. In: Ehlers J (ed) Glacial deposits in North-West Europe. Balkema, Rotterdam, pp 61–73
Warwick RM, Clarke KR (1993) Increased variability as a symptom of stress in marine communities. J Exp Mar Biol Ecol 172:215–226
Weeks WF (1994) Possible roles of sea ice in the transport of hazardous material. Arct Res US 8:34–52
Weeks WF, Barnes PW, Rearic DM, Reimnitz E (1984) Some probabilistic aspects of ice gouging on the Alaskan shelf of the Beaufort Sea. In: Barnes PW, Schell DM, Reimnitz E (eds) The Alaskan Beaufort Sea: ecosystems and environments. Academic Press, Orlando, pp 213–236
Westermeier R, Gómez I, Rivera PJ, Müller DG (1992) Macroalgas marinas antárticas: distribución, abundancia y necromasa en Isla Rey Jorge, Shetland del Sur, Antártica. Ser Cient INACH 42:21–34
Whitaker TM (1977) Sea ice habitats of Signy Island (South Orkneys) and their primary productivity. In: Llano GA (ed) Adaptations within Antarctic ecosystems. Gulf, Houston, pp 75–82
White MG (1973) Aspects of the biological significance of ice in the marine environment. Proc Challenger Soc 4:145–146
White MG (1984) Marine benthos. In: Laws RM (ed) Antarctic ecology, vol 2. Academic Press, London, pp 421–461
White PS, Pickett STA (1985) Natural disturbance and patch dynamics: an introduction. In: Pickett STA, White PS (eds) The ecology of natural disturbance and patch dynamics. Academic Press, Orlando, pp 3–13
Wlodarska M, Weslawski JM, Gromisz S (1996) A comparison of the macrofaunal community structure and diversity in two arctic glacial bays – a "cold" one and a "warm" one off Spitsbergen. Oceanologia 38:251–283
Woodworth-Lynas CMT, Josenhans HW, Barrie JV, Lewis CFM, Parrott DR (1991) The physical processes of seabed disturbance during iceberg grounding and scouring. Cont Shelf Res 11:939–961
Zamorano JH (1983) Zonación y biomassa de la macrofauna bentónica en Bahía South, Archipiélago de Palmer, Antártica. INACH Ser Cient 30:27–38
Zubov NN (1943) United States, Naval Oceanographic Office; Arctic ice. Technical report, AFGL-TR-79034 – United States Air Force, Geophysics Laboratory. NTIS, Springfield, Va

Polar Biol (2001) 24: 486–496
DOI 10.1007/s003000100243

David N. Thomas · Hilary Kennedy · Gerhard Kattner
Dieter Gerdes · Carl Gough · Gerhard S. Dieckmann

Biogeochemistry of platelet ice: its influence on particle flux under fast ice in the Weddell Sea, Antarctica

Abstract An array of four sediment traps and one current meter was deployed under a well-developed platelet layer for 15 days in the Drescher Inlet in the Riiser Larsen ice shelf, in February 1998. Traps were deployed at 10 m (just under the platelet layer), 112 m (above the thermocline), 230 m (below thermocline) and 360 m (close to sea floor). There was a substantial flux of particulate organic material out of the platelet layer, although higher amounts were collected in the traps either side of the thermocline. Material collected was predominantly composed of faecal pellets containing diatom species growing within the platelet layer. The size classes of these pellets suggest they derive from protists grazing rather than from larger metazoans. Sediment trap material was analysed for particulate organic carbon/nitrogen/phosphorus (POC/PON/POP) and $\delta^{13}C_{POC}$ (carbon isotopic composition of POC). These were compared with organic matter in the overlying platelet layer and the water column. In turn, the biogeochemistry of the platelet layer and water column was investigated and the organic matter characteristics related to inorganic nutrients (nitrate, nitrite, ammonium, silicate, phosphate), dissolved organic carbon/nitrogen (DOC/DON), pH, dissolved inorganic carbon (DIC), oxygen and $\delta^{13}C_{DIC}$ (carbon isotopic composition dissolved inorganic carbon).

Please note: When citing this article, please use the reference to the original source in the journal Polar Biology as given on the top of this page.

D.N. Thomas (✉) · H. Kennedy
School of Ocean Sciences, University of Wales-Bangor,
Menai Bridge, Anglesey, LL59 5EY, Wales
E-mail: d.thomas@bangor.ac.uk
Fax: +44-1248-382878

G. Kattner · D. Gerdes · G.S. Dieckmann
Alfred Wegener Institute for Polar and Marine Research,
27570 Bremerhaven, Germany

C. Gough
Department of Zoology, University of Aberdeen,
Tillydrone Ave, Aberdeen, AB24 3TZ, Scotland

Introduction

A key aim of the SCAR programme "Ecology of the Antarctic Sea Ice Zone" (EASIZ) is "to determine the role of the sea ice zone on Antarctic marine systems and in the control of global biogeochemical and energy exchanges". A major question being asked by the programme is "what physical, chemical and biological factors determine patterns of production, sedimentation and recycling, and the major elemental budgets, of the Antarctic coastal and shelf system" (SCAR 1994). A critical factor in understanding what happens to the biomass produced within the sea ice is the fate of sea-ice algae and the role they play in the pelagic and in bentho-pelagic coupling. Sea ice provides a stable environment for primary production and a refuge for metazoans and protozoans which graze on the high algal biomass within the ice or at the ice/water interface (see review by Ackley and Sullivan 1994 and citations therein). On ice break-up, algae contained within the ice may seed the ice-edge bloom after their release into the water column, or can be swiftly transported downwards by the formation of aggregates (Riebesell et al. 1991) and/or packaging into rapidly sinking faecal pellets (González 1992; González et al. 1994).

Although there have been several studies that have linked high productivity in sea ice with increased particle flux and sediment accumulation in regions underlying ice cover (Leventer 1998), we are as yet unable to fully interpret the importance of sea ice in the overall bentho-pelagic coupling of the Southern Ocean. This applies particularly to the shallow coastal areas where the influence of seasonal inputs of organic matter to the sediments may be highly significant (Gutt et al. 1998 and citations therein).

Diatom frustules compose a large fraction of the material in Antarctic sediments (Zielinski and Gersonde 1997; Leventer 1998), and the heavy silicification of some diatom species such as *Fragilariopsis* species ensures that frustules are well preserved in the sedimentary record (Smetacek 1999a). The importance of the durability of the frustules of these species for palaeoenvironmental indicators is discussed in the comprehensive review by Leventer (1998). Two species, *Fragilariopsis curta* and *F. cylindrus*, have been shown to be very successful in exploiting sea ice and ice-edge regions and, as such, provide useful proxies for sea-ice distribution. Stable carbon isotopic values of diatoms have been proposed as another paleoenvironmental indicator of sea-ice conditions. Several studies have illustrated an enrichment of ^{13}C in sea-ice diatoms, and have linked this to carbon dioxide limitation within sea-ice assemblages. Enrichment of ^{13}C in sediments has been suggested as a proxy for past carbon dioxide concentrations in surface waters (Rau et al. 1989, 1991a). However, laboratory investigations by Gleitz et al. (1996), and field investigations (Dunbar and Leventer 1992; Gibson et al. 1999) have shown that ^{13}C enriched carbon of sea-ice diatoms may confound any reconstruction of past surface carbon dioxide concentrations in seasonally ice-covered regions.

In some Antarctic coastal regions, fast ice persists well into the summer. This ice can often support high algal standing stocks with extremely productive bottom ice assemblages (McMinn et al. 1999). However, often an additional layer of accumulated platelet ice under the fast-ice cover enhances the biological activity. These layers form a semi-enclosed system with, at times, restricted exchange with the underlying water column (Dieckmann et al. 1992; Arrigo et al. 1995; Günther and Dieckmann 1999; Günther et al. 1999a). Although light and inorganic nutrient limitation are known to occur within platelet-ice systems, these habitats are well known as supporting the highest algal biomass ever measured in sea ice (1 mg Chl*a* l^{-1}, Arrigo et al. 1995). At the Drescher Inlet in the Riiser Larssen ice shelf, Günther et al. (1999b) investigated the biogeochemical processes taking place during growth and accumulation of algal biomass within platelet ice. As well as showing that inorganic nutrients become limiting and that there are shifts in the carbon metabolism of the diatoms, Günther et al. (1999a) also showed the platelet layer to be a rich grazing ground for a wide range of metazoans dominated by small copepods and amphipods.

The objective of the present study was to measure the export and the nature of particulate organic matter sedimenting from the platelet layer in Drescher Inlet, and assess the flux to the underlying sediments. The measurements were made in late summer, when the maximum flux from the platelet layer would be predicted. Another objective of the study was to compare the composition and chemistry of the particle flux with that in the platelet layer and water column. An array of sediment traps was deployed under a platelet-ice layer within the inlet and a suite of biogeochemical parameters within the platelet ice and water column measured concurrently. The campaign was curtailed by premature break up of ice within the inlet, although these unforeseeable events did offer the opportunity to study the effect of intensive ice break-up on particle flux. A focus of the study was the stable carbon isotope composition of the sedimenting organic material as a tracer for ice algae. These measurements, together with other parameters, would help in the clarification of bentho-pelagic coupling and the role of sea-ice systems in this coupling.

Materials and methods

Study site

The study was conducted during a 5-week field campaign at the Drescher Inlet in austral summer 1998 (Dieckmann and Thomas 1999). The inlet, a 20-km-long funnel-shaped crack in the Riiser-Larsen ice shelf (72°52′S, 19°25′W) in the eastern Weddell Sea (Fig. 1), is flanked by floating ice cliffs approximately 30 m above and 150 m below the sea surface. The inlet has an irregular topography with water depths varying between 360 and 400 m (Plötz and Bornemann 1999). The inlet is characterised by a stable fast-ice cover lasting throughout the summer and an underlying platelet-ice layer, which can reach thicknesses of >20 m. From 7 to 13 February, the weather was good and the sea-ice cover in the inlet, stable. From 14 to 23 February, weather conditions deteriorated, and there was intensive ice break-up, with three-quarters of the inlet being ice-free by the end of the field campaign.

Although the ice at the mouth of the inlet was 1 year old ice no more than 2 m thick, towards the back of the inlet the ice was multi-year ice of 6 m or greater thickness. At the study site (Fig. 1), the sea ice was approximately 4 m thick with a highly variable snow cover ranging from 0 to 2 m thick. There was no evidence of significant biological activity within the overlying sea ice.

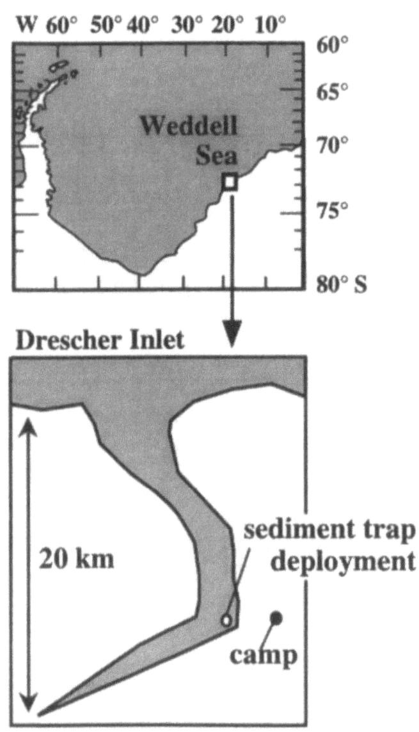

Fig. 1 Location of the Drescher Inlet in the Riiser-Larson Ice Shelf, eastern Weddell Sea. In the *lower panel* the inlet and the sediment trap array deployment site are shown

Sampling

Water column and platelet-ice layer

On three occasions, discrete water samples were collected at 50, 150 and 400 m under the ice, and on one occasion from the ice edge using a 5-l Niskin bottle. Twelve CTD profiles down to 400 m were measured using a newly calibrated Seabird SBE 19 CTD profiler. The Niskin bottle and CTD casts were deployed with a modified Valeport electric winch.

The platelet ice was sampled using an Adonis-type sampler (Dieckmann et al. 1992; Günther and Dieckmann 1999). The device allows the high-resolution discrete sampling of interstitial water from between the ice platelets. Recent tests by Günther et al. (1999b) have shown this system to be suitable for the collection of samples for dissolved gas analyses, as well as biological composition and inorganic nutrient chemistry.

The interstitial platelet water and water column samples were filtered in the field laboratory through precombusted GF/F filters (Whatmann, 450°C, 3 h). Samples for Chlorophyll *a* (Chl*a*), particulate organic carbon, nitrogen, phosphorus (POC/PON/POP) concentrations and stable carbon isotope composition of POC ($\delta^{13}C_{POC}$) were collected on filters and stored frozen until analysis in the home laboratories. Filtrates were poisoned with $HgCl_2$ and stored at 4°C in 50-ml PE bottles for later inorganic nutrient and dissolved organic nitrogen (DON) analyses (Kattner 1999). Additional filtrate samples were stored frozen (unpoisoned) in 50-ml precombusted (450°C, 3 h) glass ampoules for later dissolved organic carbon (DOC) determination.

Sub-samples for oxygen concentration, pH and alkalinity were taken immediately after retrieval, avoiding atmospheric contamination (Gleitz et al. 1995; Günther et al. 1999b). Additional samples were collected for the carbon isotopic composition of dissolved inorganic carbon ($\delta^{13}C_{DIC}$). They were immediately filtered through in-line 0.45-μm polycarbonate filters avoiding contamination with air bubbles. The filtrates were poisoned with $HgCl_2$ and sealed under an atmosphere of nitrogen in pre-weighed glass ampoules.

From each sample, 100-ml aliquots were preserved with hexamine-buffered formaldehyde (final concentration ca. 0.5%) for microscopic examination.

Sediment trap deployment

An array of four sediment traps and an Aanderaa RCM 8 current meter (Fig. 2) were deployed (by hand) on 2 February. The array was anchored to the overlying sea ice and deployed through a large crack in the ice. Two types of Hydrobios cylindrical sediment traps were deployed, both with opening areas of 0.015 m^2. The trap openings were covered by a plastic grid 40 mm thick, with 40 mm^2 holes. The trap-collecting cylinders were 560 mm long.

The "multiple" traps (Hydrobios-MST 6) had six collecting bottles, each programmed to collect samples over 6 days. The "single" traps were equipped with only one collecting bottle. Prior to deployment, all collecting cups were filled with slightly hypersaline water and poisoned with $HgCl_2$.

Multiple traps were positioned at 10 m, just below the platelet layer, and at 360 m close to the seafloor (approximately 400 m). Single-bottle traps were placed above (115 m) and below (230 m) a sharp thermocline determined by a CTD profile. However, it should be noted that the by the end the thermocline extended down to 280 m (see Fig. 4 and later discussion on hydrography). The array was deployed for 15 days, and had to be recovered due to the impending break-up of the ice in the inlet. The upper multiple trap worked as programmed, giving samples of 6, 6 and 3 days collection. However, the bottom multiple trap yielded only two samples of 6 and 9 days collection. After recovery of the traps, samples from the trap cups were split in the field laboratory and sub-samples taken for POC/PON, Chl*a* and $\delta^{13}C_{POC}$ determinations and for microscopic examination as described above.

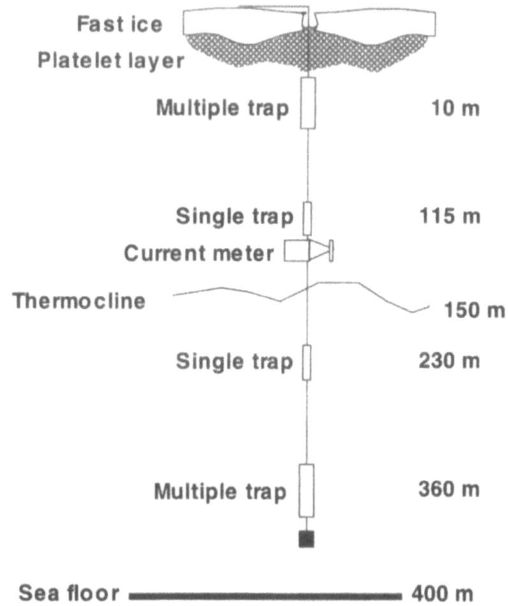

Fig. 2 Sediment trap and current meter array deployed on 2 February 1998 at Drescher Inlet under a fast-ice cover with a well-developed underlying platelet layer. The position of the thermocline was determined by CTD profiles (see Fig. 4). "Multiple" traps had multiple collecting bottles whereas "Single" traps were equipped with single sampling bottle

Analytical methods employed

In the field laboratory

Oxygen concentrations were determined using the Winkler method (Grasshoff 1983). The pH and total alkalinity were measured as described by Gleitz et al. (1995) and Günther et al. (1999b). The carbonate system was calculated from pH and alkalinity using the carbonic acid dissociation constants of Mehrbach et al. (1973). Concentrations of dissolved carbon dioxide and pH of the platelet interstitial water were calculated for a temperature of 0°C, and for the water column samples, for the in situ temperature determined from the CTD casts. Salinity was measured at laboratory temperature using a WTW microprocessor conductivity meter.

In Bremerhaven and Bangor

Chlorophyll *a* was determined using a Turner fluorometer, after overnight extraction in 90% acetone in the dark at 4°C (Evans et al. 1987). Inorganic nutrient analyses (nitrate, nitrite, silicate, ammonium, phosphate) were performed using standard autoanalyser methods (Kattner and Becker 1991). DON was analysed following persulphate wet oxidation (Kattner and Becker 1991), and DOC by high temperature oxidation using an MQ1001 TOC Analyser (Qian and Mopper 1996).

Filters for POC and $\delta^{13}C_{POC}$ were acid fumed (concentrated HCl) overnight to remove carbonate, dried at 40°C and then stored in a desiccator prior to analysis. POC and PON concentrations were determined with a Europa Scientific CHN analyser, using acetanilide as a standard. POP was measured essentially after Kattner and Brockmann (1980). Samples for $\delta^{13}C_{POC}$ analyses were processed, and subsequently analysed using a VG SIRA II isotope ratio mass spectrometer as described by Kennedy and Robertson (1995). Samples for $\delta^{13}C_{DIC}$ were processed as described by McCorkle (1987). The carbon isotope ratios are expressed in the standard $\delta^{13}C$ (‰) notation relative to the Pee Dee Belemite, with a precision (including sample collection and extraction) of ± 0.07‰.

The formaldehyde-fixed samples were examined by a combination of light and electron microscopy techniques. To facilitate the examination, fractions of different densities were taken following a range of settling periods up to 24 h. The major component of the material in the sediment traps was faecal pellets, which were classified by size using calibrated graticules in inverted microscopes.

Results

Hydrography

The automatic weather station, ARGOS (maintained by the Alfred Wegener Institute), at the camp location at Drescher (Fig. 1) recorded significant changes in the weather during the deployment period (Fig. 3): between 2 and 12 February the mean wind speed was 2.28 m s^{-1} (max. 5.43 m s^{-1}), increasing from 13 to 18 February up to 4.88 m s^{-1} (max. 11.75 m s^{-1}). Water currents at 120 m during the same period were mainly orientated to the south, with a mean speed of approximately 2 cm s^{-1}. A northeastward-directed water current was preceded by the maximum wind speed recorded on 14 February. The current peaked 2 days later with maximum velocities of 6 cm s^{-1} (Fig. 3). The high wind speed coupled with high water current speeds were followed by the massive sea-ice break-up within the inlet that was most intensive on 17 February, when the deployment was obviously ended.

These changes in wind and water current speed were also reflected by changes in the hydrography of the inlet (Fig. 4). The temperature and salinity profiles in the inlet were characterised by a stable sharp thermo- and pycnocline between 150 and 180 m water depth. On 3 February, near the beginning of the sediment trap deployment, the surface waters were also cooled down, forming a distinct shallow thermocline and pycnocline at 30 m. The lower temperature and reduced salinity in the first 30 m are clearly influenced by the platelet layer, which was observed to extend down to 25 m in places in the inlet.

During the ice break-up, warm deep water was entrained due to surface water export caused by high wind speeds and the northeastward-directed current on 16 February (Fig. 3). This resulted in a warming up of the upper water column (Fig. 4), which will have promoted platelet and sea-ice melt. The thermo- and pycnoclines also became more diffuse, extending from 150 to 280 m water depth (Fig. 4).

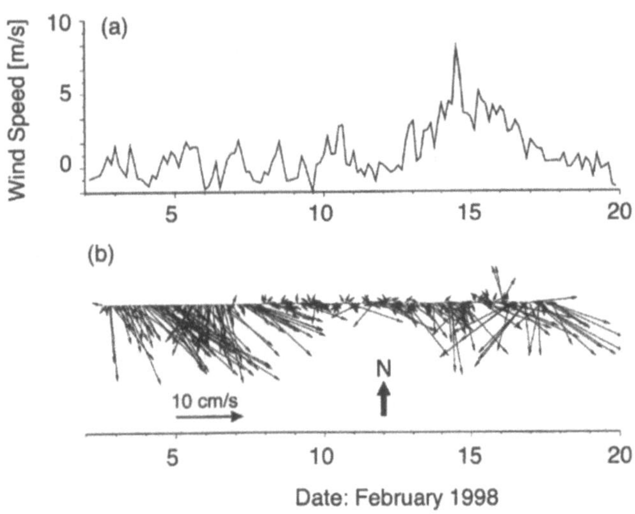

Fig. 3 Wind speed (**a**) and water current meter data (**b**) collected during the 15-day sediment trap deployment. The water current vectors indicate direction and current velocity on a 60-min integrated scale. The wind speed data were collected by the automatic weather station ARGOS at the Drescher camp (see Fig. 1 for location)

Platelet ice and water column

There were obvious differences in the biology and chemistry of the platelet layer compared with the underlying water column (Table 1). It should be noted that just taking interstitial water samples from the platelet layer underestimates the biomass of ice algae living within this layer. Grossmann et al. (1996) showed that the larger percentage of *Fragilariopsis*-dominated assemblages were attached to the ice platelets rather than in the interstitial water. Similarly, Dieckmann et al. (1992) reported that ca. 47% of the Chl*a* in the platelet ice is associated with the platelet ice crystals. During this campaign, a similar comparison was made on two occasions. When the concentration of POC was compared, the interstitial water levels were 20 and 24% of that on the surrounding platelets. However, it is not unreasonable to compare chemical characteristics of the platelet interstitial waters with the underlying water, since the chemistry of the interstitial water is a reflection of the biological activity of both components of the platelet layer, and not simply the interstitial assemblage.

Although other species were present, the platelet-layer assemblages were dominated by *Fragilariopsis* species, predominantly *F. cylindrus* and *F. curta*. Comparing mean values, there was 7 times more Chl*a* in the platelet interstitial water compared with the water column, although this enrichment was only 4 times for POC and PON. There was significantly greater enrichment of POP in the platelet interstitial water, mean concentrations being 12 times higher than those in the underlying water (Table 1). This differential chemical enrichment led to very different elemental ratios between the organic matter in the platelet interstitial water and the water column. The mean POC:PON ratios of the particulate phases were similar in both platelet interstitial water and the underlying water column, although the lower end of the range for the interstitial water was lower than that measured in the water column samples. Both the POC:POP and PON:POP ratios were lower in the interstitial water than in the water column. Although there was considerable variation in the data, the PON:POP and POC:POP ratios in the water column had mean values closer to Redfield values of 106:1 and 16:1,

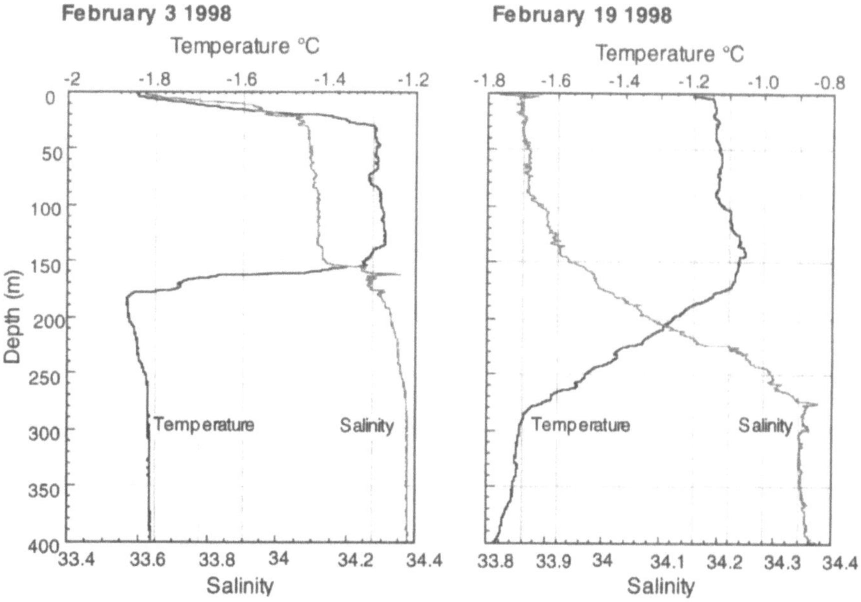

Fig. 4 Temperature and salinity profiles from CTD casts at Drescher Inlet on 3 and 19 February 1998, corresponding to the beginning and end of the sediment trap deployment

respectively, compared with those from the platelet interstitial water.

Differences in biomass were reflected in the inorganic nutrients within the two systems (Table 1). The concentrations of nitrate and silicate were significantly lower in the platelet interstitial water, and ammonium levels considerably elevated. Although the mean phosphate concentrations were not significantly different, the minimum levels recorded were very much less in the platelet interstitial water than in the underlying water. Nitrite concentrations were similar in both interstitial water and the water column. The concentrations of inorganic nutrients of the underlying water are typical of Weddell Sea water (Schröder et al. 1994). Concentrations of DOC and DON were twice as high, with a lower DOC:DON ratio, in the platelet interstitial water compared with those measured in the underlying water column (Table 1).

Table 1 Summary of chemical and biological parameters measured in interstitial water of the platelet-ice layer and the underlying water column. Mean values are given and the range of values (minimum-maximum) are given *in parentheses*. Samples were taken between 0.2 and 1.8 m in the platelet layer, and at 50, 150 and 400 m in the water column (*DIC* dissolved inorganic carbon, *Chla* chlorophyll, *POC/PON/POP* particulate organic carbon/nitrogen/phosphorus, *DOC/DON* dissolved organc carbon/nitrogen, $\delta^{13}C_{DIC}$ carbon isotopic composition of DIC, $\delta^{13}C_{POC}$ carbon isotopic composition of POC)

Parameter	Platelet interstitial water	Water column	Number of measurements	
			Platelet layer	Water column
Salinity	33.0 (31.1–33.9)	34.1 (33.9–34.4)	13	8
pH	8.46 (8.25–8.90)	8.23 (8.13–8.29)	10	8
DIC (μM)	2064 (1723–2199)	2258 (2227–2346)	10	8
CO_2 (aq) (μM)	12.3 (2.9–18.3)	20.1 (17.1–25.9)	10	8
O_2 (μM)	385.5 (325.5–514.0)	330.5 (253.6–373.9)	14	8
Chl*a* (μg/l)	3.0 (0.1–9.9)	0.4 (0.0–1.7)	13	8
POC (μM)	47.1 (8.8–155.1)	12.0 (5.7–19.3)	13	8
PON (μM)	5.5 (0.7–19.8)	1.26 (0.5–1.9)	13	7
POP (μM)	1.1 (0.1–4.0)	0.08 (0.0–0.2)	7	7
POC:PON	9.8 (6.2–12.2)	9.5 (7.1–12.1)	13	7
PON:POP	8.4 (4.2–13.6)	13.7 (8.0–22.0)	7	6
POC:POP	75.9 (39.1–96.0)	123.8 (75.7–189.7)	7	6
POC:PON:POP	106:12:2.4	106:11:0.8		
$\delta^{13}C_{POC}$ (‰)	–24.0 (–20.9 to –26.7)	–25.6 (–24.6 to –26.3)	11	8
$\delta^{13}C_{DIC}$ (‰)	1.4 (0.4–3.8)	0.7 (0.1–1.2)	13	7
Nitrate (μM)	15.7 (0.4–24.3)	25.4 (21.3–32.7)	13	7
Nitrite (μM)	0.1 (0.0–0.2)	0.1 (0.07–0.1)	13	7
Ammonium (μM)	4.2 (1.2–9.0)	1.4 (0.93–1.8)	13	7
Silicate (μM)	40.3 (4.1–62.2)	65.2 (59.48–71.1)	13	7
Phosphate (μM)	1.17 (0.2–1.8)	1.3 (1.04–1.4)	13	7
DOC (μM)	146.8 (100.8–200.6)	72.4 (55.4–85.6)	12	7
DON (μM)	24.5 (8.8–56.3)	3.7 (2.0–5.2)	12	6
DOC:DON	8.2 (2.3–20.5)	21.4 (15.3–27.1)	11	5

Elevated oxygen concentrations were measured in the platelet interstitial water, which were linked to lower concentrations of dissolved inorganic carbon (DIC) and CO_2 (aq). Likewise, elevated pH values were measured in the platelet interstitial layer, typical of ice environments with active algal assemblages with somewhat restricted water exchange. The decrease in DIC concentration was reflected in the more positive carbon isotope values of $\delta^{13}C_{DIC}$ compared with those typically found in seawater. This enrichment of ^{13}C in DIC samples is also reflected in a corresponding enrichment in the $\delta^{13}C_{POC}$ values in platelet interstitial assemblages (Table 1).

Particle flux

It is striking that the highest faecal pellet, Chl*a* and phaeopigment fluxes were at 230 m just below the thermocline (Table 2). The POC flux was only slightly higher below the thermocline than above, which was opposite to that for PON. In contrast, the POP flux was clearly higher above the thermocline (Table 2). The POC:PON ratios had a small range in values, although there was a difference between samples collected either side of the thermocline. Above the thermocline, the mean for all samples was 6.6, close to the Redfield ratio for fresh phytoplankton material, whereas in the samples collected below it was somewhat higher at 7.4. POC:POP and PON:POP ratios were more variable, but still showed lower ratios (i.e. high phosphorus content) above the thermocline.

There was little trend in the $\delta^{13}C_{POC}$ values with depth of sediment trap. The values above the thermocline were more variable ($\delta^{13}C_{POC}$ –25. 5, SD = 0.07, $n = 4$) than below ($\delta^{13}C_{POC}$ –26.6, SD = 0.4, $n = 3$), but not significantly different. In the samples where the flux of faecal pellets was less than 1×10^5 pellets m^{-2} day^{-1}, the $\delta^{13}C_{POC}$ values were more negative than –26‰ (first 6 days at 10 m and both sampling periods at 360 m), whereas at greater faecal pellet flux they were more positive than –26‰ (Table 2).

Comparing the pigment concentrations, it is pertinent to consider a pigment degradation ratio derived by dividing the phaeopigment flux by the sum of the chlorophyll plus phaeopigment flux (Table 2). There was significant degradation of chlorophyll in material in the top three traps (degradation ratios of 70–79). Since in the 10-m trap the ratios varied from 70 to 79, no trend of increasing pigment degradation with depth was recognisable down to 230 m. At 360 m, the pigment degradation ratios did increase to between 87 and 91, indicating that there was increased pigment degradation in the deepest trap compared with those above (Table 2).

The sediment trap samples were dominated by faecal pellets (Fig. 5). The contents of the pellets, independent of pellet size, were mostly diatom frustules almost exclusively from the genus *Fragilariopsis*, and mostly

Table 2 Summary of biochemical parameters measured on sediment trap material collected on a 15-day deployment of traps at 10 m, 115, 230 and 360 m under a platelet-ice layer/fast-ice cover at Drescher Inlet in February 1998. The traps at 10 m and 360 m collected multiple samples during the deployment, whereas the traps at 115 and 230 m collected over the whole 15 days (*Chla* chlorophyll, *POC/PON/POP* particulate organic carbon/nitrogen/phosphorus, $\delta^{13}C_{POC}$ carbon isotopic composition of POC)

Depth of trap	Number of days collected	Chla (µg m^{-2} day^{-1})	Phaeopigments (µg m^{-2} day^{-1})	Pigment degradation ratio[a]	POC (mmol m^{-2} day^{-1})	PON (mmol m^{-2} day^{-1})	POP (mmol m^{-2} day^{-1})	POC:PON	POC: PON:POP	$\delta^{13}C_{POC}$ (‰)	Faecal pellets (nos. m^{-2} day^{-1})
10 m	6	37.5	112.4	75	4.75	0.62	0.11	7.6	106:14:2.5	–26.5	0.4×10^5
	6	37.0	143.6	79	5.47	0.85	0.06	6.5	106:16:1.2	–25.4	1.3×10^5
	3	73.1	170.4	70	5.77	0.90	0.13	6.4	106:16:2.4	–25.0	3.1×10^5
115 m	15	112.6	364.6	76	8.71	1.49	0.11	5.8	106:18:1.3	–25.2	5.7×10^5
230 m	15	261.7	997.1	79	9.29	1.31	0.05	7.1	106:15:0.6	–25.8	14.0×10^5
360 m	6	76.5	509.3	87	4.12	0.51	0.02	8.0	106:13:0.5	–26.6	0.8×10^5
	9	77.5	745.6	91	3.65	0.51	0.02	7.2	106:15:0.6	–26.2	0.9×10^5

[a] Pigment degradation ratio = phaeopigments/(phaeopigments + Chl*a*)

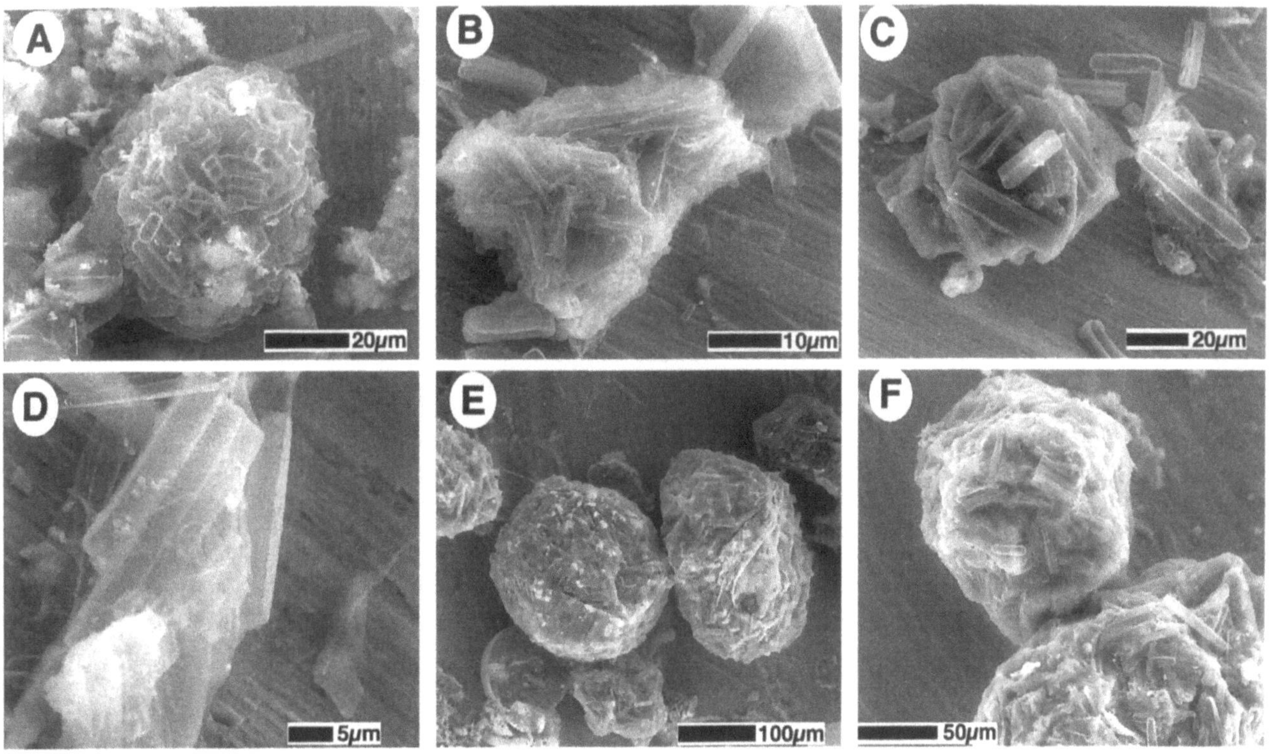

Fig. 5 Scanning electron micrographs of faecal pellets collected in sediment traps at 112 m (**A**), 122 m (**E**) and 230 m (**B**, **C**, **D**, **F**) depth. Note the difference in scale bars for each micrograph

F. cylindrus and *F. curta*. It was striking that these diatoms were largely intact and often in chains several cells long. The faecal pellets were of either undefined shape or very compact spherical pellets. No oval/elipsoid-shaped or krill pellets were seen in any of the samples. The pellets were size-classed, and in all trap samples 50% of the pellets were less than 97 μm in diameter (Fig. 6). The bottom trap had the greatest proportion of large pellets, and up to 20% of the pellets were greater than 244 μm in diameter. Although their origin is not certain, it is thought that they are probably produced by amphipods (U. Bathmann, personal communication). Amphipods were collected in traps at all depths, as were calanoid and cyclopoid copepods. An unidentified polychaete was found in both the 10-m and 115-m traps, and an unidentified salp at 115 m (metazoans were removed before samples were filtered for chemical analyses). The smaller pellets (up to 97 μm in diameter) are thought to be

Fig. 6 Size distribution of faecal pellets collected at 10 m, 115 m, 230 m and 360 m at Drescher Inlet. The *key* gives the size classes measured. At 10 m and 360 m multiple collections were made, whereas at 115 and 200 m only single collections were made over the 15-day deployment period

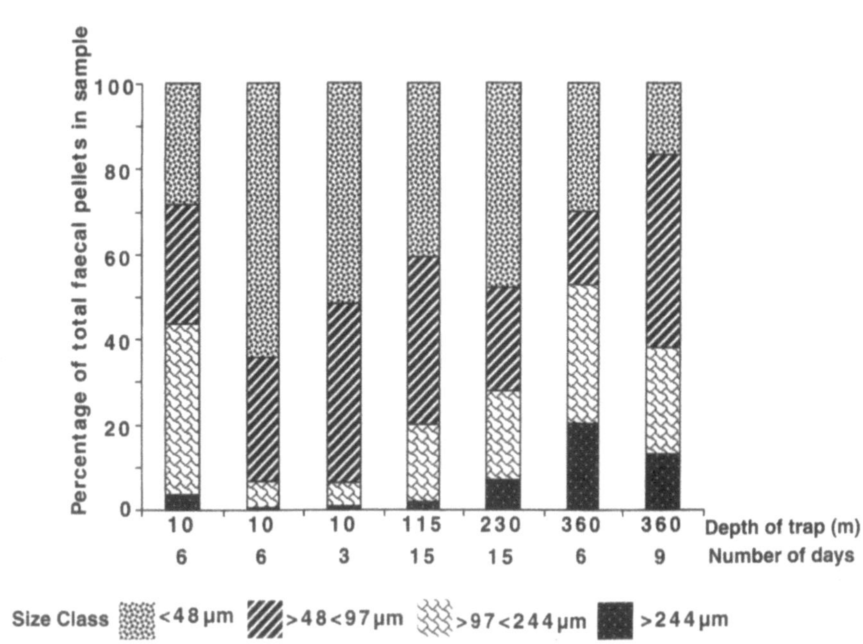

predominantly produced by protozoans (U. Bathmann, personal communication). Although the faecal pellets from the 10-m, 115-m and 230-m traps' collections were uniformly intact, it was noticeable that the pellets from 360-m were more fragmented. There was little evidence of bacteria on the surfaces of any of the pellets or debris in the traps. Sponge spicules were noted in the 360-m and, to much less extent, the 200-m trap samples. This is evidence that there may have been a degree of resuspension from the benthos.

Temporal trends in flux rates were discernable in the sediment trap deployed at 10 m. There was an increasing flux of faecal pellets, Chl*a*, phaeopigments, POC and PON, although not of POP (Table 2). This trend is most likely due to the increasing wind and surface water current velocities and the gradual increase in ice break-up in the inlet. The nature of the flux also changed with increasing $\delta^{13}C_{POC}$ values. The daily flux of faecal pellets, Chl*a*, POC, PON and POP in the 360-m trap did not change significantly with time, and there were no temporal changes in $\delta^{13}C_{POC}$. The pigment degradation ratios indicate no evidence of change with time in the degree of pigment degradation in either the 10-m or 360-m trap.

A comparison between the nature of the material collected in the traps with that found in the platelet interstial water and underlying water column showed considerable differences. The POC:PON ratios measured in the sediment trap material were lower (mean 6.1, Table 2) compared with the mean ratios of the platelet interstitial water and water column (9.8 and 9.5, respectively, Table 1). Only in a few samples from the platelet layer were the POC:PON ratios similar to those in the sediment traps. The variability in the POP measured in the ice and water column makes it impossible to compare the data with those obtained in the sediment traps. No evidence of ^{13}C enrichment in the $\delta^{13}C_{POC}$ in the sediment trap material was measured. The range of $\delta^{13}C_{POC}$ values in the sediment trap material (Table 2) is within the ranges measured for the platelet interstitial water and water column, although within the platelet ice, more ^{13}C-enriched material ($^{13}C_{POC}=-20.9‰$) was present.

Discussion

The magnitude of the daily POC flux during this study is not exceptional, and is within the ranges of fluxes recorded at other ice-covered sites for the same time of year: Ross Sea (Dunbar et al. 1998; Smith and Dunbar 1998; Asper and Smith 1999), McMurdo Sound (Dunbar and Leventer 1986), Kapp Norvegia (Bathmann et al. 1991), Lutzow-Holm Bay (Matsuda et al. 1987), Terra Nova Bay (Pusceddu et al. 1999) and Halley Bay (González et al. 1994). As in the present study diatoms, and in particular ice algae, were found in all of these studies to be the principal contributors to the organic flux collected by the sediment traps.

A flux of ice diatoms from an overlying ice cover to the sediment is well recorded in previous sediment trap studies carried out in ice-covered regions of the Southern Ocean (Leventer and Dunbar 1987; Fischer et al. 1988; Dunbar et al. 1998). The major diatoms in the traps were *F. curta* and *F. cylindrus*, two species consistently identified as dominating sea-ice assemblages and marginal ice-edge zones. The dominance of these two species has led to their use as sea-ice proxies in palaeoenvironmental studies (see review by Leventer 1998). Although both species were present in the water column, the biomass was so low that it is reasonable to conclude that the source of this algal flux was from the platelet layer underlying the fast ice in the inlet, mediated by a packaging in the form of faecal pellets. A similar highly significant contribution of faecal pellets to summer particle flux was recorded by Dunbar et al. (1998), who noted pellets being primarily composed of unfragmented diatom frustules, mainly *F. curta*.

One of the most striking observations about the faecal pellets was the integrity of the frustules. Only on a very few occasions were broken frustules observed, and apparently the diatoms are effectively transported to the benthos intact. Smetacek (1999a, b) discussed how heavy silicification of diatom frustules enhances their transfer to depth, and hence why diatoms dominate the biological pump. In his discussion he also points to how protists, in particular dinoflagellates, are voracious grazers on the diatoms, maybe even more so than the larger metazoans such as copepods and krill. The producers of the faecal pellets collected in this study cannot be determined, although we speculate that the majority of the pellets up to 97 μm were produced by protozooplankton such as heterotrophic dinoflagellates, ciliates, foraminiferans and radiolarians (Nöthig and von Bodungen 1989; González 1992). These accounted for over 50% of the trap material, and in the traps at 10, 115 and 230 m, up to 90% of the faecal pellets present. Spindler and Dieckmann (1986) reported high numbers of foraminiferans in sea ice from Drescher Inlet, and there are various reports of high levels of small copepods living in the platelet ice in the inlet (Schnack-Schiel et al. 1995; Günther et al. 1999a). The rest of the faecal material is again of unknown origin, but will have been produced by metazoan grazers, the largest pellets being produced most probably by amphipods. No krill faeces were found, which contrasts with many other sites in the Southern Ocean where such material has represented the major component of sediment trap collections (see citations in Nöthig and von Bodungen 1989).

One of the characteristics of the present study was that the traps were deployed under thick multi-year fast ice, with little evidence of ice algal activity within the ice itself. The only algal growth that was taking place at this site was within the underlying platelet ice layer. Even within the platelet layer, the algal standing stock was low compared with other late summer studies conducted at other coastal fast-ice sites (Dieckmann et al. 1992; Arrigo et al. 1993; Günther and Dieckmann 1999;

Günther et al. 1999b). This is a reflection of the thick fast ice overlying a thick platelet ice layer that had probably accumulated over many more months than is usual at other sites. The resulting low irradiance limits the growth of diatoms considerably. In spite of this, the standing stock of diatoms in the interstitial water was almost an order of magnitude higher than that measured in the underlying water column.

Diatom growth in the platelet layer, whether in the interstitial water or attached to platelets, resulted at times in depletion of inorganic nutrients and, in the case of nitrate, phosphate and silicate, almost to exhaustion. In contrast, as described in several other studies (Dieckmann et al. 1991; Thomas et al. 1998; Guglielmo et al. 2000), ammonium levels in the platelet layer were high, as were levels of dissolved organic carbon and nitrogen (DOC 146 μM, DON 25 μM). These levels of dissolved organic matter are twice as high as those reported for Antarctic surface waters (Kähler et al. 1997; Wedborg et al. 1998), which together with the high ammonium levels are indicative that remineralisation processes are high (Dieckmann et al. 1992; Arrigo et al. 1995; Günther and Dieckmann 1999). The ratio of DOC:DON indicates significant enrichment of nitrogen (cf. Thomas et al. 1995) compared with values in the open-water samples, which were typical for open-water ratios (Williams and Druffel 1988). In fact, the values of DOC:DON are of a similar magnitude to that of POC:PON in the platelet interstitial water, whereas in the water column the ratios of the dissolved phase are double that of the particulate phase. These measurements confirm the hypothesis that concentrations of dissolved organic matter are high within platelet ice systems (Grossmann et al. 1996; Günther et al. 1999b), although not as high as in other sea-ice habitats (Thomas et al. 1998).

The elemental composition of diatoms is governed by many factors, although primarily by inorganic nutrient (including inorganic carbon) supply (Burkhardt and Riebesell 1997 and citations therein). Phosphorus in the particulate material of the platelet layer was enriched compared to carbon using the Redfield ratio as benchmark. This enrichment in phosphorus was also measured in the trap material above the thermocline. In contrast, phosphorus was depleted in the particulate material in the water column and considerably so in the contents of the two deep trap samples. In both the platelet interstitial water and water column, there was evidence that the nitrogen content of the particulate material was depleted. There was no evidence of this depletion in the trap samples above the thermocline, although material collected below the thermocline did show slight nitrogen depletion compared with carbon. Phosphorus is already remineralised to ortho-phosphate to a considerable extent above the thermocline, and the trap samples clearly indicate that material reaching the sediment is poor in phosphorus and slightly reduced in nitrogen compared to carbon.

Increased biological activity within the platelet-ice layer led to the higher pH, lower dissolved inorganic carbon and CO_2 (aq) and elevated oxygen concentrations within the interstitial water. Similar magnitudes of oxygen enrichment and DIC depletion were found in platelet layers by Günther et al. (1999b) and in sea-ice brines by Gleitz et al. (1995). The decrease in DIC was also reflected in the more positive carbon isotope values of $\delta^{13}C_{DIC}$, which result from the production of isotopically depleted organic carbon during photosynthesis (Smith and Kroopnick 1981). The ^{13}C enrichment in DIC is also reflected in a corresponding enrichment in the $\delta^{13}C_{POC}$ values in platelet-ice assemblages. Enriched ^{13}C in $\delta^{13}C_{POC}$ values up to –11‰ have been reported by McMinn et al. (1999), and values of –8‰ in sea-ice cores by Dunbar and Leventer (1992). Fischer (1991) and Rau et al. (1991b) measured enrichment of ^{13}C in $\delta^{13}C_{POC}$ collected from ice cores and it has been concluded that this results from the ice restricting gas exchange, in combination with high rates of algal growth resulting in the enrichment of ^{13}C in the remaining DIC. Highest biomass of algae in the sea ice results in the most ^{13}C-enriched values and they are assumed to be due to the limited CO_2 exchange with the external seawater.

Oxygen enrichment and DIC depletion in sea-ice brines obtained by sack-hole sampling (Gleitz et al. 1995) confirm that sea-ice algal communities modify the brine chemistry. These field measurements were extended to the laboratory, where Gleitz et al. (1996) compared diatom growth, enrichment in $PO^{13}C$ with changing DIC, and pH and oxygen trends in sea ice, simulating closed incubations. To our knowledge, the measurements in the study reported here are the first of $\delta^{13}C_{DIC}$ to be made in any sea-ice system and confirm the results that enrichment in $\delta^{13}C_{POC}$ takes place within the semi-enclosed ice matrix.

If POC enriched in ^{13}C is being produced in sea ice, it is reasonable to predict that material sinking out from sea ice may result in ^{13}C-enriched sediments, irrespective of regional surface water variations in the partial pressure of carbon dioxide (Dunbar and Leventer 1992). Wada et al. (1987) concluded that high values of –20‰ originated from deposition of ice algae grown under CO_2-limited conditions.

One of the major aims of this present study was to trace the source of the organic matter to the underlying sediments and link enrichment of ^{13}C in the organic matter produced in the platelet ice with that in sedimenting material from the ice. Stable isotopes are commonly used to trace organic matter sources and to elucidate the degree to which benthic consumers are coupled to pelagic primary production (Hobson et al. 1995). Gibson et al. (1999), working at a near-shore site in Prydz Bay, measured pronounced seasonal variation in carbon isotopes. Winter values were nearly constant around –20‰, increasing to above –15‰ after inputs of sea-ice algae in spring. The platelet-ice system, however, does not form such an isolated habitat as sea-ice sheets. Yet, as already stated, platelet-ice layers are systems where rates of exchange and replenishment of dissolved inorganic nutrients and gases may be restricted. The data from the interstitial waters of the platelet layer above the traps confirm that this was the case during the time of sampling. However, in the sediment traps there was no evidence of this material being transported to depth as there were no significant differences in the chemistry of trap material compared with that from the water column. In fact, most positive $\delta^{13}C$ values for the open-water samples are higher than in any sample collected in the traps.

One possible explanation for this apparent discrepancy is that the algae that were being grazed and sinking from the platelet ice were those that were on the outer edges of the platelet layer, and hence located in regions where exchange with surrounding seawater was possible. This explanation is also supported by the elemental composition of the sediment trap material, which does not have the elevated POC:PON ratios measured within the platelet layer, but rather a ratio more akin to diatoms growing in inorganic nutrient-replete conditions.

The short field campaign in which this study took place by default results in our obtaining only a brief "snap-shot" of the nature and flux of the organic matter from the platelet layer to the underlying benthos. The effects of carbon limitation of the stable carbon isotope signature of ice algae, and the subsequent incorporation into sediments is clearly not as straightforward as has been reported elsewhere. Before the use of ^{13}C-enriched sediments as a proxy for sea-ice cover can be employed, more extensive seasonal studies must be conducted, and more rigorous attention paid to collecting efficiency and characterising the biogeochemical processes taking place in sea ice and platelet layers. However, the results have raised important questions that need to be addressed if we are to understand the importance of sea-ice primary production and the benthic-pelagic coupling in Antarctic waters and, in particular, for shallow coastal regions such as the Drescher Inlet.

Acknowledgements We would like to thank the captain and crew of RV "Polarstern" and the pilots and crew of the "Wasserthal" helicopter service. J. Plötz and H. Bornemann provided vital help in the field, without which we could not have done the work. Discussions with S. Schiel, R.W. Crawford and U. Bathmann helped us greatly with the interpretation of the results. We thank D.P. Kennedy for running the stable carbon isotope analyses. M. Stürcken-Rodewald (Bremerhaven), R. Delahunty and T. Bently (Bangor) gave excellent technical assistance. The work was partially funded by the Natural Environment Research Council (GT9/2894, GR9/3309, GR3/E0068), British Council/DAAD and the Nuffield Foundation. This is publication 1749 of the Alfred Wegener Institute.

References

Ackley SF, Sullivan CW (1994) Physical controls on the development and characteristics of Antarctic sea ice biological communities – a review and synthesis. Deep Sea Res 41:1583–1604

Arrigo KR, Robinson DH, Sullivan CW (1993) A high resolution study of the platelet ice ecosystem in McMurdo Sound, Antarctica: photosynthetic and bio-optical characteristics of a dense microalgal bloom. Mar Ecol Prog Ser 98:173–185

Arrigo KR, Dieckmann GS, Robinson DH, Fritsen CH, Sullivan CW (1995) A high resolution study of the platelet ice ecosystem in McMurdo Sound, Antarctica: biomass, nutrient and production profiles within a dense microalgal bloom. Mar Ecol Prog Ser 127:255–268

Asper VL, Smith WO Jr (1999) Particle flux during austral spring and summer in the southern Ross Sea, Antarctica. J Geophys Res C3 104:5345–5359

Bathmann U, Fischer G, Müller PJ, Gerdes D (1991) Short-term variations in particulate matter sedimentation off Kapp Norvegia, Weddell Sea, Antarctica: relation to water mass advection, ice cover, plankton biomass and feeding activity. Polar Biol 11:185–195

Burkhardt S, Riebesell U (1997) CO_2 availability affects elemental composition (C:N:P) of the marine diatom *Skeletonema costatum*. Mar Ecol Prog Ser 155:7–76

Dieckmann GS, Thomas DN (1999) Organic matter production within, and export from Antarctic sea ice platelet layers. In: Arntz WE, Gutt J (eds) The Expedition ANTARKTIS XV/3 (EASIZ II) of RV "Polarstern" in 1998. Ber Polarforsch 301:118–123

Dieckmann GS, Lange MA, Ackley SF, Jennings JC Jr (1991) The nutrient status in sea ice of the Weddell Sea during winter: effects of sea ice texture and algae. Polar Biol 11:449–456

Dieckmann GS, Arrigo K, Sullivan CW (1992) A high resolution sampler for nutrient and chlorophyll a profiles of the sea ice platelet layer and underlying water column below fast ice in polar oceans: preliminary results. Mar Ecol Prog Ser 80:291–300

Dunbar RB, Leventer A (1986) Opal and carbon fluxes beneath ice-covered regions of McMurdo Sound. Antarct J US 21:132–134

Dunbar RB, Leventer A (1992) Seasonal variation in carbon isotopic composition of Antarctic sea ice and open water plankton communities. Antarct J US:27:79–81

Dunbar RB, Leventer AR, Mauciarone DA (1998) Water column sediment fluxes in the Ross Sea, Antarctica: atmospheric and sea ice forcing. J Geophys Res C13 103:30741–30759

Evans CA, O'Reilly JE, Thomas JP (1987) A handbook for the measurement of chlorophyll a and primary production. Biological investigations of Marine Antarctic Systems and Stocks. BIOMASS 8. Texas A&M University, College Station, Texas

Fischer G (1991) Stable carbon isotope ratios of plankton carbon and sinking organic matter from the Atlantic sector of the Southern Ocean. Mar Chem 35:581–596

Fischer G, Fütterer D, Gersonde R, Honjo S, Ostermann D, Wefer G (1988) Seasonal variability of particle flux in the Weddell Sea and its relation to ice cover. Nature 335:426–428

Gibson JAE, Trull T, Nichols PD, Summons RE, McMinn A (1999) Sedimentation of C-13 rich organic matter from Antarctic sea ice algae: a potential indicator of past sea ice extent. Geology 27:331–334

Gleitz M, Loeff MR vd, Thomas DN, Dieckmann GS, Millero FJ (1995) Comparison of summer and winter inorganic carbon, oxygen and nutrient concentrations in Antarctic sea ice brine. Mar Chem 51:81–91

Gleitz M, Kukert H, Riebesell U, Dieckmann GS (1996) Carbon acquisition and growth of Antarctic sea ice diatoms in closed bottle incubations. Mar Ecol Prog Ser 135:169–177

González HE (1992) Distribution and abundance of minipellets around the Antarctic Peninsula. Implications for protistan feeding behaviour. Mar Ecol Prog Ser 90:223–236

González HE, Kurbjeweit F, Bathmann UV (1994) Occurrence of cyclopoid copepods and faecal material in the Halley Bay region, Antarctica, during January-February 1991. Polar Biol 14:331–342

Grasshoff K (1983) Determination of oxygen. In: Grasshoff K, Ehrhardt M, Kremling K (eds) Methods of seawater analysis. Verlag Chemie, Weinheim, pp 61–72

Grossmann S, Lochte K, Scharek R (1996) Algal and bacterial processes in platelet ice during late austral summer. Polar Biol 16:623–633

Guglielmo L, Carrada GC, Catalano G, Dell'Anno A, Fabiano M, Lazzara L, Mangoni O, Pusceddu A, Saggiomo V (2000) Structural and functional properties of sympagic communities in the annual sea ice at Terra Nova Bay (Ross Sea, Antarctica). Polar Biol 23:137–146

Günther S, Dieckmann GS (1999) Seasonal development of algal biomass in snow-covered fast ice and the underlying platelet layer in the Weddell Sea, Antarctica. Antarct Sci 11:305–315

Günther S, George KH, Gleitz M (1999a) High sympagic metazoan abundance in platelet layers at Drescher Inlet, Weddell Sea, Antarctica. Polar Biol 22:82–89

Günther S, Gleitz M, Dieckmann GS (1999b) Biogeochemistry of Antarctic sea ice: a case study on platelet ice layers at Drescher Inlet, Weddell Sea. Mar Ecol Prog Ser 177:1–13

Gutt J, Starmans A, Dieckmann G (1998) Phytodetritus deposited on the Antarctic shelf and upper slope: its relevance for the benthic system. J Mar Syst 17:435–444

Hobson KA, Ambrose WG Jr, Renaud PE (1995) Sources of primary production, benthic-pelagic coupling, and trophic relationships within the Northeast Water Polynia: insights from $\delta^{13}C$ and $\delta^{15}N$ analysis. Mar Ecol Prog Ser 128:1–10

Kähler P, Bjørnsen PK, Lochte K, Antia A (1997) Dissolved organic matter and its utilisation by bacteria during spring in the Southern Ocean. Deep Sea Res 44:341–353

Kattner G (1999) Storage of dissolved inorganic nutrients in seawater: poisoning with mercuric chloride. Mar Chem 67:61–66

Kattner G, Becker H (1991) Nutrients and organic nitrogenous compounds in the marginal ice zone of Fram Strait. J Mar Syst 2:385–394

Kattner G, Brockmann UH (1980) Semi-automated methods for the determination of particulate phosphorus in the marine environment. Fresenius Z Anal Chem 301:14–16

Kennedy H, Robertson J (1995) Variations in the isotopic composition of particulate organic carbon in surface waters along an 88°W transect from 67°S to 54°S. Deep Sea Res 42:1109–1122

Leventer A (1998) The fate of Antarctic "sea ice diatoms" and their use as paleoenvironmental indicators. In: Lizotte MP, Arrigo KR (eds) Antarctic sea ice biological processes. American Geophysical Union. Antarct Res Ser 73:121–137

Leventer A, Dunbar RB (1987) Diatom flux in McMurdo Sound, Antarctica. Mar Micropaleontol 12:49–64

Matsuda O, Ishikawa S, Kawaguchi K (1987) Seasonal variation of downward flux of particulate organic matter under the Antarctic fast ice. Proc NIPR Symp Polar Biol 1:23–34

McCorkle DC (1987) Stable carbon isotopes in deep sea pore waters: modern geochemistry and palaeoceanographic applications. PhD Thesis, University of Washington

McMinn A, Skerratt J, Trull T, Ashworth C, Lizotte M (1999) Nutrient stress gradient in the bottom 5 cm of fast ice, McMurdo Sound, Antarctica. Polar Biol 21:220–227

Mehrbach C, Culberson CH, Hawley JE, Pytkowicz RM (1973) Measurement of the apparent dissociation constant of carbonic acid in seawater at atmospheric pressure. Limnol Oceanogr 18:897–907

Nöthig E-M, Bodungen B von (1989) Occurrence and vertical flux of faecal pellets of probably protozoan origin in the southeastern Weddell Sea (Antarctica). Mar Ecol Prog Ser 56:281–289

Plötz J, Bornemann H (1999) Studies at the Drescher Inlet: seals, fish, sea ice and hydrography. In: Arntz WE, Gutt J (eds) The Expedition ANTARKTIS XV/3 (EASIZ II) of RV "Polarstern" in 1998. Ber Polarforsch 301:94–102

Pusceddu A, Cattaneo-Vietti R, Albertelli G, Fabiano M (1999) Origin, biochemical composition and vertical flux of particulate organic matter under the pack ice in Terra Nova Bay (Ross Sea, Antarctica) during late summer 1995. Polar Biol 22:124–132

Qian J, Mopper K (1996) Automated high-performance, high-temperature combustion total carbon analyzer. Anal Chem 68:3090–3097

Rau GH, Takahashi T, Des Mareis DJ (1989) Latitudinal variations in plankton $\delta^{13}C$: implications for CO_2 and productivity in past oceans. Nature 341:516–518

Rau GH, Froelich PN, Takahashi T, Des Marais DJ (1991a) Does sedimentary organic record variations in $\delta^{13}C$ record variations in quaternary ocean (CO2 (aq))? Paleoceanogr 6:335–347

Rau GH, Sullivan CW, Gordon Li (1991b) $\delta^{13}C$ and $\delta^{15}N$ variations in Weddell Sea particulate organic matter. Mar Chem 35:355–369

Riebesell U, Schloss I, Smetacek V (1991) Aggregation of algae released from melting sea ice: implications for seeding and sedimentation. Polar Biol 11:239–248

SCAR (1994) Coastal and Shelf Ecology of the Antarctic Sea-Ice Zone (CS-EASIZ): a SCAR programme of marine research for the coastal and shelf ecosystem of Antarctica. Science plan and implementation plan. SCAR Report no. 10

Schnack-Schiel SB, Thomas DN, Dieckmann GS, Eicken H, Gradinger R, Spindler M, Weissenberger J, Mizdalski E, Beyer K (1995) Life cycle strategy of the Antarctic calonoid copepod *Stephos longipes*. Prog Oceanogr 36:45–75

Schröder M, Wisotzki A, Brunken N, Hansen I, Heil P, Kreyscher M, Moschner S, Steiner N, Sterr U, Richter K-U, Diedrich H (1994) Physical oceanography and nutrients. In: Lemke P (ed) The expedition ANTARKTIS X/4 of RV "Polarstern" in 1992. Ber Polarforsch 140:15–34

Smetacek V (1999a) Bacteria and silica cycling. Nature 397:475–476

Smetacek V (1999b) Diatoms and the ocean carbon cycle. Protistologica 150:25–32

Smith SV, Kroopnick P (1981) Carbon-13 isotopic fractionation as a measure of aquatic metabolism. Nature 294:252–253

Smith WO Jr, Dunbar RB (1998) The relationship between new production and vertical flux on the Ross Sea continental shelf. J Mar Syst 17:445–457

Spindler M, Dieckmann GS (1986) Distribution and abundance of the planktonic foraminifer *Neogloboquadrina pachyderma* in sea ice of the Weddell Sea (Antarctica). Polar Biol 5:185–191

Thomas DN, Lara RJ, Eicken H, Kattner K, Skoog A (1995) Dissolved organic matter in Arctic multi-year sea ice during winter: major components and relationship to ice characteristics. Polar Biol 15:477–483

Thomas DN, Lara RJ, Haas C, Schnack-Schiel SB, Dieckmann GS, Kattner G, Nöthig E-M, Mizdalski E (1998) Biological soup within decaying summer sea ice in the Amundsen Sea, Antarctica. In: Lizotte MP, Arrigo KR (eds) Antarctic sea ice biological processes. American Geophysical Union. Antarct Res Ser 73:161–171

Wada E, Terazaki M, Kabaya Y, Nemoto T (1987) ^{15}N and ^{13}C abundances in the Antarctic Ocean with emphasis on the biogeochemical structure of the food web. Deep Sea Res 34:829–841

Wedborg M, Hoppema M, Skoog A (1998) On the relation between organic and inorganic carbon in the Weddell Sea. J Mar Syst 17:59–76

Williams P, Druffel ERM (1988) Dissolved organic matter in the ocean: comments on a controversy. Oceanogr Mag 1:14–17

Zielinski U, Gersonde R (1997) Diatom distribution in Southern Ocean surface sediments (Atlantic sector): implications for paleoenvironmental reconstructions. Palaeogeogr Palaeoclimatol Palaeoecol 129:213–250

Polar Biol (2001) 24: 724–728
DOI 10.1007/s003000100273

Sigrid B. Schnack-Schiel · Gerhard S. Dieckmann
Rolf Gradinger · I.A. Melnikov · Michael Spindler
David N. Thomas

Meiofauna in sea ice of the Weddell Sea (Antarctica)

Abstract Sea-ice meiofauna was studied during various cruises to the Weddell Sea. Foraminifers dominate (75%) the sea-ice community in terms of numerical abundance while turbellarians dominate (45%) in terms of biomass. Distribution of organisms is patchy and varies considerably between cruises but also between sampling sites within one cruise. The bulk of the meiofauna is concentrated in the lowest parts of the sea ice, especially during winter and autumn. However, in porous summer sea ice, sympagic organisms also occur in high densities in upper and intermediate layers of sea ice. Proto- and metazoans associated with Antarctic sea ice include organisms actually living in sea ice, as well as those on the underside of floes and in the underlying water. The sea-ice habitat serves as a feeding ground, as well as an important nursery for juveniles, providing energy-rich food resources. The ice also constitutes a shelter from predators.

Please note: When citing this article, please use the reference to the original source in the journal Polar Biology as given on the top of this page.

S.B. Schnack-Schiel (✉) · G.S. Dieckmann
Alfred-Wegener-Institut für Polar-
und Meeresforschung,
Postfach 120161,
27515 Bremerhaven, Germany
E-mail: sschiel@awi-bremerhaven.de

R. Gradinger · M. Spindler
Institut für Polarökologie, Universität Kiel,
24148 Kiel, Germany

I.A. Melnikov
P.P. Shirshov Institute of Oceanology,
Russian Academy of Sciences,
Moscow 117857, Russia

D.N. Thomas
School of Ocean Sciences,
University of Wales-Bangor,
Menai Bridge, Anglesey,
LL59 5EY, UK

Introduction

The annual cycle of sea-ice growth and melt plays a pivotal role in polar ecosystems. Productivity, behaviour, ecophysiology and life-history patterns of a diverse group of organisms ranging from bacteria to whales are governed by this phase separation in the pelagic zone. The complexity of sea ice as a habitat, or rather range of habitats, is now well documented (e.g. Horner 1985) and microalgae are known to reach very high standing stocks (e.g. Ackley et al. 1979; Dieckmann et al. 1998). However, the heterotrophic component of sea-ice assemblages, other than bacteria, has received less attention. This is especially true for the metazoans. High algal biomass in sea ice clearly serves as the basis of a dynamic food web within and associated with the ice, particularly during winter when phytoplankton in the water column is scarce. Protozoans and metazoans associated with Antarctic sea ice include heterotrophic flagellates, ciliates, foraminifers, turbellarians and copepods (e.g. Spindler et al. 1990; Gradinger 1999). Nematodes have for the first time been found in an Antarctic pack-ice sample (Blome and Riemann 1999). Many organisms appear to spend their entire life-cycle, or at least a major part thereof, within the sea ice (e.g. Dahms et al. 1990; Schnack-Schiel et al. 1995; Tanimura et al. 1996). The sea-ice peripheries or ice/water interface, which include nooks and crannies, larger gaps and enclosures, as well as floe edges, represent a tremendous surface area for algal growth. These zones are frequented by larger grazers, particularly krill and amphipods, which feed directly on ice algae (e.g. Richardson and Whitaker 1979; Marschall 1988; Stretch et al. 1988; Daly 1990). The sea ice is an important nursery ground for juveniles of many species, providing an abundance of food as well as shelter from predators. It therefore plays a significant role in the survival of overwintering zooplankton. Recent studies have shown that even porous summer sea ice may act as an important refuge for life-history stages of several organisms (Thomas et al. 1998).

Table 1 Weddell Sea studies selected for analysis

	Date	Ship	References
ANT III/3	Jan/Feb 1985	Polarstern	Spindler et al. (1990); Schnack-Schiel et al. (1995)
ANT V/3	Oct/Dec 1986	Polarstern	Spindler et al. (1990); Schnack-Schiel et al. (1995)
ANT VIII/2	Sep/Oct 1989	Polarstern	Gradinger (1999)
ANT IX/3	Jan/Mar 1991	Polarstern	Gradinger (1999)
ANT X/3	Apr/May 1992	Polarstern	Schnack-Schiel et al. (1995); Gradinger (1999)
ISW-1	Feb/May 1992	Drift station	Menshenina and Melnikov (1995)
ANT XV/3	Jan/Mar 1997	Polarstern	Schnack-Schiel et al. (in press)

The objective of this study was to summarize results obtained during various studies in the Weddell Sea and to show that the meiofauna is a significant component of sea-ice biota.

Materials and methods

Sampling was carried out during several expeditions to the Weddell Sea (Table 1). Ice cores were drilled with a 7.5-cm or 10-cm-diameter ice auger and subsequently cut into subsections of between 1 and 20 cm length. The ice was melted after addition of 200 ml 0.2-μm filtered seawater per 1 cm core length (Garrison and Buck 1986). After complete melting, the volume was determined and ice meiofauna was concentrated over 20 μm gauze. The meiofauna (here defined as heterotrophs > 20 μm) was identified and organisms counted alive under a dissecting microscope immediately afterwards. Heterotrophic flagellates, although occurring frequently in sea ice, were not enumerated during all Weddell Sea cruises and hence are not included in this study. Under-ice sampling was carried out with an NIPR-I net (Fukuchi et al. 1979), a pump (Schnack-Schiel et al. 1995) or by scuba-diving (Menshenina and Melnikov 1995). All samples were preserved in 2–5% formaldehyde. Copepods were identified and counted in the entire samples or in subsamples (1/10–1/100) depending on the numbers. Biomass (expressed as carbon) was estimated according to Gradinger (1999). For further information of the sampling procedure and analyses of the samples, see Spindler et al. (1990), Menshenina and Melnikov (1995) and Gradinger (1999).

Results and discussion

Foraminifers dominated the sea-ice meiofauna by numbers (48%), ciliates, turbellarians, harpactoid and calanoid copepods making up between 7 and 23% (Fig. 1). However, in terms of carbon biomass, turbellarians were dominant (53%), followed by copepods (calanoids and harpacticoids each 20%). In contrast to Arctic sea ice, rotifers have not been recorded in the Antarctic (Gradinger 1999) while nematodes have only been seen once (Blome and Riemann 1999).

Data from different cruises have shown the occurrence of organisms in sea ice to be highly variable (Tables 2, 3, 4), but not necessarily attributable to seasonal changes. Even during the same cruise, abundances were patchy and differed between sampling sites, which is reflected in the high ranges observed. Similar spatial variations of sympagic Antarctic organisms have been described by other authors (e.g. Garrison and Close 1993; Scott et al. 1994; Swadling et al. 1997; Schnack-Schiel et al. 1998).

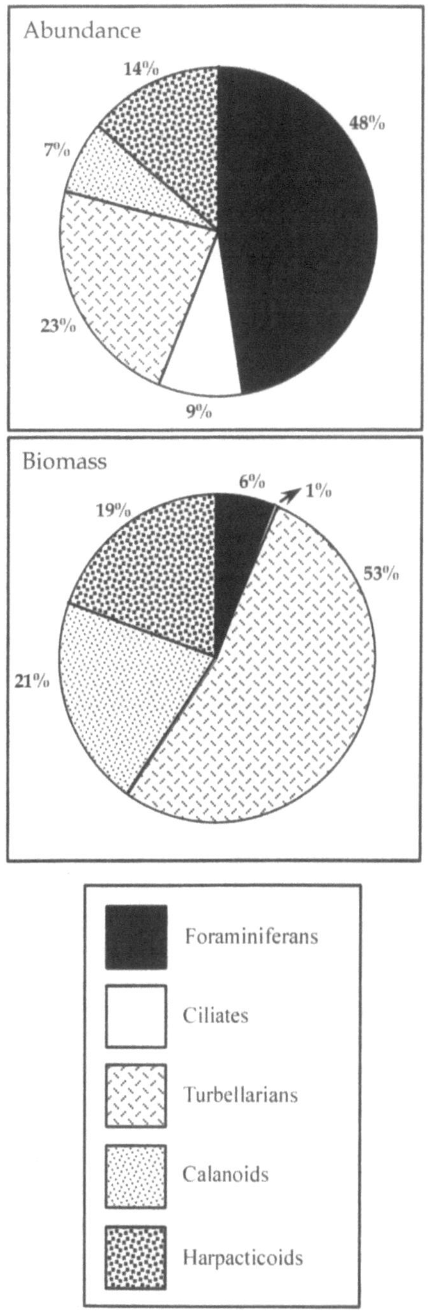

Fig. 1 Relative contribution of ice meiofauna taxa to mean abundance and biomass

Table 2 Summary of ice thickness and mean integrated chlorophyll *a* standing stock

	Year	Month	No. of cores	Average core length (cm)	Range of core length (cm)	Mean integrated Chl *a* (mg m^{-2})	Range of mean integrated Chl *a* (mg m^{-2})
ANT VIII	1989	Oct/Dec	32	75	21–170	4.17	0.14–18.40
ANT V	1986	Sep/Oct	29	216	190–270	5.48	0.01–35.43
ANT III	1985	Jan/Feb	13	209	80–285	131.66	5.90–435.3
ANT IX	1991	Jan/Mar	11	113	67–174	24.60	0.76–77.34
ANT XV	1997	Jan/Mar	17	107	16–224	3.87	0.31–26.79
ANT X	1992	Apr/May	17	58	20–180	7.95	0.08–32.24

Table 3 Mean integrated abundance of sea-ice meiofauna (– not counted)

Cruise	Foraminiferans		Ciliates		Turbellarians		Calanoid copepods		Harpacticoid copepods	
	Ind. × 10^3 m^{-2}	Range	Ind. × 10^3 m^{-2}	Range	Ind. × 10^3 m^{-2}	Range	Ind. × 10^3 m^{-2}	Range	Ind. × 10^3 m^{-2}	Range
ANT VIII	14.83	0–89	1.74	0–10	3.02	0–45	0.61	0–7	4.65	0–49
ANT V	47.85	0–440	10.36	0–67	–	–	2.02	0–21	0.32	0–2
ANT III	39.23	0–159	2.02	0–26	–	–	0.02	0–0.2	5.62	0–67
ANT IX	11.69	0–62	7.13	0.5–19	5.38	0–20	21.98	0.2–124	10.17	0–97
ANT XV	–	–	–	–	–	–	0.24	0–2	3.60	0–29
ANT X	33.5	0–383	–	–	34.17	0–436	2.13	0–20	26.68	0–226

Table 4 Mean integrated biomass of sea-ice meiofauna (– not counted)

Cruise	Foraminiferans		Ciliates		Turbellarians		Calanoid copepods		Harpacticoid copepods	
	mg C m^{-2}	Range	mg C m^{-2}	Range	mg C m^{-2}	Range	mg C m^{-2}	Range	mg C m^{-2}	Range
ANT VIII	0.46	0–3	0.02	0–0.1	1.18	0–17	0.55	0–8	1.79	0–15
ANT V	1.48	0–14	0.11	0–0.7	–	–	4.01	0–42	0.17	0–1
ANT III	1.22	0–5	0.02	0–0.3	–	–	0.02	0–0.2	3.37	0–40
ANT IX	0.36	0–2	0.08	0–0.2	2.09	0–8	8.38	0.01–44	2.51	0–24
ANT XV	–	–	–	–	–	–	3.2	0–33	0.93	0–9
ANT X	1.04	0–12	–	–	20.5	0–261	2.74	0–17	8.49	0–69

The largest proportion of Antarctic sea ice is annual ice, resulting in the loss of organisms to the water when it melts in summer. The organisms therefore undergo a pelagic phase during their life-cycle. Ice formation and organism-incorporation or migration into an already existing ice cover transfer organisms between habitats with extremely different environmental constraints. Hence, the sympagic community must be able to tolerate a wide range of rapidly changing environmental conditions. The intricate sea-ice structure comprising ice crystals and brine channels forms a labyrinth of pockets and tubes ranging in diameters from a few micrometres to several centimetres (Weissenberger et al. 1992). The volume and degree of branching of the brine channels vary with temperature and ice texture. Temperature and salinity of the ice near the ice/water interface are similar to those of the underlying water (e.g. Bartsch 1989). This is one reason why the bulk of sympagic organisms frequents the peripheries of the sea ice, where living conditions are more benign (Hoshiai and Tanimura 1986; Schnack-Schiel et al. 1995). Figure 2 shows an example of the distribution within ice at two stations in the Weddell Sea in late winter where the meiofauna was concentrated in the lowest regions of ice floes. However, ice algae as well as heterotrophic organisms also occur in upper and intermediate layers of sea ice (Bartsch 1989; Schnack-Schiel et al. 1995). This is especially true in porous summer sea ice where brine channels and pockets are larger (Fig. 3). The higher temperatures result in a phase transition in permeability, permitting movement and exchange of seawater, as well as organisms, throughout the ice (Golden et al. 1998). This is in agreement with results from the Bellingshausen and Amundsen Seas, where peak abundances occurred in rotten slush layers flooded with seawater, and hence were closely associated with higher nutrient and dense algal concentrations (Schnack-Schiel et al. 1998; Thomas et al. 1998).

Strong interactions between the ice biota and pelagic animals exist during periods of complete ice coverage. The high algal biomass at the ice/water interface clearly serves as a food source for the diverse heterotrophic organisms. Many specialized metazoans begin to feed on ice algae during late winter and early spring when phytoplankton in the water column is scarce. These include amphipods and euphausiids, which have been recorded in high numbers during winter, spring and autumn at the underside of ice floes (e.g. Richardson and Whitaker

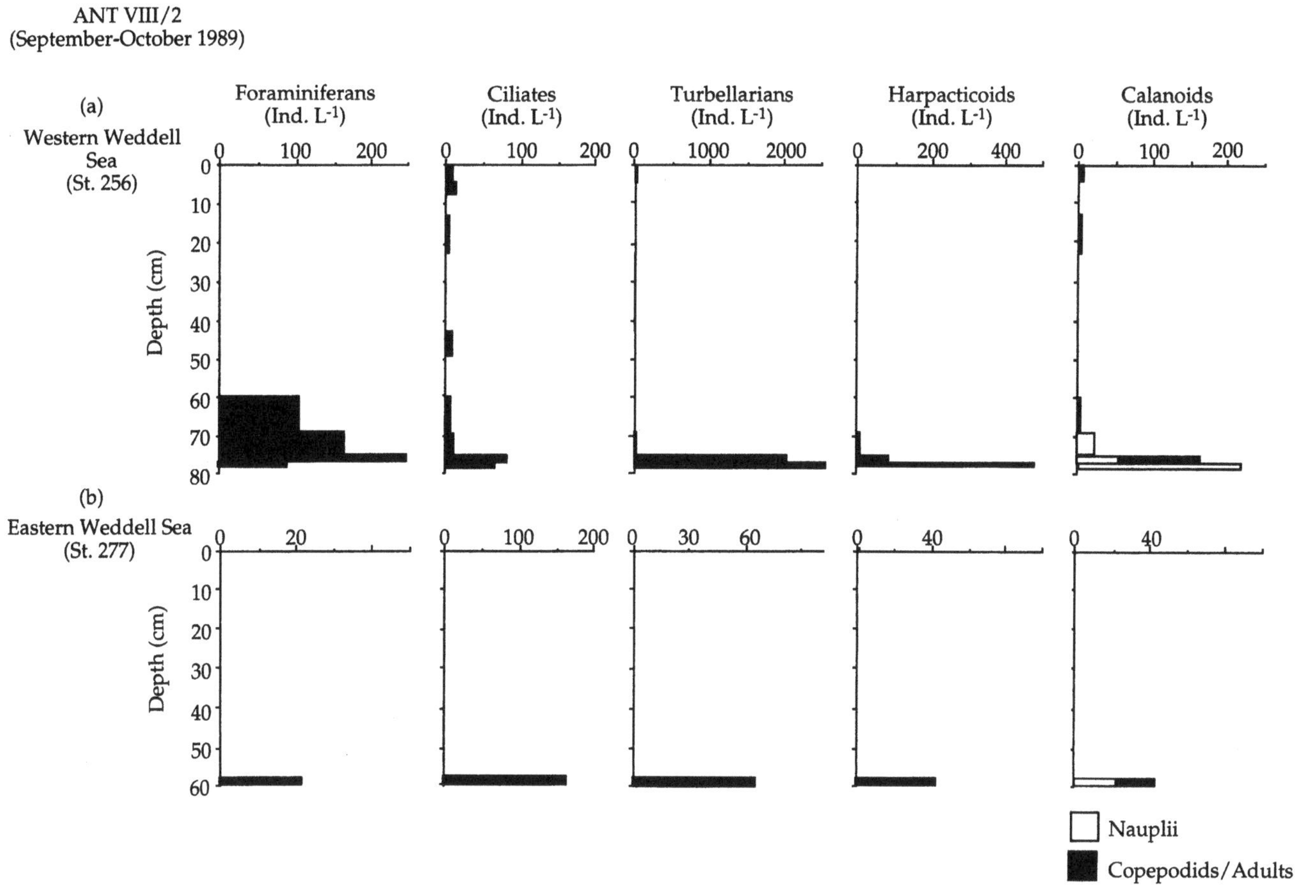

Fig. 2a, b Depth distribution of ice meiofauna taxa in sea-ice cores in late winter (September/October 1989) in the Weddell Sea

1979; Marschall 1988; Menshenina and Melnikov 1995; Melnikov and Spiridonov 1996). Many amphipods associated with Antarctic sea ice are benthic species that appear to seasonally colonize the undersides of the floes from the benthos below. Moving along the bottom of the ice, they feed directly on the bottom community and use brine channels and pressure ridges between the floes as shelter against possible predators such as fish and birds (Richardson and Whitaker 1979).

Many sympagic heterotrophs live permanently or temporarily within the sea ice or at the ice/water interface. For example, the two dominant ice-associated copepod species in the Weddell Sea, *Drescheriella glacialis* and *Stephos longipes* (Dahms et al. 1990; Kurbjeweit et al. 1993; Schnack-Schiel et al. 1995; Günther et al. 1999), appear to differ in their life-cycle strategies. The harpacticoid, *D. glacialis*, seems to spend most of its life in the ice (Dahms et al. 1990). In contrast, the life-cycle of the calanoid, *S. longipes*, is strongly associated with

Fig. 3 Depth distribution of ice meiofauna taxa in a sea-ice core in summer (January/March 1991) in the eastern Weddell Sea

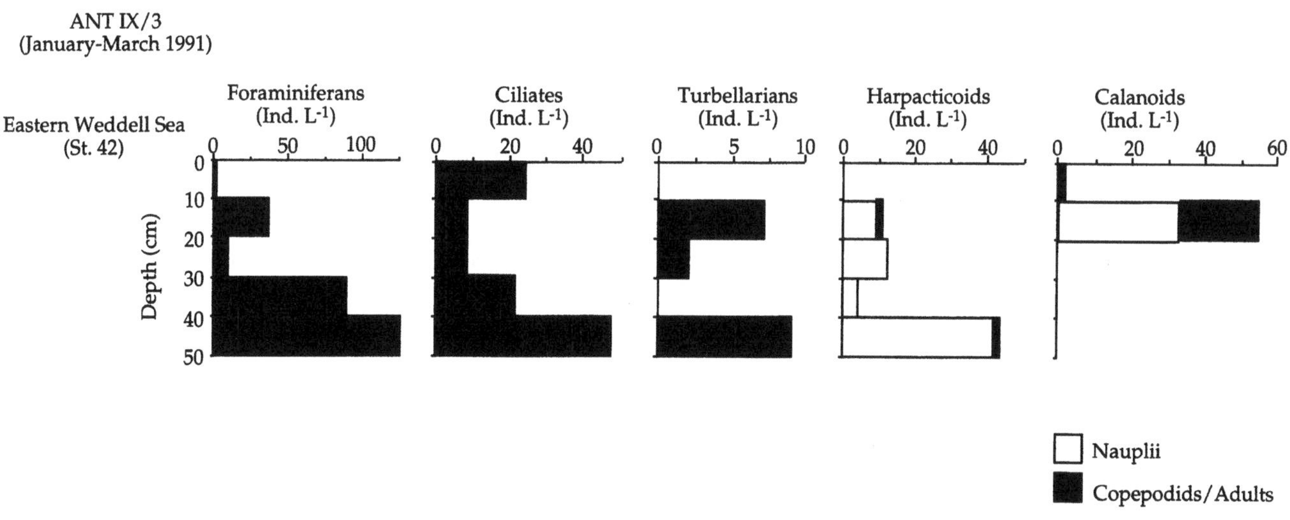

the seasonal fluctuation of sea-ice cover. In late winter/early spring and autumn, when sea ice predominates, *S. longipes* was seldom found in the water column, whereas high numbers occurred in the sea ice and in the ice/water interface (Menshenina and Melnikov 1995; Schnack-Schiel et al. 1995). In summer, when ice cover was minimal, *S. longipes* was very abundant in the water column particularly in the surface layers. According to Schnack-Schiel et al. (1995), *S. longipes* seems to have two alternative strategies for overwintering: a young population overwinters in and just below the sea ice, while an older population migrates deep into the water column and may reach the sea bed. At some stations in the Weddell Sea, the calanoid species *Paralabidocera antarctica*, a typical sympagic species in east Antarctic coastal areas (Hoshiai and Tanimura 1986; Tanimura et al. 1996), may also reach high concentrations in the sea ice and in platelet-ice layers. Platelet layers underlying fast ice adjacent to ice shelves generally attain high algal standing stocks (e.g. Arrigo et al. 1995) and, hence, are an important food source for herbivorous zooplankton over the continental shelf.

The seasonal ice melt in spring/summer has a pronounced effect on biological processes as ice algae are released to the water column where they seed phytoplankton blooms or sink to the sea bed. Ice algae also provide an important food source for a variety of heterotrophic organisms in the water column (Garrison and Mathot 1996). By release of faecal pellets to the water, the zooplankton serves as mediator for particulate organic matter from the sea ice to the water column and contributes to the particle flux in ice-covered seas.

Acknowledgements This is contribution no. 1812 of the Alfred-Wegener-Institut für Polar- und Meeresforschung, Bremerhaven.

References

Ackley SF, Buck KR, Taguchi S (1979) Standing crop of algae in the sea ice of the Weddell Sea region. Deep Sea Res 26A: 269–281

Arrigo KR, Dieckmann GS, Gosselin M, Robinson DH, Fritsen CH, Sullivan CW (1995) A high resolution study of the platelet ecosystem in McMurdo Sound, Antarctica: biomass, nutrient and production profiles within a dense microalgal bloom. Mar Ecol Prog Ser 127:255–268

Bartsch A (1989) Die Eisalgenflora des Weddellmeeres (Antarktis): Artenzusammensetzung und Biomasse sowie Ökophysiologie ausgewählter Arten. Ber Polarforsch 63:109

Blome D, Riemann F (1999) Antarctic sea ice nematodes, with description of *Geomonhystera glaciei* sp. nov. (Monhysteridae). Mitt Hamb Zool Mus Inst 96:15–20

Dahms HU, Bergmann M, Schminke HK (1990) Distribution and adaptations of sea ice inhabiting Harpacticoida (Crustacea, Copepoda) of the Weddell Sea (Antarctica). Mar Ecol 11: 207–226

Daly KL (1990) Overwinter development, growth and feeding of larval *Euphausia superba* in the Antarctic marginal ice zone. Limnol Oceanogr 35:1564–1576

Dieckmann GS, Eicken H, Haas C, Garrison DL, Gleitz M, Lange M, Nöthig E-M, Spindler M, Sullivan CW, Thomas DN, Weissenberger J (1998) A compilation of data on sea ice algal standing crop from the Bellingshausen, Amundsen, and Weddell Seas from 1983 to 1994. Antarct Res Ser 73:85–92

Fukuchi M, Tanimura A, Hoshiai T (1979) "NIPR-I": a new plankton sampler under sea ice . Bull Plankton Soc Jpn 26: 104–109

Garrison DL, Buck KR (1986) The biota of Antarctic pack ice in the Weddell Sea and Antarctic Peninsula regions. Polar Biol 10:211–219

Garrison DL, Close AR (1993) Winter ecology of the sea-ice biota in Weddell Sea pack ice. Mar Ecol Prog Ser 96:17–31

Garrison DL, Mathot S (1996) Pelagic and sea ice microbial communities. Antarct Res Ser 70:155–172

Golden KM, Ackley SF, Lytle VI (1998) The percolation phase transition in sea ice. Science 282:2238–2240

Gradinger R (1999) Integrated abundance and biomass of sympagic meiofauna in Arctic and Antarctic pack ice. Polar Biol 22:169–177

Günther S, George KH, Gleitz M (1999) High sympagic metazoan abundance in platelet layers at Drescher-Inlet, Weddell Sea, Antarctica. Polar Biol 22:82–89

Horner R (1985) Sea ice biota. CRC Press, Boca Raton, Fla

Hoshiai T, Tanimura A (1986) Sea ice meiofauna at Syowa Station, Antarctica. Mem Natl Inst Polar Res Spec Issue 44:118–124

Kurbjeweit F, Gradinger R, Weissenberger J (1993) The life cycle of *Stephos longipes* – an example for cryopelagic coupling in the Weddell Sea (Antarctica). Mar Ecol Prog Ser 98:255–262

Marschall P (1988) The overwintering strategy of Antarctic krill under the pack-ice of the Weddell Sea. Polar Biol 9:129–135

Melnikov IA, Spiridonov VA (1996) Antarctic krill under perennial sea ice in the western Weddell Sea. Antarct Sci 8:323–329

Menshenina LL, Melnikov IA (1995) Under-ice zooplankton of the western Weddell Sea. Proc NIPR Symp Polar Biol 8:126–138

Richardson MG, Whitaker TM (1979) An Antarctic fast-ice food chain: observations on the interaction on the amphipod *Pontogeneia antarctica* Chevreux with ice-associated micro-algae. Br Antarct Surv Bull 47:107–115

Schnack-Schiel SB, Thomas D, Dieckmann GS, Eiken H, Gradinger R, Spindler M, Weissenberger J, Mizdalski E, Beyer K (1995) Life cycle strategy of the Antarctic calanoid copepod *Stephos longipes*. Prog Oceanogr 36:45–75

Schnack-Schiel SB, Thomas D, Dahms HU, Haas C, Mizdalski E (1998) Copepods in Antarctic sea ice. Antarct Res Ser 73:173–182

Schnack-Schiel SB, Thomas D, Haas C, Dieckmann GS, Alheit R (in press) The occurrence of the copepods *Stephos longipes* (Calanoida) and *Drescheriella glacialis* (Harpacticoida) in summer sea ice in the Weddell Sea, Antarctica. Antarct Sci

Scott P, McMinn A, Hosie G (1994) Physical parameters influencing diatom community structure in eastern Antarctic sea ice. Polar Biol 14:507–517

Spindler M, Dieckmann GS, Lange MA (1990) Seasonal and geographical variations in sea ice community structure of the Weddell Sea, Antarctica. In: Kerry KR, Hempel G (eds) Antarctic ecosystems, ecological change and conservation. Springer, Berlin Heidelberg New York, pp 129–135

Stretch JJ, Hamner PP, Hamner WM, Michel WC, Cook J, Sullivan CW (1988) Foraging behavior of Antarctic krill *Euphausia superba* on sea ice microalgae. Mar Ecol Prog Ser 44:131–139

Swadling KM, Gibson JAE, Ritz DA, Nichols PD (1997) Horizontal patchiness in sympagic organisms of the Antarctic fast ice. Antarct Sci 9:399–406

Tanimura A, Hoshiai T, Fukuchi M (1996) The life cycle strategy of the ice-associated copepod *Paralabidocera antarctica* (Calanoida, Copepoda), at Syowa Station, Antarctica. Antarct Sci 8:257–266

Thomas DN, Lara RJ, Haas C, Schnack-Schiel SB, Dieckmann GS, Kattner G, Nöthig EM, Mizdalski E (1998) Biological soup within decaying summer sea ice in the Bellingshausen Sea. Antarct Res Ser 73:161–171

Weissenberger J, Dieckmann GS, Gradinger R, Spindler M (1992) Sea ice: a cast technique to examine and analyze brine pockets and channel structure. Limnol Oceanogr 37:179–183

Polar Biol (2001) 25: 72–80
DOI 10.1007/s003000100314

L.J. Trenerry · A. McMinn · K.G. Ryan

In situ oxygen microelectrode measurements of bottom-ice algal production in McMurdo Sound, Antarctica

Abstract In-situ estimates of fast-ice algal productivity at Cape Evans, McMurdo Sound, in 1999 were lower than at the same site in previous years. Under-ice irradiance was between 0 and 8 μmol photons m^{-2} s^{-1}; the ice was between 1.9 and 2.0 m thick and the algal biomass averaged 150 mg chl *a* m^{-2}, although values as high as 378 mg chl *a* m^{-2} were recorded. Production on 11 and 12 November was between 0.053 and 1.474 mg C m^{-2} h^{-1}. When the data from 11 November were fitted to a hyperbolic tangent function, a multilinear regression gave estimates for P_{max} of 0.571 nmol O_2 cm^{-2} s^{-1}, an α of 0.167 nmol O_2 cm^{-2} s^{-1} μmol^{-1} photons m^{-2} s^{-1} and an E_k of 3.419 μmol photons m^{-2} s^{-1}. A P_{max} of 2.674 nmol O_2 cm^{-2} s^{-1}, an α of 0.275 nmol O_2 cm^{-2} s^{-1} μmol^{-1} photons m^{-2} s^{-1}, *r* of 0.305 nmol O_2 cm^{-2} s^{-1} and an E_k of 9.724 μmol^{-1} photons m^{-2} s^{-1} were estimated from the 12 November data. The sea-ice algal community was principally comprised of *Nitzschia stellata*, *Entomoneis kjellmanii* and *Berkeleya adeliensis*. Other taxa present included *N. lecointei*, *Fragilariopsis* spp., *Navicula glaciei*, *Pleurosigma* spp. and *Amphora* spp. Variations in the method for estimating the thickness of the diffusive boundary layer were not found to significantly affect the measurements of oxygen flux. However, the inability to accurately measure fine-scale variations in biomass is thought to contribute to the scatter of the *P* versus *E* data.

Please note: When citing this article, please use the reference to the original source in the journal Polar Biology as given on the top of this page.

L.J. Trenerry · A. McMinn (✉)
Institute of Antarctic and Southern Ocean Studies,
University of Tasmania, Box 252-77,
Hobart 7001, Tasmania, Australia
E-mail: andrew.mcminn@utas.edu.au

K.G. Ryan
Industrial Research Ltd, P.O. Box 31310,
Lower Hutt, New Zealand

Introduction

Sea-ice microbial communities are a critical component of polar marine ecosystems (Bunt and Lee 1970; Palmisano and Sullivan 1983; Sullivan et al. 1983; Grossi et al. 1984; Horner 1985; Garrison et al. 1986; Palmisano and Garrison 1993; Arrigo et al. 1995; Robinson et al. 1997) and contribute approximately 25% of the annual primary production in perennially ice-covered areas (Legendre et al. 1992). Palmisano and Sullivan (1983) reported that biomass was almost entirely located in the bottom 20 cm of ice, and a maximum standing crop of 309 mg chl *a* m^{-2} has been recorded in annual sea ice of McMurdo Sound. Estimates of productivity in Antarctic fast-ice algal mats, however, are inconsistent. Comparison of primary productivity data has been difficult due to differences in methodology and units used between researchers, and the high variability in biomass and in situ irradiance.

The development of microsensors for use in studies of microbial ecology has provided in situ information on the metabolic processes of microorganisms and their interactions with their physical and chemical microenvironment. Oxygen microelectrodes have revealed the structure (Jørgensen and Revsbech 1985) and the importance of a diffusive boundary layer (DBL), which may impede the flux of O_2, CO_2 and nutrients to benthic microbial communities (Gundersen and Jørgensen 1990; Jørgensen and Des Marais 1990; Glud et al. 1994; Revsbech 1994). The microelectrodes have a fine tip diameter (< 20 μm), fast response times (90% response in < 1 s) to changes in oxygen tension, low stirring sensitivity (< 1%) and relatively good long-term stability, and are an excellent tool for studying the spatial and temporal coupling between production and consumption of O_2 in these communities (Glud et al. 1999).

The use of oxygen microelectrodes to determine oxygen production within sediment and benthic algal communities has been well documented over the past 20 years (Jørgensen et al. 1983; Revsbech and Ward

1983; Revsbech et al. 1983; Jørgensen and Revsbech 1985; Lorenzen et al. 1995; Klimant et al. 1997). Microelectrode techniques are also suited to the study of some sea-ice microbial communities as, like those in sediment, they are adhered to the substratum. Most previous studies of sea-ice productivity have applied methods used for phytoplankton. This has involved melting the ice to extract the algae and then resuspension in either an incubator with artificial light or in incubation bottles placed in the water column. Both methods cause massive disturbance to the sea-ice microbial community. McMinn and Ashworth (1998) were the first to use oxygen microelectrodes to determine the productivity of sea ice. Due to the difficulty of inserting the fragile electrodes into sea ice, a method based on measuring the oxygen flux through the DBL (Jørgensen and Revsbech 1985) was used to calculate the net oxygen production of the whole mat. This method, however, modified the natural conditions experienced by the ice algae, by extracting an ice core, inverting it in an incubator and using artificial illumination. McMinn et al. (1999a) used the same equipment but used natural sunlight provided by a heleostat. Further development of this technique by McMinn et al. (2000) has enabled under-ice deployment and in situ measurements of photosynthetic rates in sea-ice algal mats in Antarctic fast ice. By positioning the microelectrodes beneath the sea ice, measurements of the algal community production are recorded under natural illumination and under a range of natural processes, such as grazing, nutrient limitation and current strength. Importantly, the algal community is never touched and there is no modification of the natural environment. Given these natural conditions, oxygen profiles measured over short time periods approximate a steady state. On a longer time scale, the oxygen flux is in dynamic equilibrium with diurnal changes in light levels. Our aim here is to determine the primary productivity of fast ice in McMurdo Sound (1999) using oxygen microelectrodes. Unlike the study in McMurdo Sound in 1997 (McMinn et al. 2000), this study presents results from a more typical ice thickness (1.8 m rather than 0.9 m) and from earlier in the year (late October/early November rather than late November/early December). These results are then compared with productivity measurements obtained by more conventional methods. Our results provide a clearer insight into processes affecting primary production in undisturbed sea-ice microbial communities.

Materials and methods

Field studies were conducted between 28 October and 12 November 1999 on annual fast ice above 30 m of water, 200 m north of Cape Evans, McMurdo Sound (Fig. 1). This site was selected because of the absence of ice platelets at the ice-water interface (McMinn et al. 1999a). This was necessary because the oxygen electrodes are unable to penetrate them without being damaged. Snow cover in the area was variable, ranging from < 1 cm up to 10 cm. Ice thickness was between 1.9 and 2 m over the experimental area. The weather during this period was generally overcast, with several days of strong wind and blowing snow, although cloud-free weather was experienced on a number of days.

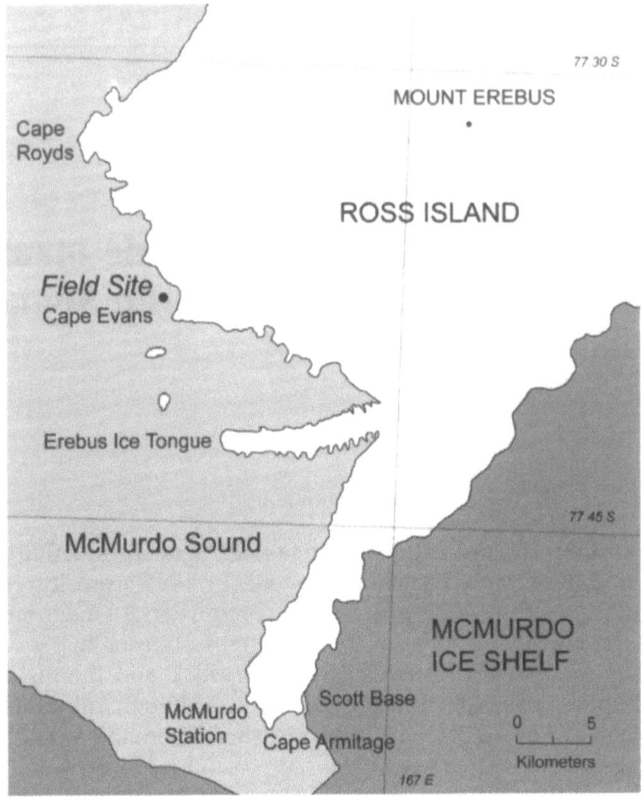

Fig. 1 Map showing Cape Evans field site in McMurdo Sound, Antarctica

Under-ice net primary production was measured using the oxygen flux across the diffusive boundary layer (DBL) method (Jorgensen and Revsbech 1985; McMinn et al. 2000). Using this technique it is necessary to measure both the flux of oxygen from beneath the algal mat and the thickness of the DBL. Measurement of both the DBL thickness and the oxygen flux was achieved by stepping the oxygen microelectrode at 10-μm intervals through the first few millimetres of water under the ice. The in situ under-ice deployment and measuring equipment was purpose-built to measure the oxygen flux across the DBL at the ice-water interface (Fig. 2). The equipment and micro-oxygen electrodes (Precision Measurement Engineering, Encinitas) were the same as used by McMinn et al. (2000). The oxygen microelectrodes used had a tip diameter of approximately 40 μm, a 90% response time of approximately 1 s and a stirring sensitivity of 1–2%. They were calibrated by obtaining values for air-saturated seawater (saturated by bubbling with a standard aquarium pump for 20 min) and for deoxygenated seawater (deoxygenated with sodium sulphite). A fibre-optic light sensor was constructed after Kühl et al. (1994) and calibrated with a Biospherical QSP 200 4p PAR sensor. All light measurements were made beneath the algal mat.

The measured O_2 gradients over the DBL were used to calculate the net oxygen production in light. The top of the DBL is defined as the ice-water interface. Recognition of the top of the DBL was by repeated O_2 values or a fall in O_2 concentration associated with touching the ice. The bottom of the DBL is defined as the position at which there is a measurable increase in O_2 concentration (referred to as the "true diffusive boundary layer"). The limit of the true DBL can be poorly defined and a more precise and practical parameter can be defined as the depth at which there had been an increase of 10% of the total change in oxygen concentration (i.e. the oxygen export equivalent of the 90% of "free-flow concentration" of

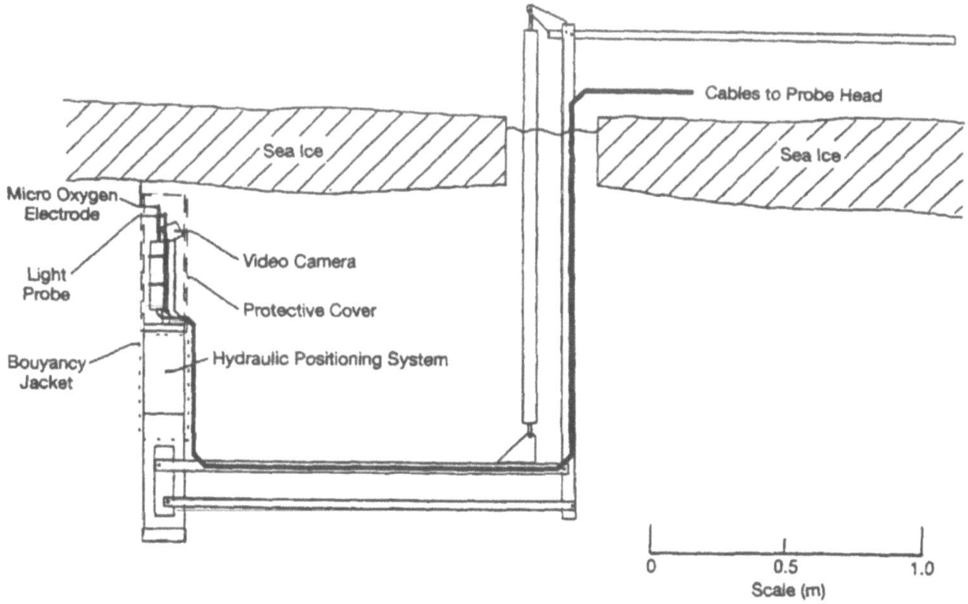

Fig. 2 The in situ under-ice deployment and measuring equipment used to determine productivity of bottom-ice algal communities (from McMinn et al. 2000)

Jørgensen and Des Marais 1990). This is referred to as the "effective diffusive boundary layer" (Jørgensen and Revsbech 1985).

The oxygen diffusion flux (J) across the DBL is equivalent to the net productivity of the algal mat and was calculated using the 1-dimensional version of Fick's first law of diffusion (Revsbech and Jørgensen 1986):

$$J = D_0 \frac{d[O_2]}{dx} \quad (1)$$

where D_0 = molecular diffusion coefficient (at −1.9°C = 1.11×10^{-5} $cm^2\ s^{-1}$; Broecker and Peng 1974), dx = effective DBL thickness or 90% level, $d[O_2]$ = change in oxygen concentration across the DBL. The photosynthetic quotient was assumed to be 1.03 (Satoh and Watanabe 1988). J is derived for both the true and effective DBLs. The oxygen flux at the effective DBL is equivalent to the net effective productivity of the algal mat.

The basic description of a P versus E relationship requires a non-linear mathematical function to account for the light saturation effect (Falkowski and Raven 1997). Several formulations have been employed to provide estimates of the photosynthetic parameters: P_{max} (maximum photosynthetic rate, in nmol $O_2\ cm^{-2}\ s^{-1}$); E_k (the minimum irradiance required to saturate photosynthesis, in μmol photons $m^{-2}\ s^{-1}$); α (the light-limited photosynthetic efficiency, in nmol $O_2\ cm^{-2}\ s^{-1}$ μmol^{-1} photons $m^{-2}\ s^{-1}$) and r (respiration constant, in nmol $O_2\ cm^{-2}\ s^{-1}$) (Jassby and Platt 1976; Aalderink and Jovin 1997). Different functions have been found to better fit phytoplankton data from different communities (Jassby and Platt 1976; Falkowski and Raven 1997). Systat was used to estimate the values of P_{max}, α and r. E_k was calculated by dividing P_{max} by α.

The non-linear regression of the production data against a hyperbolic tangent function from Jassby and Platt (1976) was used:

$$P = P_{max} \tanh(\alpha \times E/P_{max}) - r \quad (2)$$

where $\tanh(\alpha \times E/P_{max}) = \left(e^{(2E\alpha/P_{max})} - 1\right)/\left(e^{(2E\alpha/P_{max})} + 1\right)$

The application of an exponential function to the analysis of the photosynthesis-irradiance relationship used by Webb and colleagues (1974) (as cited in Jassby and Platt 1976) is tested:

$$P = P_{max}\left(1 - \exp\left(\frac{-\alpha \times E}{P_{max}}\right)\right) \quad (3)$$

Also used were the rectangular hyperbola from Falkowski and Raven (1997):

$$P = P_{max}\left(\frac{\alpha E}{P_{max} + \alpha E}\right) \quad (4)$$

and the quadratic equation of Smith (1936) (as cited in Jassby and Platt 1976):

$$P = P_{max}\alpha E / \left[(P_{max})^2 + (\alpha E)^2\right]^{1/2} \quad (5)$$

Chl*a* measurements were taken from five replicate ice cores from as close as possible to the site of the productivity measurements; this was always within 1 m. Ice cores were obtained by drilling to 1.5 m and then coring with an SIPRE corer to the bottom. The bottom 10 cm of the core sample was removed and trimmed to a block 7×7×10 cm and melted into filtered seawater. The block trimming avoided errors introduced by using the whole core, as small amounts of algae inevitably broke off the edge of the core during sampling. A 2.5-ml subsample was preserved in glutaraldehyde for later species identifications and cell counts (Ryan et al. in press). Melted sample (100 ml) was filtered onto a Whatman GF/F 47-mm-diameter filter and extracted in 20 ml methanol; 10 ml of this was used for chl*a* analysis. A Turner 10 AU fluorometer was used to measure chl*a* biomass following the acidification method of Strickland and Parsons (1972). The fluorometer was calibrated against a Sigma Chemicals chl*a* standard and regularly checked against a secondary standard. Under-ice currents were measured with a mini-air 2 flowmeter (Schiltknecht) with an 18-mm impeller chamber.

Results

The ice algal mat was comprised of *Nitzschia stellata* (average abundance 31%), *Entomoneis kjellmanii* (23%) and *Berkeleya adeliensis* (17%). Other taxa present included *N. lecointei*, *Fragilariopsis* spp., *Navicula glaciei*, *Pleurosigma* spp. and *Amphora* spp. Chlorophyll biomass during the oxygen electrode experimentation was 212.6 ± 15.1 and 288.7 ± 82.1 mg chl *a* m^{-2} for 9 and 12 November, respectively. The highest single chlorophyll biomass value recorded was 378 mg chl *a* m^{-2}, which is higher than previous standing-crop estimates reported by Palmisano and Sullivan (1983) for bottom-ice communities in McMurdo Sound. The water temperature during this time remained at –1.9°C and there was no evidence that the ice thickness was either increasing or decreasing. Under-ice currents at the Cape Evans field site were low and beneath the detection limit of 0.02 m s^{-1}.

Oxygen profiles from the sea-ice DBL were obtained between 9 and 12 November, although only profiles for 11 and 12 November were successfully recorded over a full range of diurnal irradiances. Midday surface irradiances were up to 1,610 μmol photons m^{-2} s^{-1}, while under-ice irradiances up to 8 μmol photons m^{-2} s^{-1} were recorded. An example of an under-ice oxygen profile is given in Fig. 3. Both estimates (true and effective) of the DBL were used to determine net oxygen flux from Eq. 1 and there was a strong correlation ($r^2 = 0.928$) between oxygen fluxes calculated from both measures of DBL thickness.

Data presented in Fig. 4 show the correlation between productivity, effective productivity and irradiance during part of a day for 11 and 12 November. Net oxygen export (at the true DBL) increased from 0.0001 nmol O_2 cm^{-2} s^{-1} at an irradiance of 0.2 μmol photons m^{-2} s^{-1} to over 0.0012 nmol O_2 cm^{-2} s^{-1} at an irradiance of 6 μmol photons m^{-2} s^{-1} on 11 November. This is equivalent to a carbon-based productivity range of 0.053–0.523 mg C m^{-2} h^{-1} (assuming a photosynthetic quotient of 1.03 for fast-ice algal communities proposed by Satoh and Watanabe 1988). Assimilation numbers for 11 November are between 0.0002 and 0.0021 mg C (mg chl $a)^{-1}$ h^{-1}. At the effective DBL, production ranged between 0.050 and 0.602 mg C m^{-2} h^{-1} and assimilation numbers from 0.0002 to 0.0024 mg C (mg chl $a)^{-1}$ h^{-1}.

On 12 November (Fig. 4b), net oxygen flux ranged from 0.0012 to 0.0034 nmol O_2 cm^{-2} s^{-1} at irradiances of 3–7 μmol photons m^{-2} s^{-1}. Carbon-based production on this day ranged from 0.51 to 1.42 mg C m^{-2} h^{-1} and assimilation from 0.002 to 0.0057 mg C (mg chl $a)^{-1}$ h^{-1}. Effective production was between 0.63 and 1.47 mg C m^{-2} h^{-1} and assimilation ranged from 0.0025 to 0.0059 mg C (mg chl $a)^{-1}$ h^{-1}.

All the data obtained for each day are summarised on *P* versus *E* curves (Figs. 5, 6) and in Table 1. The productivity data for the true DBL from 11 November (Fig. 5a), when fitted to a hyperbolic tangent function (Jassby and Platt 1976), produced a value for P_{max} of 0.54 nmol O_2 cm^{-2} s^{-1}, an α of 0.12 nmol O_2 cm^{-2} s^{-1} μmol^{-1} photons m^{-2} s^{-1} and an E_k of 4.3 μmol photons m^{-2} s^{-1}. Determination of respiration was attempted for 11 November, but there was no significant result. *P* versus *E* curves were also plotted using the effective DBL, given as effective productivity in Fig. 5b. For 11 November, P_{max} was 0.57 nmol O_2 cm^{-2} s^{-1}, α 0.16 nmol O_2 cm^{-2} s^{-1} μmol^{-1} photons m^{-2} s^{-1}, and E_k 3 μmol photons m^{-2} s^{-1}.

P versus *E* data for 12 November fitted to a hyperbolic tangent function gave a value for P_{max} of 2.67 nmol O_2 cm^{-2} s^{-1}, an α of 0.27 nmol O_2 cm^{-2} s^{-1} μmol^{-1} photons m^{-2} s^{-1}, an *r* of 0.31 nmol O_2 cm^{-2} s^{-1} and an E_k of 9 μmol photons m^{-2} s^{-1} (Fig. 6a). This was repeated for effective production data (Fig. 6b), where P_{max} was 2.56 nmol O_2 cm^{-2} s^{-1}, α 0.39 nmol O_2 cm^{-2} s^{-1} μmol^{-1} photons m^{-2} s^{-1}, *r* 0.54 nmol O_2 cm^{-2} s^{-1} and E_k 6 μmol photons m^{-2} s^{-1}. Results for P_{max}, E_k and α determined from fitting data to exponential, rectangular, quadratic and hyperbolic tangent functions (Jassby and Platt 1976) are given in Table 1.

Discussion

The sea-ice environment supports significant primary and secondary production (Horner 1985); however, the

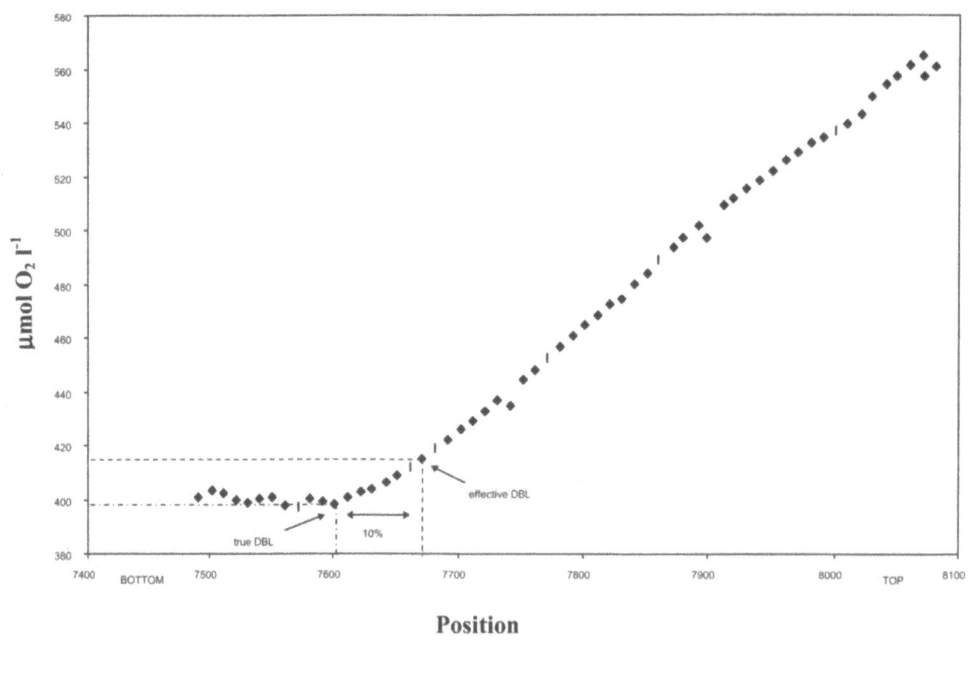

Fig. 3 Oxygen profile for 12 November 1999. Position on the *x-axis* are 10-μm units from an arbitrary starting position, and the *y-axis* shows calibrated oxygen values in μmol O_2 l^{-1}. This profile shows the difference in net oxygen flux determined from the true diffusive boundary layer (DBL)and from the effective DBL

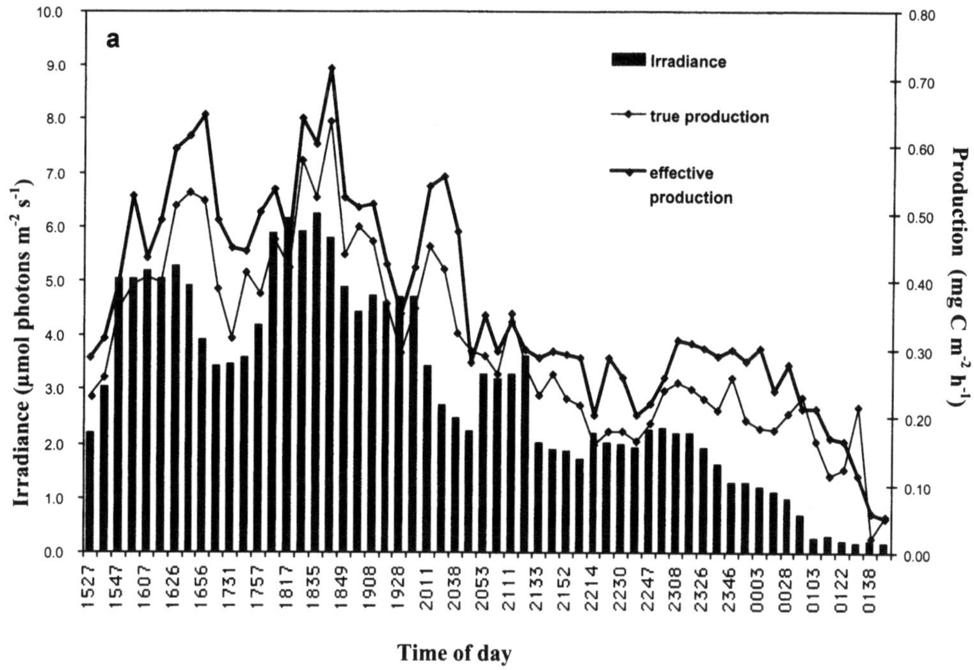

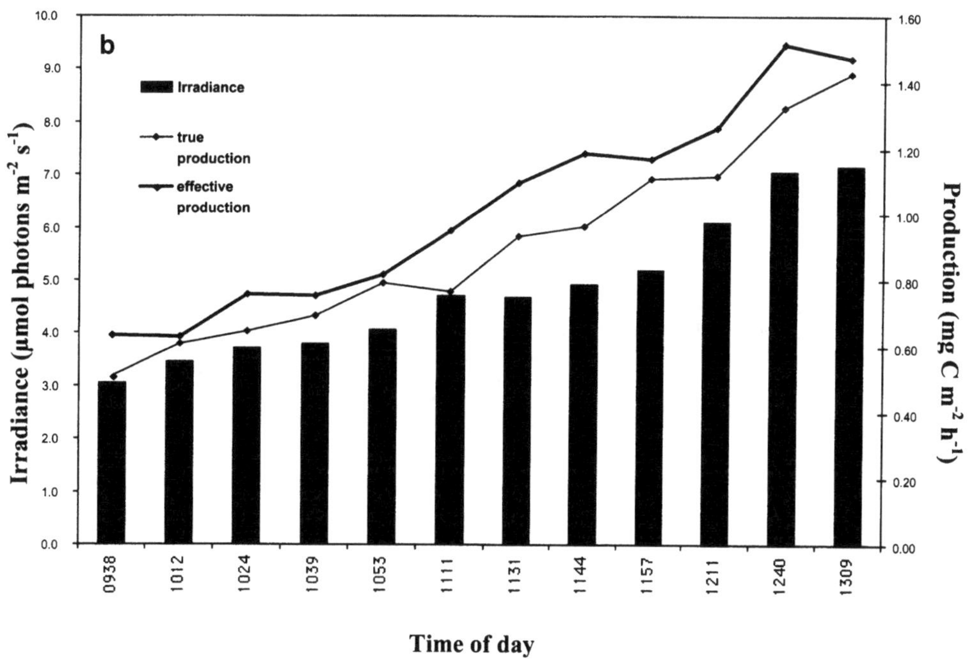

Fig. 4a, b Change in production (mg C m^{-2} h^{-1}) with irradiance (µmol photons m^{-2} s^{-1}) and time of day (hour, minute). Time of day is on the *x-axis*; production (*thin line*) and effective production (*solid line*) on the *right y-axis* and irradiance (*solid bars*) with the scale on the *left y-axis*. **a** 11 November. **b** 12 November

extent of ice-based production is still questionable. Inherent spatial and temporal variability, as well as methodological inconsistency, has prevented more reliable estimates or comparison between estimates. Here we present additional production data from the bottom interstitial algal community in McMurdo Sound using a method that avoids many of the errors caused by handling. Production by sea-ice communities in McMurdo Sound has been reported previously from the surface (Robinson et al. 1997), the platelet community (Arrigo et al. 1995) and the bottom (Palmisano and Sullivan 1983; Palmisano et al. 1985a, b). Photosynthetic parameters previously reported from McMurdo Sound bottom interstitial algal communities are summarised in Table 2. Carbon-based production and assimilation numbers from fast-ice algal communities in McMurdo Sound in 1997 (Table 2, McMinn et al. 2000) are much higher than those from the same site and time of year in 1996 (Table 2, McMinn et al. 1999a) and also 1999 (herein). Under-ice PAR irradiances during mid-November and early December 1997 were between 3 and 55 µmol photons m^{-2} s^{-1} (McMinn et al. 2000), whereas irradiances presented here for early to mid-November 1999 were between only 0.2 and 7 µmol photons m^{-2} s^{-1}. Midday under-ice irradiances in 1996 were between 0.2 and 5.0 µmol photons m^{-2} s^{-1}, which are within a similar

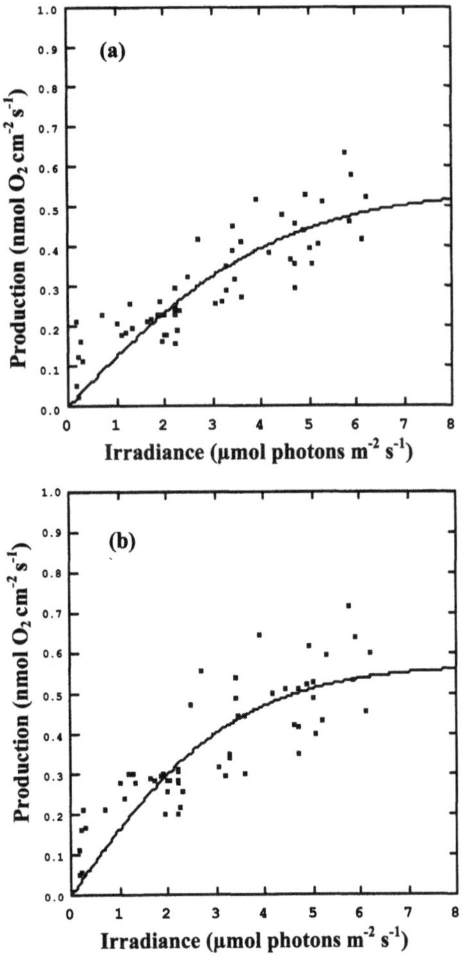

Fig. 5a, b Productivity (nmol O_2 cm^{-2} s^{-1}) versus irradiance (μmol photons m^{-2} s^{-1}) curve for 11 November. Data have been fitted to a hyperbolic tangent function to obtain P_{max} (maximum photosynthetic rate), E_k (irradiance of the onset of saturation) and α (photosynthetic efficiency) (Jassby and Platt 1976). **a** Production data determined for true DBL. **b** Effective production

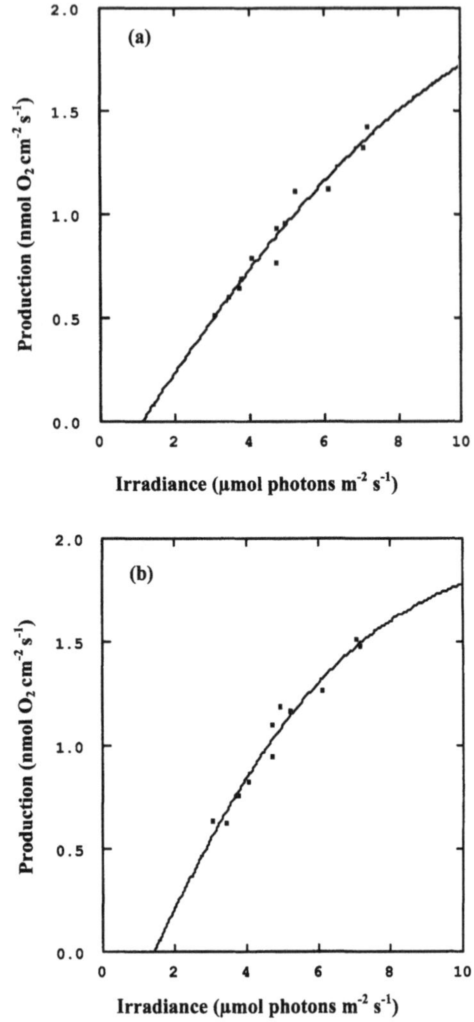

Fig. 6a, b Productivity (nmol O_2 cm^{-2} s^{-1}) versus irradiance (μmol photons m^{-2} s^{-1}) curve for 12 November. Data have been fitted to a hyperbolic tangent function (Jassby and Platt 1976). **a** Production data determined for true DBL. **b** Effective production

range to those reported here. The differences in productivity and assimilation numbers can be attributed to both differences in under-ice light and algal mat biomass, as well as ice thickness. The very high mat biomass in 1999, up to 378 mg chl*a* m^{-2}, was responsible for significant self shading, which would have greatly reduced the photosynthetic output per unit chlorophyll. McMinn et al. (2000) recorded ice at 90 cm thick in 1997, whereas the ice thickness recorded here (1999) was up to 2 m, and in 1996 was 1.7 m thick (McMinn et al. 1999a). From the *P* versus *E* data presented by McMinn et al. (2000), production in 1997 was approximately 0.01 nmol O_2 cm^{-2} s^{-1} at an irradiance of 7 μmol photons m^{-2} s^{-1}. This is considerably higher (0.0012 nmol O_2 cm^{-2} s^{-1}) than the production recorded here at a similar irradiance.

McMinn et al. (2000) suggested that sea-ice productivity was also affected by DBL thickness and, hence, under-ice currents. Jørgensen and Des Marais (1990) reported DBLs 0.16–0.6 mm thick over a cyanobacterial mat, measured at different velocities of water flow over the mat. They found that an increase in flow velocity caused a thinner DBL, steeper gradients, higher diffusion through the DBL, a higher oxygen concentration at the mat surface and a deeper penetration of oxygen into the mat. Here we recorded diffusive boundary layers of 0.49–5.3 mm thickness over a range of irradiances. Under-ice currents at McMurdo Sound were low and beneath the detection limit of 0.02 m s^{-1}. This would partially account for the unusually thick DBL. McMinn et al. (2000) suggested that sea-ice algal production could be more than doubled if there were strong currents present. The low under-ice current velocities at Cape Evans in 1999, which caused the considerably thicker DBLs, are likely to have contributed to the lower productivity in that year. It is often difficult to measure the precise thickness of the DBL. Jørgensen and Des Marais (1990) adopted a thickness measure where the concen-

Table 1 Photosynthetic parameters calculated from exponential, rectangular, quadratic and hyperbolic tangent functions for bottom-ice algae, McMurdo Sound, with P_{max} in nmol O_2 cm^{-2} s^{-1}, α in nmol O_2 cm^{-2} s^{-1} (μmol photons m^{-2} s^{-1})$^{-1}$ and E_k in μmol photons m^{-2} s^{-1}. For 12 November, *r* is respiration (nmol O_2 cm^{-2} s^{-1})

DBL	Function	P_{max}	α	E_k	R^2	r
11 November 1999						
True	Hyperbolic tangent	0.544	0.125	4.352	0.952	
True	Exponential function	0.635	0.151	4.205	0.954	
True	Rectangular hyperbola	0.944	0.165	5.721	0.954	
True	Quadratic function	0.614	0.128	4.797	0.952	
Effective	Hyperbolic tangent	0.571	0.167	3.419	0.951	
12 November 1999						
True	Hyperbolic tangent	2.674	0.275	9.724	0.996	0.305
True	Exponential function	3.521	0,346	10.176	0.996	0.406
True	Rectangular hyperbola	5.675	0,373	15.2U	0.996	0.441
True	Quadratic function	3.143	0.279	11.265	0.996	0.315
Effective	Hyperbolic tangent	2.561	0.390	6.567	0.997	0.546

Table 2 Photosynthetic parameters of fast-ice bottom algae, McMurdo Sound, Antarctica [PAR^S midday surface irradiance, in μmol photons $m^{-2}s^{-1}$); PAR^{UI} under-ice irradiance, μmol photons $m^{-2}s^{-1}$; ice thickness in metres (*m*); P_{max} maximum photosynthetic rate, in mg C (mg chl*a*)$^{-1}$ h^{-1}; E_k the minimum irradiance required to saturate photosynthesis, in μmol photons m^{-2} s^{-1}; α the light-limited photosynthetic efficiency, in nmol O_2 cm^{-2} s^{-1} μmol^{-1} photons m^{-2} s^{-1}]

Reference	PAR^S	PAR^{UI}	Ice thickness (m)	P_{max}	α	E_k
Palmisano et al. (1985a)	1410–1543	0.5–2.9	1.5	0.062; 0.079–0.13	0.019; 0.02–0.026	3.5; 4.2–5.4
Palmisano et al. (1985b)		0.5–7.4	2.3	0.35		
Cota and Sullivan (1990)	1500–1700	17.7	1.75	0.22 ± 0.10	0.024 ± 0.019	10.3 ± 4.4
McMinn et al. (1999b)	2050	0.2–5.0	1.75	0.05	0.0019	25
McMinn et al. (2000)	1600	3–55	0.9	0.29–2.01	0.003	14
Present study	1610	8	1.9–2.0	0.01	0.167–0.275	3.149–9.724

tration had dropped below its free-flow concentration to 90% of its total gradient within the DBL. The DBL thickness defined in this way was found to deviate < 10% from its precise thickness determined from whole gradients. A similar method is applied here but as there is a strong correlation ($r^2 = 0.928$) between estimates of production using both methods of DBL measurement, we see no reason to persist with the 90% change as suggested by Jørgensen and Des Marais (1990).

A high degree of low-light adaptation has been reported for algal assemblages that develop under the low photon flux densities beneath the annual fast ice in McMurdo Sound (Palmisano et al. 1985b; Cota and Sullivan 1990). These assemblages are characterised by nutrient-replete and shade-adapted cells (Lizotte and Sullivan 1992; Kirst and Wiencke 1995). Photosynthetic characteristics typical for low-light adapted algae are saturation at low irradiance (i.e. low E_k), high α values and photoinhibition of photosynthesis at low irradiances (Lizotte and Sullivan 1991; Falkowski and Raven 1997). Cota (1985) proposed that sea-ice algae in the high Arctic are obligately shade adapted; that is, the microalgae could not adapt over time to photosynthesise at increased irradiance upon exposure. Obligately shade-adapted microalgae would reach their maximum photosynthetic rate at low irradiances, and exposure to higher irradiance would result in a decrease in the maximum photosynthetic rate, i.e. photoinhibition. Barlow et al. (1988), however, concluded that the Hudson Bay ice-algae assemblage was not obligately shade adapted because E_k was consistently higher than the in situ irradiance and because ice algae could adapt gradually to higher in situ irradiances by increasing E_k and P_{max}. Our algal communities were never light saturated (i.e. they were always light limited), as they had E_k values of 4–9 μmol photons m^{-2} s^{-1}, which were similar to the highest irradiance that the algal mats were exposed to. McMinn et al. (2000) measured an E_k of 14 μmol photons m^{-2} s^{-1} at a maximum irradiance of 55 μmol photons m^{-2} s^{-1} and suggested that the McMurdo Sound algal community was able to adapt to higher light levels.

Furthermore, there is some evidence to suggest that there is a community succession response, with a change in species composition towards more shade-adapted species in lower light. The bottom interstitial algal communities studied here were dominated by *Nitzschia stellata* and *Entomoneis kjellmanii*, which are species characteristic of low-light environments (McMinn 1998). In higher light environments, it is expected there would be a shift toward communities dominated by *Berkeleya adeliensis* (McMinn 1998).

The *P* versus *E* data were fitted to the hyperbolic tangent, rectangular hyperbola, quadratic and exponential functions (Jassby and Platt 1976). Generally, there was little significant difference in the fit of the data to the functions. On 11 November, the rectangular hyperbola gave the best fit, while on 12 November, the fits were identical. It is also interesting to note that the different methods of determining DBL thickness did not significantly influence the fit of the data. However, there was considerable scatter in the data and to unambiguously determine the best function to fit the data, a larger data set with less scatter is required.

To obtain measures of photosynthesis, such as the assimilation number, it is necessary to measure the photosynthesising biomass. With our microelectrode method, we have had to take a large sample (i.e an ice core) and assume that the biomass distribution is uniform. Swadling et al. (1997) showed high variances in fast-ice biomass distribution on several scales above 1 m, but microscale variations have not previously been assessed. Biomass here was estimated from a 7×7×10 cm section cut from the bottom 10 cm of an ice core, whereas the electrodes measured O_2 flux on a 40 μm scale. Our microelectrode equipment provides point measurements of oxygen flux at the base of the ice column and it is difficult to normalise chlorophyll on such a small scale. This contributes to the scatter of points about the *P* versus *E* curves (because each point cannot be individually normalised) and detracts from the precision in determining photosynthetic parameters. Two of the limitations of the oxygen microelectrode method of measuring photosynthesis are the small area over which the measurements are taken (i.e electrode tips are often < 10 μm) and the large spatial heterogeneity of the sea-ice algal mat. In future, many spatial replicates should be taken to characterise this variance. Furthermore, it is probable that not all the biomass is contributing to the net oxygen flux but we have not yet been able to determine the non-contributing proportion. This is why we mostly present our data as production, i.e. unnormalised, rather than as assimilation numbers.

Oxygen electrodes and diffusive boundary layer techniques provide an alternative, non-invasive method of determining sea-ice algal productivity. Our results demonstrate the usefulness of the method in typical McMurdo Sound fast-ice conditions. They reveal production values and assimilation numbers at the bottom end of previous estimates for an algal community that is constantly light limited at low (i.e. < 10 μmol photons m^{-2} s^{-1}) ambient irradiance. Our low productivity values probably indicate overestimation by other methods, but it is also possible that the assimilation numbers may be low as they incorporate algae (in the chl*a* estimate) that are not contributing to production.

Acknowledgements We would like to thank the Australian Research Council for providing financial support for Andrew McMinn, and the FORST Foundation (CO 8817) for funding Ken Ryan. We thank Antarctic New Zealand for providing logistical support (event K136), and Dave Clement, Kevin Mitchell and Scott Thomsen for assistance in the field. We would also like to thank our anonymous reviewers for their constructive criticisms.

References

Aalderink RH, Jovin R (1997) Estimation of the photosynthesis/irradiance (P/I) curve parameters from light and dark bottle experiments. J Plankton Res 19:1713–1742

Arrigo KR, Dieckmann G, Gosselin M, Robinson DH, Fritsen CH, Sullivan CW (1995) High resolution study of the platelet ice ecosystem in McMurdo Sound: biomass, nutrient and production profiles within a dense microalgal bloom. Mar Ecol Prog Ser 127:255–268

Barlow RG, Gosselin M, Legendre L, Therriault J-C, Demers S, Mantoura RFC, Llewellyn CA (1988) Photoadaptive strategies in sea-ice microalgae. Mar Ecol Prog Ser 45:145–152

Broecker WS, Peng TH (1974) Gas exchange rates between air and sea. Tellus 26:21–35

Bunt JS, Lee CC (1970) Seasonal primary production in Antarctic sea ice at McMurdo Sound in 1967. J Mar Res 28: 304–320

Cota GF (1985) Photoadaptation of high Arctic ice algae. Nature 315:219–222

Cota GF, Sullivan CW (1990) Photoadaptation, growth and production of bottom ice algae in the Antarctic. J Phycol 26:399–411

Falkowski PG, Raven JA (1997) Aquatic photosynthesis. Blackwell, Mass

Garrison DL, Sullivan CW, Ackley SF (1986) Sea ice microbial communities in Antarctica. BioScience 36:243–250

Glud RN, Gundersen JK, Revsbech NP, Jørgensen BB (1994) Effects on the benthic diffusive boundary layer imposed by microelectrodes. Limnol Oceanogr 39:462–467

Glud RN, Kühl M, Kohls O, Ramsing NB (1999) Heterogeneity of oxygen production and consumption in a photosynthetic microbial mat as studied by planar optodes. J Phycol 35: 270–279

Grossi SM, Kottmeier ST, Sullivan CW (1984) Sea ice microbial communities. III. Seasonal abundance of microalgae and associated bacteria, McMurdo Sound, Antarctica. Microb Ecol 10:231–242

Gundersen JK, Jørgensen BB (1990) Microstructure of diffusive boundary layers and the oxygen uptake of the sea floor. Nature 345:604–607

Horner RA (1985) Ecology of sea ice microalgae. In: Horner RA (ed) Sea ice biota. CRC Press, Boca Raton, pp 83–103

Jassby AD, Platt T (1976) Mathematical formulation of the relationship between photosynthesis and light for phytoplankton. Limnol Oceanogr 21:540–547

Jørgensen BB, Des Marais DJ (1990) The diffusive boundary layer of sediments: oxygen microgradients over a microbial mat. Limnol Oceanogr 35:1343–1355

Jørgensen BB, Revsbech NP (1985) Diffusive boundary layers and the oxygen uptake of sediments and detritus. Limnol Oceanogr 30:111–122

Jørgensen BB, Revsbech NP, Cohen Y (1983) Photosynthesis and structure of benthic microbial mats: microelectrode and SEM studies of four cyanobacterial communities. Limnol Oceanogr 28:1075–1093

Kirst GO, Wiencke C (1995) Ecophysiology of polar algae. J Phy col 31:181–199

Klimant I, Kühl M, Glud RN, Holst G (1997) Optical measurement of oxygen and temperature in microscale: strategies and biological applications. Sensors Actuators B 38–39:29–37

Kühl M, Lassen C, Jørgensen BB (1994) Light penetration and light intensity in sandy marine sediments measured with irradiance and scalar irradiance fibre-optic microprobes. Mar Ecol Prog Ser 105:139–148

Legendre L, Ackley SF, Dieckmann GS, Gulliksen B, Horner R, Hoshiai T, Melnikov IA, Reeburgh WS, Spindler M, Sullivan CW (1992) Ecology of sea ice biota. 2. Global significance. Polar Biol 12:429–444

Lizotte MP, Sullivan CW (1991) Rates of photoadaptation in sea ice diatoms from McMurdo Sound, Antarctica. J Phycol 27:367–373

Lizotte MP, Sullivan CW (1992) Photosynthetic capacity in microalgae associated with Antarctic pack ice. Polar Biol 12:497–502

Lorenzen J, Glud RN, Revsbech NP (1995) Impact of microsensor-caused changes in diffusive boundary layer thickness on O2 profiles and photosynthetic rates in benthic communities of microorganisms. Mar Ecol Prog Ser 119:237–241

McMinn A (1998) Species succession in fast ice algal communities; a response to UV-B radiation? Korean J Polar Res 8:47–52

McMinn A, Ashworth C (1998) The use of oxygen microelectrodes to determine the net production by an Antarctic sea ice algal community. Antarct Sci 10:39–44
McMinn A, Ashworth C, Ryan KG (1999a) Growth and productivity of Antarctic sea ice algae under PAR and UV irradiances. Bot Mar 42:401–407
McMinn A, Skerratt J, Trull T, Ashworth C, Lizotte M (1999b) Nutrient stress gradient in the bottom 5 cm of fast ice, McMurdo Sound, Antarctica. Polar Biol 21:220–227
McMinn A, Ashworth C, Ryan KG (2000) In situ net primary productivity of an Antarctic fast ice bottom algal community. Aquat Microb Ecol 21:177–185
Palmisano AC, Garrison DL (1993) Microorganisms in Antarctic sea ice. In: Friedmann EI (ed) Antarctic microbiology. Wiley, New York, pp 167–218
Palmisano AC, Sullivan CW (1983) Sea ice microbial communities (SIMCO). 1. Distribution, abundance and primary production of ice microalgae in McMurdo Sound, Antarctica in 1980. Polar Biol 2:171–177
Palmisano AC, Kottmeier ST, Moe RL, Sullivan CW (1985a) Sea ice microbial communities. IV. The effect of light perturbation on microalgae at the ice-seawater interface in McMurdo Sound, Antarctica. Mar Ecol Prog Ser 21:37–45
Palmisano AC, SooHoo JB, Sullivan CW (1985b) Photosynthesis-irradiance relationships in sea ice microalgae from McMurdo Sound, Antarctica. J Phycol 21:341–346
Revsbech NP (1994) Analysis of microbial mats by use of electrochemical microsensors: recent advances. In: Stal LJ, Caumette P (eds) Microbial mats: structure, development and environmental significance. Institute NAS NATO ASI Series G: Ecological sciences. Springer, Berlin Heidelberg New York, pp 135–147
Revsbech NP, Jørgensen BB (1986) Microelectrodes: their use in microbial ecology. In: Marshall KC (ed) Advances in microbial ecology. Plenum, New York, pp 293–352
Revsbech NP, Ward DM (1983) Oxygen microelectrode that is insensitive to medium chemical composition: use in an acid microbial mat dominated by *Cyanidium caldarium*. Appl Environ Microbiol 45:755–759
Revsbech NP, Jørgensen BB, Blackburn TH, Cohen Y (1983) Microelectrode studies of the photosynthesis and O_2, H_2S and pH profiles of a microbial mat. Limnol Oceanogr 28:1062–1074
Robinson DH, Kolber Z, Sullivan CW (1997) Photophysiology and photoacclimation in surface sea ice algae from McMurdo Sound Antarctica. Mar Ecol Prog Ser 147:243–256
Ryan KG, McMinn A, Trenerry LJ, Clement D (in press) In situ effects of elevated UV radiation on Antarctic fast ice algal communities. Antarct Sci
Satoh H, Watanabe K (1988) Primary productivity in the fast ice area near Syowa Station, Antarctica, during spring and summer 1983/84. J Oceanogr Soc Jpn 44:287–292
Strickland JDH, Parsons TR (1972) A practical handbook of seawater analysis, 2nd edn. Bull Fish Res Board Can 167:1–311
Sullivan CW, Palmisano AC, Kottmeier S, Grossi SM, Moe R, Taylor GT (1983) The influence of light on development and growth of sea ice microbial communities in McMurdo Sound. Antarct J US 18:177–179
Swadling KM, Gibson JAE, Ritz DA, Nichols, PD (1997) Horizontal patchiness in sympagic organisms of the Antarctic fast ice. Antarct Sci 9:399–406

Polar Biol (2001) 24: 918–925
DOI 10.1007/s003000100300

Hee Joong Lee · Sandra Vanhove · Lloyd S. Peck
Magda Vincx

Recolonisation of meiofauna after catastrophic iceberg scouring in shallow Antarctic sediments

Abstract A series of 10 samples from sediment in and adjacent to a shallow coastal iceberg scour at Signy Island, Antarctica, were taken by hand coring from 17 December 1993 until 23 August 1994. Scouring by the iceberg led to more than a 95% decrease in meiofaunal abundance and to a certain degree of reduction in diversity. Nematodes were always the most dominant group of meiofauna. The return of major meiofauna groups to control levels was accomplished in 30 days, although a decrease in abundance on the 50th day made interpretation difficult. The pioneering meiofaunal colonisers were copepods and ostracods, followed by nematodes. *Microlaimus* sp.1 was dominant among the nematodes throughout the whole period. Epistratum feeders and non-selective deposit feeders were highly dominant over selective deposit feeders and predators/omnivores. The Maturity Index, a measure for stress within nematode communities, was relatively low at all times and in controls, which indicates that *r*-strategists prevail in this community. In spite of the catastrophic destruction, nematode community structure was not affected by the iceberg impact, and there was no evidence of succession during recovery. This suggests that the nematode community in the shallow subtidal coast at Signy Island is well adapted to ice disturbance.

Please note: When citing this article, please use the reference to the original source in the journal Polar Biology as given on the top of this page.

H.J. Lee (✉) · S. Vanhove · M. Vincx
Marine Biology Section, Department of Biology,
University of Ghent, K.L. Ledeganckstraat 35,
9000 Ghent, Belgium
E-mail: Lee.HeeJoong@rug.ac.be
Fax: +32-9-2645344

L.S. Peck
Natural Environment Research Council,
British Antarctic Survey, High Cross,
Madingley Road, Cambridge, CB3 0ET, UK

Introduction

Benthic organisms in shallow Antarctic coasts are exposed to ice scouring. This includes potential scouring by floes in summer and also anchor-ice disturbance in winter (Picken 1984; Dayton 1990), together with scouring by icebergs. Frequency of iceberg scouring at Signy Island was estimated at once every 50–75 years (Peck and Bullough 1993). Global warming may increase the breaking up of ice shelves in Antarctica (Doake and Vaughan 1991; Gammie 1995), which would result in a more frequent abrasion of the Antarctic sea-floor habitats. Few studies have focused on the structure of shallow-water Antarctic benthic communities under ice disturbances (Dayton et al. 1969, 1970 in Cape Armitage and Hut Bay at McMurdo; Lenihan and Oliver 1995 at Signy Island; Peck et al. 1999 at Signy Island).

All of these above studies were concerned primarily with macrobenthos and none of them treated meiofaunas in detail. Meiofaunas are, however, useful in assessing environmental disturbance. They have short life-cycles, which make their response to environmental change rapid. Due to the lack of planktonic stages in their life-cycle, the population changes depend more on reproduction than recruitment (Heip 1980; Platt and Warwick 1980; Vincx and Heip 1991). Reflecting this, a number of experimental studies have focused on the effects of various disturbances (see Schratzberger and Warwick 1999). The general conclusions of those studies are that the resilience of meiofauna is strong; nevertheless, the response depends on the intrinsic resilience of the particular community and the intensity of disturbance.

Antarctic benthic organisms have been exposed to ice-disturbance stress since the major continental glaciation, which probably started from early Oligocene (Arntz et al. 1994). Its long geological history suggests that ice disturbance is possibly a major structuring element in the adaptation and evolution of benthic communities around Antarctic coasts. The evaluation of responses of meiofaunal communities to iceberg scouring should, therefore, give an insight into community adaptation and evolution and, hence, their resilience. An enhanced understanding of the ecology of Antarctic meiofauna should also help the prediction of community reaction to potential future environmental changes.

The aims of this study were: (1) to assess the intensity of scouring by an iceberg and its short-term effect on a meiofauna community; (2) to investigate the recolonisation process or succession of such a community; (3) to estimate resilience, and (4) to study the long-term influence of iceberg scouring on a meiofauna community, with special emphasis on nematodes, in a shallow coast in Antarctica.

Materials and methods

Sampling area

In mid-December 1993, an iceberg was observed grounded in Factory Cove near the British Antarctic Survey base at Signy Island, Antarctica (Fig. 1). The site affected was at about 8–9 m depth on a soft-sediment bottom graded into solid rock at the landward side. A few days later, on 17 December 1993, the iceberg drifted off leaving a crater of approximately 10 m wide and 50 cm to 1 m deep, with furrows showing the direction of its arrival and departure. The sediment in the scour was relatively well-sorted fine sand (median size: about 150 μm).

Sampling and preparation

Duplicate samples from the sediment in and adjacent to the iceberg scour were taken with a hand corer (ø 5.5 cm) by scuba divers between 17 December 1993 and 23 August 1994. In total, 10 samples were taken from the inside of the scour mark: on 0 (hereafter D0), 10 (D10), 20 (D20), 30 (D30), 50 (D50), 80 (D80), 110 (D110), 150 (D150), 200 (D200) and 250 (D250) days after impact. Two samples were taken from adjacent sediments as controls: 1 on day 0 (hereafter C0) and the other on day 170 (C170). The sediment in the core was preserved in 4% warm neutral formaldehyde solution. Sediments were decanted, and sieved using 2 different mesh sizes (1,000 μm and 38 μm) before centrifugation. The animals passed through the 1,000-μm sieve and retained on the 38-μm sieve were regarded as meiofauna. Final extraction of meiofauna retained on the 38-μm sieve was achieved using the LUDOX centrifugation flotation technique (McIntyre and Warwick 1984). The metazoan meiofauna was counted after staining with Rose Bengal. Approximately 200 nematodes (all, when there were less than 200 individuals in a sample) per replicate were randomly picked out and dehydrated in a series of glycerin-alcohol solutions. The dehydrated nematodes were mounted on slides with anhydrous glycerin medium and sealed with paraffin wax. Observations were carried out under a Wild M20 light microscope. Morphospecies were determined based on morphological characters. For each specimen, its developmental stage and sex were categorised into four different groups: juvenile, non-gravid female, gravid female and male. Nematode genera were classified into four feeding groups: selective deposit feeders (1A), non-selective deposit feeders (1B), epistratum feeders (2A) and predators/ omnivores (2B).

Statistical analysis

Multivariate methods

The species abundance matrix was analysed by using TWINSPAN (Two-Way Indicator Species Analysis; Hill 1979) from pooled and non-pooled nematode abundance data after 4th root transformation. Classification-clustering based on the Bray-Curtis similarity index was carried out using the same data.

Univariate methods

Nematode species diversity was measured using Hill's diversity indices: N_0 (the number of species), N_1 (the exponential of Shan-

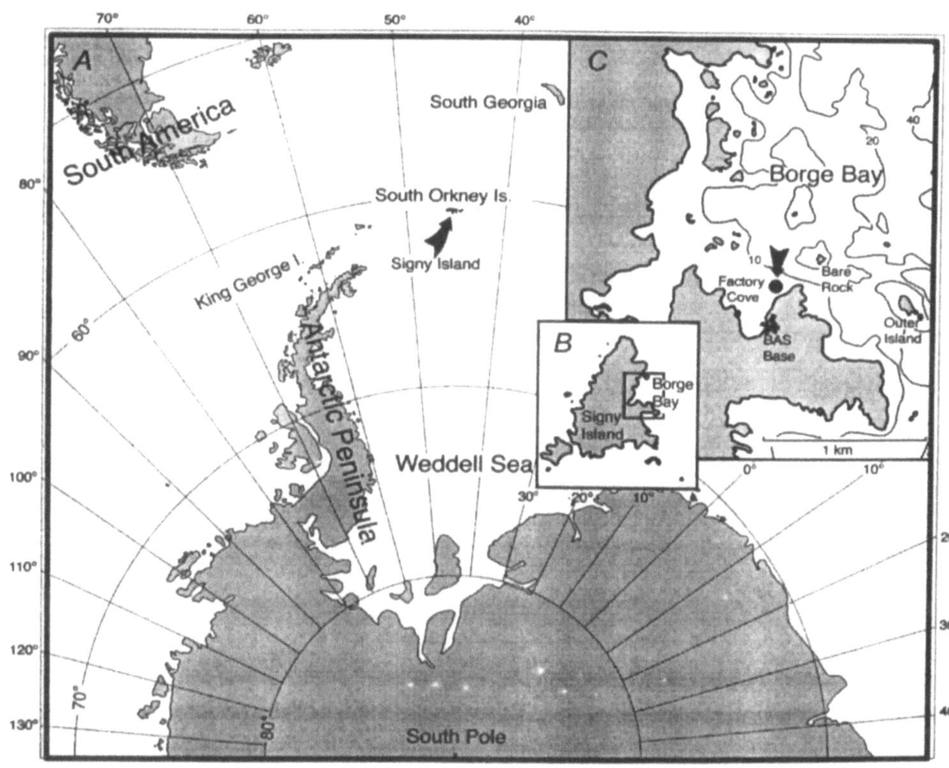

Fig. 1 Maps showing the location of Signy Island in the Antarctic (**A**), Borge Bay in Signy Island (**B**) and Factory Cove and British Antarctic Survey (BAS) station south of Borge Bay and the location of sampling station (*filled circle*) with details of surrounding (**C**)

non's index), N_2 (the reciprocal of Simpson's index) and N_∞ (the reciprocal of the relative abundance of the most dominant species) (Hill 1973). The maturity index (MI: Bongers 1990; Bongers et al. 1991) was used to characterise the life-style of nematode communities. Although the MI is generally used for terrestrial nematode communities, it has been proposed as a useful measure for the study of disturbance effects on marine communities (Bongers et al. 1991).

Results

Meiofaunal abundance

In total, 14 meiofaunal groups (for pragmatic reasons all nauplii were considered as a single group) were recovered and their abundance varied throughout the study (Table 1). The least number of meiofaunal groups was found on D0 and D10, whilst the maximum was found on D50, D110, D150, D250 and C0. The iceberg scouring removed about 93.5% of total metazoan density. In general, the early samples contained lower numbers of meiofaunal groups. Nematodes were the most abundant group (overall mean 1,180 ind./10 cm^2). An exception was on D50 when nematode abundance was low and numbers of nauplii exceeded that of nematodes. Meiofaunal crustaceans (copepods, ostracods and nauplii) were the second most abundant group (overall mean 92, 31 and 202 ind./10 cm^2, respectively).

The earliest recolonisation of meiofauna was achieved by copepods, which recovered to control levels on D20. On D30, all meiofaunal major taxa exceeded control levels, except for ostracods whose initial number at the control site was extraordinarily high. Overall meiofauna abundance decreased after D30, mainly due to the abrupt fall in nematode numbers, which only reached stable levels in 80 days.

Changes of nematode community

Abundance

The most obvious change caused by iceberg scouring was an abrupt decrease of nematode abundance (Table 1). The average density of nematodes at the scour mark on the initial day (D0) was about 54.0 ± 23.9 ind./10 cm^2, which was only 4.5% of that of the control site (C0, 1,208.6 ± 555.8 ind./10 cm^2). The number stayed low until D20 (29.3 ± 6.3 ind./10 cm^2). A conspicuous increase was found by D30 with a density of 2,351.7 ± 93.4 ind./10 cm^2, which is almost double the density of the control site on the initial day. This, however, decreased on D50 to 183.1 ± 130.1 ind./10 cm^2. A larger fluctuation was found on D150 when the density reached the highest value (4,125.3 ± 1,319.3 ind./10 cm^2) during the study. Except for this instance, it seemed that the nematode community was fairly stable from D80 on, with overall density between 587.6 ± 39.7 ind./10 cm^2 and 1,215.8 ± 36.7 ind./10 cm^2.

Table 1 Changes of meiofaunal mean abundance of two replicates (ind./10 cm^2) (*SD* standard deviation, – not applicable)

Days	D0		D10		D20		D30		D50		D80		D110		D150		D200		D250		C0		C170	
Date	17/12/93		28/12/93		06/01/94		17/01/94		05/02/94		08/03/94		09/04/94		16/05/94		05/07/94		23/08/94		17/12/93		05/06/94	
	Mean	SD	Mean	SD	Mean	SD	Mean	SD	Mean	SD	Mean	SD	Mean	SD	Mean	SD	Mean	SD	Mean	SD	Mean	SD	Mean	SD
Nematoda	54.0	23.9	19.0	6.6	29.3	6.3	2351.7	93.4	183.1	130.1	1001.5	867.9	975.5	439.8	4125.3	1319.3	587.6	39.7	1215.8	36.7	1208.6	555.8	2415.2	299.0
Copepoda	10.8	13.4	2.7	2.7	38.6	42.7	70.3	77.3	72.6	36.4	472.2	23.3	10.1	1.8	76.4	55.5	100.6	81.5	156.1	10.7	38.0	17.9	51.9	22.7
Ostracoda	8.2	2.7	11.0	1.8	3.8	4.8	89.9	29.8	20.5	2.1	43.9	43.6	28.1	2.1	13.3	0.3	11.2	0.9	11.4	4.2	108.4	77.0	20.9	1.5
Nauplii	12.7	13.7	0.2	0.3	1.9	2.1	85.0	34.9	435.4	514.4	400.8	65.0	0.6	0.9	121.9	20.9	201.1	34.9	1055.7	58.5	50.0	24.8	63.5	42.1
Bivalvia	0	–	1.3	0	1.7	0	22.4	11.9	3.4	0	3.2	1.5	3.8	5.4	0	–	0	–	0.4	0.6	13.9	2.4	0	–
Amphipoda	1.3	0.6	0	–	3.6	3.3	1.5	2.1	1.3	1.2	1.3	0.6	0	–	0.2	0.3	1.3	1.8	3.6	3.9	8.9	4.8	2.7	0.9
Polychaeta	0.2	0.3	0	–	0.2	0.3	0.6	0.9	0.2	0.3	0	–	0	–	0.6	0.3	0.2	0.3	8.4	11.9	0.4	0.6	0	–
Oligochaeta	0	–	0	–	0	–	0	–	0	–	0	–	0	–	1.3	0.6	0	–	0.6	0.3	0.2	0.3	0	–
Rotifera	0	–	0	–	0	–	0	–	0	–	0	–	0.2	0.3	0.2	0.3	0	–	0	–	3.0	1.2	0.6	0.9
Halacarida	0	–	0.2	0.3	1.1	1.5	0	–	0.2	0.3	0.4	0.6	0.4	0.6	0.2	0.3	0.2	0.3	0	–	0	–	0	–
Gastropoda	0	–	0	–	0	–	0	–	0.2	0.3	0.4	0.6	0.6	0.3	0	–	0	–	0	–	0.2	0.3	0	–
Tanaidacea	0	–	0	–	0	–	0	–	0	–	0	–	0.8	1.2	0	–	0	–	0	–	0	–	0	–
Kinorhyncha	0	–	0	–	0	–	0	–	0.4	0.6	0	–	0.2	0.3	0.2	0.3	0.6	0.9	0.4	0.6	0	–	0.4	0.6
Priapulida	0	–	0	–	0	–	0.6	0.9	0	–	0	–	0	–	0	–	0.2	0.3	12.2	0.6	0	–	0.8	1.2
Total	87.1	48.6	34.4	5.7	80.2	40.6	2621.9	90.1	717.3	680.3	1923.6	955.3	1020.5	442.5	4339.7	1242.4	903.0	5.4	2464.8	124.4	1431.6	674.3	2556.1	365.2

Diversity

Hill's diversity indices calculated at nematode species level decreased initially after the impact (Fig. 2). In general, the lowest values occurred 30 days after the scouring, and recovery to initial levels occurred on day 50. Among Hill's indices, the average species number (N_0) showed a slightly different trend from N_1, N_2 and N_∞ which showed remarkably similar patterns. All indices were initially similar to the control site on the initial day (C0). N_1, N_2 and N_∞ all showed, after the decline to minimum values on D30, a sharp rise to high maximum levels on D50, which then gradually declined again to low levels on D110. This second set of low values was followed by a plateau of moderate to high values until the end of the study.

The complex *K*-dominance curves from all samples (Fig. 3a) made it difficult to discriminate any particular day from the others. Since more dramatic changes were expected in earlier days, all curves after D50 were eliminated (Fig. 3b), and the result showed better resolution. There were two clearly different tendencies. The *K*-dominance curves of earlier days moved upwards from D0 to D30 (highest position). On D50, the original position of D0 was restored. The curve for C0 was closer to the curves for D0 and D50 than the curves for other earlier days but somewhat higher.

Taxonomic composition

In total, 65 nematode species of 49 genera in 19 families were found. Only 14 species made up the majority of the community (more than 93% on average, with each species occurring in at least half the samples). Dominance and abundance rankings were calculated for the ten most abundant species (Table 2). *Microlaimus* sp.1 was the most dominant through the whole period. *Sabatieria* sp. was the second most dominant species, followed by *Daptonema* sp.1, *Pomponema* sp. and *Odontophora* sp.1. This rank was fairly consistent through the whole study period, including controls. In total, four orders were

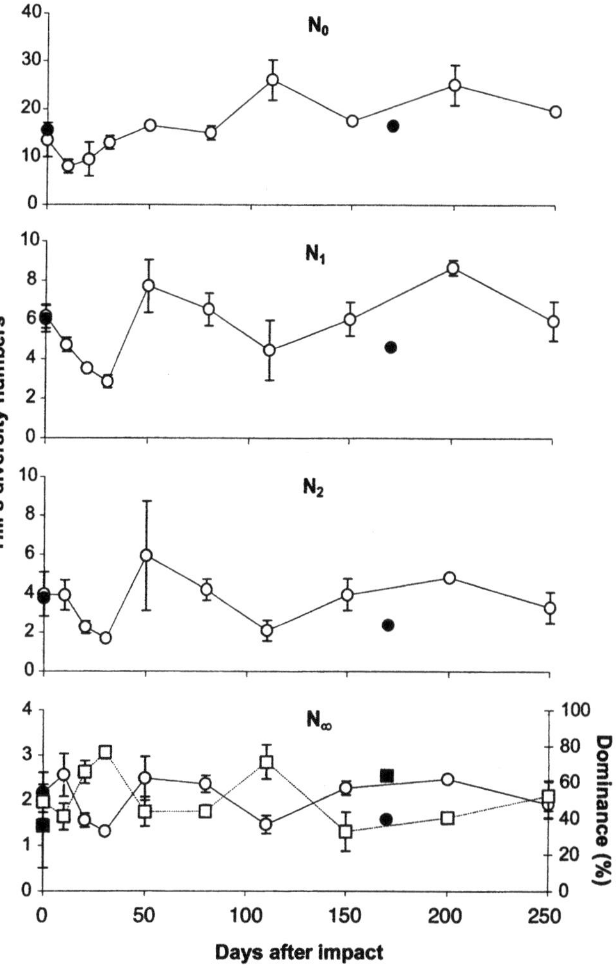

Fig. 2 Dynamics of Hill's diversity numbers of nematode communities (*error bar*: standard deviation; *unfilled circles*: diversity on corresponding days of scoured site; *filled circles*: diversity on corresponding days of control sites; *unfilled squares*: dominance of *Microlaimus* sp. 1 on corresponding days of scoured site; *filled squares*: dominance of *Microlaimus* sp. 1 on corresponding days of control site). The dominance (%) of *Microlaimus* sp. 1 is plotted together with N_∞ for comparison. High dominance of *Microlaimus* sp. 1 was responsible for the low diversity of day 30 and day 110

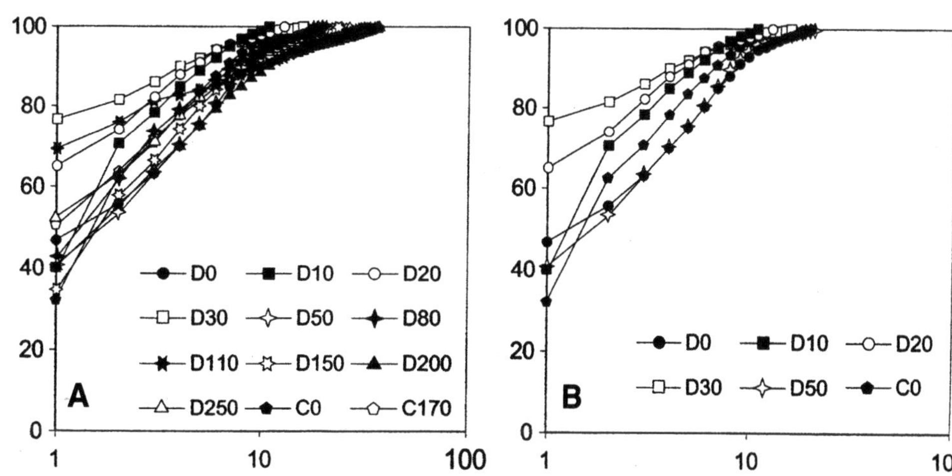

Fig. 3 *k*-dominance curves for nematode species from the mean of replicates of each sample (**A**) and from the data for early days and the corresponding control (C0) (**B**)

Table 2 Dominance rank and relative dominance of the ten most abundant nematode species through the whole period. *Numbers* on the *left side* indicate the dominance rank, and *asterisks* indicate relative dominance: * < 1%, ** 1–3%, *** 3–10%, **** 10–30%, ***** > 30%, – absent

Sample	D0	D10	D20	D30	D50	D80	D110	D150	D200	D250	C0	C170	Total
Date	17/12/93	28/12/93	06/01/94	17/01/94	05/02/94	08/03/94	09/04/94	16/05/94	05/07/94	23/08/94	17/12/93	05/06/94	
Microlaimus sp. 1	1 *****	1 *****	1 *****	1 *****	1 *****	1 *****	1 *****	1 *****	1 *****	1 *****	1 *****	1 *****	1 *****
Sabatieria sp.	3 ***	2 *****	4 ***	3 ***	2 ****	3 ****	3 ***	3 ***	2 ****	3 ***	2 ****	3 ***	2 ****
Daptonema sp. 1	2 ***	5 ***	3 ***	2 ***	3 ****	2 ****	14 *	10 **	4 ***	2 ****	7 **	6 ***	3 ***
Pomponema sp.	4 ***	4 ***	2 ***	4 ***	4 ***	4 ***	6 **	4 ***	5 ***	4 ***	5 ***	2 ***	4 ***
Odontophora sp. 1	9 **	6 **	7 **	6 **	5 ***	5 ***	2 ***	6 ***	3 ***	6 ***	6 ***	8 **	5 ***
Neochromadora sp.	5 ***	8 **	5 ***	13 *	7 ***	6 **	7 **	14 *	6 ***	9 *	3 ***	5 ***	6 ***
Eleutherolaimus sp.	6 ***	9 **	10 *	8 **	8 ***	10 **	11 *	7 ***	10 **	7 ***	4 ***	7 **	7 **
Microlaimus sp. 2	– –	– –	– –	– –	– –	– –	21 *	2 ****	20 *	17 *	12 *	17 *	8 **
Prochromadorella sp.	11 **	– –	6 ***	10 *	6 ***	9 **	5 **	8 **	8 **	8 **	10 **	4 ***	9 **
Odontophora sp. 2	– –	3 ***	8 *	9 *	10 **	7 **	10 *	5 ***	7 ***	10 *	9 **	13 **	10 **

present. The largest order, in species number, was Monhysterida, comprising 33 species, followed by Chromadorida with 27 species. Only three species of Enoplida were found and two species of Tylenchida were also present. In terms of individual abundance, however, Chromadorida accounted for the majority (72.4%) while Monhysterida represented 27.3%. Enoplida (0.3%) and Tylenchida (0.1%) were negligible in terms of individual abundance.

Age and sex composition

Juveniles dominated the age structure of the nematode community throughout the whole period (average 49.2%, Table 3). The relative proportion of juveniles dipped to minimum values on D20 and D30 (38.0% and 28.7%, respectively). The highest proportion (67.3%) of juveniles was observed on D200. On all other days, the values were between 45.0% and 54.9%. The fluctuation of non-gravid female numbers was less than other groups. They ranged from 8.2% to 22.2% and averaged 16.7% of the population. Gravid females were a smaller group (average 9.2%). Their proportion was high at first and became extremely low on D110, D200 and C170. A high proportion (37.8%) of males was observed on D30. On other days, however, the values varied between 20.6% and 30.1%, except on D10 (18.3%). The fecundity, expressed as the proportion of gravid females over total females, was relatively higher (> 40%) in the earlier days (D0–D20).

Trophic composition

The epistratum feeders (2A) dominated the nematode community at this site, accounting for 48.5–80.5% of the nematodes (Table 3). Together with the non-selective deposit feeders (1B), the two groups composed more than 90% of the community in most samples. The maximum percentage of these two groups was on D110, showing 97.5%. The smallest feeding group was the selective deposit feeders (1A), which were not detected in some samples. The maximum representation of this group was on D0 (3.5%). The proportion of the selective deposit feeders (1A) and the predators/omnivores (2B) fluctuated without a clear pattern during the period of investigation. The non-selective deposit feeders (1B) decreased conspicuously in the early days after scouring, especially on D20 and D30. In contrast, the epistratum feeders (2A) increased on those days. However, the maximum proportion of 1B was found on D10, while the minimum proportion of the 2A feeding group was found on the same day.

Maturity index (MI)

The mean MI (Table 3) of the nematode communities through the study period was 2.14 (total range = 2.02–

Table 3 Integrated table showing the composition of the nematode community as life-history stage (%), fecundity, nematode feeding type (%) and maturity index (MI, Bongers et al. 1991). [*Female–* non-gravid female; *Female+* gravid female; *Fecundity* relative proportion (%) of gravid females to total females; *SD* standard deviation ($n = 2$); – not applicable]

Samples	D0		D10		D20		D30		D50		D80	
Date	17/12/93		28/12/93		06/01/94		17/01/94		05/02/94		08/03/94	
	Mean	SD	Mean	SD	Mean	SD	Mean	SD	Mean	SD	Mean	SD
Generation												
Juvenile	49.1	7.9	50.0	11.4	38.0	11.0	28.7	3.9	49.6	4.4	53.4	5.2
Female –	17.9	5.4	18.3	6.1	15.5	0.4	22.2	0.8	18.4	3.8	15.3	2.1
Female +	12.4	0.9	13.5	0.8	19.1	6.5	11.3	2.3	7.4	2.6	9.0	5.3
Male	20.6	1.6	18.3	6.1	27.4	17.1	37.8	5.4	24.6	1.9	22.3	2.1
Fecundity	41.5	5.6	43.2	9.6	54.3	9.2	33.5	5.4	28.2	3.0	35.3	11.0
Feeding type												
1A	3.5	5.0	0	–	2.2	0.7	0	–	0.3	0.4	0	–
1B	30.2	1.9	45.6	3.9	18.3	2.0	14.6	1.9	36.9	0.9	43.3	0.6
2A	60.2	3.3	49.7	1.9	69.9	5.4	80.5	1.6	55.6	0.7	51.7	1.0
2B	6.1	3.6	4.7	2.0	9.6	8.0	4.9	0.3	7.3	0.2	5.0	0.4
MI	2.12	0.04	2.18	0.01	2.20	0.15	2.10	0.01	2.18	0.05	2.12	0.03

Samples	D110		D150		D200		D250		C0		C170	
Date	09/04/94		16/05/94		05/07/94		23/08/94		17/12/93		05/06/94	
	Mean	SD	Mean	SD	Mean	SD	Mean	SD	Mean	SD	Mean	SD
Generation												
Juvenile	52.2	5.8	52.8	9.9	67.3	1.4	45.0	1.8	49.3	3.6	54.9	3.3
Female –	18.0	2.7	15.2	0.3	8.2	1.5	13.5	3.4	13.9	4.5	16.1	2.4
Female +	1.7	0.4	7.9	5.4	3.1	0.7	11.4	1.5	9.8	0.3	3.8	1.8
Male	28.1	3.5	24.2	4.8	21.4	0.6	30.1	6.7	27.0	1.3	25.1	0.9
Fecundity	9.0	3.0	32.4	16.2	27.5	8.2	45.9	2.9	42.3	7.4	18.7	5.0
Feeding type												
1A	1.5	0	0	–	0.4	0.6	0.8	0.3	0	–	0.8	0.4
1B	21.8	7.2	29.7	14.5	23.4	5.6	36.3	6.6	44.2	36.3	22.6	11.5
2A	75.7	6.6	62.2	20.2	70.1	9.0	57.1	8.0	48.5	37.3	70.6	11.0
2B	1.0	0.7	8.1	5.8	6.1	2.8	5.8	1.0	7.4	1.0	6.0	0.9
MI	2.02	0.01	2.19	0.16	2.12	0.10	2.09	0.09	2.15	0.03	2.17	0

2.20). Unexpectedly, there was no significant change in MI between sampling periods.

The result of TWINSPAN (Fig. 4) suggested that the nematode populations in samples taken on D0, D10, D20 and D50 were different from the rest. The separation of these days was made based on the disappearance of *Monhystera* sp. 2. The rest of the samples were split into 4 different groups: in the second division D110 and D200 were grouped together due to the presence of *Desmolaimus* sp. 1; in the third division D150 was separated due to the presence of *Daptonema* sp. 3; in the fourth and final division D80 and C0 were differentiated, based on the presence of *Perepsilonema* sp. 1, from D30, D250 and C170. D30 was a borderline negative in the fourth division.

Discussion

Initial reduction

Disturbance generally results in a reduction in meiofauna (e. g. Schratzberger and Warwick 1999), although there are some exceptions. The iceberg scouring in the present study produced a dramatic decrease of meiofaunal abundance. The abrupt decrease of meiofaunal abundance in the initial days after scouring compared to the control site emphasises the severity of the effect of iceberg scouring.

Surprisingly, such catastrophic iceberg scouring did not directly affect nematode community structure. Although the diversity decreased in the early recovery phase, there were no remarkable changes in species composition, dominance rank, maturity index or trophic structure after the scouring. The decreased diversity indices were mainly because of the low species numbers and the dominance of *Microlaimus* sp.1, together with *Sabatieria* sp. in some cases.

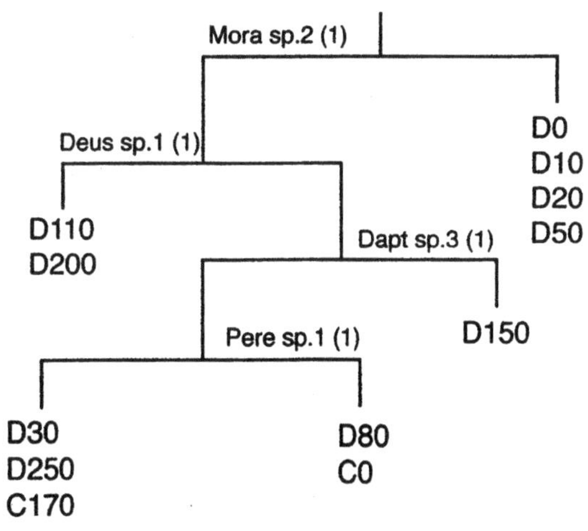

Fig. 4 TWINSPAN dendrogram based on 4th root-transformed nematode mean abundance data (*Mora Monhystera*; *Deus Desmolaimus*; *Dapt Daptonema*; *Pere Perepsilonema*)

Diversity indices, especially species number (N_0), are dependent on sample size (Soetaert and Heip 1990). High dominance and small sample size bias species numbers by reducing the chance of encountering rare species. For this reason, in the early days when the total nematode density in the sample was less than 200 and dominated by 1 or 2 species, species numbers were probably underestimated.

The results of the TWINSPAN analysis discriminated the early days from the later ones because of the absence of *Monhystera* sp.2. However, this analysis failed to group D30 with earlier days when the nematode abundance was high, because *Monhystera* sp.2 appeared in this day. In fact *Monhystera* sp.2 was a rare species throughout the whole study period. The presence or absence of this nematode was probably related to nematode abundance in samples rather than the scouring effect.

Recovery

Densities

In order to investigate the recovery of nematode communities after scouring, it is necessary to have an idea of natural temporal variations for comparisons. A study of Vanhove et al. (2000) provides data on temporal variations in a subtidal nematode community for two winter seasons and the intervening summer (from April 1991 till November 1992) at a station adjacent to our study site at Signy Island. This study shows that the temporal fluctuation of nematode density in this area seems to be closely related to food availability in the water column, and the annual pattern can be greatly different from year to year. Meiofauna appears to bloom during summer and just after pack-ice formation. The exceptionally low densities of nematodes in earlier days after iceberg scouring in our study were outside the range of seasonal fluctuation, even considering the patchy distribution of nematodes.

The recovery potential of meiofauna is known to be high. In most cases of disturbance at temperate sites, meiofauna recovers in a few weeks (Alongi et al. 1983; Brylinsky and Gibson 1994; Danovaro et al. 1995) or in days or even hours in some cases (Sherman and Coull 1980; Sun and Fleeger 1994). The recovery of the major meiofauna groups occurred in 30 days after the impact, although there was an abrupt decrease of their abundance on day 50. Peck et al. (1999), who studied macrofaunal recovery from the same scoured site, attributed the high abundance occurring on D30 to a storm on the 23rd day after scouring. A storm event influences the recovery of meiofaunal communities by redistributing organisms, but also indirectly by improving sediment condition.

It would be expected that the abundance of deeper-living nematodes such as *Sabateria* sp. would remain low after the storm. The comparison of the actual density of *Sabatieria* sp. between D20 and D30 (1.9 and 110.0 ind./10 cm^2, respectively) revealed an increase of >×50 during 10 days. The increase of total nematode abundance on D30 therefore cannot be explained solely by the storm. Most nematodes observed in this area are primary colonisers, and the rapid reproduction of nematodes in the community must have contributed to the high abundance on D30. The decrease of the nematode abundance on D50, however, is not easily explained. Possible causes include a patchy distribution, depletion of food resources caused by the overpopulation on D30, or sampling error.

Nematode community structure

Species dominance rank stayed fairly stable, showing no evidence of succession during the whole study period. *Microlaimus* sp.1 was the most dominant all the time. *Sabatieria* sp., the second most dominant species in overall abundance, kept second and third rank, except on D20 when it became the fourth most important species. *Daptonema* sp.1, the third most dominant species in overall abundance, was more important in earlier days than later, probably because of response to seasonal food item changes (Vanhove et al. 1998). The low proportions of juveniles on D20 and D30 show that juveniles are more sensitive to physical disturbance. The increase in the number of gravid females in the early days shows that reproduction was stimulated by the low community density after the scouring.

The trophic composition of the nematode population in this study was similar to that found by Vanhove et al. (2000). The dominance of 2A group in our study reflects the very high numbers of *Microlaimus* sp.1. Nematodes belonging to feeding type 1A are probably not successful in a frequently disturbed habitat. The frequent sediment agitation may limit microbial growth, and thereby limit the population of feeding group 1A.

The Maturity Index, MI, of Bongers (1990) is basically an extension of *r*- and *K*-selection theory. Neilson et al. (1996) found this maturity index to be robust in a marine ecological context, although it is not yet as popular in marine as in terrestrial nematode studies. For this reason, the direct comparison of MI for environmental stress assessment between different marine habitats or ecosystems must be interpreted with caution.

There are as yet no MI data from polar marine systems available to compare with our MI values; it is therefore necessary to compare with studies from other seas. The MI in the present study was comparable with that in an intertidal station close to a sewage outfall in Tay Estuary, Scotland (Neilson et al. 1996) and also with the polluted Ems estuary (Essink and Romeyn 1994). These comparisons indicate that the MI in the present study is higher than in seriously polluted areas, but lower than normal marine habitats around Europe. The relatively low and constant values during our study periods further imply that the MI per se is not directly affected by iceberg scouring in the short term.

Long-term adaptation of a nematode community to an ice-induced disturbance

The colonisation capacity of nematodes is defined by their life-style rather than their mobility. Colonisers (*r*-strategists) must have more advantages than persisters (*K*-strategists) in such a circumstance. This may imply that natural selection pressure has acted during the long history of Antarctic glaciation in this area and led to the relatively low but constant MI in our study area. Iceberg scouring in a deeper coastal shelf in the Weddell Sea also resulted in a low MI and a low proportion of the 1A feeding group, showing a resemblance of community properties with those of this study (Lee et al., 2001). The nematode community in the shallow coast at Signy Island seemed to be forced towards 1B, and 2A feeding types and *r*-strategists. The regularly disturbed sediments may contain very little old detritus and dissolved organic matter, which microbes can easily utilise, compared with undisturbed stable sediment, and thus the new fresh detritus and benthic diatoms are more common food items in this area. This, together with indirect influence caused by a low MI, may explain the extremely low representation of the 1A group.

Acknowledgements The authors are grateful to the British Antarctic Survey for providing the samples. Thanks go to Mrs. M. Beghyn for helping to treat the samples. We also thank Dr. A. Vanreusel and N. Smol for extensive critical comments on our draft and useful discussions. Dr. J. Mees assisted with statistical analysis. This study was funded by the Scientific Research Programme on Antarctica – Phase IV, from the Belgian State-Prime Minister's Federal Office for Scientific, Technical and Cultural Affairs, DWTC (contract: A4/DD/B01), and by the University of Ghent, Belgium (contract: BOF 98-03, 12050398).

References

Alongi DM, Boesch DF, Diaz RJ (1983) Colonization of meiobenthos in oil-contaminated subtidal sands in the lower Chesapeake Bay. Mar Biol 72:325–335

Arntz WE, Brey T, Gallardo VA (1994) Antarctic zoobenthos. Oceanogr Mar Biol Annu Rev 32:241–304

Bongers T (1990) The maturity index: an ecological measure of environmental disturbance based on nematode species composition. Oecologia 83:14–19

Bongers T, Alkemade R, Yeates GW (1991) Interpretation of disturbance-induced maturity decrease in marine nematode assemblages by means of the Maturity Index. Mar Ecol Prog Ser 76:135–142

Brylinsky M, Gibson J (1994) Impacts of flounder trawls on the intertidal habitat and community of the Minas Basin, Bay of Fundy. Can J Fish Aquat Sci 51:650–661

Danovaro R, Fabiano M, Vincx M (1995) Meiofauna response to the Agip Abruzzo oil spill in subtidal sediments of the Ligurian Sea. Mar Pollut Bull 30:133–145

Dayton PK (1990) Polar benthos. In: Smith WO (ed) Polar oceanography, Part B. Chemistry, biology and geology. Academic Press, London

Dayton PK, Robilliard GA, Vries ALD (1969) Anchor ice formation in McMurdo Sound, Antarctica, and its biological effects. Science 163:273–275

Dayton PK, Robilliard CA, Paine RT (1970) Benthic fauna zonation as a result of anchor ice at McMurdo Sound, Antarctica. In: Holdgate MW (ed) Antarctic ecology. Academic Press, London, pp 244–258

Doake CSM, Vaughan DG (1991) Rapid disintegration of the Wordie Ice Shelf in response to atmospheric warming. Nature 350:228–330

Essink K, Romeyn K (1994) Estuarine nematodes as indicators of organic pollution; an example from the Ems estuary (the Netherlands). Neth J Aquat Ecol 28:213–219

Gammie F (1995) Breakaway iceberg "due to warming". Nature 374:108

Heip C (1980) Meiobenthos as tool in the assessment of marine environmental quality. Rapp P V Reun Cons Int Explor Mer 179:182–187

Hill H (1973) Diversity and evenness: a unifying notation and its consequences. Ecology 54:427–432

Hill H (1979) TWINSPAN – A FORTRAN program for detrended correspondence analysis and reciprocal averaging. Cornell University, Ithaca, NY

Lee HJ, Gerdes D, Vanhove S, Vincx M (2001) Meiofauna to iceberg distrubance on the Antarctic continental shelf at Kapp Norvegia (Weddell Sea). Polar Biol 24:918–925

Lenihan HS, Oliver JS (1995) Anthropogenic and natural disturbances to marine benthic communities in Antarctica. Ecol Appl 5:311–326

McIntyre AD, Warwick RM (1984) Meiofauna technique. In: Holme NA, McIntyre AD (eds) Methods for the study of marine benthos. Blackwell, Oxford, pp 217–224

Neilson R, Boag B, Plamer LF (1996) The effect of environment on marine nematode assemblages as indicated by the maturity index. Nematologica 42:232–242

Peck LS, Bullough LW (1993) Growth and population structure in the infaunal bivalve *Yoldia eightsi* in relation to iceberg activity at Signy Island, Antarctica. Mar Biol 117:235–241

Peck LS, Brockington S, Vanhove S, Beghyn M (1999) Community recovery following catastrophic iceberg impacts in a soft-sediment shallow-water site at Signy Island, Antarctica. Mar Ecol Prog Ser 186:1–8

Picken GB (1984) Marine habitats-benthos. In: Bonner WN (ed) Key environments: Antarctica. Pergamon Press, Oxford, pp 154–172

Platt HM, Warwick RM (1980) Significance of free-living nematodes to the littoral ecosystem. In: Price JH, Irvine DEG, Farnham WF (eds) The shore environment. Academic Press, London, p 729

Schratzberger M, Warwick RM (1999) Differential effects of various types of disturbances on the structure of nematode assemblages: an experimental approach. Mar Ecol Prog Ser 181:227–236

Sherman KM, Coull BC (1980) The response of meiofauna to sediment disturbance. J Exp Mar Biol Ecol 46:59–71

Soetaert K, Heip C (1990) Sample-size dependence of diversity indices and the determination of sufficient sample size in a high-diversity deep-sea environment. Mar Ecol Prog Ser 59:305–307

Sun B, Fleeger JW (1994) Field experiments on the colonization of meiofauna into sediment depressions. Mar Ecol Prog Ser 110:167–175

Vanhove S, Lee HJ, Beghyn M, Gansbeke DV, Brockington S, Vincx M (1998) The metazoan meiofauna in its biogeochemical environment: the case of an Antarctic coastal sediment. J Mar Biol Assoc UK 78:411–434

Vanhove S, Beghyn M, Van Gansbeke D, Bullough LW, Vincx M (2000) The seasonal varying biotope of Signy Island: implication for meiofaunal structure. Mar Ecol Prog Ser 202:13–25

Vincx M, Heip C (1991) The use of meiobenthos in pollution monitoring studies: a review. ICES 16:50–67

Polar Biol (2001) 24: 926–933
DOI 10.1007/s003000100301

Hee Joong Lee · Dieter Gerdes · Sandra Vanhove
Magda Vincx

Meiofauna response to iceberg disturbance on the Antarctic continental shelf at Kapp Norvegia (Weddell Sea)

Abstract The impact of iceberg scouring on meiofauna communities, especially nematodes, was studied on the Kapp Norvegia shelf in the Weddell Sea, Antarctica. Three stations with different stages of recolonisation following scour were selected on the basis of seafloor video images, sediment characteristics and faunal occurrences. These stations comprised a fresh scour, an older scour, and an undisturbed control site where a sponge spicule mat covered the sediment with dense epifauna. Meiofaunal abundance and taxonomic diversity of meiofauna groups were significantly reduced in the fresh scour. The highest abundance and diversity were found in the older scour as compared with the undisturbed site. The abundance and diversity of nematodes also decreased due to scouring. The abundance in the older scour recovered to the level of the undisturbed site whereas the diversity remained low. Scouring also changed the nematode community structure, with the suborders Desmoscolecina and Leptolaimina as the most sensitive groups. In addition, scouring resulted in the decrease of selective deposit feeders and the Maturity Index. The low diversity and the change in nematode generic composition in the older scour compared with the undisturbed site, despite the complete recovery in terms of abundance, suggest that the deep continental shelf nematode community in this area is sensitive to iceberg disturbance.

Please note: When citing this article, please use the reference to the original source in the journal Polar Biology as given on the top of this page.

H.J. Lee (✉) · S. Vanhove · M. Vincx
Marine Biology Section, Department of Biology,
University of Ghent, K.L. Ledeganckstraat 35,
9000 Ghent, Belgium
E-mail: Lee.HeeJoong@rug.ac.be
Fax: +32-9-2645344

D. Gerdes
Alfred Wegener Institute for Polar and Marine Research,
Columbusstrasse, 27568 Bremerhaven, Germany

Introduction

Iceberg scouring is a catastrophic event that physically disturbs benthic ecosystems in polar regions. Scouring affects wide areas in both polar systems, in depths down to about 70 m in the Arctic and 500 m in the Antarctic (Gutt et al. 1996). Iceberg scouring is thus an ecologically important structuring factor in polar benthic ecosystems (Arntz and Gallardo 1994), and although its long history must have controlled the ecology and evolutionary adaptation of polar benthic organisms, few studies have been published (for a review see Gutt, 2001). Most of these studies focus on shallow coastal communities.

There have been few ecological studies of Antarctic meiofauna (Dahms et al. 1990; Herman and Dahms 1992; Vanhove et al. 1995, 1998, 1999; Fabiano and Danovaro 1999; Lee et al., 2001). Although ice scour has been identified as an important structuring force in deep-water nematode distribution (Vanhove et al. 1999), the present study is the first to consider the effect of iceberg scour on deep-water meiofauna.

Materials and methods

Sampling area and methods

The continental shelf off Kapp Norvegia in the eastern Weddell Sea is one of the areas where intensive biological studies have been conducted during the EPOS (European "Polarstern" Study) and the first and second EASIZ (Ecology of Antarctic Sea Ice Zone) cruises. Our knowledge about the benthic fauna in this area is substantial, and it is one of the places where iceberg scouring activity is very high. Therefore this area was selected for our study (Fig. 1).

During the second EASIZ cruise from 13 January to 26 March 1998, samples were taken by means of a multi-box corer (Gerdes 1990) in order to study the impact of iceberg scouring on meiobenthic communities. Three contrasting stations were selected for this study: 225 (water depth, 278 m), 187 (water depth, 255 m) and 228 (water depth, 298 m). These stations were regarded as a very fresh scour, a relatively older scour and an undisturbed site, respectively. This discrimination was based on the combination of

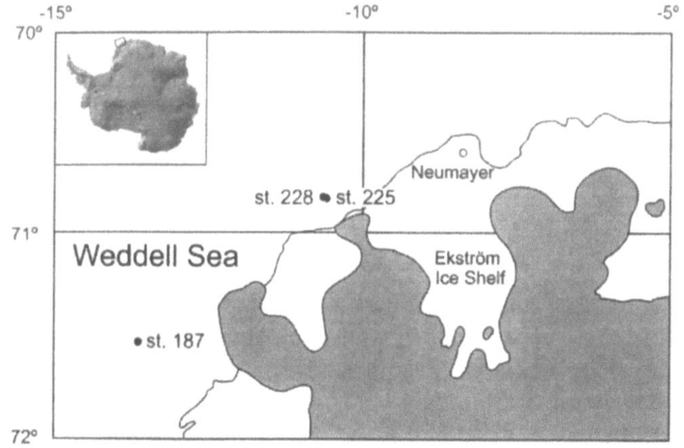

Fig. 1 Map showing the sampling sites (stations, *filled circles*) for this study during EASIZ II (ANT XV/III). Station 225 was a fresh scour; station 187 was an older scour and station 228 was an undisturbed site

in situ observations of the bottom scenery produced by a video camera attached to the multi-box corer, the sediment textual conditions and the macro- and epifauna occurrence (Table 1). The video images at station 225 showed a very fresh scour, based on the lack of epifauna and, especially, on the very fluid sediment condition. The ploughing and pumping activities of the iceberg must resuspend the sediment (Lien et al. 1989), and this resuspended sediment will settle down, especially in the depression of the scour. This sediment will be more fluid, because it contains more interstitial water than under normal conditions. After some time, however, water currents will redistribute the fluid fine sediment and underlying compact sediments will remain in scour marks. The fluid sediment at station 225 can thus be considered as evidence of a very fresh scour mark.

The sediment from station 187 was not as fluid as at station 225, and the presence of a sessile hydrozoan found in 1 sediment sample from this station, *Symplectoscyphus plectilis* (about 1 month old; J-M Gili, personal communication), suggests that the scour mark at this station is at least older than 1 month. Although the actual age of this scour remains uncertain, it is considered that it is not older than 1 year. The undisturbed reference station, station 228, showed a typical dense cover of epifauna on the seabed, in contrast to the barren surface at the scoured sites. The sediment samples from this station were also different from the sediments of the other stations, in that they were covered with a sponge spicule mat about 1 cm thick.

Three standard meiofauna hand-cores (10 cm^2 surface area) for the meiofauna, and a large hand-core (diameter about 6 cm) for sediment analyses were taken from one box core of each station.

Sample treatment

The sediment cores were sliced into 5 layers (0–1, 1–3, 3–5, 5–10 cm and the rest) immediately after the samples were recovered on board. Only the 3 top layers up to 5 cm, where the majority of meiofauna dwells, were used for this study. Meiofauna samples were preserved with 4% neutral hot (60°C) formaldehyde solution on board before further studies in the home laboratory. Sediments were decanted and sieved over 1,000-μm and 32-μm mesh sizes. Animals passing the 1,000-μm sieve and retained on the 32-μm sieve were regarded as meiofauna. Final extraction of meiofauna was achieved using the LUDOX centrifugation flotation technique (McIntyre and Warwick 1984). The number of all metazoan meiofauna was counted after staining with Rose Bengal. Approximately 100 nematodes (all, in samples with less than 100 individuals) per replicate were randomly picked out and dehydrated in a series of glycerin-alcohol solutions. The dehydrated nematodes were mounted on slides with anhydrous glycerin medium and sealed with paraffin wax. Observations were carried out under a Wild M20 light microscope. Identification to genus level was based on morphological characters (Platt and Warwick 1988). Developmental stages and sex of specimens were categorised into four different groups: juvenile, non-gravid female, gravid female and male. The feeding types were classified into four categories; selective deposit feeders (1A), non-selective deposit feeders (1B), epistratum feeders (2A) and predators/omnivores (2B).

Sediment analysis was performed with a Coulter-Counter (the sponge spicule mats on the top of sediments from station 228 were removed prior to analysis).

Statistical analysis

Analysis of variance (ANOVA) was used to determine significant differences ($P < 0.05$) for the following parameters: diversity, feeding type, age guild, maturity index and dominance of nematode genera between stations. Subsequent post-hoc comparison (Tukey HSD) was used on stations.

Nematode diversity at genus level was measured using the suite of diversity indices proposed by Hill (1973; Fig. 2), where N_0 is the number of genera, N_1 is the exponential of the Shannon index, N_2 is the reciprocal of the Simpson index and N_∞ is the reciprocal of the relative abundance of the most dominant genus. The maturity index (MI; Bongers 1990; Bongers et al. 1991) was used to characterise the life-style of nematode communities.

Results

Sediment composition

The sediments from the fresh scour and the undisturbed site consisted of silt (median grain size 43.4 μm and 32.4 μm, respectively), while that of the older scour was very fine sand (median grain size 101 μm) (Table 1). Although the grain size distributions of the fresh scour and the undisturbed site were similar to each other, there were clear differences in sediment condition. The sediment from the fresh scour was very fluid, with more interstitial water than that of the older scour or the

Table 1 Coordinates and description of the sampling stations in the Weddell Sea, Antarctica

Scour condition	Station no.	Depth (m)	Latitude	Longitude	Sediment median size (μm)	Sediment condition	Epifauna on the sediment samples
Fresh	225	278	70°50.1S	10°35.2W	43.4	Very fluid	None
Older	187	255	71°32.3S	13°31.7W	32.4	Normal	One hydrozoan
Undisturbed	228	298	70°49.8S	10°38.0W	101.0	Sponge spicule mat	Diverse

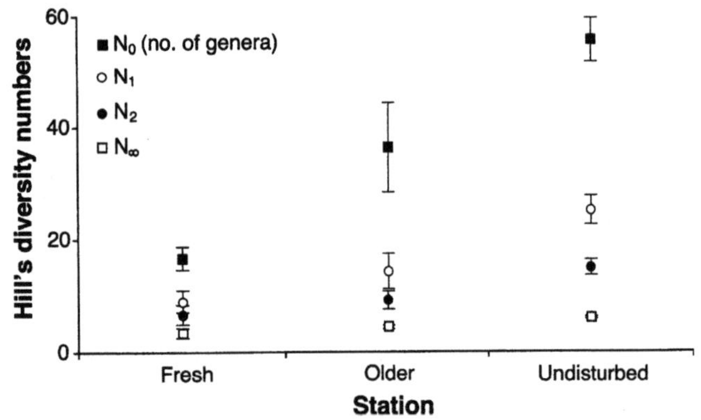

Fig. 2 Hill's diversity numbers of the nematode communities (*error bar*, standard deviation)

undisturbed site. The sediment from the undisturbed site differed from the other stations in having a surface sponge spicule mat forming the top 1 cm of the sample.

Meiofauna

A total of 20 different meiofauna groups were recovered from the 3 stations (Table 2). There were large differences in density between the fresh scour and the other stations (fresh scour, 120 ± 15.4 ind./10 cm^2; older scour, 1,326 ± 287.5 ind./10 cm^2; undisturbed site, 1,342 ± 70.8 ind./10 cm^2). The number of meiofauna groups was also very low in the fresh scour (7 groups) compared with the older scour (16 groups) and the undisturbed site (13 groups).

Nematodes were the most dominant meiofaunal group at all stations, and their relative abundance in-

Table 2 Mean density (ind./10 cm^2) of the meiofauna with standard deviation in parentheses ($n = 3$)

Station	Fresh	Older	Undisturbed
Nematoda	72.7 (9.3)	1028.7 (217.5)	1234.3 (69.3)
Copepoda	21.0 (11.3)	89.7 (23.1)	30.7 (10.6)
Ostracoda	1.3 (1.5)	12.0 (10.1)	5.7 (4.7)
Nauplii	22.7 (8.0)	162.0 (25.2)	49.7 (17.1)
Priapulida	0	2.0 (1.0)	0
Kinorhyncha	0	0.7 (1.2)	4.3 (4.9)
Tardigrada	0	0.7 (0.6)	4.7 (0.6)
Turbellaria	0	2.7 (3.1)	0.3 (0.6)
Tanaidacea	0.3 (0.6)	0	0
Rotifera	1.3 (1.5)	0.7 (0.6)	3.0 (1.0)
Isopoda	0	0.3 (0.6)	0
Amphipoda	0	0.7(1.2)	0
Polychaeta	0.7 (0.6)	22.0 (10.8)	5.3 (5.9)
Oligochaeta	0	1.0 (1.0)	0
Bivalvia	0	1.0 (1.0)	0
Sipuncula	0	0.3 (0.6)	0
Hydrozoa	0	0	1.0 (1.0)
Acari	0	1.7 (2.9)	2.0 (1.0)
Aplacophora	0	0	0.3 (0.6)
Bryozoa	0	0	1.0 (1.0)
Total	120 (15.4)	1326.0 (287.5)	1342.3 (70.8)

creased from the fresh scour (60.6%) to the older scour (77.6%) and the undisturbed site (92.0%). Crustaceans (copepods, ostracods and nauplii) were the second most dominant group. These 2 groups made up more than 97% of community density at all stations.

Nematode communities

The abundance of nematodes was significantly lower (73 ± 9.3 ind./10 cm^2) in the fresh scour compared with the older scour (1,029 ± 217.5 ind./10 cm^2) and the undisturbed site (1,234 ± 69.3 ind./10 cm^2), which were comparable (Tables 2, 3).

The average genus number expressed as N_0 of Hill's diversity numbers increased significantly with the stage of recolonisation (fresh scour, $N_0 = 16.7 \pm 2.0$; older scour, $N_0 = 36.3 \pm 8.0$; undisturbed site, $N_0 = 55.7 \pm 4.0$) (Fig. 2). All other Hill diversity numbers showed the same tendency, although the differences of other indices between the fresh scour and the older scour were not significant. In the case of N_∞, only the difference between the fresh scour and the undisturbed site was significant.

There were clear differences in genus composition between all three stations (Fig. 3, Table 3). The genus *Monhystera* was the most dominant at all stations, but the dominance decreased towards the undisturbed site (26.1% in the fresh scour, 18.8% in the older scour and 16.7% in the undisturbed site). Next to *Monhystera*, *Neochromadora* and *Daptonema* were the subdominant genera in the fresh and the older scour, but of less importance in the undisturbed site. *Desmoscolex* and *Leptolaimus* were the second and third dominant genera in the undisturbed site, although they were less important in the older scour and absent or very rare in the fresh scour. The most unexpected observation was the rarity of the suborders Leptolaimina and Desmoscolecina in the fresh scour. Leptolaimina were represented only by *Aegialoalaimus* (1.3%), *Prismatolaimus* (2.7%) and *Teratocephalus* (0.7%), and the suborder Desmoscolecina was entirely absent from the samples.

The genus number and abundance of these suborders increased with the age of the scour (older scour: 9.3%, 6 genera; the undisturbed site: 34.2%, 14 genera); neither *Prismatolaimus* nor *Teratocephalus* occurred in the undisturbed site. *Sabatieria* was one of the common genera at all stations, although this genus was less important in the fresh scour compared with the other stations. The dominance of *Desmoscolex*, *Leptolaimus*, *Acantholaimus*, and *Tricoma* showed an increasing tendency from the fresh scour to the undisturbed site (Fig. 3a). *Sabatieria*, *Halalaimus*, *Amphimonhystrella* and *Sphaerolaimus* showed no constant pattern (Fig. 3b). The proportions of *Monhystera*, *Neochromadora*, *Daptonema*, *Cervonema* and *Prismatolaimus* tended to increase at the more recently scoured site (Fig. 3c). A similar trend was found at a coarser taxonomic level (mostly family level), except for the family Comesomatidae. Table 4 shows the dominance rank of families with a

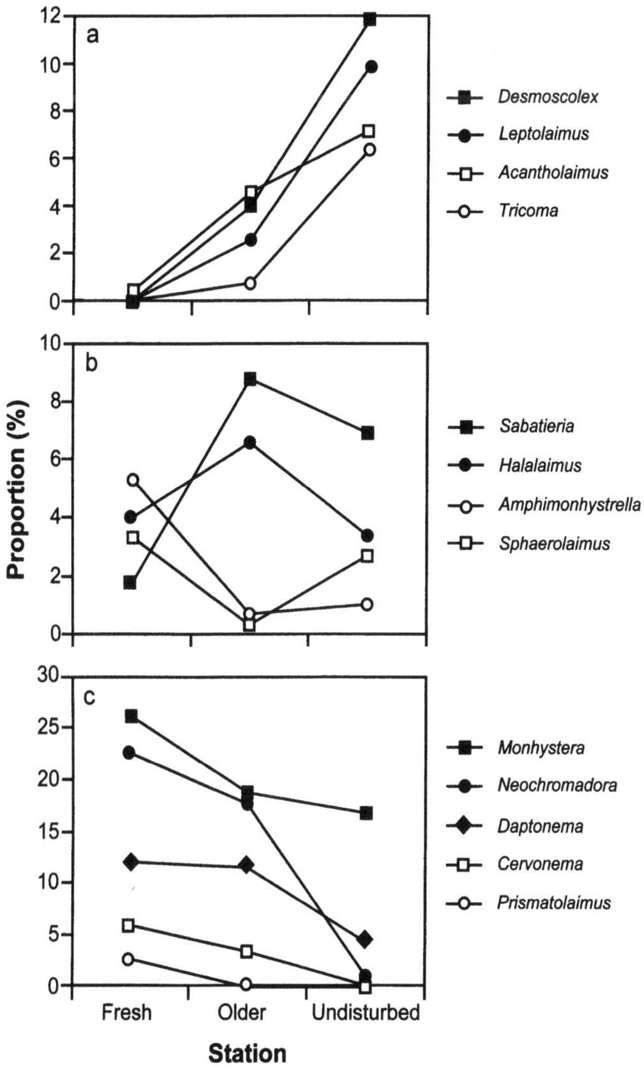

Fig. 3a–c Different tendencies of major nematode genera expressed as the relative proportion according to different degrees of freshness of scours. The nematodes that proportionally decreased due to scouring were regarded as persisters (**a**), and those that increased after scouring were regarded as colonisers (**c**). The intermediate group (**b**) did not show a clear trend

density higher than 0.5%. The families Monhysteridae, Chromadoridae and Xyalidae were commonly dominant at all stations. The families Comesomatidae and Oxystominidae were also common and subdominant at all three stations.

Juvenile stages were predominant at all stations (average: 64.0–69.3%) followed by males (average: 13.7–17.7%), but the differences between stations were not significant (Fig. 4). The non-gravid females, the third dominant group (average: 7.9–19.7%), were significantly ($P < 0.05$) lower in the fresh scour compared with the undisturbed site, but differences between the samples were not significant. The gravid females were least abundant (average: 1.3–6.0%) and their proportion was significantly lower in the undisturbed site compared with the other stations.

Table 3 Mean density (ind./10 cm^2, $n = 3$) of the important nematode genera (> 1% at any station)

Genus	Fresh	Older	Undisturbed
Monhystera	19.7	192.4	206.1
Neochromadora	17.0	181.9	13.2
Daptonema	9.0	119.4	52.6
Sabatieria	1.3	90.4	85.4
Leptolaimus	–	25.1	121.6
Acantholaimus	0.3	45.7	88.0
Halalaimus	3.0	67.5	42.4
Tricoma	–	7.6	78.4
Desmoscolex	–	39.9	145.8
Cervonema	4.3	35.0	1.9
Amphimonhystrella	2.5	3.3	33.1
Molgolaimus	–	9.4	24.9
Aegialoalaimus	1.0	14.6	15.9
Prochromadorella	0.5	19.3	6.4
Sphaerolaimus	4.0	7.0	11.7
Odontanticoma	–	18.5	1.8
Anticoma	0.3	17.0	2.5
Actinonema	–	7.3	12.6
Diplolaimella	–	18.5	–
Tylenchidae genus	1.3	6.1	10.9
Spilophorella	–	11.2	6.7
Camacolaimus	–	5.6	12.0
Microlaimus	–	0.3	15.0
Oxystomina	1.0	2.3	3.0
Pseudosteineria	1.3	3.3	–
Chromadoridae genus	1.5	–	1.9
Prismatolaimus	2.0	0.3	–
Other (No. of other genera)	2.7 (11)	78.8 (39)	240.4 (67)
Total	72.7	1028.7	1234.3

Among the four feeding guilds, the 1A group showed increased importance from the fresh scour to the undisturbed site (Fig. 5). The proportion of this group was significantly higher in the undisturbed site compared with the other stations. The feeding group 1B showed an opposite tendency to the 1A group. The proportion of 1B group was significantly lower only in the undisturbed site compared with the fresh scour. No significant difference between the stations was observed for the 2A group. Amongst the feeding guilds, the 2B group was the smallest. The proportion of this group was significantly higher in the fresh scour compared with the other stations.

The maturity index showed a significant decreasing tendency with the freshness of disturbance (Fig. 6).

Discussion

Influence of iceberg scouring on meiofauna

The very low meiofaunal abundance in the fresh scour (< 10% that in the older scour and the undisturbed site) is similar to the 93% reduction of meiofauna abundance after iceberg scouring in the shallow coastal sediments of Signy Island (Lee et al., 2001). Iceberg scouring also decreased the range of meiofauna taxa. Initially it removed most sessile and some motile animals, e.g.

Table 4 Relative importance, feeding type (*FT*) and *c-p* (coloniser-persister) value of the nematodes (Bongers et al. 1991) grouped into families, except for Comesomatidae where the different genera showed different tendencies (* >0.5%, ** 2.5–5%, *** 5–10%, **** 10–20%, ***** >20%, ? unknown)

Nematode group	Fresh	Older	Undisturbed	FT	c-p value
Monhysteridae	*****	*****	****	1B	1
Chromadoridae	*****	*****	****	2AB	3
Xyalidae	****	****	****	1B	2
Oxystominidae	***	***	***	1A	4
Comesomatidae *Cervonema*	***	**	*	1A	2
Sphaerolaimidae	***	*	*	2B	3
Prismatolaimidae	**			1B	?
Rhabditidae	*	*		1A	1
Teratocephalidae	*	*		1A	3
Comesomatidae *Sabatieria*	*	***	***	1B	2
Trefusiidae			*	1A	4
Epsilonematidae			*	1A	4
Microlaimidae			*	2A	3
Ceramonematidae		*	*	1A	3
Diplopeltidae		*	*	1A	3
Cyatholaimidae		*	*	2A	3
Desmodoridae		*	*	2B	3
Meyliidae		*	***	1A	4
Leptolaimidae		**	****	1A	3
Desmoscolecidae		**	****	1A	4

hydrozoans, bryozoans, kinorhynchs, tardigrades, turbellarians and acarines. The first immigrants into the scour were motile organisms such as amphipods and isopods, which can be considered as early colonisers. The absence of some lesser motile burrowers, e.g. priapulids, bivalves, sipunculids and oligochaetes, in the undisturbed site in spite of their presence in the older scour may be related to the presence of sponge spicule mats covering the sediment.

Previous studies of disturbance have shown time scales of meiofaunal recovery ranging from days to years (e.g. Coull 1969; Sherman and Coull 1980; Danovaro et al. 1995). A general conclusion is that the recovery of meiofauna is relatively fast although it may depend on the type, frequency and scale of disturbance. In our study, the meiofaunal abundance in the older scour had already recovered to the level of the undisturbed site and the taxon number in the older scour even exceeded that of the undisturbed site (Table 2). Unfortunately, we cannot provide a good estimate of the recovery time for the meiofauna community in the current deep-water study. Recolonisation of the shallow-coast meiofauna community at Signy Island occurred between 30 and 80 days (Lee et al., 2001), but as there are major differences in the ecological conditions between shallow coastal and deep-water habitats, one might expect that recolonisation of meiofauna in deep water will occur at a slower rate. Complete recovery of the nematode community in this area may take some years.

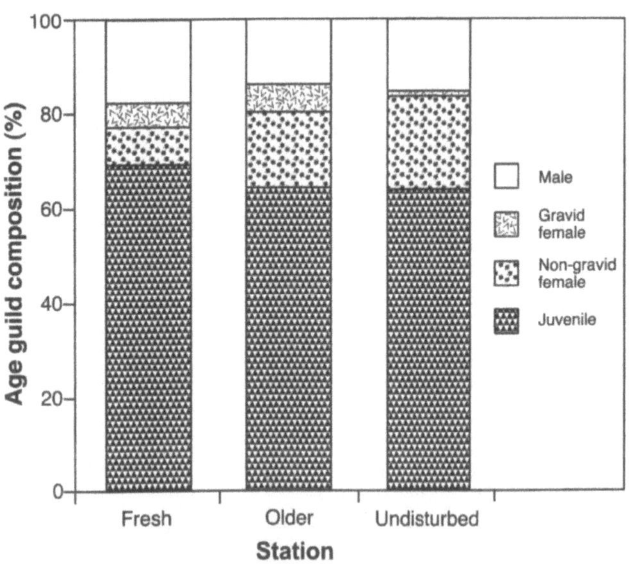

Fig. 4 Age and sex composition of the nematode community in the three stations with different stages of recovery

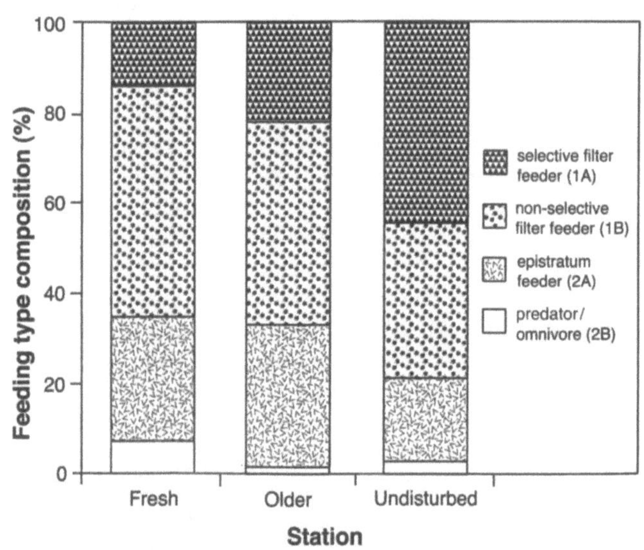

Fig. 5 Trophic guild composition of the nematode community in the three stations with different stages of recovery

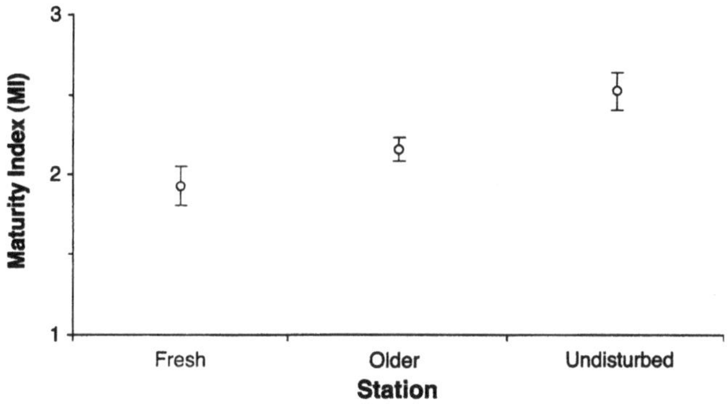

Fig. 6 Maturity Index (MI) of the nematode community (*error bar*, standard deviation)

Influence of iceberg scouring on the nematode community

Abundance

The pattern of nematode abundance between samples was similar to that of total meiofauna abundance. The extremely low density in the fresh scour shows the severity of iceberg scouring impact in the initial stage. Although the nematode density in the older scour recovered to the level of the undisturbed site, their proportions among meiofauna were still low in the scours, implying that nematodes recolonise more slowly than other major meiofauna, e.g. copepods and ostracods. This is probably because of the slower dispersal of nematodes.

Diversity

Nematode diversity was greatly influenced by scouring. The relatively high diversity of nematodes in terms of genus number in the undisturbed site ($N_0 = 55.7 \pm 4.0$) was in accordance with the results of Vanhove et al. (1999) from Kapp Norvegia and Halley Bay ($N_0 = 56 \pm 5.0$ and 52 ± 10.8, respectively) in the Weddell Sea. However, it was very low in the fresh scour of this study ($N_0 = 16.7 \pm 2.0$). The nematode genus number in the older scour site was, unlike the abundance, still low ($N_0 = 36.3 \pm 8.0$) compared with the undisturbed site, which means that the restoration of nematode community structure is slower than the recovery in abundance.

Iceberg scouring also resulted in the increase of the dominance of a few nematode genera. Each of the three most abundant genera in the fresh and the older scour, *Monhystera*, *Neochromadora* and *Daptonema*, composed more than 10% of the population, while only *Monhystera* exceeded 10% in the undisturbed site. These three dominant genera made up 60.6% in the fresh scour and 48.1% in the older scour. The three most dominant genera in the undisturbed site, *Monhystera*, *Desmoscolex* and *Leptolaimus*, composed only 38.4% of the community.

Composition

The nematode genus composition of the fresh scour was more similar to the community in the older scour than that in the undisturbed community, despite the fresh scour being located much closer to the undisturbed site than to the older scour. Three different trends of nematode response to scouring effects were found (Fig. 3). A group of nematodes showed a higher relative abundance in the fresh scour, reflecting their recolonisation ability. The common feature of this nematode group was that most of them were non-selective deposit feeders (1B). The second group, which showed a higher relative abundance in the undisturbed site, was represented by selective deposit feeders (1A). *Acantholaimus* (2A, an exceptional genus in this group) seems to be a typical deep-sea nematode (Soetaert and Heip 1995; Vanaverbeke et al. 1997) and prefers physically more stable environments. It is not clear which parameters influence the sensitivity of *Acantholaimus* to iceberg scouring. The last group, which was composed of nematodes with various feeding types and different reproductive strategies, showed intermediate properties. Among these three groups, the second group seemed to be most sensitive to iceberg scouring. The sensitivity of nematodes in this group is probably caused by their restricted food preference and/or due to their reproductive strategy, because these nematodes were either selective deposit feeders (1A) or extreme colonisers.

In general, the feeding guilds in the Antarctic deeper water are more or less evenly shared between 1A, 1B and 2A groups, with a slight dominance of the 2A group, while the 2B group appears to be the least abundant group (Vanhove et al. 1999). The results of the present study mainly confirmed this tendency, although it was biased towards the 1A group in the undisturbed site and towards the 1B group in the other stations.

Sex and age

Reproductive activity at the scoured sites was higher than at the undisturbed site. One of the reasons for the high proportion of gravid females in the scour samples was the change of nematode generic composition in the communities. The dominant genera in the undisturbed site, *Desmoscolex*, *Leptolaimus* and *Tricoma*, showed a relatively lower fecundity. Therefore the low proportion of *Leptolaimus* and Desmoscolecina genera in the scours must have influenced the proportion of gravid females. However, the lower density of the population and therefore the lower competition in the scours may have stimulated the reproduction of nematodes.

Shallow-water community versus continental-shelf community in the Antarctic

The influence of iceberg scouring has also been investigated at a shallow-coast site at Signy Island (Lee et al.,

2001). The comparison of the results from Signy Island and the Weddell Sea, from this study, provides further insight into the impact of iceberg scouring on the meiofauna and nematode communities. Both areas are under the influence of catastrophic physical disturbances of iceberg scouring. However, the frequency of disturbance is different. The study site on the shelf off Kapp Norvegia is deeper and an occasional iceberg scouring happens at each square metre once every 340 years on average (Gutt, 2001), whereas scouring can happen much more frequently in the case of the coast at Signy Island (50–75 years, Peck and Bullough 1993). Between scouring events, the seabed of Kapp Norvegia remains more or less constant, whereas the seabed close to Signy Island is under the continuous disturbance of wave action and frequent ice-induced disturbances. Therefore a combination of several physical disturbances structures the nematode community at shallow Antarctic coasts, whereas only occasional iceberg scour disturbs the nematode community at Kapp Norvegia. The nematode communities of undisturbed habitats in the two areas are also very different. At Signy Island, they are characterised by a low diversity, a high dominance, a low MI and a low proportion of the 1A feeding group. The undisturbed nematode community at Kapp Norvegia shows opposite characteristics. However, iceberg scouring leads to a higher similarity between the characteristics of the two nematode communities at Kapp Norvegia and Signy Island. Similarly, Conlan et al. (1998) found that the scour communities of macrofauna closely resembled inshore shallow-water benthos despite the remoteness of several hundred metres.

Conclusions

It is likely that a significant threat for the Antarctic deep-water communities comes from increased iceberg scouring as a result of global warming (Doake and Vaughan 1991; Gammie 1995). Gutt et al. (1996) concluded that a slight increase in iceberg scour could be accommodated by the macrobenthic system because of its adaptation to such disturbance. Meiofaunal communities have a strong natural capacity for recovery from all kinds of disturbance including iceberg scour, but a large contrast is observed between shallow- and deep-water communities. Shallow meiofauna seems to be strongly adapted to iceberg scouring as it is frequently faced with different kinds of physical constraints characteristic of shallow-water environments. However, the structural recovery of the meiofauna, and more specifically the nematodes, from the deeper continental shelf is a slower process which can take some years, in contrast to the rapid recovery of abundance. This indicates that the communities are much more fragile and that they do not show similar adaptation to shallow-water communities. Probably this is mainly because of the relative constancy of the Antarctic deep-water environment. Hence, the increasing frequency of iceberg scouring due to global warming might have much greater effects on a deep-water meiofauna community, as compared with a shallow community.

Acknowledgements The authors are grateful to the Alfred Wegener Institute for Polar and Marine Research, Germany for providing the opportunity to join the EASIZ II Antarctic expedition. The first author is grateful to Professor W. Arntz and Captain E.-P. Greve for their efforts to organise the limited sampling schedule efficiently, reflecting all the demands of small working groups, and for eventually leading the successful and safe sampling cruise. Thanks go to the officers and crew of RV "Polarstern" for their assistance in sampling and also their kindness during the cruise. Thanks also go to J. Van de Velde and D. Van Gansbeke for technical assistance. We thank Dr. A. Vanreusel and N. Smol for comments on our draft and useful discussions. This study was funded by the Scientific Research Programme on Antarctica – Phase IV from the Belgian State Prime Minister's Federal Office for Scientific, Technical and Cultural Affairs, DWTC (contract: A4/DD/B01) and by the University of Ghent, Belgium (contract: BOF 98-03, 12050398).

References

Arntz WE, Gallardo VA (1994) Antarctic benthos: present position and future prospects. In: Hempel G (ed) Antarctic science. Springer, Berlin Heidelberg New York, pp 243–277

Bongers T (1990) The maturity index: an ecological measure of environmental disturbance based on nematode species composition. Oecologia 83:14–19

Bongers T, Alkemade R, Yeates GW (1991) Interpretation of disturbance-induced maturity decrease in marine nematode assemblages by means of the Maturity Index. Mar Ecol Prog Ser 76:135–142

Conlan KE, Lenihan HS, Kvitek RG, Oliver JS (1998) Ice scour disturbance to benthic communities in the Canadian High Arctic. Mar Ecol Prog Ser 166:1–16

Coull BC (1969) Hydrographic control of meiobenthos in Bermuda. Limnol Oceanogr 14:953–957

Dahms HU, Herman RL, Schockaert E (1990) Meiobenthos on the Halley Bay and Kapp Norvegia transects. Ber Polarforsch 68:91–96

Danovaro R, Fabiano M, Vincx M (1995) Meiofauna response to the Agip Abruzzo oil spill in subtidal sediments of the Ligurian Sea. Mar Pollut Bull 30:133–145

Doake CSM, Vaughan DG (1991) Rapid disintegration of the Wordie Ice Shelf in response to atmospheric warming. Nature 350:228–330

Fabiano M, Danovaro R (1999) Meiofauna distribution and mesoscale variability in two sites of the Ross Sea (Antarctica) with contrasting food supply. Polar Biol 22:115–123

Gammie F (1995) Breakaway iceberg 'due to warming'. Nature 374:108

Gerdes D (1990) Antarctic trials with the multi-box corer, a new device for benthos sampling. Polar Rec 26:35–38

Gutt J (2001) On the direct impact of ice on benthic communities, a review. Polar Biol 24:553–564

Gutt J, Starmans A, Dieckmann G (1996) Impact of iceberg scouring on polar benthic habitats. Mar Ecol Prog Ser 137:311–316

Herman RL, Dahms HU (1992) Meiofauna communities along a depth transect off Halley Bay (Weddell Sea-Antarctica). Polar Biol 12:313–320

Hill H (1973) Diversity and evenness: a unifying notation and its consequences. Ecology 54:427–432

Lee HJ, Vanhove S, Peck LS, Vincx M (2001) Recolonisation of meiofauna after a catastrophic iceberg scouring in shallow Antarctic sediments. Polar Biol 24:918–925

Lien R, Solheim A, Elverhøi A, Rokoengen K (1989) Iceberg scouring and sea bed morphology on the eastern Weddell Sea shelf, Antarctica. Polar Res 7:43–57
McIntyre AD, Warwick RM (1984) Meiofauna technique. In: Holme NA, McIntyre AD (eds) Methods for the study of marine benthos. Blackwell, Oxford, pp 217–224
Peck LS, Bullough LW (1993) Growth and population structure in the infaunal bivalve *Yoldia eightsi* in relation to iceberg activity at Signy Island, Antarctica. Mar Biol 117:235–241
Platt HM, Warwick RM (1988) Free-living marine nematodes. Part 2. British chromadorids. Pictorial key to world genera and notes for the identification of British species. Brill, Leiden
Sherman KM, Coull BC (1980) The response of meiofauna to sediment disturbance. J Exp Mar Biol Ecol 46:59–71
Soetaert K, Heip C (1995) Nematode assemblages of deep-sea and shelf break sites in the North Atlantic and Mediterranean Sea. Mar Ecol Prog Ser 125:171–183
Vanaverbeke J, Soetaert K, Heip C, Vanreusel A (1997) The metazoan meiobenthos along the continental slope of the Goban Spur (NE Atlantic). J Sea Res 38:63–107
Vanhove S, Wittoeck J, Desmet G, Vandenberghe B, Herman RL, Bak RPM, Nieuwland G, Vosjan JH, Boldrin A, Rabitti S, Vincx M (1995) Deep-sea meiofauna communities in Antarctica – structural analysis and relation with the environment. Mar Ecol Prog Ser 127:65–76
Vanhove S, Lee HJ, Beghyn M, Gansbeke DV, Brockington S, Vincx M (1998) The metazoan meiofauna in its biogeochemical environment: The case of an Antarctic coastal sediment. J Mar Biol Assoc UK 78:411–434
Vanhove S, Arntz W, Vincx M (1999) Comparative study of the nematode communities on the southeastern Weddell Sea shelf and slope (Antarctica). Mar Ecol Prog Ser 181:237

Polar Biol (2001) 24: 615–619
DOI 10.1007/s003000100263

Julian Gutt · Andreas Starmans

Quantification of iceberg impact and benthic recolonisation patterns in the Weddell Sea (Antarctica)

Abstract Video transects in the eastern Weddell Sea were used to classify the mega-epibenthos into stages of recolonisation after iceberg impact and unaffected fauna. Three site categories differing in bottom topography and concentration of grounded icebergs were analysed. At small iceberg banks and on a comparatively plain seabed, 52–60% of undisturbed seafloor was found, and below 20% at a large iceberg bank. The impact was calculated as a function of values for recently disturbed areas and an estimated recovery time. The results show that, statistically, the Antarctic benthos never reaches peak maturity and that iceberg scouring is among the five most significant disturbances that any large ecosystem on earth experiences.

Introduction

The impact of ice disturbance on benthic communities is currently the subject of considerable interest (for further literature, see Conlan et al. 1998; Gutt 2001). The devastation of life on the seafloor caused by icebergs was even referred to as an "Armageddon" and the most destructive natural disaster that any living community on earth can experience (BBC News Online). Investigations on such phenomena are helpful to improve our understanding of how ecosystems function, as well as providing a basis for applied ecology. However, so far, most studies on the effects of ice scour have been only qualitative.

It has been known for some time that Antarctic tabular icebergs impact on the seabed at depths ≤ 600 m whereas the effects of sea-ice keels and small growlers are restricted to shallower waters. The ecological consequences of such disturbances generally depend on their space and time architecture (Pickett and White 1985), biological attributes of the assemblages and their "normal" conditions. The recolonisation of a scour mark is characterised by the initial invasion of mobile animals, fast and often massive recruitment of pioneers and the subsequent appearance of species with other life strategies. It appears that the recovery of these assemblages is not deterministic but follows highly complex pathways of succession (see, e.g., Gray 1977; Thiery 1982; Sousa 1984).

The aim of this paper is to provide information on the descriptors of the iceberg disturbance regime in the Weddell Sea and on the proportion of successional stages after iceberg impact. The use of a Remotely Operated Vehicle (ROV) for obtaining kilometre-long video transects made such analyses possible. The results are also used to test whether the extreme scenario mentioned, which was drawn from a small shallow site at Signy Island (Peck et al. 1999), is of general validity. Moreover, our results enable us to estimate the significance of iceberg disturbance on a global scale by comparison with other natural catastrophes.

Materials and methods

Investigations were carried out aboard R/V Polarstern during the expeditions ANT XIII/3 (1996) and ANT XV/3 (1998). For cruise reports, see Arntz and Gutt (1997, 1999). Sampling was carried out along an approximately 200-km-long coastline in the Weddell Sea between Atka Bay (70°30′S, 08°00′W) and southwest of Kapp Norvegia (71°30′S, 12°20′W). A total of 39 stations were classified:

a. Level plateau. Relatively even seafloor between 152 and 366 m depth, without concentrations of grounded icebergs.
b. Small iceberg bank. Area of 10×10 km between 114 and 230 m depth, with concentrations of grounded icebergs.
c. Large iceberg bank. Relatively deep area (232–315 m) approximately 30×10 km with many grounded icebergs off Auståsen, 70°43′S, 10°18′W.

Please note: When citing this article, please use the reference to the original source in the journal Polar Biology as given on the top of this page.

J. Gutt (✉) · A. Starmans
Alfred Wegener Institute for Polar and Marine Research,
Columbusstrasse, 27568 Bremerhaven, Germany
E-mail: jgutt@awi-bremerhaven.de
Tel.: +49-471-4831333
Fax: +49-471-4831149

Table 1 Definitions used to classify putative scour marks at different stages of recolonisation (*R0–R2* recolonisation stages, from younger to older; *ID* indirectly disturbed; *UD* undisturbed; *??* area of uncertain status). All criteria are obligatory if not indicated to be "facultative" or when "or" separates alternatives

Label	Criteria	Possible errors
R0, R1, R2	Peculiarities in bottom topography cannot be explained by other phenomena (facultative) At least one abrupt change (within 3 m) of megabenthic species composition visible	Physical impact other than iceberg scouring generated these structures Scour marks with two non-abrupt zones at their margins are not included Abrupt change is due to other reasons (poison, disease, chemical defence, attraction)
R0	Abundance of sessile mega-epibenthos in the affected area < 10% of outside	
R1	Less dominant species. No. of species making up 66% of abundance in the affected areas is < 1/3 compared to that at non-disturbed sites, *or* ≤ 5 dominant species	
R2	Abrupt change in species composition but higher no. of dominant species, higher no. of total species and higher abundance than in R0 and R1, *and* missing large hexactinellid sponges (≥20 cm tall) in distinct part of the transect	When dispersion of large hexactinellids is very patchy or abundance of hexactinellids is low, areas without hexactinellids considered as R2 although they are UD
ID	> 33% of large hexactinellids (≥20 cm tall) on the side in a distinct part of the transect *or* > 66% of large hexactinellid sponges (≥20 cm tall) inclined (≥30°)	
UD	No abrupt change in community structure	Abrupt change not visible because it is situated outside the observed transect See R2
??	Criteria for R0, R1 or R2 not completely fulfilled but no convincing evidence for UD situation	

Most stations comprised 60–110 min of video observation of the seafloor using an ROV. In order to achieve straight transects, the ROV was left suspended in mid-water and the ship allowed to drift. Co-ordinates of the transects were taken every minute using GPS. The transects had an average length of ~1.5 km and a width of ~50 cm, which was indicated by two parallel laser beams acting as a scale on the image. All organisms ≥1 cm were identified. In addition, ≥100 underwater photographs were taken by the ROV at each station to identifying the taxa.

Six stages of recolonisation were determined at 1-s intervals from the video transects. Theoretically, it is easy to imagine a homogenous benthic structure in which grounding icebergs have opened delimited scours. In reality, the variability in mature communities and the recovery stages will soon disabuse anyone of that ideal. Therefore, we defined scour marks at different stages of recolonisation (Table 1).

To estimate the significance of iceberg scouring in comparison to other types of disturbance, we added data to a table compiled by Garwood et al. (1979) using the following parameters:

- Standardised rate of disturbance (RD). Percentage of an area disturbed over a period of 100 years; can be calculated from any data that contain information on the proportion of a disturbed area and the period to which the disturbance refers.
- Recovery time (RT). Time required to return to the pre-disturbance community structure.
- Impact factor = total area disturbed ($TD = RD/100 \times RT$).

Results

In total, almost 50 km of video transects were analysed. The number of areas where each stage was found ranged between 7 for R1 and "indirectly disturbed" (ID) at "level plateau", and 82 for R2 at "small iceberg bank".

Figure 1 shows the proportions of the stages of recolonisation. At "level plateau", without concentrations of icebergs, undisturbed conditions (UD) were found for

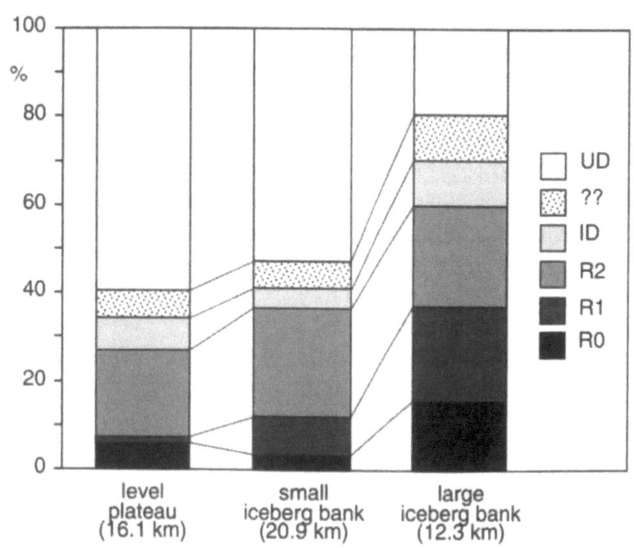

Fig. 1 Relative proportions of the different recolonisation stages per site category with distances observed in parentheses (*R0–R2* recolonisation stages, from younger to older; *ID* indirectly disturbed; *UD* undisturbed; *??* area of uncertain status)

about 60% of the transects, whereas disturbed areas (R0, R1, R2, ID) accounted for 34%. The proportion of disturbed areas (R0, R1, R2, ID) was higher (42%) for "small iceberg banks", with stage R2 forming a major component (26%). However, the most recent scour marks (R0) were only found on almost 4% of the transects. The effects of iceberg scouring were dramatic at the "large iceberg bank" off Auståsen. Over 70% of the analysed area was found to be disturbed (R0, R1, R2, ID), with only 20 % undisturbed.

Taking the maximum age of recent scour marks, R0 and R1, to be 15 years and their proportion of 7.3 and 37.3% in this period, a standardised rate of 50 and 250% area disturbed in 100 years can be calculated for the "large iceberg bank" and the "level plateau", respectively (Table 2). This estimate is based on a growth study of one pioneer, the bryozoan, *Cellaria incula* (Brey et al. 1999). Further assuming that the recovery time is 250–500 years, based on the presence of large hexactinellid sponges and the slow growth rate of other invertebrates (see Gutt 2000), the total percentage of the area disturbed by icebergs off Auståsen is between 625 and 1250%. This is more than twice that of tree falls in a tropical forest in Panama (270–640%) or fire and tree falls in temperate forests (150–600%). For "level plateau" areas without concentrations of icebergs, the total area disturbed amounted to 125–250%.

Discussion

The aims of this study were to assess the impact of grounding Antarctic icebergs to the benthic assemblage and to compare its effect at a global scale. To draw general conclusions, the results can be assumed to be representative of the entire investigation area and, to a certain degree, also the entire Antarctic shelf. There are sufficient numbers of recolonised and fresh scour marks, as well as undisturbed parts of the transects, for each of the different categories to represent this part of the shelf. The total length (49.2 km) of the video transects is relatively long compared to the extent of the investigation area (200×40 km), and the opportunity to obtain data for each metre of the transect provides extremely detailed results compared to traditional sampling. Deeper areas (> 350 m) were not sampled because the probability of iceberg scouring is low. At this depth, only 3 out of the 20 stations videotaped or photographed between 1988 and 1991 showed scours. A true mapping would yield even more precise ecological information; however, the applied stratified random sampling strategy provides relatively high efficiency especially in waters where the probability of ice cover is high.

The study area is generally typical of the eastern Antarctic. It comprises a narrow shelf with both small and large banks and inner-shelf depressions (500–800 m). A coastal current superimposed by the tides flows over the area, being bordered by an ice-shelf coast, grounded ice walls or glacier cliffs. As a consequence, there is not much shallow water (< 150 m). It is theoretically possible for an iceberg to circumnavigate the entire continent before disintegrating, if a half-life expectancy of 2–4 years and a net drift in the coastal current of < 2 km day^{-1} is assumed (Gutt 2001). This means that the long-term pattern of iceberg grounding might be determined to a greater extent by the bottom topography, current regime and sea-ice cover than by the sites of intense calving activity.

The classification of the sea floor and its megafauna into different stages of recolonisation and undisturbed areas shows an unexpected result. The proportions of the early stages R0 and R1 found on the "level plateau" without iceberg concentrations were not much lower than those on the "small iceberg bank". However, the two bank categories differed considerably from each other. The following explanation seems to be most likely (Fig. 2). If the seafloor is relatively even, icebergs make contact with the bottom and continue to drift, leaving relatively long scour marks behind them which are randomly distributed. In contrast, when elevations have relatively steep slopes and the minimum water depth is quite shallow, these structures can act as obstacles for icebergs of all sizes. If they run aground, they build up a certain concentration called "small iceberg bank", but the seafloor is not intensively disturbed due to the small size of individual scours. At large icebergs banks, such as Auståsen, the effects described for both categories are combined. In addition, if the water is relatively deep, and assuming that there is a correlation between the draught and horizontal extent of an iceberg, the probability of generally large scours is high. Consequently, the percentages of recently disturbed areas (R0 and R1) are 3–4 times higher than at the "level plateau" or "small iceberg bank".

Comparison of the large-scale processes in different biomes is one of the great challenges facing modern ecosystem research. The reason why, so far, there are only a few comparable results is that such an approach requires the measurement of disturbance descriptors which are not yet generally realised. Nevertheless, proxies of the significance of various impacts are available for a number of large ecosystems. Impact factors above 100, as calculated in Table 2, indicate that statistically the ecosystem will never reach a mature stage. In these systems, the only chance for the biota to develop undisturbed over a longer period, is if the disturbance is patchy. Thus, it is important to specify the size of the area the results refer to, and how large the ecosystem is for which the results are considered to be representative.

Results taken from an intensively scoured area, such as an Antarctic "large iceberg bank", show that icebergs seem to cause the most intensive impact that has so far been investigated in any ecosystem. Apart from the use of the term "Armageddon" (see Introduction), it is justified to classify such a phenomenon as extreme. The more general situation, represented by the "level plateau", is still among the five most significant disturbances, with only tree falls in the tropical forests of Panama being clearly more significant. The high ranking of iceberg scouring is mainly due to the assumed slow

Table 2 Natural disturbances in marine and terrestrial ecosystems: rates of disturbance and recovery times. "% area disturbed" refers to "Period (years)"; standardised rate of disturbance (RD) = percent of area disturbed in 100 years. *Impactfactor* $= RD/100 \times RT$

Type of disturbance	Scource	% area disturbed	Period (years)	Std. rate of disturbance	Recovery time, RT (years)
Large iceberg bank, megafauna, Antarctica	This study	37	15	250	250-500
Tree falls, tropical forest, Panama	Brokaw (1985)	100	62-114	88-161	300-400
Fire, tree falls, temperate forests	Runkle (1985)	0.5-2.0	1	50-200	300
Cyclones, coral reefs, Belize	Pearson (1981)	100	30	330	60-100
Tree falls, tropical forest, Panama	Hubbell et al. (1999)	72	13	554	20-50[a]
Icebergs, level plateau, megafauna, Antarctica	This study	7.2	15	50	250-500
Ice scouring, macrofauna, shallow Antarctic	Peck and Bullough (1993)	100	50-75	133-200	65
Sea-ice, exposed rocks, Massachusetts, USA	Wethey (1985)	100	5	2000	4
Earthquake landslides, tropical forest, New Guinea	Garwood et al. (1979)	8-16	100	8-16	200-500
Foraging gray whales, macrofauna, Bering Sea	Nerini and Oliver (1983) Johnson and Nelson (1984)	9-27 5.6	1 1	560-2700	≥2
El Niño, shallow benthos, south-eastern Pacific	Arntz and Fahrbach (1991)	100	4	2500	2
Sea-ice keels, macrofauna, Canadian Arctic	Conlan et al. (1998)	40	50	80	≤50
Sea-ice, rocky shores, Halifax, Canada	McCook and Chapman (1991) Minichton et al. (1997)	100	26	385	4
Earthquake landslides, tropical forest, Panama	Garwood et al. (1979)	2	100	2	200-500
Ice, meiofauna, shallow Antarctic	Peck et al. (1999)	100	50-75	143-200[b]	0.25

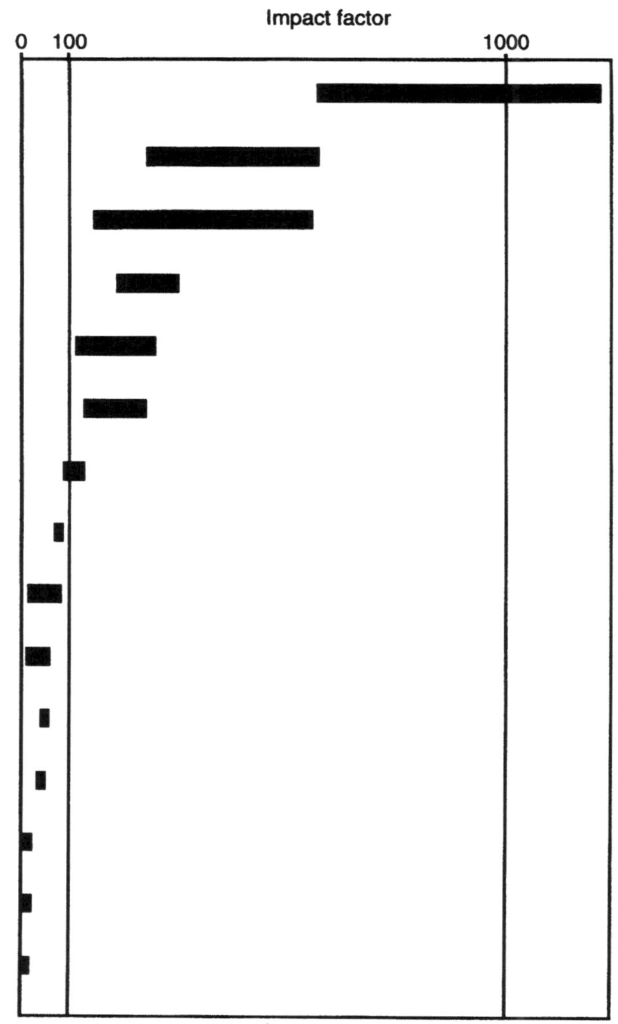

[a] From Garwood et al. 1979
[b] From Peck and Bullough 1993

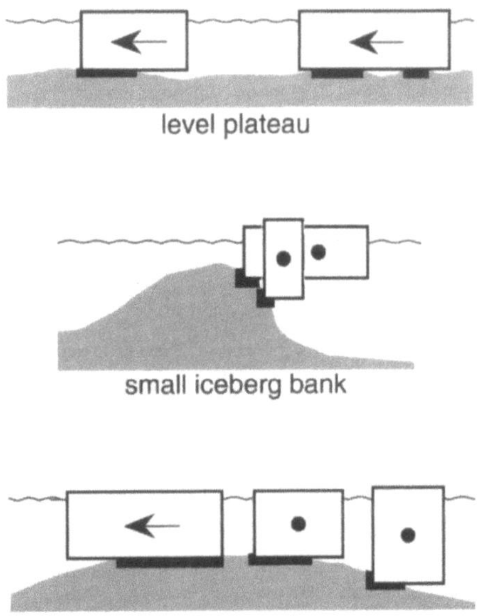

Fig. 2 Schematic presentation of the three different categories of iceberg scouring in relation to bottom topography

regeneration, whereas its rate belongs to the three lowest values listed for the various ecosystems. Other disturbances in the marine realm are cyclones in coral reefs, where the disturbance rate is clearly higher but the recovery is faster. All other marine disturbances are clearly less severe. Also, the results found by Peck et al. (1999) for shallow Antarctic waters are lower than those calculated for our "level plateau" category, which might be representative of most of the entire high-Antarctic shelf.

Future work might amend our calculations for the following paradoxon. The percentage of young recolonisation stages is too high in relation to that of the later stage (R2), assuming that the above-mentioned recovery period is relatively accurate. Two reasons for this discrepancy are possible, both of which would lead to a decrease in impact significance: (1) a false classification, R1 instead of R2, since these discrete definitions in fact represent a continuous gradient of a varying faunistic development; (2) the age of a mature assemblage, indicated by the presence of large hexactinellid sponges, is less than 250–500 years. This assumption is based on Dayton and Robilliard (1971) "...almost all the sponges are certainly much older than ... *Odontaster validus* which ... could reach ages of over 100 years". However, it is not only the presence of the extremely slow-growing sponges that is necessary to reach a relatively mature benthic community. Associated animals have to re-establish themselves after the sponges have reached a certain size. There is also an argument that the impact factor is underestimated. The 15 years indicate only the minimum age for stage R1; it is also possible that the bryozoan, *Cellaria incula*, on whose age (Brey et al. 1999) this estimate was based, actually started growing several years after the pristine sediment became available for recolonisation. Furthermore, it cannot be ignored that climatic shifts or local phenomena may have led to a variability in the grounding rate of icebergs in the past decades or centuries.

References

Arntz WE, Fahrbach E (1991) El Niño – Klimaexperiment der Natur. Birkhäuser, Basel

Arntz WE, Gutt J (eds) (1997) The Expedition ANTARKTIS XIII/3 (EASIZ I) of RV "Polarstern" to the eastern Weddell Sea in 1996. Ber Polarforsch 249:1–148

Arntz WE, Gutt J (eds) (1999) The expedition ANTARKTIS XV/3 (EASIZ II) of RV "Polarstern" in 1998. Ber Polarforsch 301: 1–229

Borkaw NVL (1985) Treefalls, regrowth, and community structure in tropical forests. In: Pickett STA, White PS (eds) The ecology of natural disturbance and patch dynamics. Academic Press, Orlando, pp 53–69

Brey T, Gerdes D, Gutt J, Mackensen A, Starmans A (1999) Growth and age of the Antarctic bryozoan *Cellaria incula* on the Weddell Sea shelf. Antarct Sci 11:408–414

Conlan KE, Lenihan HS, Kvitek RG, Oliver JS (1998) Ice scour disturbance to benthic communities in the Canadian High Arctic. Mar Ecol Prog Ser 166:1–16

Dayton PK, Robilliard GA (1971) The benthic community near McMurdo station. Antarct J US 6:54–56

Garwood NC, Janos DP, Brokaw N (1979) Earthquake-caused landslides: a major disturbance to tropical forests. Science 205:997–999

Gray JS (1977) The stability of benthic ecosystems. Helgol Wiss Meeresunters 30:427–444

Gutt J (2000) Some "driving forces" structuring communities of the sublittoral Antarctic macrobenthos. Antarct Sci 12:297–313

Gutt J (2001) On the direct impact of ice on marine benthic communities, a review. Polar Biol 24:553–564

Hubbell SP, Foster RB, O'Brien ST, Harms KE, Condit R, Wechsler B, Wright SJ, Loo de Lao S (1999) Light-gap disturbances, recruitment limitation, and tree diversity in a neotropical forest. Science 283:554–557

Johnson KR, Nelson CH (1984) Side-scan sonar assessment of gray whale feeding in the Bering Sea. Science 225:1150–1152

McCook LJ, Chapman ARO (1991) Community succession following massive ice-scour on an exposed rocky shore: effects of *Fucus* canopy algae and of mussels during late succession. J Exp Mar Biol Ecol 154:137–169

Minichton TE, Scheibling RE, Hunt HL (1997) Recovery of an intertidal assemblage following a rare occurrence of scouring by sea ice in Nova Scotia, Canada. Bot Mar 40:139–148

Nerini MK, Oliver JS (1983) Gray whales and the structure of the Bering Sea benthos. Oecologia 59:224–225

Pearson RG (1981) Recovery and recolonization of coral reefs. Mar Ecol Prog Ser 4:105–122

Peck LS, Bullough LW (1993) Growth and population structure in the infaunal bivalve *Yoldia eightsi* in relation to iceberg activity at Signey Island, Antarctica. Mar Biol 117:235–241

Peck LS, Brockington S, Vanhove S, Beghyn M (1999) Community recovery following catastrophic iceberg impacts in a soft-sediment shallow-water site at Signy Island, Antarctica. Mar Ecol Prog Ser 186:1–8

Pickett STA, White PS (1985) The ecology of natural disturbance and patchy dynamics. Academic Press, Orlando

Runkle JR (1985) Disturbance regimes in temperate forests. In: Pickett STA, White PS (eds) The ecology of natural disturbance and patch dynamics. Academic Press, Orlando, pp 17–33

Sousa WP (1984) The role of disturbance in natural communities. Annu Rev Ecol Syst 15:353–391

Thiery RG (1982) Environmental instability and community diversity. Biol Rev 57:691–710

Wethey DS (1985) Catastrophe, extinction, and species diversity: a rocky intertidal example. Ecology 66:445–456

Polar Biol (2001) 24: 502–507
DOI 10.1007/s003000100246

M. Brenner · B.H. Buck · S. Cordes · L. Dietrich
U. Jacob · K. Mintenbeck · A. Schröder · T. Brey
R. Knust · W. Arntz

The role of iceberg scours in niche separation within the Antarctic fish genus *Trematomus*

Abstract Species of the Antarctic fish genus *Trematomus* occupy different trophic niches. It is not clear, however, whether small-scale variability in benthic community structure affects niche separation. Therefore abundance and biomass of fish were determined and stomach content and food composition were compared in areas affected by iceberg scours and unaffected areas in the Weddell Sea. *Trematomus eulepidotus*, *T. lepidorhinus* and *T. scotti* dominate undisturbed areas, whereas *T. nicolai* and especially *T. pennellii* dominate disturbed areas. Total stomach content and number of prey taxa per fish are higher in preferred than in non-preferred areas. These findings indicate that small-scale horizontal patterns caused by iceberg scours play a distinct role in *Trematomus* niche separation.

Introduction

Iceberg grounding has been reported from many Antarctic shallow-water and shelf bottoms down to several hundred metres (Barnes et al. 1984; Lien et al. 1989; Woodworth-Lynas et al. 1991; Dowdeswell et al. 1993). Type and strength of physical iceberg impact on the benthos vary (e.g. Gruzov et al. 1968; Kauffmann 1974), but it is estimated that 5% of the Antarctic shelf shows detectable iceberg scours (Gutt et al. 1999). Iceberg impact causes various degrees of damage to the benthos up to local extinction of the fauna (Dayton et al. 1974; Brey and Clarke 1993; Arntz et al. 1994; Peck et al. 1999), and disturbed bottoms are subsequently recolonised (Starmans 1997; Gutt et al. 1998). Temporal distribution of iceberg scouring and time-scales of recolonisation are still poorly understood, but seem to be an intrinsic feature of Antarctic shallow-water and shelf benthic community dynamics, leading to a spatial and temporal patchwork of benthic recovery. Hence, mobile non-benthic species exploiting benthic resources such as food or shelter may have adapted to this pattern, e.g. by specialising on certain recovery stages. This study investigates whether such adaptations are detectable in demersal and bentho-pelagic fish species of the genus *Trematomus*.

Please note: When citing this article, please use the reference to the original source in the journal Polar Biology as given on the top of this page.

M. Brenner · B.H. Buck · S. Cordes · L. Dietrich
U. Jacob · K. Mintenbeck · A. Schröder
T. Brey (✉) · R. Knust · W. Arntz
Alfred Wegener Institute for Polar and Marine Research,
P.O. Box 120161, 27515 Bremerhaven, Germany
E-mail: tbrey@awi-bremerhaven.de

Materials and methods

Sampling

Trematomus species were caught during the RV "Polarstern" expedition ANT XV/3 (EASIZ II, 1998) into the eastern Weddell Sea, using a 140-ft bottom trawl (GSN) and an Agassiz trawl (AGT) with 3-m mouth opening and 10-mm mesh size in the cod end. Disturbed and undisturbed bottom patches were identified by underwater video. Subsequently, eight GSN hauls were taken in undisturbed areas and two GSN hauls in disturbed areas; additionally, some stations were trawled by AGT (all hauls between 200 and 400 m water depth). Fishes caught were identified according to Gon and Heemstra (1990), counted, measured to the lower centimetre and weighed. Abundance and biomass from GSN samples were standardised to 15 min trawling time for numerical comparisons. AGT catches were used for stomach content analysis only.

Stomach content analysis

Prey organisms found in the stomachs of 296 fish (Table 1) were identified as exactly as possible, counted and weighed. In the case of fragmented animals, we used typical parts such as heads or eyes to infer the number of individuals. If taxonomic identification but neither counting nor weighing was possible, the number was set to 1 and the weight was set to 0.01 g for further statistical calculations.

The effect of fish size was removed from further analysis by standardising stomach content weights to a 50-g standard fish for each species using the following procedure:

Table 1 Fish used for stomach content analysis. Distribution of numbers among species and stations (*np* not present; *Depth* trawl on ground – start of recovery). For further information on stations, see Anonymous (1999)

Station no.	48/071	48/078	48/062	48/194	48/222	48/220
Region	Kap Norvegia (North)	Drescher Inlet	Kap Norvegia (North)	Kap Norvegia	Kap Norvegia	Kap Norvegia
Condition	Undisturbed	Undisturbed	Undisturbed	Undisturbed	Disturbed	Disturbed
Gear	GSN	GSN	AGT	AGT	GSN	GSN
Depth (m)	281–301	391–390	241–253	244–246	267–234	272–236
Time	1549–1610 hours	0654–0711 hours	1738–1746 hours	0931–0941 hours	1427–1437 hours	1057–1106 hours
Species						
Trematomus lepidorhinus	8	4	2	4	7	1
T. eulepidotus	np	27	np	np	7	3
T. pennellii	np	np	np	np	52	29
T. nicolai	np	3	np	np	25	9
T. scotti	45	51	2	16	np	1

1. A linear regression between fish weight (WF) and empty stomach weight (WES) was established:

$$WES = a + b^* WF$$

2. Subsequently, number ($NP_{j,i}$) and weight ($WP_{j,i}$) of each prey item j found in the stomach of a fish i of weight WF_i was standardised to a 50-g standard fish (SF) by:

$$NP(SF)_{j,i} = NP_{j,i}{}^*(a + b^* 50)/(a + b^* WF_i)$$

$$WP(SF)_{j,i} = WP_{j,i}{}^*(a + b^* 50)/(a + b^* WF_i)$$

The sum of all $NP(SF)_{j,i}$ and $WP(SF)_{j,i}$ is defined as standardised total prey number ($STPN_i$) and standardised total prey weight ($STPW_i$) in this study. Based on these standardised data, the main food index (MFI) of major prey taxa for each fish species was computed to compare the diet of the different species:

$$\mathrm{MFI} = \sqrt{\frac{A_I + F}{2} * A_G}$$

where A_I is numerical dominance (in % of mean STPN), F is presence (in % of all stomachs analysed), and A_G is weight dominance (in % of mean STPW). Effects of area and species on total stomach content and number of prey taxa per fish were analysed by ANOVA with subsequent Games-Howell post-hoc test on differences between means.

Results

Abundance, biomass and size of fish

In total, 689 *Trematomus* spp. of 7 species were collected by 10 GSN hauls (undisturbed 8 hauls, 500 fish; disturbed 2 hauls, 189 fish). The majority of the catch consisted of *Trematomus scotti* (Boulenger, 1907) (N=247), *T. pennellii* (Regan, 1914) (N=140), *T. eulepidotus* (Regan, 1914) (N=120) and *T. lepidorhinus* (Pappenheim, 1911) (N=111), whereas *T. nicolai* (Boulenger, 1902) (N=36), *T. loennbergii* (Regan, 1913) (N=16), *T. hansoni* (Boulenger, 1902) (N=17) and juveniles (N=2) were less abundant. The last three species and juveniles were excluded from further analysis. The GSN samples indicate distinct differences in abundance and biomass between disturbed and undisturbed areas (Fig. 1); *T. eulepidotus*, *T. lepidorhinus* and *T. scotti* dominated undisturbed areas, whereas *T. nicolai* and, especially, *T. pennellii* dominated disturbed areas. The pooled length-frequency distribution of all *Trematomus* species differs significantly between disturbed and undisturbed areas (Kolmogoroff-Smirnoff test, $P < 0.05$), indicating that there are fewer small but more large individuals present in disturbed areas.

Stomach contents

Because of empty cells in the areaxspecies matrix (*T. pennellii* present in disturbed areas only; only 1 individual of *T. scotti* in disturbed areas), we started with a two-way ANOVA, including the remaining three species (*T. eulepidotus*, *T. lepidorhinus*, *T. nicolai*) to analyse effects of area and species on total stomach content. There are significant ($P < 0.05$) effects of species and species-area interaction on total stomach content, but no significant effects of area (Tables 2, 3). One-way ANOVAs of total stomach content versus area and versus species, respectively, detected significant ($P < 0.05$) effects of area as well as of species. Total stomach content was significantly higher in undisturbed areas (1.432 g WM; SE range 1.344–1.523 g compared to 0.675 g WM; SE range 0.617–0.722 g) and in the species *T. pennellii* and *T. scotti* (Tables 4, 5).

Food composition

The total number of different taxa consumed by the several *Trematomus* species is distinctly different between disturbed and undisturbed areas (Fig. 2). *T. eulepidotus*, *T. lepidorhinus* and *T. scotti* consumed a higher number of taxa in undisturbed than in disturbed areas, whereas *T. nicolai* seemed to feed on more taxa in disturbed areas. The two-way ANOVA with the three species occurring in both areas indicated significant effects of species-area interactions, but neither of area nor of species. One-way ANOVAs including all species

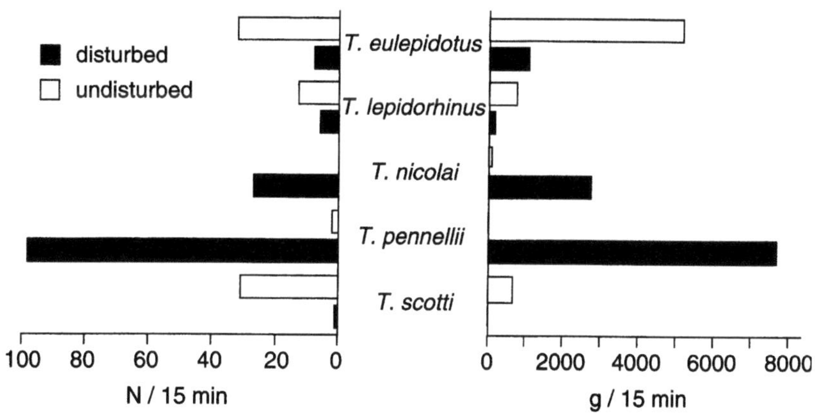

Fig. 1 Average abundance (N 15 min^{-1}) and biomass (g 15 min^{-1}) of all *Trematomus* species caught in bottom trawls (GSN). Samples from disturbed (■) and undisturbed (□) areas. *Trematomus loennbergii*, *T. hansoni* and juveniles are neglected because of extremely low numbers

Table 2 Analysis of differences in standardised total prey weight (STPW; log (1 + gWM)) among *T. eulepidotus*, *T. lepidorhinus* and *T. nicolai* and between disturbed and undisturbed areas by two-way ANOVA

ANOVA				
Source	*df*	Mean square	*F*-value	*P*-value
Area	1	0.059	3.775	0.055
Species	2	0.075	4.820	0.010
Area×Species	2	0.085	5.411	0.006
Residual	94	0.016		

showed no significant difference between disturbed and undisturbed areas ($P = 0.414$); on average, each fish had fed 1.9 (min. 0, max. 7) different taxa. Among species, total number of food taxa per fish was significantly higher in *T. pennellii* and *T. scotti* (Tables 6, 7).

Figure 3 shows the main food index (MFI) of the six most frequent prey taxa for the five *Trematomus* species and for all species combined, in disturbed as well as undisturbed areas. Regarding all *Trematomus* species, MFIs of amphipods and isopods are higher in disturbed areas, whereas the MFI of polychaetes is lower. In terms of number and biomass per standard stomach, polychaetes (mean $N = 9.27$, mean $B = 0.16$ g), amphipods (mean $N = 7.14$, mean $B = 0.14$ g), euphausiaceans (mean $N = 0.88$, mean $B = 0.09$ g) and isopods (mean $N = 1.37$, mean $B = 0.02$ g) were the most important prey items. A two-way ANOVA of number (or biomass) versus area (disturbed and undisturbed) and *Trematomus* species (five species) showed significant differences among *Trematomus* species with respect to all prey taxa, but only one significant effect of area: biomass of euphausiaceans was significantly ($P < 0.05$) lower in disturbed areas (0.016 g stomach^{-1} compared to 0.155 g stomach^{-1}).

In summary, the diet of *T. scotti* and *T. pennellii* is quite similar. Both feed almost exclusively on benthic organisms, with highest MFI values for amphipods and polychaetes. *T. nicolai* and *T. lepidorhinus* feed on benthic (amphipods, isopods, polychaetes) as well as pelagic organisms (euphausiaceans and others). Pelagic euphausiaceans and amphipods dominate the diet of *T. eulepidotus*.

Discussion

Methods

Our study suffers from several methodical shortcomings, which may have introduced some bias in the data. The separation of benthic communities into disturbed and undisturbed according to visual inspection neglects the various stages of benthic community succession. Hence, variability within these two groups may obscure differ-

Table 4 Analysis of differences in standardised total prey weight (STPW; log (1 + gWM)) of fish among species by ANOVA

ANOVA				
Source	*df*	Mean square	*F*-value	*P*-value
Species	4	1.034	45.013	< 0.001
Residual	290	0.023		

Table 3 Mean standardised total prey weight (STPW) of *T. eulepidotus*, *T. lepidorhinus* and *T. nicolai* in disturbed and undisturbed areas. SE range: mean ± one standard error

Mean STPW (g WM)						
Species	Disturbed			Undisturbed		
	Mean STPW	SE range	*N*	Mean	SE range	*N*
Trematomus eulepidotus	0.442	0.365–0.524	10	0.816	0.675–0.968	27
T. lepidorhinus	0.026	0.000–0.052	8	0.629	0.515–0.750	18
T. nicolai	0.442	0.396–0.496	34	0.140	0.084–0.199	3

Table 5 Games-Howell post hoc test on differences between mean STPW (based on ANOVA shown in Table 4; nsd: not significantly different; *: significantly different)

Games-Howell post-hoc test on differences between means ($a = 0.050$)

Species *N*	Mean STPW	*T. eule* 37	*T. lepi* 26	*T. nicol* 37	*T. penn* 80	*T. scott* 115
Trematomus eulepidotus	0.706	/	nsd	nsd	nsd	*
T. lepidorhinus	0.413		/	nsd	*	*
T. nicolai	0.413			/	*	*
T. pennellii	0.910				/	*
T. scotti	1.818					/

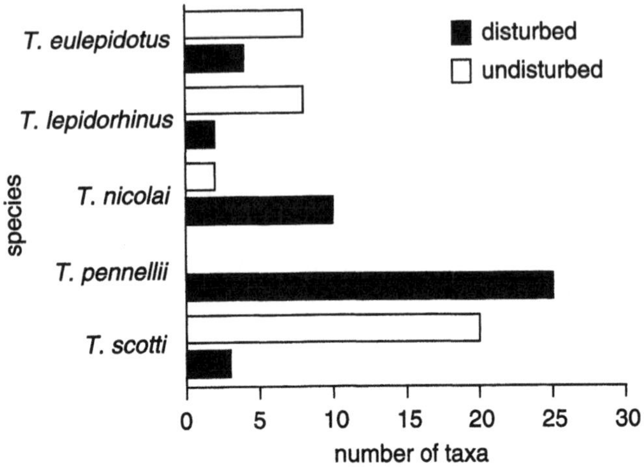

Fig. 2 Total number of taxa consumed by each *Trematomus* species in disturbed (■) and undisturbed (□) areas

Table 6 Analysis of differences in the average number of prey taxa per fish among fish species by ANOVA

ANOVA

Source	*df*	Mean square	*F*-value	*P*-value
Species	4	9.893	7.310	<0.001
Residual	290	1.353		

ences between them. The number of GSN hauls from undisturbed ($N = 8$) and disturbed ($N = 2$) areas does not allow for statistical comparison of fish abundance and biomass. Variability of trawl efficiency is unknown and may depend on area type. Hauls were carried out at different hours of the day, which may affect stomach contents. Finally, neither the benthic nor the pelagic community was sampled; i.e. we have no information on selectivity of feeding by the five *Trematomus* species. The consistency of our results, however, indicates that the database is of sufficient quality, especially when taking into account that this is the first study of this type carried out in polar waters.

Food composition

Regarding feeding mode, our results confirm most findings of Ekau (1988) and Schwarzbach (1988). *T. scotti* and *T. pennellii* are demersal benthos feeders; *T. lepidorhinus* is intermediate between benthic and bentho-pelagic and *T. eulepidotus* is a pelagic feeder. Only *T. nicolai*, formerly characterised as a pure nekton feeder (Schwarzbach 1988), was found to feed in a rather bentho-pelagic manner in this study.

Overall diet composition of each species seems to be little affected by iceberg scouring. The main food taxa (Fig. 3) are the same in disturbed and in undisturbed areas, but there are differences in diet diversity (Fig. 2). The comparison of total stomach content (STPW) or number of prey taxa per fish between areas and among species (Tables 2, 3, 4, 5) did not reveal a clear spatial pattern. However, if we re-arrange the data according to the area preferred by each fish species as inferred from its abundance (see above), it becomes obvious that amount as well as diversity of prey ingested depend on whether a certain fish species feeds in its preferred area (disturbed or undisturbed, Fig. 1) or not. STPW (1.138 g vs 0.233 g), as well as average number of prey taxa per fish (2.00 vs 0.91), are significantly higher ($P < 0.001$) in the preferred area compared to the non-preferred area.

Table 7 Games-Howell post hoc test on differences between mean number of prey taxa per fish (based on ANOVA shown in Table 6; nsd: not significantly different; *: significantly different)

Games-Howell post-hoc test on differences between means ($\alpha = 0.050$)

Species Mean *N*	*N* Taxa	*T. eul* 37	*T. lepi* 26	*T. nicol* 37	*T. penn* 80	*T. scott* 115
T. eul	1.54	/	nsd	nsd	*	*
T. lepi	1.19		/	nsd	*	*
T. nicol	1.57			/	*	nsd
T. penn	2.35				/	nsd
T. scott	2.02					/

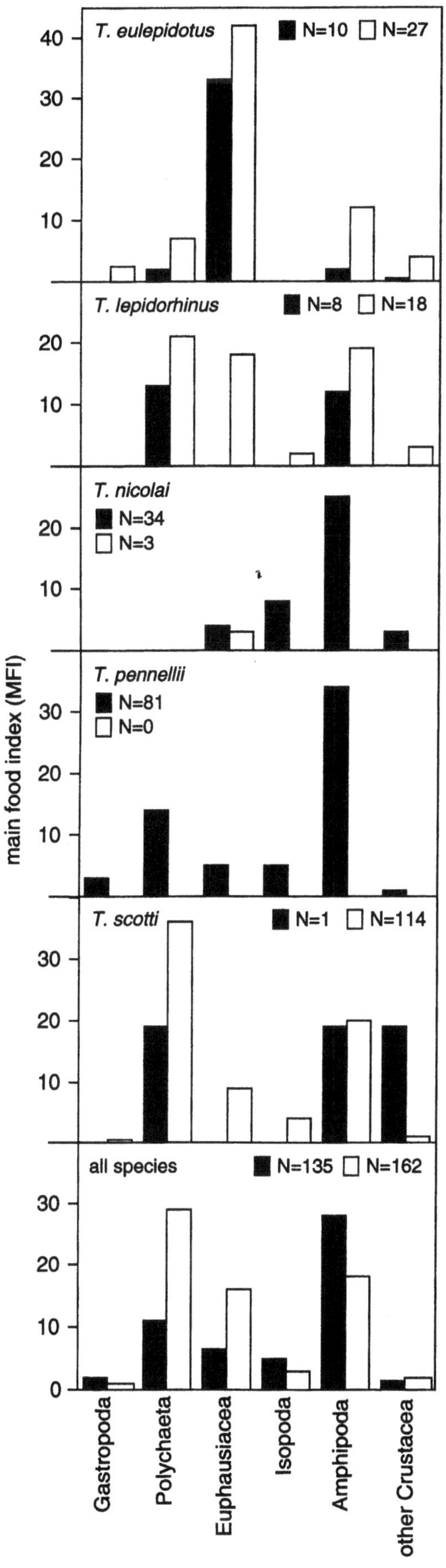

◄ **Fig. 3** Main food index (MFI) of the six most frequent prey taxa in disturbed (■) and undisturbed (□) areas

Distribution of *Trematomus* species

All Antarctic Nototheniidae including the genus *Trematomus* presumably evolved from a primarily demersal ancestor, radiating into many species with different lifestyles (Eastman 1985; Ekau 1988). Ecological niches of the Nototheniidae have been characterised mainly by large-scale geographic distribution, vertical distribution and diet (Ekau 1988; Schwarzbach 1988).

Regarding vertical distribution and diet (see above), *T. scotti* and *T. pennellii* are demersal, *T. lepidorhinus* and *T. nicolai* are bentho-pelagic, and *T. eulepidotus* is a pelagic species. Our study clearly indicates, however, that the small-scale horizontal distribution is another important feature of niche separation in *Trematomus* spp., especially between those species which occupy similar trophic niches. This is evident in the two demersal feeders, *T. scotti* and *T. pennellii*. Whereas *T. scotti* occurs in undisturbed areas only, *T. pennellii* prefers the disturbed sites, thus reducing competition by horizontal niche separation. This separation is detectable also in the bentho-pelagic species, where *T. lepidorhinus* is more abundant above undisturbed areas and *T. nicolai* is more abundant above disturbed areas. The truly pelagic *T. eulepidotus* is unlikely to be affected directly by the disturbance stage of the benthos, although its lower abundance and lower STPW over disturbed areas may indicate indirect effects via its prey taxa.

Regarding size, we found that individuals < 7 cm were clearly under-represented in disturbed areas, whereas large individuals seemed to be more abundant here. Predator-prey interactions may explain this difference. Small fish, i.e. juveniles or generally smaller species such as *T. scotti*, may use the three-dimensional structure of the intact Antarctic shelf community (Arntz et al. 1994) as shelter against predators. However, larger individuals need less shelter and may be better equipped to handle the seemingly larger and more mobile invertebrate prey in disturbed areas (T. Brey, personal observation).

Conclusions

Our study shows that niche separation among the five *Trematomus* species of the Weddell Sea shelf investigated here occurs along three interacting gradients: a trophic, a vertical and a small-scale horizontal gradient. All five species show distinct preference for either undisturbed or disturbed areas. Specialisation is most advanced in the demersal species, where predator-prey interactions may have caused divergent evolution regarding body size. These adaptations indicate that local disturbance by icebergs is a long-term intrinsic feature of Antarctic shelf communities, driving even the evolution of upper level predators.

References

Anonymous (1999) Station list. In: Arntz WE, Gutt J (eds) The expedition ANTARKTIS XV/3 (EASIZ II) of RV Polarstern. Rep Polar Res 301/99:208–223

Arntz WE, Brey T, Gallardo VA (1994) Antarctic zoobenthos. Oceanogr Mar Biol Annu Rev 32:241–304

Barnes PW, Schell D, Reimnitz E (1984) The Alaskan Beaufort Sea: ecosystem and environments. Academic Press, San Diego

Brey T, Clarke A (1993) Population dynamics of marine benthic invertebrates in Antarctic and Subantarctic environments: are there unique adaptations? Antarct Sci 8:3–6

Dayton PK, Robilliard GA, Paine RT, Dayton LB (1974) Biological accommodation in the benthic community at McMurdo Sound, Antarctica. Ecol Monogr 44:105–128

Dowdeswell JA, Villinger H, Whittington RJ, Marienfeld P (1993) Iceberg scouring in Scoresby Sund and on the East Greenland continental shelf. Mar Geol 111:37–53

Eastman JT (1985) The evolution of neutrally buoyant notothenoid fishes: their specializations and potential interactions in the antarctic marine food web. In: Siegfried RW, Condy PR, Laws RM (eds) Antarctic nutrient cycles and food webs. Springer, Berlin Heidelberg New York, pp 430–436

Ekau W (1988) Ökomorphologie nototheniider Fische aus dem Weddellmeer, Antarktis. Rep Polar Res 51/88:

Gon O, Heemstra PC (1990) Fishes of the southern ocean, 1st edn. J.L.B. Smith Institute of Ichthyology. CTP, Capetown

Gruzov YN, Propp MV, Pushkin AF (1968) Biological associations of coastal areas of the Davis Sea (based on the observations of divers). Sov Antarct Exped Inf Bull 6:523–533

Gutt J, Storch V, Arntz WE (1998) Benthosforschung im Südpolarmeer: Störung schafft Vielfalt. Biol Zeit 6/98:362–370

Gutt J, Buschmann A, Dijkstra J, Dimmler W, Piepenburg D, Teixido (1999) Study on benthic resilience of the macro- and megabenthos by imaging methods. Rep Polar Res 301/99:17–22

Kauffmann TA (1974) Seasonality and disturbance in benthic communities, Arthur Harbour, Antarctic Peninsula. Antarct J U S 9:307–310

Lien R, Solheim A, Elverhøi A, Rokoengen K (1989) Iceberg scouring and sea bed morphology on the eastern Weddell Sea, Antarctica. Polar Res 7:43–57

Peck L, Brockington S, Vanhove S, Beghyn M (1999) Community recovery following catastrophic iceberg impacts in a soft-sediment shallow-water site at Signy Island, Antarctica. Mar Ecol Prog Ser 186:1–8

Schwarzbach W (1988) Die Fischfauna des östlichen und südlichen Weddellmeeres: geographische Verbreitung, Nahrung und trophische Stellung der Fischarten. Rep Polar Res 54/88:

Starmans A (1997) Vergleichende Untersuchung zur Ökologie und Biodiversität des Mega-Epibenthos der Arktis und Antarktis. Rep Polar Res 250/97:

Woodworth-Lynas CMT, Josenhans HW, Barrie JV, Lewis CFM, Parrott DR (1991) The physical processes of seabed disturbance during iceberg grounding and scouring. Cont Shelf Res 11:939–961

IV.

Ecophysiology

Polar Biol (2001) 25: 31–40
DOI 10.1007/s003000100308

Lloyd S. Peck

Ecophysiology of Antarctic marine ectotherms: limits to life

Abstract The ecophysiology of Antarctic marine ectotherms is an area of active research. This review of recent progress covers metabolism, mitochondrial function, aerobic scope, growth and development. Energetics is shown to be a central feature of adaptation to temperature, with mitochondrial function and tissue oxygen supply important in setting limits to organismal size and performance.

Introduction

Since life was first discovered in polar seas scientists have been interested in how animals survive in what is perceived as a harsh environment. The earliest studies on polar marine invertebrates (Fox 1936; Thorson 1936) measured metabolic rates, and energetics has been at the centre of debates concerning polar adaptations ever since. In this review of recent advances in our understanding of adaptation to low temperature in marine organisms I will concentrate first on metabolic rate and energetics, then move to discuss growth and development, finishing with a brief outline of some very recent work on the relationship between temperature and animal size and distribution.

Metabolic rate

The very first measurements of the metabolic rate of polar marine ectotherms were concerned specifically with testing the hypothesis that "animals living at a very low temperature should show a relatively high level of standard metabolism at that temperature compared with others living normally at a high temperature" (Krogh 1916). Results were equivocal, with some species showing elevated metabolism, whilst others did not. Over two decades later, more sophisticated work on fish also produced variable results, but the data were interpreted as showing elevated metabolism in Antarctic species (Scholander et al. 1953; Wohlschlag 1964). Holeton (1974) suggested that polar species were sensitive to experimental conditions and that, when routine metabolism was measured carefully, rates in polar species were low. Since then data have accumulated showing that a low standard or routine metabolic rate is the norm for polar marine ectotherms (Clarke 1980; Johnston et al. 1991a; Chapelle and Peck 1995; Clarke and Johnston 1999), although a few studies have still reported elevated metabolism (e.g. Wells 1987).

Please note: When citing this article, please use the reference to the original source in the journal Polar Biology as given on the top of this page.

L.S. Peck
British Antarctic Survey, High Cross,
Madingley Road, Cambridge, CB3 0ET, UK
E-mail: lspe@pcmail.nerc-bas.ac.uk

Temperature and physiological compensation

Species living at low temperatures generally have lowered physiological rates. Antarctic ectotherm metabolism is significantly lower than for temperate and tropical equivalents. Comparisons across latitudes of the metabolism of species at their normal habitat temperatures show a monotonic increase from polar to tropical species in fish (Clarke and Johnston 1999) and bivalve molluscs (Peck and Conway 2000).

Careful investigations of short-term temperature effects on metabolism in Antarctic ectotherms have found low Q_{10} values when temperatures are within the normal range experienced, but markedly elevated values above the natural range (Davenport 1988; Peck 1989; Johnston et al. 1991a). Clarke (1991a) has discussed the use of Q_{10} coefficients in detail, and emphasised that great care should be taken in their use in metabolic studies because metabolism is a complex entity and different components of metabolism may have differing responses to temperature. This is especially so in investigations of seasonal effects as other factors including the onset of phases of reduced energy analogous to diapause in insects may produce spurious results.

It has long been recognised that an acute change in temperature can induce a sharp change in physiological rate. The different types of responses exhibited by organisms were classified in the 1950s by Precht et al. (1955) and Prosser (1958), and the four main types of response are shown in Fig. 1. These categorisations should only be used for short-term physiological responses over hours, days or possibly weeks, and should not be confused with compensation of function over evolutionary time-scales in low temperature species (Clarke 1991b). Time-course evaluations of physiological responses to temperature change in Antarctic ectotherms are however rare. Short-term studies of temperature effects on metabolism in an Antarctic eelpout (*Pachycara brachycephalum*) show little or no compensation following initial elevations in metabolism after a temperature rise (van Dijk et al. 1999). Metabolic changes with temperature in brachiopods and limpets (Peck 1989), and the bivalve molluscs *Limopsis marionensis* (Pörtner et al. 1999a) and *Laternula elliptica* (Peck, Pörtner and Hardewig, unpublished work) show partial compensation. Longer term examples of partial acclimation up to seasonal acclimation have also been found for polar species (Pörtner et al. 1998, 1999b). There are currently no known examples of over compensation from Antarctica.

Mitochondriae function

A significant component of respiratory demand is comprised of mitochondrial activity and the study of mitochondrial function at low temperatures has seen significant recent advances. Mitochondrial function appears to show only limited temperature compensation, and the number of mitochondria in fish muscle increases as temperature decreases. These studies are complicated by problems of making taxonomically valid comparisons of species with similar lifestyles. Thus motile species have different mitochondrial characteristics from sedentary ones. Comparing maximal respiration rate of mitochondria from red muscles of fish from differing thermal environments, Blier and Guderley (1993) showed there was little or no temperature compensation of function as temperature decreased. Johnston et al. (1994) demonstrated the same finding for different fish

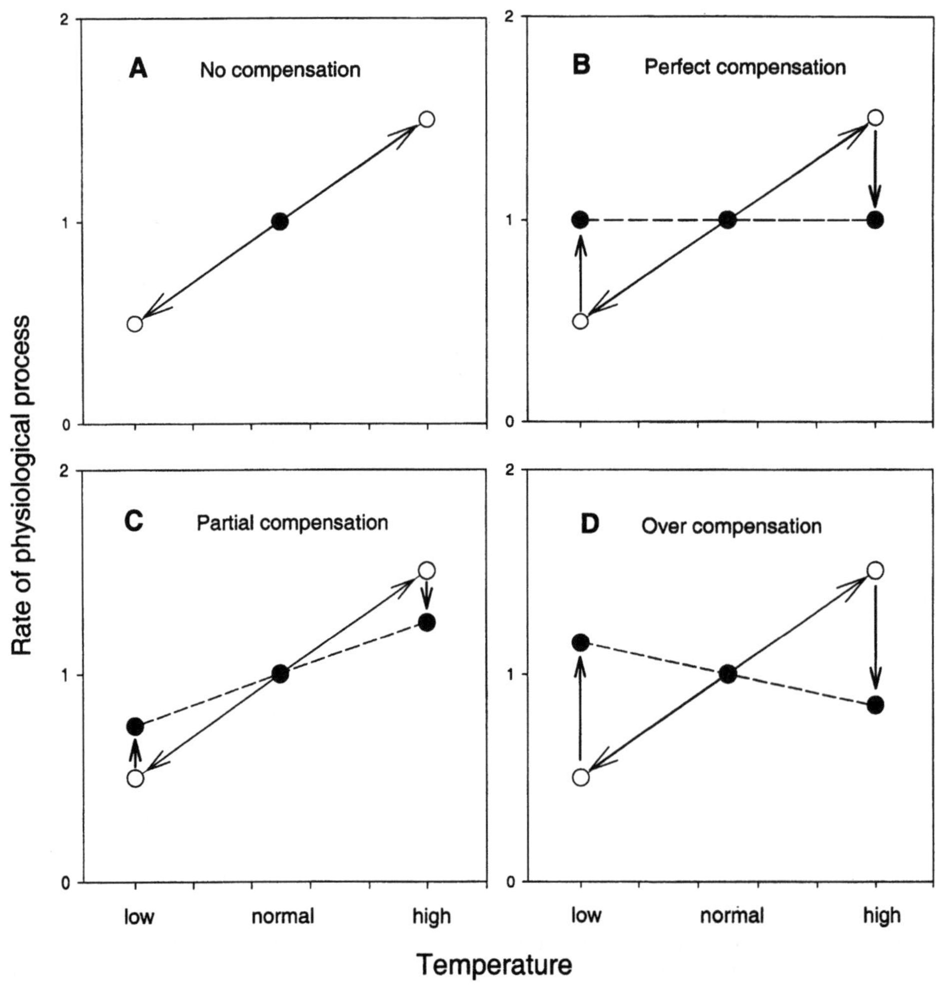

Fig. 1a–d The main types of physiological compensation exhibited with temperature change. With *no compensation* (**a**) physiological rates move to new levels with temperature change, either up or down, and stay at the new rate thereafter. *Perfect compensation* (**b**) occurs when rates return to initial levels after a temperature change. *Partial compensation* (**c**) is when rates return part of the way to previous levels, and *over compensation* (**d**) is when rates rebound beyond the initial physiological state. The fifth response is inverse acclimation, where physiological rates continue to move away from initial rates after a temperature change, instead of remaining constant or returning towards the initial rate. This response is rare, but may be seen when new physiological states, e.g. hibernation, are instigated by temperature change. These definitions follow Precht et al. (1955)

species across latitudinal zones, and these data have recently been summarised (Fig. 2; Clarke 1998; Guderley 1998). Johnston et al. (1998) refined this work by studying perciform fish from different latitudes with similar activity levels and swimming mode (predominantly labriform). Maximum mitochondrial oxygen consumption rate doubled with each 10°C temperature rise (Q_{10} =2). Combining all data for perciform fish produced a continuous relationship between mitochondrial respiration and temperature. Because the low temperature species were consistent with data from lower latitudes, Johnston et al. (1998) argued that this indicated no compensation at low temperatures.

In a recent examination of data on mitochondrial function Pörtner et al. (1999c) found significant differences between stenotherms and eurytherms in their adaptation to low temperatures. Seasonal low temperature acclimation of eurythermal fish causes an increase in the specific aerobic capacity of mitochondria (Guderley 1998). Intraspecific investigations of mitochondrial function in the eurythermal polychaete *Arenicola marina* in the North Sea and White Sea found that all functional states were 2 times higher in the White Sea than the North Sea (Sommer and Pörtner 1999). Whole animal resting metabolic rates were also higher in the colder White Sea than the North Sea. From this within-species study it was concluded that invertebrate life is more costly at higher latitudes, because whole animal oxygen consumption, mitochondrial densities and mitochondrial capacities were all higher. The differences between these findings and those of Johnston et al. (1998) may be explained by the differing thermal regimes of the organisms in the studies. Antarctic ectotherms are stenothermal, living in an ocean with an annual fluctuation of 3°C or less. The other fish used by Johnston et al. (1998)

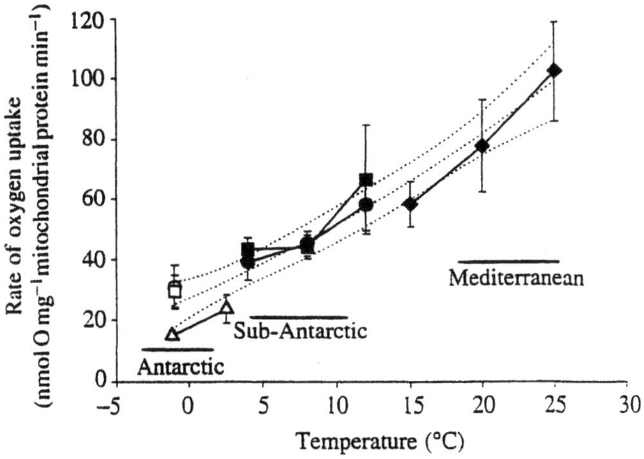

Fig. 2 Maximum mitochondrial respiration rate (state III) for fish species from Antarctica (*open symbols*), the sub-Antarctic (*closed circles* and *squares*) and the Mediterranean Sea (*diamonds*). Mitochondrial respiration declines by around 5 times as temperature is reduced from 25 C to 0°C. However, the data are best fitted by a second order polynomial equation ($rate = 27.21 + 2.16T + 0.28T^2$), shown as the *central dotted line*. *Outer dotted lines* are 95% confidence intervals. From Johnston et al. (1998)

were from the sub-Antarctic and Mediterranean where seasonal temperature fluctuations are around 5°C (Pörtner et al. 1999c). The polychaetes in the North Sea-White Sea comparison experience annual temperature fluctuations of over 15°C. Pörtner et al. (1999c) suggested that eurythermal species have enhanced costs at low temperatures, in line with theoretical expectations, but that in stenothermal species enhanced mitochondrial capacity is negligible. Because of this, whole animal aerobic capacities are lower in stenotherms (Pörtner et al. 1999c), which is in accord with the observations of low maximum aerobic capacities in Antarctic ectotherms (Peck 1998). It may also be associated with a loss of ability to acclimate to temperature change (e.g. Weinstein and Somero 1998).

Muscle function and temperature

Some low-temperature species can compensate maximum sustainable muscle capacity, and maximum sustainable swimming speeds are similar in Antarctic and temperate fish. Moderately active fish from both environments maintain speeds of 1.4 to 2.6 body lengths s^{-1} (Johnson and Bennet 1995; van Dijk et al. 1998). Data for burst swimming, i.e. the maximum instantaneous capacities of fish, are limited but do not appear to show compensation (Johnston et al. 1991b).

Sustainable swimming in fish is powered by red muscle which is predominantly aerobic, whereas burst swimming is powered by white fast twitch muscle which relies greatly on anaerobic glycolysis. Whilst aerobic swimming in low to moderate activity Antarctic teleosts appears well compensated, data on anaerobic capacity shows little compensation (Egginton and Davison 1998), and burst activity is limited by the pool size of high energy phosphates and the ability to remove and process end products of anaerobiosis. There is also evidence that aerobic capacity in active fish is markedly reduced in Antarctic species (Johnston et al. 1998), although data for zoarcid eelpout produce a different conclusion (Hardewig et al. 1998). Thus overall aerobic and anaerobic swimming capacities of Antarctic fish are limited, but good compensation has been achieved in the case of low to moderate activity aerobic requirements.

There are few assessments of activity in Antarctic marine invertebrates. Pelagic communities do contain moderately active species, but their activity rates have been suggested to be relatively low in polar taxa (Ikeda 1974). The limpet *Nacella concinna* has a 2–20 times lower capability for sticking to rock surfaces than comparable temperate limpets (Davenport 1988). Peck et al. (1993) found that *N. concinna* produced similar amounts of mucus to temperate limpets of similar lifestyle, whereas its walking speed (8–15 mm min^{-1} for routine walking and 20–60 mm min^{-1} at maximum, M. Palmer, personal communication) is below the range reported for temperate limpets (Branch and Marsh 1978). Similar speeds of locomotion were measured for

N. concinna at South Georgia Island (54–55°S, 36–38 W) by Davenport (1997), who also found that speed declined with temperature from 0°C to 12°C. More recently, burrowing rates in the Antarctic anemone *Heilcampoides* sp. were shown to be 1.6–4 times slower than the temperate *Peachia hastata* (Ansell and Peck 2000), and burrowing in the large infaunal bivalve *Laternula elliptica* also shows low compensation for temperature (A.D. Ansell, personal communication). Although more data are needed, the overall pattern appears to be one of little temperature compensation of activity in Antarctic marine invertebrates.

Compensation of sustainable swimming in Antarctic fish is achieved by an increase in numbers and also possibly internal surface area, of mitochondria in the muscles (Johnston et al. 1988, 1998; Archer and Johnston 1991). Within certain limits the power generated by aerobic muscle is related to the proportion of mitochondria in the muscle. The lack of compensation in the function rate of mitochondria at low temperatures in Antarctic fish muscle is thus partially overcome by increasing mitochondrial density. Mitochondrial density is also higher in active fish compared with slow moving species. In low activity Antarctic fish mitochondrial density is around 29–33%, whereas in Mediterranean fish with similar activity the value is 8–13% (Fig. 3, Johnston et al. 1998). The highest reported mitochondrial density in Antarctic fish is 56% for *Pleuragramma antarcticum*, although icefish species also have values around 50%, and haemoglobinless species tend to have higher densities than red-blooded species (Johnston et al. 1988, 1998). These are fairly slow moving species, yet values of 40% and over in warmer water habitats are restricted to active fish, with values of 46% and 29% being recorded for European anchovies and tuna respectively (Johnston 1982; Moyes et al. 1992). Values > 50% are also close to the theoretical limit for muscle function (Johnston et al. 1998).

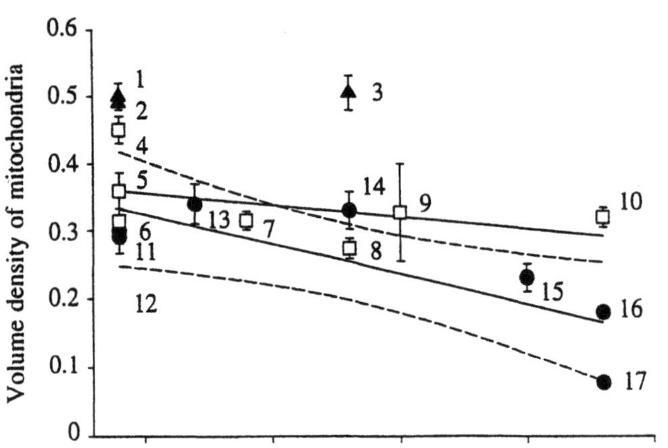

Fig. 3 Relationship between volume density of mitochondria and habitat temperature for red pectoral fin adductor muscle in perciform fish. Data are for three Channichthyid species (*triangles*), seven active pelagic species (*squares*) and demersal or only moderately active mid-water species (*circles*). Data for low activity species decline from around 0.35 at 0°C to 0.2 at 15–20°C. From Johnston et al. (1998)

Tyler and Sidell (1984) suggested that increasing mitochondrial density in muscles may overcome diffusion difficulties by reducing inter-mitochondrial distances. However, Hubley et al. (1997) combined nuclear magnetic resonance measures with a model of phosphate reaction/diffusion for goldfish acclimated to 5°C and 25°C to show that metabolite gradients did not change with temperature, thermal acclimation or mitochondrial volume density. These findings suggest that limitation of intracellular diffusion of high energy phosphates is not the stimulus for mitochondrial proliferation, at least in short term acclimation. Johnston et al. (1998) postulated that diffusion restrictions in other metabolites may be important. In contrast Pörtner et al. (1999c) suggested that adjustments of mitochondrial density and capacity are cost determined.

The second proposed mechanism for compensation of muscle function is increased internal surface area of mitochondria at low temperatures. Internal mitochondrial surface area is a function of internal membrane folding and is expressed as internal area per unit volume of mitochondrion. The highest values reported are for tuna red muscle (63–70 $\mu m^2\ \mu m^{-3}$) and hummingbird flight muscle (58 $\mu m^2\ \mu m^{-3}$). Values for Antarctic red blooded fish (36–37 $\mu m^2\ \mu m^{-3}$) are not significantly different from temperate and tropical perciformes with similar lifestyles (Johnston et al. 1998).

Metabolic scope and specific dynamic action of feeding

A second area where any low temperature limitation of mitochondrial function might be expected to be seen is in metabolic scope, which is the difference between energy utilised under maximum aerobic workload compared with the minimum needed to maintain life. It can be measured as the actual increment in energy utilisation, or as a factorial scope expressed as the ratio of the two extremes. The concept is usually understood in terms of maximum sustainable locomotor activity compared with minimum resting metabolism and measured as the relative rates of oxygen consumption. Increasing the numbers of mitochondria in muscles, as seen in Antarctic fish, should increase aerobic capacity, because it will increase the power available per unit body mass.

However, the limited data available suggest that polar fish have low metabolic scopes, with the exception of the zoarcid eelpout *Pachycara brachycephalum* which has a considerable aerobic reserve (Hardewig et al. 1998). Atlantic and Greenland cod have the same factorial scopes (×2.5) despite inhabiting cool temperate and polar localities respectively. Factorial scopes of 3.9 to 5.7 were found for *Pagothenia borchgrevinki* at Scott Base, Antarctica (Forster et al. 1991), and a scope of 5.7

was also recorded for *Notothenia coriiceps* at Signy Island, South Orkney Islands (Johnston et al. 1991a). These scopes are well within the range for temperate demersal species, and are much lower than active fish (e.g. salmonids) which have values from ×10 to ×12 (Clarke 1983).

These comparisons are all of factorial scopes (maximum metabolic rate/standard metabolism). Because standard metabolic rates are ×2–×10 lower in Antarctic ectotherms compared with temperate or tropical species, the energy available to an Antarctic species is much lower than for warm water equivalents (Fig. 4). A Q_{10} of 2 means that an animal at 20°C has 4 times as much energy available for work compared with a similar species at 0°C, and a Q_{10} of 3 means 9 times as much power is available at 20 C, even though the factorial scopes remain the same.

Data on metabolic scopes in Antarctic species are limited, and generally confined to fish. However, it has been argued that for sessile or very slow moving organisms demands for work other than locomotion push metabolism to its limit, and give a measure of metabolic scope. Thus the same maximum metabolic rate was found following temperature increases and feeding large meals in Antarctic brachiopods and limpets by Peck (1998), who argued that these measures could be taken as a measure of maximum metabolic scope. Post-prandial elevations of metabolism (specific dynamic action, SDA) in Antarctic marine ectotherms are typically low and of long duration, and elevations in post-prandial nitrogen excretion are of longer duration than oxygen consumption (Peck 1998). Thus for brachiopods and nemerteans peak elevations are between ×1.5 and ×2 standard metabolic rates, and SDA durations last up to 25 days (Fig. 5). Factorial relationships between standard and SDA peak metabolism appear constant across polar, temperate and tropical regimes (Johnston and Battram 1993; Peck 1998) but, as for locomotory metabolic scope, the actual energy available is less at low temperature because standard metabolism is reduced. This leads to an increase in time required to complete an SDA by between ×2 and ×10 compared with that for temperate species.

The proportion of the energy content of a meal used in the SDA (the SDA coefficient) varies with meal size and composition (Jobling 1994). In Antarctic species this coefficient varied from <1% in the nemertean *Parborlasia corrugatus* fed meals of 20% to 110% of its own body mass (Clarke and Prothero-Thomas 1997), to 55% in the plunderfish *Harpagifer antarcticus* fed meals of 2.5% of their own body mass. Using meal sizes of about 10% body mass Johnston and Battram (1993) measured SDA coefficients in demersal, low activity fish from different latitudes. Values were 15%–23% throughout, suggesting SDA coefficients are independent of ambient temperature.

Growth and development

Growth is also a major energetic sink and data on the growth and development of Antarctic marine ectotherms have been collected for three decades. In benthic species growth rates are, with a few exceptions, very slow (Everson 1977; Clarke 1983; Arntz et al. 1994). The exceptions are two sponges (Dayton et al. 1974), three

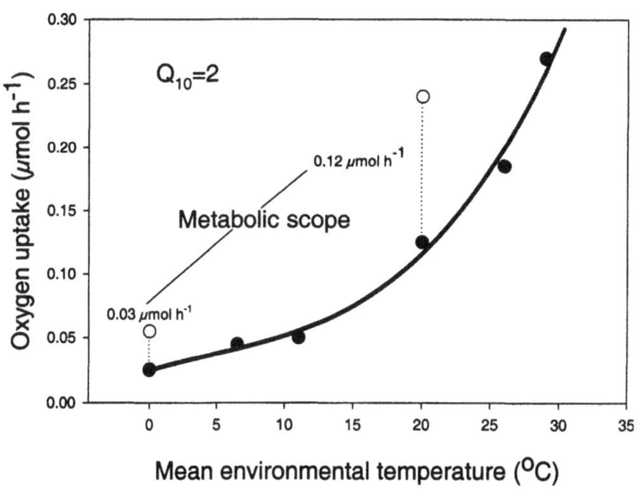

Fig. 4 Diagrammatic representation of changes in metabolic scope with temperature. A Q_{10} of 2 is assumed. Data for standard metabolism in crustacean species (*closed circles*) recalculated from Clarke (1983), based on previous data from Ivleva (1980). For a factorial metabolic scope of 2 a species at 0°C has a maximum possible rise in metabolism of 0.03 μmol O_2 h^{-1}, whereas at 20°C the maximum rise is 0.12 μmol O_2 h^{-1}, a value 4 times that at 0°C

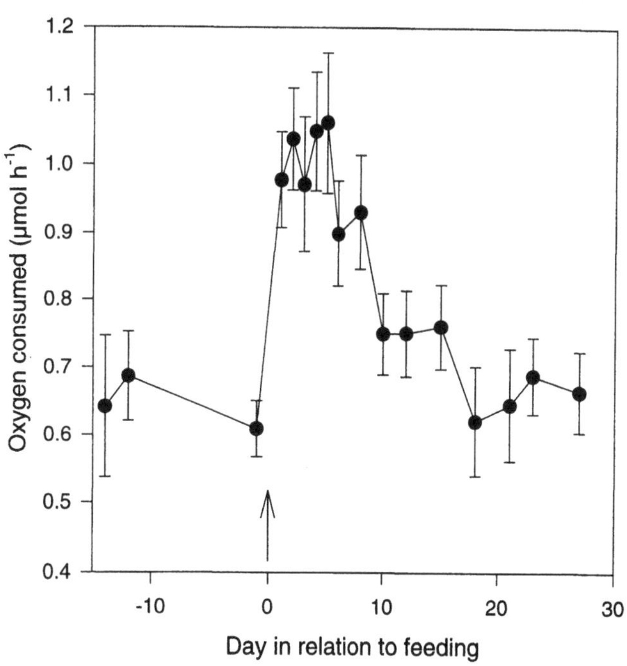

Fig. 5 The post-prandial rise in metabolism following feeding (specific dynamic action, SDA) in the Antarctic brachiopod *Liothyrella uva*. The peak is around ×1.6 that of the preceding standard metabolic rate, and oxygen consumption remains elevated for around 18 days after feeding. From Peck (1996)

ascidians (Rauschert 1991) and a bryozoan (Barnes 1995). These exceptions, plus the fact that many Antarctic pelagic invertebrates grow quickly in summer months, have been used to argue that low temperature does not limit growth (Clarke 1991a). It has also been argued that resource limitation constrains growth rates, and that intense seasonality is the mechanism limiting resources (Clarke 1988; Clarke and Peck 1991). An implication of this hypothesis is that growth would be expected to be highly seasonal and linked to periods of food supply. This is generally the case for pelagic species (Clarke and Peck 1991) but good assessments of seasonality of growth in benthos are rare and data are variable. Barnes (1995) found that growth in some Antarctic bryozoans continued at the same rate throughout the year, whereas in others there was strong seasonality. Peck et al. (2000) found that shell growth in juveniles of the infaunal bivalve *Yoldia eightsi* proceeded at the same rate in winter and summer, whereas tissue mass increased in the summer and decreased in winter. Remarkably, Peck et al. (1997) found up to 12 times faster shell growth in winter in the brachiopod *Liothyrella uva* (Table 1). Peck and Holmes (1989) showed that *L. uva*, in common with *Y. eightsi* has a tissue mass cycle in phase with the summer bloom. Thus either shell and tissue growth are decoupled, with tissue growth occurring in summer, or all growth is decoupled from the phytoplankton bloom with consumed resources being stored for use throughout the year. This is clearly an area where further research is needed.

Larval development in Antarctic marine ectotherms is slowed dramatically compared with temperate species. As yet no exceptions to this rule have been reported. The first Antarctic studies were in the 1960s (Pearse 1962, 1965). Pearse's group have since compiled a comprehensive data set for echinoderm development worldwide indicating rates for Antarctic echinoderms 5–10 times slower than temperate or tropical species (Fig. 6, Bosch et al. 1987). These data were later confirmed by Stanwell-Smith and Peck (1998), who also found a temperature window between 0.1°C and 1.7°C where development rate was optimal and larval mortality minimal for the urchin *Sterechinus neumayeri* (Fig. 7). Reproduction of field populations of this species at Signy Island matched these temperatures, with gamete release and subsequent development occurring when temperatures were between 0°C and 2°C.

Development rates of encapsulated embryos and both feeding (planktotrophic) and non-feeding (lecithotrophic) larvae are similarly slowed. Hoegh-Guldberg and Pearse (1995) showed that rates of echinoderm development are near their maximum for a given ambient

Table 1 Liothyrella uva: daily growth increments (mm day^{-1}) for specimens of 5, 10, 20 and 40 mm length, for summer and winter periods in 1992 and 1993 (Figures in parentheses denote SE, * denotes brachiopods of 40 mm in length and above, which did not increase in length during the austral summer of 1993). From Peck et al. (1997)

Shell length	1992		1993	
	Summer	Winter	Summer	Winter
5	2.67 (0.53)	9.25 (1.05)	1.89 (0.83)	6.88 (0.68)
10	2.33 (0.44)	8.13 (0.89)	1.42 (0.70)	6.17 (0.58)
20	1.63 (0.31)	5.88 (0.60)	0.46 (0.47)	4.75 (0.39)
40	0.25 (0.35)	1.39 (0.62)	*	1.91 (0.39)

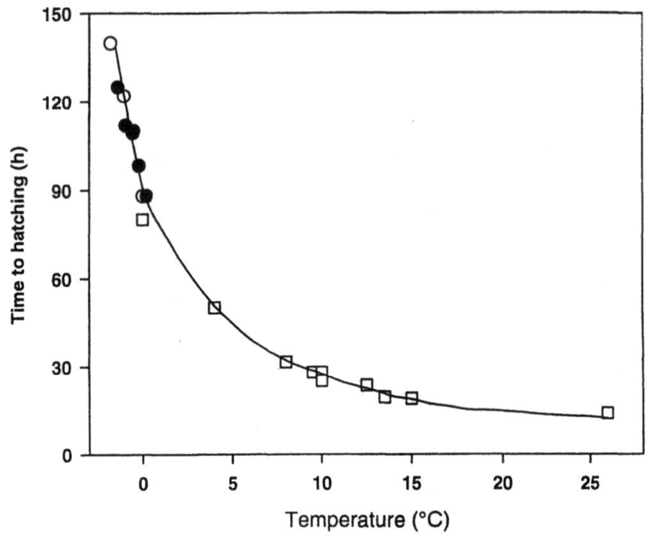

Fig. 6 Duration of embryonic development, as time to hatching, for echinoids at different latitudes. Data are for tropical and temperate species (*open squares*), *Sterechinus neumayeri* from McMurdo Sound (*open circles*) and Signy Island (*closed circles*). All species were held at naturally occurring temperatures. Note the much longer times to hatching as temperatures approach 0°C. Modified from Bosch et al. (1987), and Stanwell-Smith and Peck (1998)

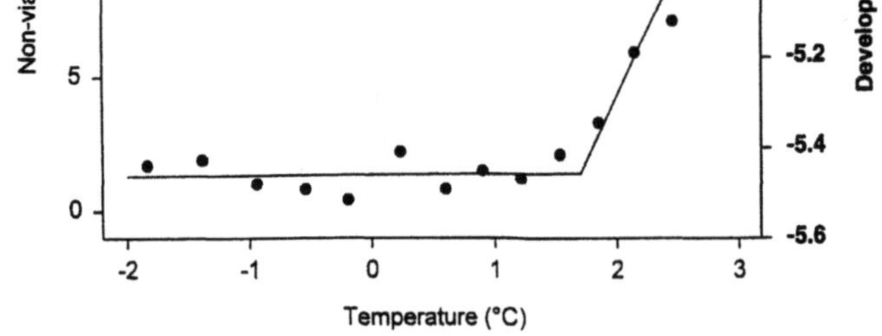

Fig. 7 Development rate (*triangles*) and embryo viability (*circles*) for *Sterechinus neumayeri* over a range of temperatures from –2°C to +2.5°C. Data are for a population from Signy Island, and show an optimum temperature window between 0.2°C and 1.7°C where development rate is maximum and embryo viability optimum. Development rate is shown as the natural logarithm of the inverse of the time between fertilisation and hatching of 50% of the embryos. Modified from Stanwell-Smith and Peck (1998)

temperature at any site worldwide, and that temperature is the main regulator of development rate. Similarly slowed development rates have been found in other groups of Antarctic marine invertebrates. Nemerteans are slowed around 5 times compared to temperate species (Peck 1993), and brachiopods by 7–100 times (Peck and Robinson 1994). The groups so far documented with the most dramatically extended developmental periods are brooding crustaceans and gastropod molluscs. Among the isopods *Glyptonotus antarcticus* broods embryos for 20 months (7–20 times temperate species; White 1984). In amphipods development times decrease exponentially with temperature (Arntz et al. 1994), and the giant *Eusirus perdentatus* takes over 14 months to reach the juvenile stage (Klages 1993). For small gastropods at Signy Island Picken (1980) found development periods of 3–6 months, and broods of *Trophon shackletoni* and *Marseniopsis mollis* require 7–9 months to hatch at Rothera station (Chapman, personal communication). However, the slowest development was found in the large gastropods *Neobuccinum eatoni* and *Trophon* sp which took 25 months for broods to hatch (Hain 1991).

Upper lethal limits and critical temperatures

The stenothermal nature of Antarctic marine ectotherms is now well documented (Peck and Conway 2000). Several fish species live within a narrow temperature window, between −1.86°C and +4°C (Somero and DeVries 1967; Eastman 1993; Somero et al. 1998). Many invertebrate species are also stenothermal, with Antarctic bivalve molluscs inhabiting temperature ranges 2–5 times smaller than temperate and tropical species. The brachiopod *Liothyrella uva* has an upper lethal temperature between 3.0°C and 4.5°C and the limpet *Nacella concinna* of 9°C (Peck 1989). The most stenothermal species found so far is the bivalve *Limopsis marionensis*, which has a short-term upper temperature of 4°C and a long-term upper limit of around 2°C (Pörtner et al. 1999a). Some Antarctic species however are less limited. The starfish *Odontaster validus* survives up to 12°C (T. Hill, personal communication) and the eelpout *Pachycara brachycephalum* to 9–10°C (van Dijk et al. 1999). Davenport (1997) found the limpet *N. concinna* could survive temperatures up to 15°C for 12 h. Exposure time clearly affects the upper lethal temperatures measured. Long-term exposures are of greatest value in defining a species' ecological limits. Thus, many Antarctic marine ectotherms live within a 6–7°C temperature window, compared with 15–25°C for most temperate species.

Studies of temperate species have suggested that upper temperature limits may be set by oxygen supply mechanisms and aerobic energy production (Pörtner et al. 1998, 1999a). For example, in *Arenicola marina* populations from the North Sea and the White Sea, seasonal and latitudinal adaptation to lower environmental temperatures led to lowering of both the upper and lower lethal temperatures (Fig. 8). Both temperature limits were characterised by the accumulation of anaerobic end products of metabolism (e.g. acetate and succinate) and the temperatures at which aerobic metabolism could no longer be sustained were termed the upper and lower critical temperatures (T_{CI} and T_{CII}; Sommer et al. 1997; Pörtner et al. 1998). These temperatures marked regimes where oxygen demand outstripped supply. Evolutionary adaptation to low stable temperatures also appears to lead to a greater reduction in the upper critical temperature than the lower one (Fig. 8).

Recent studies on Antarctic invertebrates have supported this theory. Anaerobic end products of metabolism were accumulated at, and above the long-term upper critical temperature by *Limopsis marionensis* (Pörtner et al. 1999a), and oxygen consumption and ventilation rate declined at T_{CII} in the fish *Lepidonotothen*

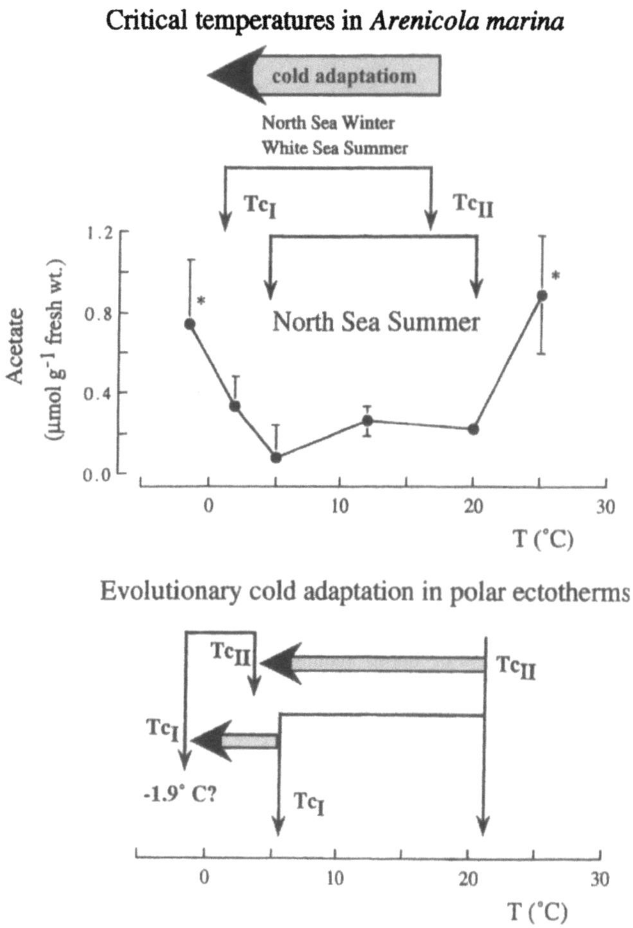

Fig. 8 Critical temperatures in the lugworm *Arenicola marina* are reduced with seasonal acclimatisation to winter temperatures (*upper*). The critical temperatures are characterised by accumulation of anaerobic metabolic products (eg. acetate). Both critical upper and lower temperatures (Tc_I and Tc_{II}) are lowered. They are also lower in populations living at higher latitudes. In evolutionary adaptation to low temperature both Tc_I and Tc_{II} are reduced, but Tc_{II} is more markedly reduced and the viable temperature window is smaller. From Pörtner et al. (1998)

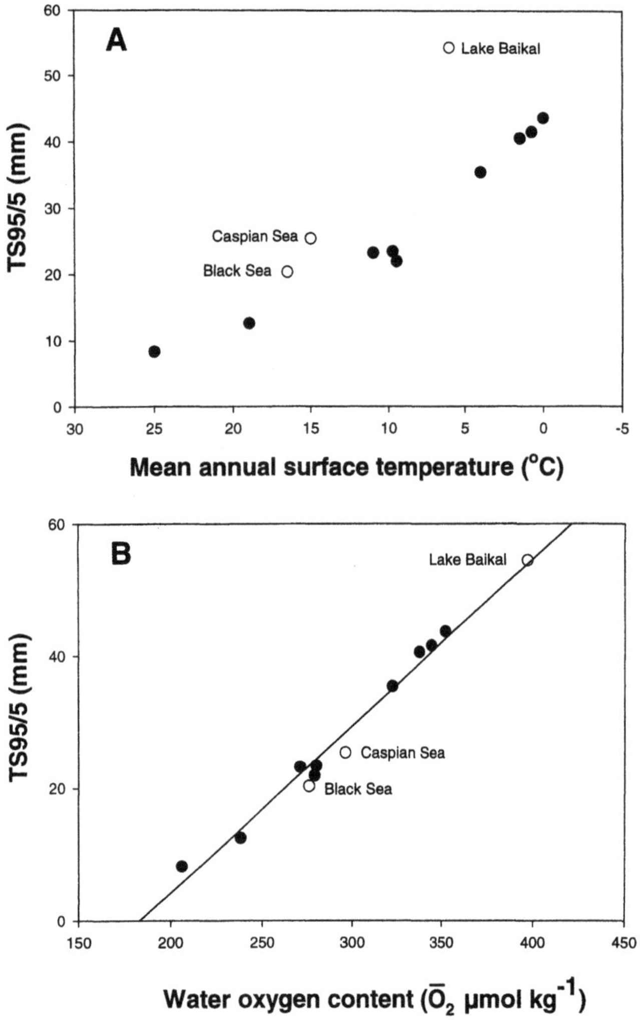

Fig. 9 **A** Maximum amphiped size in relation to ambient temperature and **B**. The relationship between maximum amphipod size and water oxygen content. All data fit a consistent relationship between maximum potential size and water oxygen content. These data strongly support the hypothesis that oxygen availability and not environmental temperature controls maximum size at any site for amphipod crustaceans. Data were collected from 12 sites worldwide, ranging from Madagascar in the tropics to the Antarctic (for details see Chapelle and Peck 1999) Analyses were restricted to 250 m depth (continental shelf depth) except in Antarctica, which included species to 500 m, because of continental shelf depression by the Antarctic icecap (n = number of species, T = mean annual water temperature). From Chapelle and Peck (1999)

nudifrons (Hardewig et al., submitted for publication). Oxygen consumption and haemolymph oxygen content have also been shown to decline at T_{CII} in the bivalve *Laternula elliptica* (Peck, Pörtner and Hardewig, unpublished data) and the echinoid *Sterechinus neumayeri* (Brockington 1999). These recent studies suggest that Antarctic marine ectotherms live close to the limits of their aerobic scopes. If this is a general requirement for adaptation to low temperature it may be one reason that species with lifestyles requiring high levels of activity and energy utilisation are absent in polar regions (Clarke 1998).

Gigantism and oxygen supply

The finding that oxygen supply mechanisms are important in setting upper temperature limits for Antarctic marine species has come at the same time as the demonstration that oxygen availability sets the maximum size for amphipod crustaceans. Chapelle and Peck (1999) plotted the largest sizes which amphipods reached at 12 sites worldwide and found a fivefold increase in maximum length from tropical to high Antarctic sites. They also found that the largest freshwater species were 1.5–2 times larger than marine species at the same temperature. However, when the data were plotted against oxygen content of the water from the environments they inhabited a linear relationship was obtained which explained over 98% of the variance in the data (Fig. 9). The low levels of blood pigment and simple circulatory systems of amphipods are thought to be important in emphasising the relationship between oxygen and gigantism in this taxon. Amphipods have these characters in common with isopods and pycnogonids, the other two groups most often cited as exhibiting polar gigantism.

Conclusions

This review of recent advances in the ecophysiology of polar marine ectotherms has confirmed the central role of energetics in adaptation to temperature. It is now clear that polar ectotherms have low resting metabolic rates and that this may be related to mitochondrial function at low temperature. The low resting metabolic rates are also associated with slow growth and development rates, and low aerobic scopes. Oxygen supply to the tissues may be important in determining organismal thermal limits and size.

References

Ansell AD, Peck LS (2000) Burrowing in the Antarctic anemone *Edwardsia heilcampoides*. J Exp Mar Biol Ecol 252(1):45–55

Archer SD, Johnston IA (1991) Density of cristae and distribution of mitochondria in the slow muscles of Antarctic fish. Physiol Zool 64:242–258

Arntz WE, Brey T, Gallardo VA (1994) Antarctic zoobenthos. Oceanogr Mar Biol Annu Rev 32:241–304

Barnes DKA (1995) Seasonal and annual growth in erect species of Antarctic bryozoans. J Exp Mar Biol Ecol 188:181–198

Blier PU, Guderley HE (1993) Effects of pH and temperature on the kinetics of pyruvate oxidation by muscle mitochondria from rainbow trout (*Oncorhynchus mykiss*). Physiol Zool 66:474–489

Bosch I, Beauchamp KA, Steele ME, Pearse JS (1987) Development, metamorphosis and seasonal abundance of embryos and larvae of the Antarctic sea urchin *Sterechinus neumayeri*. Biol Bull 173:126–135

Branch GM, Marsh AC (1978) Tenacity and shell shape in six *Patella* species: adaptive features. J Exp Mar Biol Ecol 34: 111–130

Brockington S (2001) Ecology and physiology of *S. neumayeri* at Adelaide Island, Antarctica. PhD thesis, Open University, Milton Keynes

Chapelle G, Peck LS (1995) The influence of acclimation and substratum on the metabolism of the amphipod *Waldeckia obesa* (Chevreux 1905) and *Bovallia gigantea* (Pfeffer 1888). Polar Biol 15:225–232

Chapelle G, Peck LS (1999) Polar gigantism and oxygen. Nature 399:114–115

Clarke A (1980) A reappraisal of the concept of metabolic cold adaptation in polar marine invertebrates. Biol J Linn Soc 14:77–92

Clarke A (1983) Life in cold water: the physiological ecology of polar marine ectotherms. Oceanogr Mar Biol Annu Rev 21:341–453

Clarke A (1988) Seasonality in the Antarctic marine environment. Comp Biochem Physiol B 90:461–473

Clarke A (1991a) What is cold adaptation and how should we measure it? Am Zool 31:81–92

Clarke A (1991b) Cold adaptation. J Zool 225:691–699

Clarke A (1998) Temperature and energetics: an introduction to cold ocean physiology. In: Pörtner HO, Playle R (eds) Cold ocean physiology. Cambridge University Press, Cambridge, pp 3–30

Clarke A, Johnston N (1999) Scaling of metabolic rate and temperature in teleost fish. J Anim Ecol 68:893–905

Clarke A, Peck LS (1991) The physiology of polar marine zooplankton. In: Sakshaug E, Hopkins CCE, Øritsland NA (eds) Proceedings of the Pro-Mare symposium on polar marine ecology, Trondheim. Polar Res 10:355–369

Clarke A, Prothero-Thomas E (1997) The effect of feeding on oxygen consumption and nitrogen excretion of the Antarctic nemertean *Parborlasia corrugatus*. Physiol Zool 70:639–649

Davenport J (1988). Tenacity of the Antarctic limpet *Nacella concinna*. J Moll Stud 54:355–356

Davenport J (1997). Comparisons of the biology of the intertidal subantarctic limpets *Nacella concinna* and *Kerguelenella lateralis*. J Molluscan Stud 63:39–48

Dayton PK, Newman WA, Paine RT, Dayton LB (1974) Ecological accommodation in the benthic community at McMurdo Sound, Antarctica. Ecol Monogr 44:105–128

Dijk PLM van, Hardewig I, Pörtner H (1998) Exercise in the cold: high energy turnover in Antarctic fish. In: Prisco G di, Pisano E, Clarke A (eds) Fishes of Antarctica. A biological overview. Springer, Milan, pp 225–236

Dijk PLM van, Tesch C, Hardewig I, Pörtner H (1999) Physiological disturbances at critically high temperatures. A comparison between stenothermal Antarctic, and eurythermal temperate eelpouts (Zoarcidae). J Exp Biol 202:3611–3621

Eastman J (1993) Antarctic fish biology. Academic Press, San Diego

Egginton S, Davison W (1998) Effects of environmental and experimental stress on Antarctic fish. In: Pörtner HO, Playle R (eds) Cold ocean physiology. Cambridge University Press, Cambridge, pp 299–326

Everson I (1977) Antarctic marine secondary production and the phenomenon of cold adaptation. Philos Trans R Soc Lond Ser B 279:55–66

Forster ME, Franklin CE, Taylor HH, Davison W (1991) The aerobic scope of an Antarctic fish, *Pagothenia borchgrevinki* and its significance for metabolic cold adaptation. Polar Biol 8: 155–159

Fox HM (1936) The activity and metabolism of poikilothermic animals in different latitudes, I. Proc Zool Soc Lond 1936: 945–955

Guderley H (1998). Temperature and growth rates as modulators of the metabolic capacities of fish muscle. In: Pörtner HO, Playle R (eds) Cold ocean physiology. Cambridge University Press, Cambridge, 58–87

Hain S (1991) Life maintenance of benthic mollusks from the Eastern Weddell Sea, Antarctica. Proceedings of the 10th International Malacological Congress, Tübingen 1989, pp 339–341

Hardewig I, van Dijk PLM, Pörtner HO (1998) High energy turnover at low temperatures: recovery from exercise in antarctic and common eelpout (Zoarcidae). Am J Physiol 274:R1789–1796

Hoegh-Guldberg O, Pearse JS (1995) Temperature, food availability and the development of marine invertebrate larvae. Am Zool 35:415–425

Holeton GF (1974) Metabolic cold adaptation of polar fish: fact or artefact? Physiol Zool 47:137–152

Hubley MJ, Locke BR, Moerland TS (1997) Reaction-diffusion analysis of the effects of temperature on high-energy phosphate dynamics in goldfish skeletal muscle. J Exp Biol 200:975–988

Ikeda T (1974) Nutritional ecology of marine zooplankton. Mem Fac Fish Hokkaido Univ 22:1–97

Ivleva IV (1980) The dependence of crustacean respiration on body mass and habitat temperature. Int Rev Ges Hydrobiol 65:1–47

Jobling M (1994) Fish bioenergetics. Chapman & Hall, London

Johnston IA (1982) Quantitative analyses of ultrastructure and vascularisation of the slow muscle of the anchovy. Tissue Cell 14:319–328

Johnston IA, Battram J (1993) Feeding energetics and metabolism in demersal fish species from Antarctic, temperate and tropical environments. Mar Biol 115:7–14

Johnston IA, Camm JP, White MG (1988) Specialisations of swimming muscles in the pelagic Antarctic fish *Pleuragramma antarcticum*. Mar Biol 100:3–12

Johnston IA, Clarke A, Ward P (1991a) Temperature and metabolic rate in sedentary fish from the Antarctic, North Sea and Indo-West Pacific Ocean. Mar Biol 109:191–195

Johnston IA, Johnson TP, Battram JC (1991b) Low temperature limits burst swimming performance in Antarctic fish. In: Prisco G di, Maresca B, Tota B (eds) Biology of Antarctic fish. Springer, Berlin Heidelberg New York, pp 179–190

Johnston IA, Guderley HE, Franklin CE, Crockford T, Kamunde C (1994) Are mitochondria subject to evolutionary temperature adaptation? J Exp Biol 195:293–306

Johnston IA, Calvo J, Guderley H, Fernandez D, Palmer L (1998) Latitudinal variation in the abundance and oxidative capacities of muscle mitochondria in perciform fishes. J Exp Biol 201:1–12

Johnson TP, Bennet AF (1995) The thermal acclimation of burst escape performance in fish: an integrated study of molecular and cellular physiology and organismal performance. J Exp Biol 198:2165–2175

Klages M (1993) Biology of the Antarctic gammaridean amphipod *Eusirus perdentatus* Chevreux, 1912 (Crustacea: Amphipoda): distribution, reproduction and population dynamics. Antarct Sci 5:349–359

Krogh A (1916) Respiratory exchange of animals and man. Longmans, London

Moyes CD, Mathieu-Costello OA, Brill RW, Hochachka PW (1992) Mitochondrial metabolism of cardiac and skeletal muscles from a fast (*Katsuwonas pelamis*) and a slow (*Cyprinus carpio*) fish. Can J Zool 70:1246–1253

Pearse JS (1962) Early development of the Antarctic asteroid *Odontaster validus* Koehler. Am Zool 2:436

Pearse JS (1965) Reproductive periodicities in several contrasting populations of *Odontaster validus* Koehler, a common antarctic asteroid. (Biology of the Antarctic seas, 2) Antarct Res Ser 5:39–85

Peck LS (1989) Temperature and basal metabolism in two Antarctic marine herbivores. J Exp Mar Biol Ecol 127:1–12

Peck LS (1993) Larval development in the Antarctic nemertean *Parborlasia corrugatus* (Heteronemertea: Lineidae). Mar Biol 116:301–310

Peck LS (1996) Metabolism and feeding in the Antarctic brachiopod *Liothyrella uva*: a low energy lifestyle species with restricted metabolic scope. Proc R Soc Lond Ser B 263:223–228

Peck LS (1998) Feeding, metabolism and metabolic scope in Antarctic marine ectotherms. In: Pörtner HO, Playle R (eds) Cold ocean physiology. Cambridge University Press, Cambridge, pp 365–390

Peck LS, Conway LZ (2000) The myth of metabolic cold adaptation: oxygen consumption in stenothermal Antarctic bivalves. In: Harper E, Crame AJ (eds) The evolutionary biology of bivalve molluscs. Cambridge University Press, Cambridge, pp 441–450

Peck LS, Holmes LJ (1989) Seasonal and ontogenetic changes in tissue size in the Antarctic brachiopod *Liothyrella uva* (Broderip, 1833). J Exp Mar Biol Ecol 134:25–36
Peck LS, Robinson K (1994) Pelagic larval development in the brooding Antarctic brachiopod *Liothyrella uva*. Mar Biol 120:279–286
Peck LS, Prothero-Thomas E, Hough N (1993) Pedal mucus production by the Antarctic limpet *Nacella concinna* (Strebel, 1908). J Exp Mar Biol Ecol 174:177–192
Peck LS, Brockington S, Brey T (1997) Growth and metabolism in the Antarctic brachiopod *Liothyrella uva*. Philos Trans R Soc Lond Ser B 352:851–858
Peck LS, Colman J, Murray AWA (2000) Growth and tissue mass cycles in the infaunal bivalve *Yoldia eightsi* at Signy Island, Antarctica. Polar Biol 23:420–428
Picken GB (1980) The distribution, growth and reproduction of the Antarctic limpet *Nacella* (*Patinigera*) *concinna*. (Strebel, 1908). J Exp Mar Biol Ecol 42:71–85
Pörtner HO, Hardewig I, Sartoris FJ, Dijk P van (1998) Energetic aspects of cold adaptation: critical temperatures in metabolic, ionic and acid base regulation? In: Pörtner HO, Playle R (eds) Cold ocean physiology. Cambridge University Press, Cambridge, pp 88–120
Pörtner HO, Peck LS, Zielinski S, Conway LZ (1999a) Intracellular pH and energy metabolism in the highly stenothermal Antarctic bivalve *Limopsis marionensis* as a function of ambient temperature. Polar Biol 22:17–30
Pörtner HO, Hardewig I, Peck LS (1999b) Mitochondrial function and critical temperature in the Antarctic bivalve *Laternula elliptica*. Comp Biochem Physiol A 124:179–189
Pörtner HO, Dijk P van, Hardewig I, Sommer A (1999c) Levels of metabolic cold adaptation: tradeoffs in eurythermal and stenothermal ectotherms. In: Davison W, Williams CH (eds) Antarctic ecosystems: models for wider ecological understanding. Caxton Press, Christchurch, (in press)
Precht H, Christophersen J, Hensel H (1955) Temperatur und Leben. Springer, Berlin Heidelberg New York, 514 pp
Prosser CL (1958). General summary: the nature of physiological adaptation. In: Prosser CL (ed) Physiological adaptation. American Physiology Society, Washington, pp 167–180
Rauschert M (1991) Ergebnisse der faunistischen Arbeiten im Benthal von King George Island (Südshetlandinseln, Antarktis). Ber Polarforsch 76:1–75
Scholander PF, Flagg W, Walters V, Irving L (1953) Climatic adaptation in Arctic and tropical poikilotherms. Physiol Zool 26:67–92
Somero GN, DeVries AL (1967) Temperature tolerance of some Antarctic fishes. Science 156:257–258
Somero GN, Fields PA, Hofmann GE, Weinstein RB, Kawall H (1998) Cold adaptation and stenothermy in Antarctic notothenioid fishes: what has been gained and what has been lost? In: Prisco G di, Pisano E, Clarke A (eds) Fishes of Antarctica. A biological overview, Springer, Milan, pp 97–109
Sommer A, Pörtner HO (1999) Exposure of *Arenicola marina* (L.) to extreme temperatures: adaptive flexibility of a boreal and subpolar population. Mar Ecol Prog Ser 181:215–226
Sommer A, Klein B, Pörtner HO (1997) Temperature induced anaerobiosis in two populations of the polychaete worm *Arenicola marina*. J Comp Physiol B 167:25–35
Stanwell-Smith DP, Peck LS (1998) Temperature and embryonic development in relation to spawning and field occurrence of larvae of three Antarctic echinoderms. Biol Bull (Woods Hole) 194:44–52
Thorson G (1936) The larval development, growth and metabolism of Arctic marine bottom invertebrates compared with those of other seas. Medd Groen 100:1–155
Tyler S, Sidell BD (1984) Changes in mitochondrial disruption and diffusion distances in muscle of goldfish on acclimation to warm and cold temperatures. J Exp Zool 232:1–9
Weinstein RB, Somero GN (1998) Effects of temperature on mitochondrial function in the Antarctic fish *Trematomus bernachii*. J Comp Physiol 168:190–196
Wells RMG (1987) Respiration of Antarctic fishes from McMurdo Sound. Comp Biochem Physiol A 88:417–424
White MG (1984) Marine benthos. In: Laws RM (ed) Antarctic ecology, vol 2. Academic Press, London, pp 421–461
Wohlschlag DE (1964) Respiratory metabolism and ecological characteristics of some fishes in McMurdo Sound, Antarctica. Antarct Res Ser Am Geophys Union 1:33–62

Polar Biol (2001) 24: 598–603
DOI 10.1007/s003000100260

Ulrike H. Lüder · Jürgen Knoetzel · Christian Wiencke

Acclimation of photosynthesis and pigments to seasonally changing light conditions in the endemic Antarctic red macroalga *Palmaria decipiens*

Abstract The influence of seasonally fluctuating Antarctic daylengths on the photosynthetic apparatus of *Palmaria decipiens* was studied in culture experiments. Maximal photosynthetic activity (ETRmax) and maximal quantum yield (Fv/Fm), measured by in vivo chlorophyll fluorescence, and concentrations of pigments (phycobiliproteins and Chl *a*) were determined monthly. Fv/Fm remained constantly high between 0.62 and 0.67 during mid-autumn, winter and spring. ETRmax and pigment contents increased continuously in mid-autumn and winter and were highest in spring. A positive correlation between pigments and ETRmax was found. In summer, ETRmax, Fv/Fm and pigment levels decreased to their lowest values. *P. decipiens* acclimated by increasing phycobilisome (PBS) number and changing PBS structure, probably changing rod length and rod number. The data show that *P. decipiens* is efficiently adapted to the short period of favourable light conditions in the field. A photoperiodic control of pigment synthesis triggered by daylength is suggested.

Introduction

The environment of Antarctic benthic macroalgae was recently reviewed by Wiencke (1996). In the Antarctic sublittoral, algae are exposed to constant water temperatures and high nutrient concentrations over the entire year. Therefore, the seasonal development of Antarctic macroalgae depends mainly on the seasonal variation of daylengths varying, for example, between 5 h in winter and 20 h in summer at King George Island, South Shetlands. The underwater light climate is further affected by ice cover in winter and dim light or darkness prevail in the sublittoral. In summer, the water becomes turbid due to plankton blooms and glacial melt-water. So, optimal light conditions for sublittoral macroalgae are present only for a short time in the Antarctic spring, after break-up of sea ice. At this time the water is very clear and the sunlight penetrates deeply into the water column.

How do Antarctic macroalgae acclimate to seasonal changes in light? *Palmaria decipiens* is an excellent subject for this type of study, as there have been several previous studies on its life strategy. The alga is pseudoperennial and develops new blades during late winter/early spring even in darkness (Wiencke 1990b; Weykam et al. 1997). Growth starts in July and is, like photosynthesis, maximal in spring (October/November) (Wiencke 1990b; Weykam and Wiencke 1996). The light requirements for growth of this species are very low (Wiencke 1990b). *P. decipiens* even survives a period of 6 months of darkness by utilising stored floridean starch (Weykam et al. 1997).

In this study, we investigate the acclimation of the photosynthetic apparatus to seasonally changing light conditions. The algae were cultivated under seasonally fluctuating Antarctic daylengths. Photosynthetic rates and pigment contents, in particular the content of the major light-harvesting pigments – the phycobiliproteins – were determined monthly. The results give insight into the life strategy of this alga from a physiological point of view.

Please note: When citing this article, please use the reference to the original source in the journal Polar Biology as given on the top of this page.

U.H. Lüder (✉) · C. Wiencke
Alfred Wegener Institute for Polar and Marine Research,
Am Handelshafen 12, 27570 Bremerhaven, Germany
E-mail: ulueder@awi-bremerhaven.de
Tel.: +49-471-48311522
Fax: +49-471-48311425

J. Knoetzel
Institute of Cell Biology, Biochemistry and Biotechnology,
University of Bremen, 28334 Bremen, Germany

Materials and methods

Algal material and culture conditions

Tetraspores of *P. decipiens* (Reinsch) Ricker were collected from King George Island (South Shetland Islands, Antarctica) and since then cultivated in the laboratory under seasonally fluctuating daylengths (Fig. 1A) mimicking field conditions (Wiencke 1990a, b). Numerous young gametophytes were grown at 0 ± 1°C in twenty-four 3-l glass beakers for 2 years under simulated Antarctic daylength conditions (Fig. 1A) and a constant photon fluence rate of 25 μmol photons m^{-2} s^{-1} using daylight fluorescent tubes (Osram L58/W19). The cultures were grown in Provasoli-enriched North Sea water (Provasoli 1966, 34‰, pH 8), aerated with pressed air. To ensure sufficient nutrient supply, the culture medium was changed every 2nd week. After about 1 year we started, under Antarctic March conditions, to harvest individual plants from different glass beakers at monthly intervals for the photosynthesis and pigment measurements.

Chlorophyll fluorescence measurements

In vivo chlorophyll fluorescence was measured with a pulse-amplitude modulation fluorometer (PAM 2000, Walz, Germany) connected to a personal computer. All measurements were conducted in seawater at 0°C using eight to ten different individuals. Generally, the results of fluorescence measurements are consistent with those of oxygen measurements, as shown earlier (Hanelt 1992; Hanelt et al. 1992). The ratio of variable to maximal fluorescence (Fv/Fm) was determined in dark-acclimated plants as described by Bischof et al. (1999) and used as an indicator for maximal quantum yield. The maximal electron transport rate (ETRmax) was determined by ETR versus irradiance curves as described by Bischof et al. (1999) and used as an indicator of maximal photosynthetic activity (Pmax).

Pigment determinations

Phycobiliprotein levels were determined after Rosenberg (1981) in six different individuals. The algal material was disrupted by grinding in liquid nitrogen, and extracted in 0.1 M potassium phosphate buffer, pH 6.8. The extract was frozen and thawed several times to optimise the extraction process. The resulting extract was centrifuged and the absorbance of the supernant at 565 nm, 615 nm and 650 nm was determined and calculated using the following equations:

$$PE(phycoerythrin)(\mu g/ml) = 123.5 * OD565 - 73.5 * OD615 - 16.3 * OD650$$

$$PC(phycocyanin)(\mu g/ml) = 163.2 * OD615 - 117.1 * OD650$$

and

$$APC(allophycocyanin)(\mu g/ml) = 165.6 * OD650 - 16.4 * OD615$$

Chlorophyll *a* (Chl *a*) was extracted in *N,-N* dimethylformamide (DMF) according to Inskeep and Bloom (1985) and carried out with six to eight different individuals.

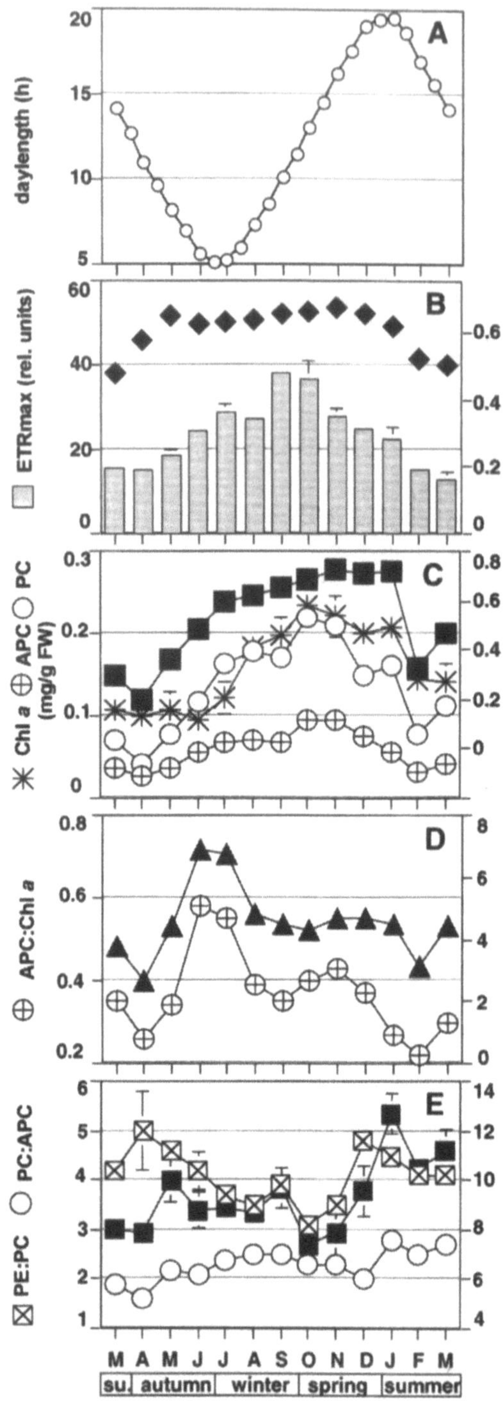

Fig. 1 Simulation of seasonal changes of daylength on King George Island, Antarctica (**A**). Seasonal changes in maximal electron transport rate (*ETRmax*) and in maximal quantum yield (*Fv/Fm*) (**B**), pigment contents (**C**), ratios of total phycobiliprotein and APC content to Chl *a* (**D**) and ratios of phycobiliprotein contents (**E**) (*Chl a* chlorophyll *a*, *APC* allophycocyanin, *PC* phycocyanin, *PE* phycoerythrin, *PBtot* total phycobiliprotein content). Means ± SD, *n* = 8–10 in **B**; *n* = 6–8 in **C** and **E**. In those cases where no error bars are given, the SD is smaller than the symbol for the mean value

Results

P. decipiens was cultivated for 1 year under seasonally fluctuating daylengths according to the conditions on King George Island, Antarctica (Fig. 1A). The maximal electron transport rate increased continuously during mid-autumn and winter (May/August) and was maximal in spring (September/October) (Fig. 1B). Lowest ETRmax values were found in summer between February and April. The maximal quantum yield Fv/Fm remained high, between 0.62 and 0.67 during mid-autumn, winter and spring, decreasing from January onwards to a value of 0.47–0.50 in March, and increasing again until May (Fig. 1B).

Seasonal changes of the light-harvesting pigments were calculated per fresh weight (FW) and are shown in Fig. 1C. All three phycobiliproteins (allophycocyanin APC, phycocyanin PC and phycoerythrin PE) started to increase in May, in parallel with ETRmax, but reached maximal values slightly later in October/November (234 μg Chl *a*, 95 μg APC, 211 μg PC and 732 μg PE per g FW). Chlorophyll *a* started to increase later in July. Lowest pigment concentrations were found in summer between February and April (99 μg Chl *a*, 31 μg APC, 71 μg PC and 202 μg PE per g FW). In autumn, winter and early spring (April/October), a positive correlation was found between ETRmax and the total phycobiliprotein levels (Fig. 2A), whereas between ETRmax and Chl *a* the correlation was weaker (Fig. 2B).

The seasonal changes in the ratios of total phycobiliproteins (PBtot) to Chl *a* (PBtot:Chl *a*) are very similar to the changes in APC:Chl *a* (Fig. 1D). Between April and September the ratios were negatively correlated to the changes in daylength. PBtot:Chl *a* and APC:Chl *a* ratios were highest in June/July when the light was minimised to 5 h per day. Ratios began to decrease when daylengths increased. Between September and January, PBtot:Chl *a* ratios were constant whereas the APC:Chl *a* ratios showed a second small peak in November, gradually falling to their lowest levels in February.

The ratios of phycobiliproteins are shown in Fig. 1E. The PE:PC ratio and (slightly) the PE:APC ratio decreased from May onwards to their lowest values in October (with a slight peak in September), whereas the PC:APC ratio was more or less constant, although there may have been a slight increase in August/September. PE:APC and PE:PC increased again to their highest values in December/January. Even the PC:APC ratio was highest in January.

Lowest dry weights (DW) of 12.6–12.7% were found between April and June (Fig. 3). From July onwards, the DW increased continuously, with a slight peak in October, until February (up to 16.5% DW).

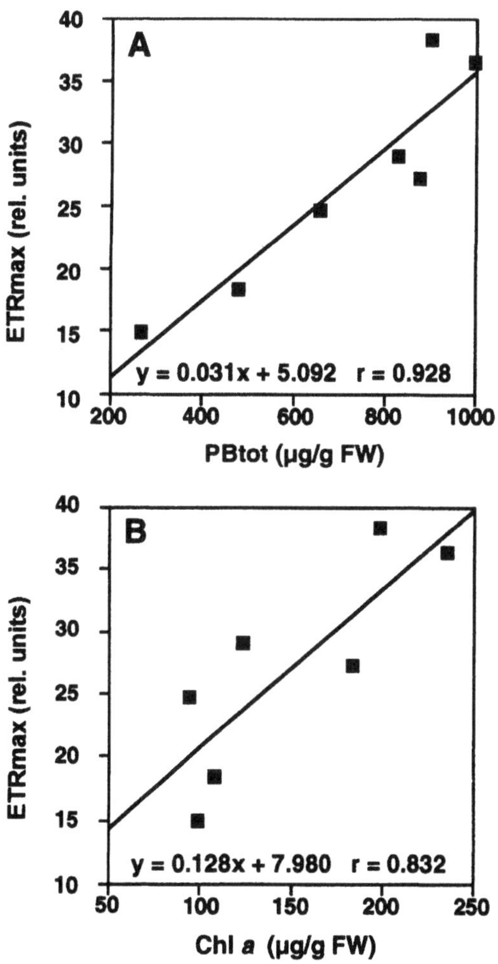

Fig. 2 Correlation of pigment contents and photosynthetic activity (*ETRmax*) between April/October: total phycobiliprotein content (*PBtot*) and ETRmax (**A**) and Chl *a* content and ETRmax (**B**)

Discussion

The results of this study demonstrate a strong relationship between the seasonal pattern of photosynthesis and pigments. The seasonal maximum of photosynthetic capacity in September/October is followed by a clear maximum in the concentrations of all pigments in October/November. This peak coincides in time with the natural sea-ice break-up in the Antarctic Peninsula region between early September and late November. Furthermore, the photosynthetic capacity increased in parallel with the phycobiliprotein content during the entire Antarctic autumn, winter and spring (from April/May to October). During this period, a positive correlation between total phycobiliprotein content and ETRmax was found, whereas a weaker correlation was found between Chl *a* and ETRmax. Weykam and Wiencke (1996) observed in a similar study with *P. decipiens* an optimum in oxygen production (per DW) 1 month later, in October/November, and maximal Chl *a* concentrations in December, with an increase during winter. The total phycobiliprotein content showed no clear seasonal pattern, even when related to FW. One reason

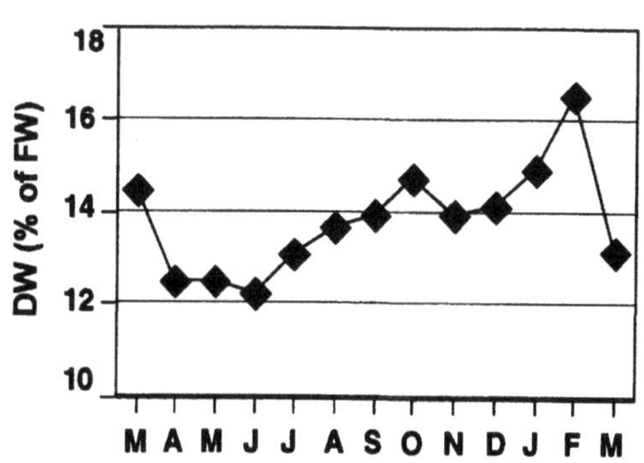

Fig. 3 Seasonal changes in dry weight (*DW*, % of fresh weight, *FW*). Means ± SD, $n = 10$. In those cases where no error bars are given, the SD is smaller than the symbol for the mean value

for differences in comparison with the present study may be the low number of replicates and the difference in age of the plants. All measurements were performed on only one old gametophyte, whereas in this study many 2-year young individuals were used. Nevertheless, changes in Chl *a* associated with changes in oxygen flux were also found in *P. decipiens* growing in different depths (Gómez et al. 1997). Similar seasonal photosynthetic patterns have been observed in several Antarctic brown macroalgae in the laboratory (Gómez and Wiencke 1997) and in the field (Gutkowski and Maleszewski 1989; Gómez et al. 1995, 1998). A clear seasonal variation in pigment pattern (Chl $a + c$), with a maximum in October coupled with an increase in photosynthetic activity, was found in *Adenocystis utricularis* (Gutkowski and Maleszewski 1989). Little seasonal variation of Chl *a*, with a peak in November, was observed in *Ascoseira mirabilis* (Gómez et al. 1995) and in *Desmarestia menziesii* (Gómez et al. 1998); however, a relationship between Chl *a* and thallus size rather than a relation between Chl *a* and light was emphasised by Gómez and Wiencke (1997).

Phycobiliprotein ratios may give further indications about changes in phycobilisome (PBS) number and size/structure. In general, PBS number and size (number of rods, rod length) can change according to acclimation processes, in response to light intensity, light quality or nutrient availability (Gantt 1990; Grossman et al. 1994; Talarico 1996). In autumn, the observed increase of APC content was due to an increase of PBS number. The decrease of PE:PC suggests a reduction of rod length and the slight increase of PC:APC suggests an increase of rod number. In summary, *P. decipiens* seems to acclimate during the autumn months by increasing the amount of "smaller" PBS (many rods with shorter rod length). In winter, the APC and PC contents were unchanged. But in September, the ratio of PC:APC was maximal and PE:PC, as well as PE:APC, increased slightly again, which suggests a slight increase in rod length. Thus, *P. decipiens* seems to acclimate during the winter months by slightly rising the PBS size (rod length). In early spring (October/November), the further increase of APC content to its highest value is due to a further increase in PBS number. From late spring (December) onwards the PBS amount seems to decline, but the high values of PE:PC and PE:APC suggest an increase in rod length of the remaining PBS. So in early spring, *P. decipiens* had the highest number of PBS, which it started to reduce in late spring, but the remaining PBS seems to be well assembled with prolonged rod length. In summer, the number of PBS further declined and the disruption of the remaining PBS started by reducing rod length, due to an decrease in PC:APC and PE:APC. These seasonal changes agree with observations in *Halopithys incurvus* (Rhodomelaceae) from the Mediterranean during an annual cycle; well-assembled PBS were found at low phycobiliprotein levels whereas small and incompletely assembled PBS occurred at high phycobiliprotein levels (Talarico and Kosovel 1983). In general, phycobiliprotein synthesis does not always lead to phycobilisome assembly (Lüning and Schmitz 1988; Foltran et al. 1996; Talarico 1996). So, the acclimation process of PBS discussed here should be proven by parallel electron microscopic studies.

The increase of phycobiliprotein content between April and June may be explained by photoacclimation to low light conditions, caused by decreasing daylength. This would result in maximal ratios of light-harvesting pigments to reaction centres and explains the maximum of PBtot:Chl *a* ratio and APC: Chl *a* ratio (Fig. 1D). The continuing increase of phycobiliprotein content from July on is presumably initiated by a specific seasonal trigger. Interestingly, it was also from July on that both the daily light dose and the Chl *a* content began to increase. The Chl *a* increase points to a rise in photosystem II reaction centres (RC II) and/or photosystem I reaction centres (RC I). The rise of Chl *a* may be coupled with an increase in thylakoid area, as shown in the red alga *Iridaea cordata* from Antarctica (Foltran et al. 1996). This species exhibits high levels of Chl *a*, combined with a thylakoid-rich chloroplast structure, when growing under the ice. In ice-free water, the chloroplasts possess a reduced thylakoid system and a lower Chl *a* content.

The increase in daylength may be a trigger for chlorophyll synthesis and a further production of phycobiliproteins. The hypothesis of a photoperiodic control of pigment synthesis in *P. decipiens* is possible, since different types of photoreceptors seem to control chlorophyll and phycobiliprotein synthesis in macroalgae (Rüdiger and López-Figueroa 1992; Häder and Figueroa 1997). López-Figueroa (1992) demonstrated that diurnal changes in the red/far-red (R:FR) ratio and in the blue:red ratio are well correlated with diurnal changes in photosynthetic pigment content in red macroalgae. Furthermore, annual variation of the end-of-day R:FR ratio is well correlated with the photoperiod, suggesting that it could act as an annual environmental light signal (Figueroa 1996). Annual changes in R:FR ratios have even been discussed as being responsible for modulating annual growth and reproduction in kelps (Lüning 1993; Häder and Figueroa 1997). The question is, do these wavelengths reach the algae growing in the sublittoral? The Antarctic water around King George Island was described as Jerlov water type III (Klöser et al. 1993), in which far-red is absorbed between 5 and 10 m depth and red light in 15 m depth. Therefore, the annual control by a red/far-red receptor is questionable; however, a control by a blue light receptor is possible. In this context, it is interesting that all phycobiliproteins show a common ultraviolet absorption band at 330 nm (Stadnichuk 1995), which may function as a photoreceptor. Furthermore, the chromophores of the phycobiliproteins are very similar to the phytochrome chromophore in higher plants, and have been suggested to control the chromatic adaptation in cyanobacteria (Grossman et al. 1994).

Another hypothesis is that the initiation of chlorophyll synthesis is triggered by growth. We know that

P. decipiens starts to grow in July, develops new blades during late winter/early spring (September) and exhibits optimal growth rates in October/November (Wiencke 1990b; Weykam and Wiencke 1996). This growth pattern is under photoperiodic control, and may have an underlying circannual rhythm, which may be triggered by daylength (Wiencke 1996). A good correlation between growth and daylength was also found in *D. menziesii* (Gómez and Wiencke 1997). Photoperiodic short-day responses are also responsible for the production of tetraspores, monospores, gametes, new blades and upright thalli in many algae, specially in red and brown macroalgae (Kain and Norton 1990; Lüning 1990).

We think that the extension of the daylength is a trigger to start growth and pigment synthesis in *P. decipiens*. The pigments increase number and/or size of light-harvesting antennae and reaction centres in order to trap as much light energy as possible, in order to optimise photosynthetic capacity and accelerate growth. This theory is supported by the fact that the pigment optimum reported in this study corresponds with the observed optimum in growth (Wiencke 1990b; Weykam and Wiencke 1996).

The continuously high Fv/Fm value of *P. decipiens* between May and December confirms that the photosynthetic apparatus remains intact during the entire mid-autumn, winter and spring. Conversely, the decrease of Fv/Fm, ETRmax and all pigments during summer suggests a degradation of the photosynthetic apparatus. The latter could be a normal photoacclimation process to high light, caused by the prolonged daylength, or also a photoinhibitory effect as demonstrated in *P. decipiens* in the field (Hanelt et al. 1994). A decrease of phycobiliproteins during summer was also found in *Gracilaria verrucosa* (Kosovel and Talarico 1979).

We conclude that *P. decipiens* is very well adapted to the short period of favourable light conditions in the field. *P. decipiens* maintains an intact photosynthetic apparatus during the entire mid-autumn, winter and spring. The alga starts to increase phycobiliprotein levels in mid-autumn, to build up new small phycobilisomes, the main light-harvesting antennae of red algae. Later, during winter, *P. decipiens* starts to increase Chl *a* levels to build up new reaction centres. This leads to a parallel increase of maximal photosynthetic activity. Subsequently, photosynthetic activity and pigment contents are maximal in Antarctic spring when the water is very clear and the sunlight penetrates deep into the water column, after break-up of sea ice and before the water becomes turbid due to plankton blooms and glacial melt-water. In contrast, during summer *P. decipiens* reduces its photosynthetic apparatus to a minimum: maximal quantum yield, maximal photosynthetic activity, and phycobiliprotein and Chl *a* levels are minimised. A photoperiodic control of pigment synthesis triggered by daylength is suggested. These results are in accordance with the life strategy of *P. decipiens* as season anticipator.

Acknowledgements This work was supported by grants from the Deutsche Forschungsgemeinschaft. We gratefully thank C. Langreder for excellent technical support, and K. Bischof and D. Hanelt for introducing the PAM technique. We acknowledge M. Schoenwaelder and I. Gómez for critically reading this manuscript. This is publication no. 1785 of the Alfred Wegener Institute.

References

Bischof K, Hanelt D, Wiencke C (1999) Acclimation of maximal quantum yield of photosynthesis in the brown alga *Alaria esculenta* under high light and UV radiation. Plant Biol 1:435–444

Figueroa FL (1996) Underwater light signals: role of algal photoreceptors in natural environment. G Bot Ital 130:608–623

Foltran A, Maranzana G, Rascio N, Scarabel L, Talarico L, Andreoli C (1996) *Iridaea cordata* from Antarctica: an ultrastructural, cytochemical and pigment study. Bot Mar 39:533–541

Gantt E (1990) Pigmentation and photoacclimation. In: Cole KM, Sheath RG (eds) Biology of the red algae. Cambridge University Press, Cambridge, pp 203–219

Gómez I, Wiencke C (1997) Seasonal growth and photosynthetic performance of the Antarctic macroalga *Desmarestia menziesii* (Phaeophyceae) cultivated under fluctuating Antarctic daylengths. Bot Acta 110:25–31

Gómez I, Wiencke C, Weykam G (1995) Seasonal photosynthetic characteristics of *Ascoseira mirabilis* (Ascoseirales, Phaeophyceae) from King George Island, Antarctica. Mar Biol 123:167–172

Gómez I, Weykam G, Klöser H, Wiencke C (1997) Photosynthetic light requirements, metabolic carbon balance and zonation of sublittoral macroalgae from King George Island (Antarctica). Mar Ecol Prog Ser 148:281–293

Gómez I, Weykam G, Wiencke C (1998) Photosynthetic metabolism and major organic compounds in the marine brown alga *Desmarestia menziesii* from King George Island (Antarctica). Aquatic Bot 60:105–118

Grossman AR, Schaefer MR, Chiang GG, Colier JL (1994) The responses of cyanobacteria to environmental conditions: light and nutrients. In: Bryant DA (ed) The molecular biology of Cyanobacteria. Kluwer, Dordrecht, pp 641–675

Gutkowski R, Maleszewski S (1989) Seasonal changes of the photosynthetic capacity of the Antarctic macroalga *Adenocystis utricularis* (Bory) Skottsberg. Polar Biol 10:145–148

Häder DP, Figueroa FL (1997) Photoecophysiology of marine macroalgae. Photochem Photobiol 66:1–14

Hanelt D (1992) Photoinhibition of photosynthesis in marine macrophytes of the South Chinese Sea. Mar Ecol Prog Ser 82:199–206

Hanelt D, Huppertz K, Nultsch W (1992) Photoinhibition of photosynthesis and its recovery in red algae. Bot Acta 105:278–284

Hanelt D, Jaramillo J, Nultsch W, Senger S, Westermeier R (1994) Photoinhibition as a regulative mechanism of photosynthesis in marine algae of Antarctica. Ser Cient INACH 44:67–77

Inskeep WP, Bloom PR (1985) Extinction coefficients of chlorophyll *a* and *b* in *N,N*-dimethylformamide and 80% acetone. Plant Physiol 77:483–485

Kain JM, Norton TA (1990) Marine ecology. In: Cole KM, Sheath RG (eds) Biology of the red algae. Cambridge University Press, Cambridge, pp 377–422

Klöser H, Ferreyra G, Schloss I, Mercuri G, Laturnus F, Curtosi A (1993) Seasonal variation of algal growth conditions in sheltered Antarctic bays: the example of Potter Cove (King George Island, South Shetlands). J Mar Syst 4:289–301

Kosovel V, Talarico L (1979) Seasonal variations of photosynthetic pigments in *Gracilaria verrucosa* (Huds.) Papenfuss (Florideophyceae-Gigartinales). Boll Soc Adr Sci 63:5–15

López-Figueroa F (1992) Diurnal variation in pigment content in *Porphyra laciniata* and *Chondrus crispus* and its relation to the

diurnal changes of underwater light quality and quantity. Mar Ecol 13:285–305
Lüning K (1990) Seaweeds. Their environment, biogeography and ecophysiology. Wiley, New York
Lüning K (1993) Environmental and intertidal control of seasonal growth in seaweeds. Hydrobiologia 260:1–14
Lüning K, Schmitz K (1988) Dark growth of the red alga *Delesseria sanguinea* (Ceramiales): lack of chlorophyll, photosynthetic capability and phycobilisomes. Phycologia 27:72–77
Provasoli L (1966) Media and prospects for the cultivation of marine algae. Proc Jpn Conf Hakone:63–75
Rosenberg G (1981) Ecological growth strategies in the seaweeds *Gracilaria foliifera* (Rhodophyceae) and *Ulva* sp. (Chlorophyceae). PhD Thesis, Yale University
Rüdiger W, López-Figueroa F (1992) Photoreceptors in algae. Photochem Photobiol 55:949–954
Stadnichuk IN (1995) Phycobiliproteins: determination of chromophore composition and content. Phytochem Anal 6:281–288
Talarico L (1996) Phycobiliproteins and phycobilisomes in red algae: adaptive responses to light. Sci Mar 60:205–222
Talarico L, Kosovel V (1983) Ficobilisomi e variazioni delle ficobiliproteine in *Halopithys incurvus* (Huds.) Batt. (Rhodomelaceae-Rhodophyta) durante il ciclo annuale. Boll Soc Adr Sci 67:69–81
Weykam G, Wiencke C (1996) Seasonal photosynthetic performance of the endemic Antarctic alga *Palmaria decipiens* (Reinsch) Ricker. Polar Biol 16:357–361
Weykam G, Thomas DN, Wiencke C (1997) Growth and photosynthesis of the Antarctic red algae *Palmaria decipiens* (Palmariales) and *Iridaea cordata* (Gigartinales) during and following extended periods of darkness. Phycologia 36:395–405
Wiencke C (1990a) Seasonality of brown macroalgae from Antarctica – a long-term culture study under fluctuating Antarctic daylengths. Polar Biol 10:589–600
Wiencke C (1990b) Seasonality of red and green macroalgae from Antarctica – a long-term culture study under fluctuating Antarctic daylengths. Polar Biol 10:601–607
Wiencke C (1996) Recent advances in the investigation of Antarctic macroalgae. Polar Biol 16:231–240

Polar Biol (2001) 25: 12–20
DOI 10.1007/s003000100306

M. Hernando · J.I. Carreto · M.O. Carignan
G.A. Ferreyra · C. Gross

Effects of solar radiation on growth and mycosporine-like amino acids content in *Thalassiosira* sp, an Antarctic diatom

Abstract The main objective of this study was to investigate the role of mycosporine-like amino acids (MAAs) in the short-term (hours) and long-term (several days) photoacclimation to solar ultraviolet radiation (UVR) of a marine Antarctic diatom (*Thalassiosira* sp.). The cultures were exposed to natural solar radiation and grown over 72 h and 6 days under three experimental irradiance treatments: (1) photosynthetic available radiation (PAR) + UVR (280–700 nm), (2) PAR + UVA (315–700 nm), and (3) PAR (400–700 nm). Two MAAs were identified and quantified in the cells: porphyra-334 and shinorine. The cultures showed an initial period of growth inhibition and the rate of synthesis of MAAs (μ) was very high in all treatments in the short-term experiment and during the first 2 days of the long-term experiment. No differences between treatments were observed in the accumulation of these compounds. After 2 days of exposure, however, the MAA content per cell was several times higher than that initially found in all treatments, and the cells exposed to the PAR + UVR treatment displayed the maximum concentration.

Please note: When citing this article, please use the reference to the original source in the journal Polar Biology as given on the top of this page.

M. Hernando (✉)
Centro Austral de Investigaciones Científicas
(CADIC-CONICET), CC 92, 9410 Tierra del Fuego,
Ushuaia, Argentina
E-mail: hernando@infovia.com.ar

J.I. Carreto · M.O. Carignan
Instituto Nacional de Investigación y Desarrollo Pesquero,
C.C.175, 7600 Mar del Plata, Argentina

G.A. Ferreyra
Instituto Antártico Argentino, Cerrito 1248,
1010 Buenos Aires, Argentina

C. Gross
Alfred Wegener Institute for Polar
and Marine Research, Columbusstrasse,
27568 Bremerhaven, Germany

Introduction

Ultraviolet-B (UVB, 280–315 nm) and ultraviolet-A (UVA, 315–400 nm) radiation are known to be harmful for living organisms. During the past decades, springtime stratospheric ozone depletion over the Antarctic and the Southern Ocean (Lubin et al.1989) has resulted in enhanced levels of UVB radiation reaching the earth's surface.

One main interest is the identification of repair and protective mechanisms that may allow Antarctic species to survive and reproduce under natural conditions of ultraviolet radiation (UVR, 280–400 nm), and particularly under an increased UVB flux scenario (Lubin et al. 1989). UVR can induce significant damage on a variety of cell targets including DNA (Karentz et al. 1991; Karentz 1994), photosystem II (Franklin et al. 1999) and Rubisco (Lesser 1996). However, growth and biomass accumulation will result from the complex interactions between harmful direct and indirect effects of UVR, and a series of counteracting repair mechanisms (Lesser et al. 1994). One of the adaptations whereby phytoplankton can prevent UVR-induced damage is the synthesis of mycosporine-like amino acids (MAAs), molecules constituted by imino-carbonyl derivatives of the mycosporine cyclohexenone chromophore which strongly absorb within the UVR range (Carreto et al. 1990a; Karentz et al. 1991; Karsten et al. 1999). MAAs, with maximum absorption from around 310–360 nm, have been hypothesized to act as sunscreens and thus to reduce harmful effects of UVR (Shibata 1969; Carreto et al.1989). In addition some MAAs may act as antioxidants to prevent cellular damage resulting from UV-induced production of toxic oxygen species (Dunlap and Yamamoto 1995). The synthesis and accumulation of these UV-absorbing compounds has been recognized as being able to minimize structural and physiological damage caused by UV exposure in tropical and temperate marine organisms (Carreto et al. 1989). Besides there is a widespread distribution of MAAs in Antarctic flora and fauna, including phytoplankton (Karentz et al. 1991; Smith et al. 1992; Helbling et al.

1996). Many reports support the view that these substances could act as a protective mechanism against UVR (Dunlap et al. 1989; Dunlap and Yamamoto 1995; Neale et al. 1998b; Karsten et al. 1999), while others show that MAAs are unable to provide full protection against UVR effects (Garcia-Pichel et al. 1993; Lesser 1996).

The main objective of the study was to assess the short (hours) and long-term (days) effects of exposure to UV radiation on growth and induction of MAA synthesis in the photoacclimation process of a commonly occurring marine Antarctic diatom (*Thalassiosira* sp.).

Materials and methods

The experiments were carried out at Potter Cove (South Shetland Islands, Antarctica, 62°14′S, 58°38′W). Surface water samples were taken with inflatable boats using a 5-l Niskin bottle at a station located in the outer part of the cove. They were transported immediately to the laboratory and maintained at a cool room temperature of 2°C. *Thalassiosira* sp. single cells were isolated with micropipets using an inverted microscope, and inoculated into 200-ml flasks with filtered seawater plus F/2 culture medium (Guillard and Ryther 1962). One inoculum of the *Thalassiosira* species isolated from the natural community was grown outdoors for 10 days exposed to photosynthetic available radiation (PAR, 400–700 nm) plus UVA radiation only. Once in exponential growth, aliquots of the culture were inoculated to a series of experimental 500-ml vessels and then exposed to solar radiation within two time scales: one short-term exposure experiment (72 h), 7–9 December 1998, and one long-term exposure experiment (6 days), 4–10 January 1999. During both experiments, cells were exposed to three irradiance treatments: PAR + UVR, PAR + UVA and PAR (control). Cylindrical UV-transparent vessels (quartz) were used for exposure of phytoplankton to full solar radiation. For the PAR + UVA treatment, the bottles were covered with Mylar foils (Dupont) and PAR controls were performed with cylindrical Plexiglass vessels (UF3; Röhm and Haas) cutting off UV radiation. Three replicate samples were included for each of the treatments and controls. Culture medium was added to the different treatments at the time 0 (t_0) of the experiments, and temperature was maintained approximately constant (around 1–2°C) by continuous pumping of surface seawater from the coast of Potter Cove. Sub-samples from the short-term experiment were taken at 3, 6, 12, 18, 24, 48 and 72 h for analyses, whereas during the long-term experiment sub-samples from each treatment were taken daily approximately at noon.

Biomass determinations

Chlorophyll-*a* analyses were made by filtering the water through Whatman fiberglass filters (GF/F, 25 mm) followed by the extraction of pigments in absolute methanol (Holm-Hansen and Riemann 1978). Sub-samples for cell counting were kept in dark bottles and fixed with formalin previously neutralized with sodium borate (final concentration 0.4%). All the samples were counted with an inverted microscope.

Growth measurements

Cell growth rate and the rate of synthesis of MAAs were determined with the following equation:

$$\mu = \ln(N_1/N_0)/(t_1 - t_0)$$

where μ is a specific rate constant (day^{-1}), t_1 is the time of measurement and t_0 is the time of the previous one, N_1 is the parameter concentration at time t_1, and N_0 is the parameter concentration at time t_0. Parametric one-way analysis of variance (ANOVA) was applied to test the significance of the observed differences between treatments, and post-hoc comparisons were used (Duncan's multiple range test; Statistica for Windows version 5). Data were checked prior to the analyses in order to verify the homoscedacity and normality requirements of the ANOVA.

MAA measurements

For identification and quantification of MAAs, 10–30 ml sample were filtered onto Whatman GF/F glass fiber filters and frozen (–20°C) until analysis. The three replicates corresponding to each treatment were pooled, in order to assure a sufficient amount of cells in the filtrates. Filters were extracted with a Vibra Cell probe sonicator (Sonics & Materials; 1 min, 100 W pulse mode, 0°C) into 4 ml 100% HPLC grade methanol. The extracts were filtered (Whatman GF/F filters) and evaporated to dryness using a centrifugal evaporator (Centrivap, Labconco). The residue was redissolved in 500 μl water and vortexed for 30 s. After passing through a 10^5-Da ultrafilter (Ultra-spin), samples were analyzed with a Shimadzu LC-10A HPLC system according to the method of Carreto et al. (in press). Individual MAAs were separated by reversed phase, gradient elution on two Alltima C18 high carbon loaded (16%), fully endcapped 5-μm columns [4.6 mm inner diameter (ID) ×150 mm] protected with an Alltech guard column cartridge (4.6 mm ID ×20 mm) filled with the same material. They were connected in series and thermostated at 20°C (Shimadzu CTO column oven). An initial isocratic hold for 8 min with 0.2% acetic acid (solvent A) was followed by a 12-min linear gradient from 0% to 40% of solvent B (methanol:acetonitrile:water, 25:25:50) and then an isocratic hold for 10 min at a flow rate of 0.8 ml min^{-1}. Peak detection was carried out using a diode array UV-Vis absorbance detector (Shimadzu SPD-M10 Avp). Individual peaks were identified by on-line absorption spectra, retention time and co-chromatography with authentic standards isolated from the red macroalgae *Porphyra* sp. Quantification was made by peak-area integration at 330 nm, using the following molar extinction coefficients: $\epsilon_{334}=44{,}700$ for shinorine (Tsujino et al. 1980) and: $\epsilon_{334}=42{,}300$ for porphyra-334 (Takano et al. 1979).

Ultraviolet radiation

Atmospheric incident solar radiation was monitored continuously using a multichannel UV spectroradiometer developed at the Alfred Wegener Institute for Polar and Marine Research and distributed by Isitec Bremerhaven. The instrument is based on a Bentham DM 150 double monochromator (Bentham), with a multichannel detector system. Data were recorded every second from 290 nm to 320 nm with a data point every 1.35 nm due to the 32 detection channels. Data were stored as 1-min average values. The 290–320 nm range was used because at wavelengths lower than 290 nm radiation is below detection limit for the solar altitudes reached at Potter Cove. UVB was weighted with the biological weighting function described in Neale et al. (1998a).

Results

Light environment: atmospheric measurements

The UVB weighted dose for each sampling interval is shown in Fig. 1. Maximum UVB irradiances of around 0.06 and 0.08 J m^{-2} were measured during the short-term experiment for the 6–12 and 24–48 h intervals, respectively (Fig. 1A). Irradiances during the long-term experiment ranged between 3 and 4.5 J m^{-2}, and maximum values were experienced at the beginning of the experiment (Fig. 1B).

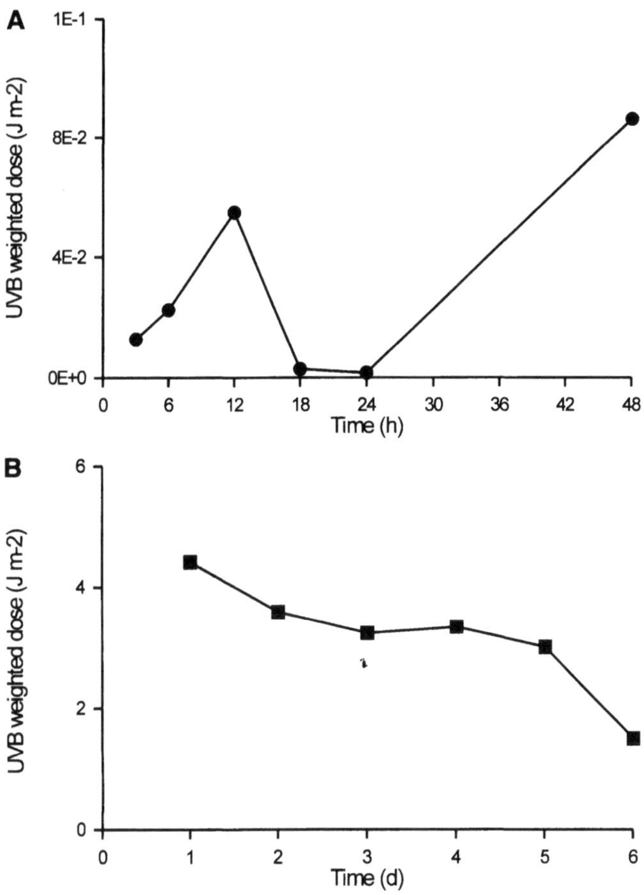

Fig. 1 Ultraviolet-B (UVB) doses (J m^{-2}) received by the experimental vessels during: **A** short-term experiment and **B** long-term experiment. It should be noted that no data were available for the last day of the long-term experiment, and values for day 6 (9 January 1999) were only measured from noon to midnight, and not up to noon of the next day. For this reason, the daily dose corresponding to that day was approximately half of the other days

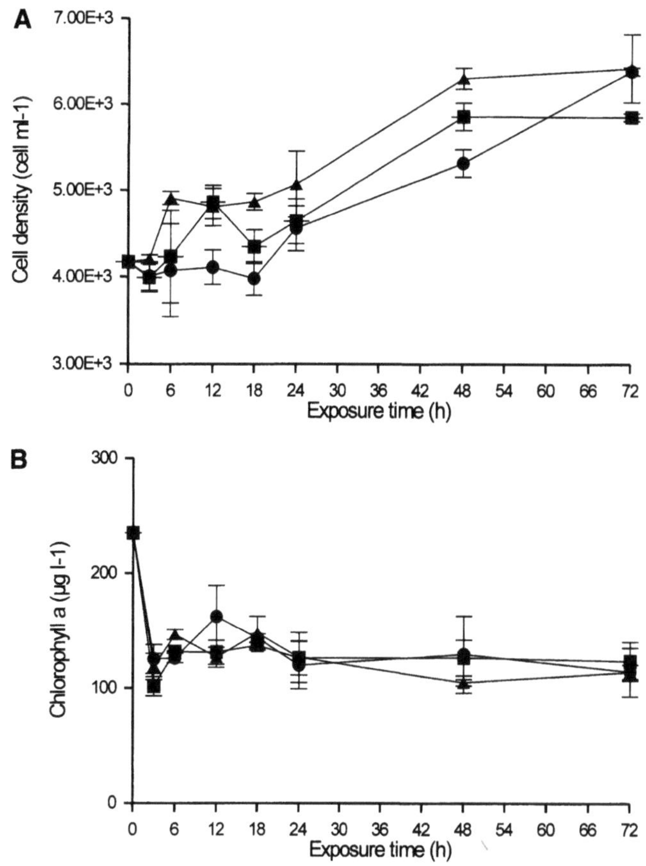

Fig. 2 Time course evolution of: **A** cell density (cell ml^{-1}) and **B** chlorophyll *a* ($\mu g\ l^{-1}$) corresponding to the short-term experiment. *Vertical bars* indicate standard deviation, *PAR* photosynthetic active radiation, *UVR* ultraviolet radiation, *UVA* ultraviolet-A, *filled circle* PAR + UVR, *filled triangle* PAR + UVA, *filled square* PAR

Short-term experiment

Biomass and growth

Cell number increased in all treatments after 24-h exposure of the cultures to solar radiation (Fig. 2A). The cell number in the UVB treatment differed significantly from the other treatments only at 12 h and 48 h ($P < 0.05$), in coincidence with the maximum UVB doses measured (Fig. 1A). Chlorophyll *a* decreased sharply during the first 3 h of the experiment, remaining approximately constant in all treatments up to the end of the observations. Moreover, chlorophyll *a* did not significantly differ between the three irradiance treatments ($P > 0.05$, Fig. 2B).

UV-absorbing compounds

The analyses of the cell extracts revealed the presence of two MAAs, identified as porphyra-334 and shinorine. The concentration of porphyra-334 exceeded that of shinorine by approximately 9 times. Both compounds increased rapidly in concentration with time, apparently independent of the light quality received by the cultures (Fig. 3A, B). This result suggests that the synthesis of MAAs in *Thalassiosira* sp. was more affected by PAR than by UVR. The kinetics of MAAs formation showed: (1) the absence of a lag phase and (2) high specific rates of synthesis for both compounds. However, the rate of synthesis of shinorine ($\mu_{shinorine} = 0.60\ day^{-1}$) was slightly lower than that measured for porphyra-334 ($\mu_{porphyra\text{-}334} = 0.86\ day^{-1}$). Contrasting with the synthesis rate, the cell accumulation of both MAAs was shown to be dependent on the light treatment (Fig. 3C, D). Highest MAA contents per cell were reached by both compounds in the cultures exposed to PAR + UVR, with maximum values at 18 h. The rate of cell accumulation for shinorine was 0.65, 0.69 and 1.05 day^{-1} for the PAR, PAR + UVA and PAR + UVR treatments, respectively. In the case of porphyra-334, these values were 0.99, 1.05 and 1.45 day^{-1}, respectively.

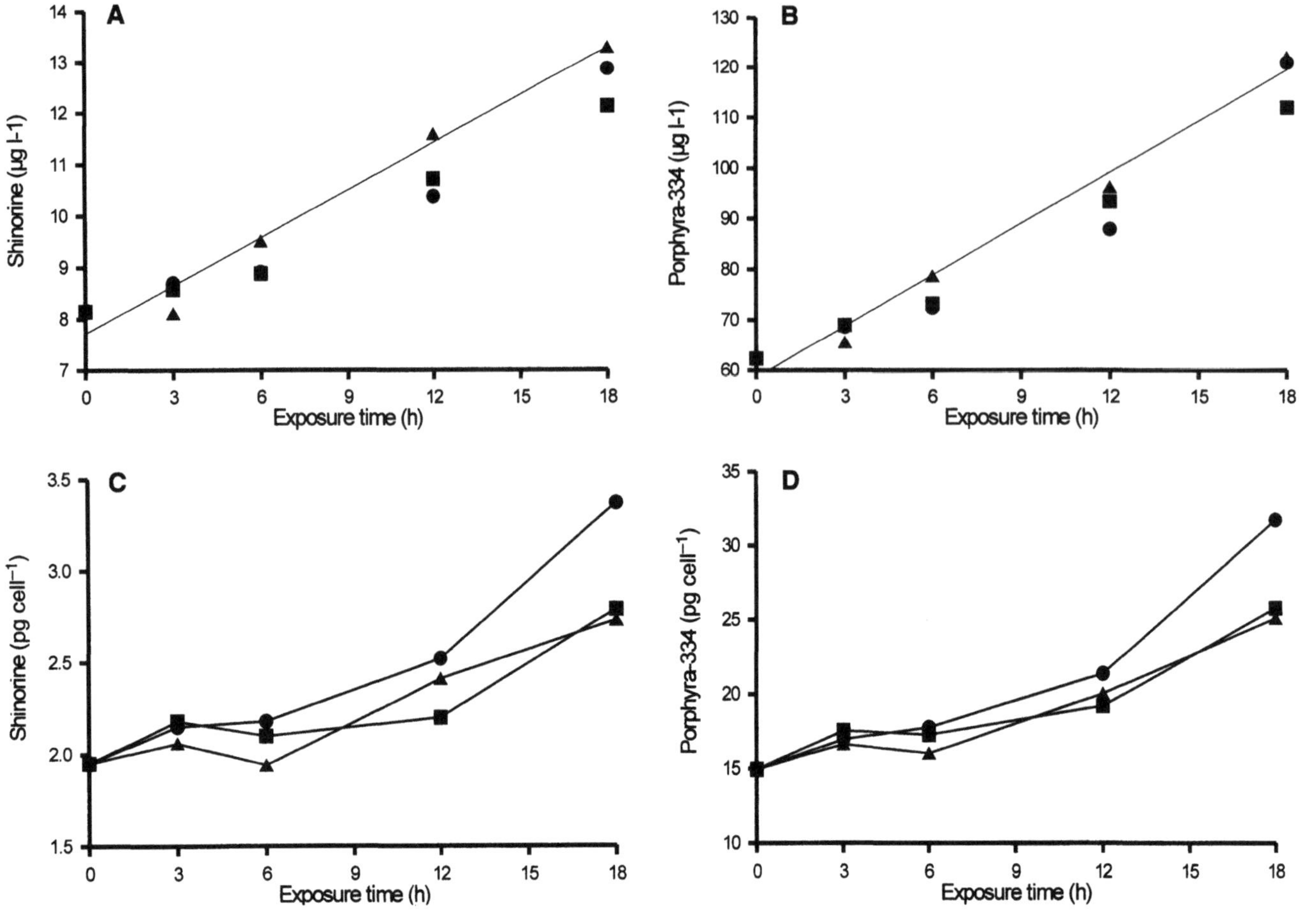

Fig. 3 Time course for the rate of synthesis of shinorine (**A**) and porphyra-334 (**B**) and cell content of shinorine (**C**) and porphyra-334 (**D**) in *Thalassiosira* sp. during the short-term experiment (*filled circle* PAR + UVR, *filled triangle* PAR + UVA, *filled square* PAR)

Long-term experiment

Biomass and growth

A significant increase in cell number began in the PAR and PAR + UVA treatments after the 1st day of acclimation to the experimental conditions and continued up to day 4, reaching constant values afterwards (Fig. 4A). In contrast, biomass accumulation in the PAR + UVR treatment started after a 2-day lag and ceased on day 5. The PAR + UVR treatment presented significant differences with the other two treatments during days 1–4. There were no significant differences between the PAR and PAR + UVA treatments in most cases (Fig. 4A). However, algae continued growing under UVB exposure and reached biomass values which did not differ significantly from the PAR and PAR + UVA treatments at the end of the experiment ($P > 0.05$). A similar pattern was evidenced for the chlorophyll-*a* concentration. In this case differences between treatments were observed only on day 4, when concentrations in the PAR + UVR treatment were significantly lower than those in the treatments where UVB radiation was excluded ($P < 0.01$, Fig. 4B). In contrast to the different cell densities between the different treatments, the chlorophyll-*a* concentration per cell remained approximately constant during the course of the experiment (15–30 pg cell^{-1}). The UVB dose responses of *Thalassiosira* sp. growth rate were consistent with the trends presented by the biomass accumulation. The algae were more affected by UVB radiation than by PAR and PAR + UVA radiation, particularly at doses above 3.6 J m^{-2}. The response of the species cultivated under PAR and PAR + UVA treatments presented μ values between 0.2–0.3 day^{-1}. In contrast, growth rates in the PAR + UVR treatment were lower, reaching values close to 0 at the highest exposure doses measured during the study (Fig. 5).

UV-absorbing compounds

In spite of this strong UVB-induced photoinhibition of growth, the specific rate of synthesis of MAA (Fig. 6A, B) was very high in the PAR + UVR treatment ($\mu_{porphyra\text{-}334} = 0.73$ day^{-1}; $\mu_{shinorine} = 0.68$ day^{-1}) but only slightly lower than that observed in PAR ($\mu_{porphyra\text{-}334} = 0.80$ day^{-1}; $\mu_{shinorine} = 0.77$ day^{-1}) and PAR + UVA ($\mu_{porphyra\text{-}334} = 0.80$ day^{-1}; $\mu_{shinorine} = 0.77$ day^{-1}) treatments. These values, calculated from

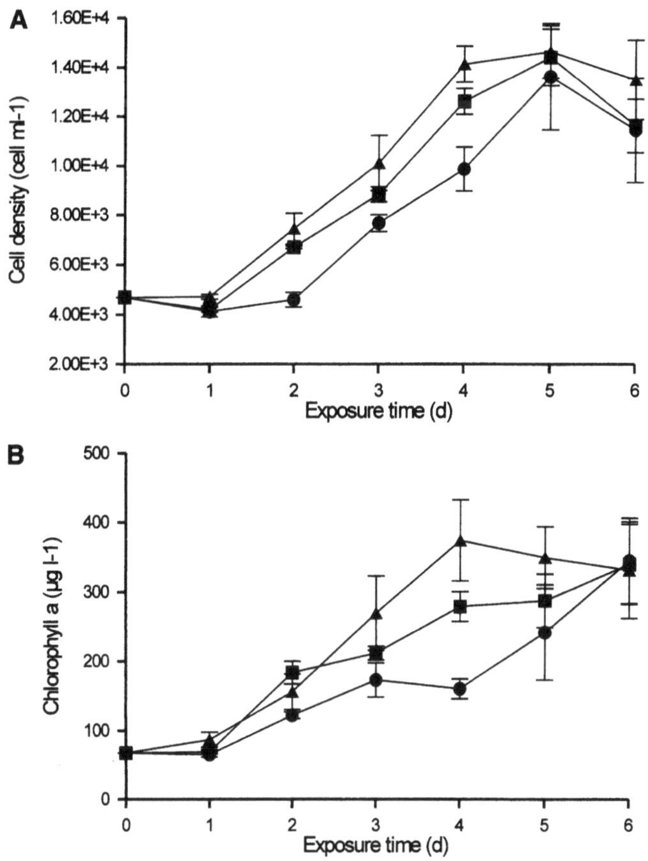

Fig. 4 Time course evolution of: **A** cell density (cell ml^{-1}) and **B** chlorophyll *a* ($\mu g\ l^{-1}$) corresponding to the long-term experiment. *Vertical bars* indicate standard deviation, *filled circle* PAR + UVR, *filled triangle* PAR + UVA, *filled square* PAR

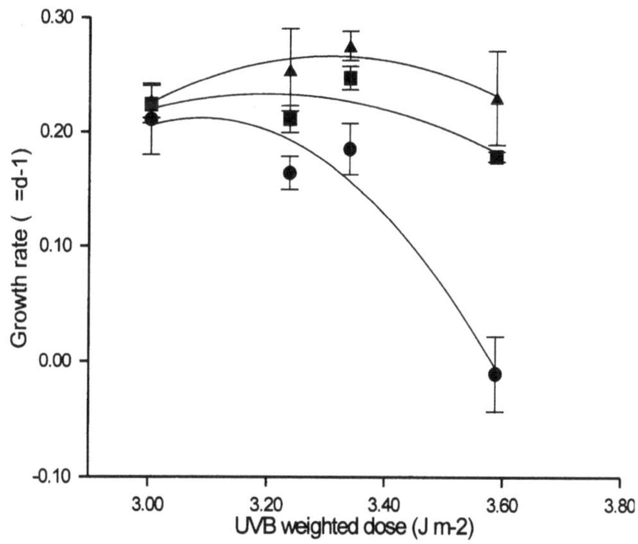

Fig. 5 Cell growth rate ($\mu = day^{-1}$) of *Thalassiosira* sp. as a function of the UVB weighted dose (J m^{-2}) during day 2–5 in the long-term experiment. *Filled triangles* PAR treatment, *filled squares* PAR + UVA, *filled circles* PAR + UVR, *vertical bars* are standard errors

two point (t_0 and t_2) extrapolation, are similar to the mean value calculated ($\mu_{porphyra\text{-}334} = 0.86\ day^{-1}$; $\mu_{shinorine} = 0.60\ day^{-1}$) in the short-term photoacclimation experiment (Fig. 3A, B). After 2-days exposure, the cell content of porphyra-334 and shinorine in all treatments was 4–5 times higher than that corresponding to t_0 (Fig. 6A, B). However, the highest MAA concentrations per cell were displayed by algae exposed to PAR + UVR (Fig. 6C, D). When comparing the cell growth rate within the different treatments with the rate of synthesis of MAAs (Fig. 7), the following can be evidenced: (1) algal growth was approximately constant in the PAR and PAR + UVA treatments (μ around 0.2 day^{-1}) and no significant differences were found between them ($P > 0.05$); (2) the PAR + UVR treatment was significantly lower than the previous treatments during days 2 to 4 ($P < 0.01$); (3) at the beginning of the experiment, $\mu_{porphyra\text{-}334}$ was around 3 times higher than cell growth in the PAR and PAR + UVA treatments and a similar trend was presented by shinorine (data not shown). An even higher difference could be observed in the PAR + UVR treatment between days 0 and 2, with the rate of synthesis of MAAs 7 times superior than that of *Thalassiosira* sp. growth rate; after this initial period, the rate of synthesis of MAAs decreased, gradually matching the cell growth rate towards the end of the experiment (Fig. 7).

The time course of the relative accumulation of the two identified MAAs in the cells was explored, in order to identify possible treatment-specific responses. A marked increase in porphyra-334 concentration relative to shinorine per cell was observed during the short-term experiment as a function of time in all treatments (Fig. 8A). In contrast during the long-term experiment (1. 8B) porphyra-334 accumulated in the cells more intensely than shinorine under PAR + UVA and PAR + UVR exposure conditions, as compared with the PAR control. The same trends are shown if the MAA synthesis rate is calculated (data not shown).

Discussion

MAAs have an important UVR photoprotective function in algae because their accumulation is correlated with exposure to UVR in many marine organisms (Shibata 1969; Dunlap et al. 1986; Carreto et al. 1989; Vernet et al. 1989). There are two important reasons for this: first, the high molar extinction coefficients in the range of environmentally relevant UVR which characterizes MAA compounds (Shibata 1969), and second, because they are photo-inducible by high light conditions (Carreto et al. 1989; Marchant et al. 1991; Karsten et al. 1998).

Experiments carried out with several species of Antarctic diatoms to identify the wavelength dependence of MAA formation (Davidson et al. 1994; Villafañe et al. 1995; Helbling et al. 1996; Riegger and Robinson 1997; Prézelin et al. 1998) showed a high variability between species. In some species MAAs are accumulated in response to UVR, in others in response to PAR.

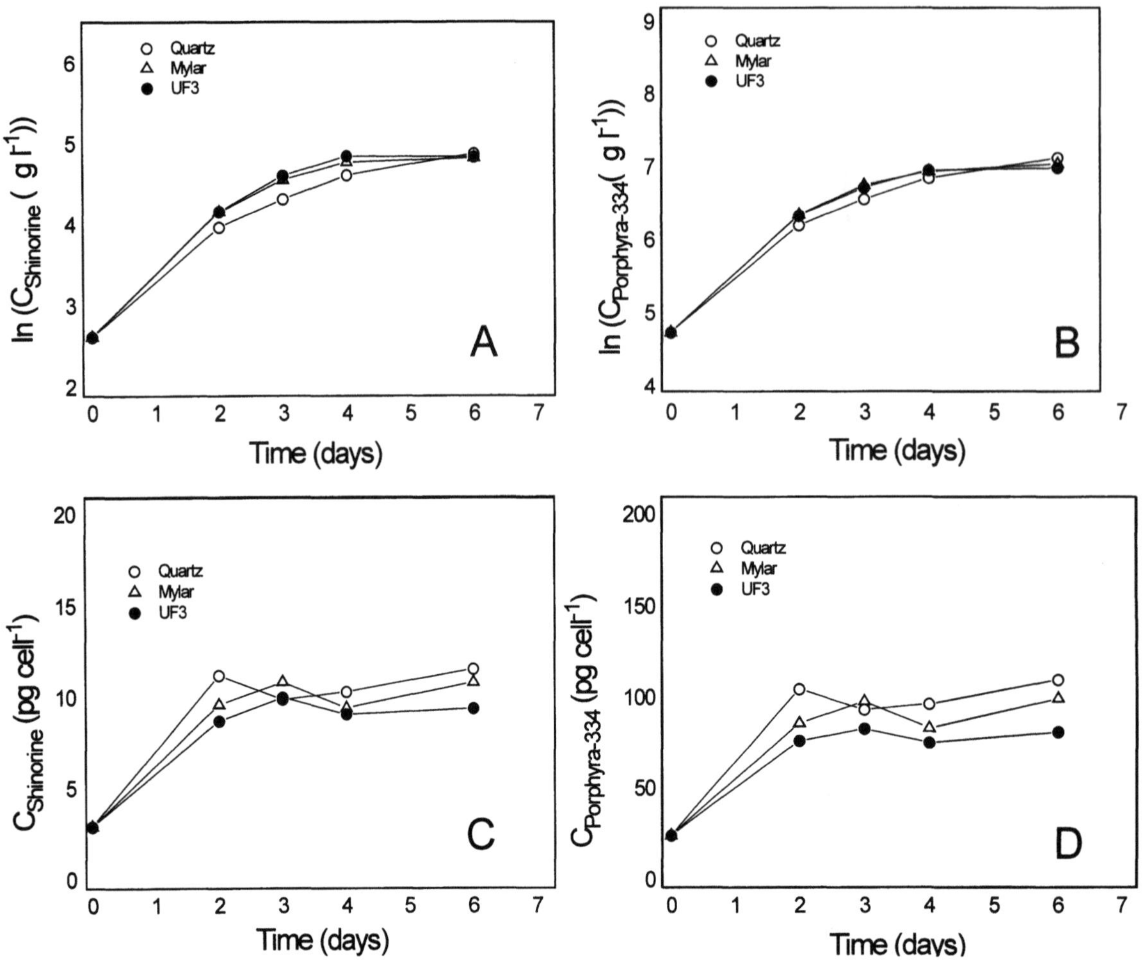

Fig. 6 Time course for the rate of synthesis of shinorine (**A**) and porphyra-334 (**B**) and cell content of shinorine (**C**) and porphyra-334 (**D**) in *Thalassiosira* sp. during the long-term experiment (*filled circle* PAR + UVR, *filled triangle* PAR + UVA, *filled square* PAR)

Our data clearly show that in this strain of *Thalassiosira* sp., the synthesis rates of porphyra-334 and shinorine were only related to PAR. This dependence on PAR irradiance reflects that photosynthesis is necessary to the formation of MAA precursors (Bandaranayake 1998). As previously shown (Carreto et al. 1990a), the synthesis of these compounds at high irradiances could be prevented by simultaneous exposure to 3-(3-4-dichlorophenyl)-1, 1-dimethylurea, an inhibitor of photosynthesis. We could not discern any specific role for UVA or UVB in stimulating the synthesis of these two MAAs. However, it may be possible that the low UVA doses present in the PAR treatment due to imperfect cutting off the UVA wavelengths could trigger the biosynthesis of MAAs (Carreto et al. 1989, 1990b; Riegger and Robinson 1997).

On the other hand, UVR can modulate the accumulation process of MAAs by modifying the cellular balance between rates of net synthesis, excretion (Vernet and Whitehead 1996) and cellular division. Our data show that during acclimation, the cellular division of *Thalassiosira* sp. was strongly inhibited by UVR. As the synthesis of MAAs was not affected by UVR, the cells exposed to full solar radiation accumulated more MAAs than in the PAR treatment. This probably means that cells have to invest a significant amount of energy in the synthesis of photoprotective compounds and repair mechanisms at the expense of the build-up of carbon skeletons. Furthermore, after this initial period the cellular content of MAAs remained nearly constant, indicating some limitations to the cellular accumulation processes. The regulatory mechanisms of these processes are largely unknown, but osmotic mechanisms were suggested for a community of halophylic cyanobacteria (Oren 1997). In contrast, the release of MAAs to the surrounding medium by the dinoflagellate *Lyngulodinium polyedra* was inversely proportional to their intracellular concentration (Vernet and Whitehead 1996). Although we have not measured the rate of MAAs excretion, the decrease of their synthesis rate during the balanced growth suggests the control of MAAs biosynthesis by feed-back regulation.

The accumulation processes seem to be closely related to increased resistance to UVR, since the inhibition of growth rates was significantly reduced when cell MAA concentrations were high. Similar results have been found in other species of marine Antarctic diatoms

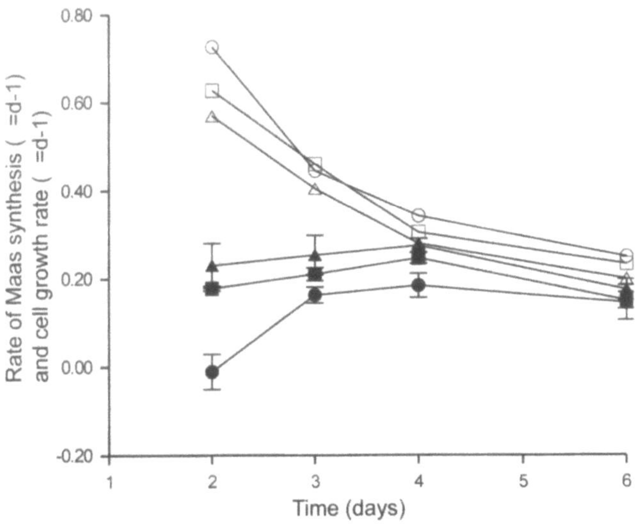

Fig. 7 Growth rate (μ = day^{-1}) of *Thalassiosira* sp. (*filled triangles* PAR treatment, *filled squares* PAR + UVA, *filled circles* PAR + UVR), and rate of synthesis of porphyra-334 (*open triangles* PAR, *open squares* PAR + UVA, *open circles* PAR + UVR) during the long-term experiment. *Vertical bars* indicate standard deviations. The *x axis* represents time intervals

(Helbling et al. 1996) and in natural phytoplankton assemblages (Vernet et al. 1994; Villafañe et al. 1995). However, some species of Antarctic diatoms which do not accumulate MAAs survive UVB exposure better than the Antarctic prymnesiophyte *Phaeocystis antarctica* that accumulates MAAs (Davidson et al. 1996; Riegger and Robinson 1997). Organisms have various strategies for minimizing UV-induced damage (Roy 2000) and we cannot determine if the increased resistance was due to the MAAs accumulation and/or to other photoprotective or repair processes which might be induced concomitantly with MAAs accumulation (Neale et al. 1998b).

In order to estimate the optical protection increase related to MAA accumulation during acclimation, a sunscreen factor (*S*; García-Pichel 1994) was calculated. We estimated $S = 0.76$, using measured MAA concentrations and the extinction coefficient of porphyra-334 (2.81×10^{-2} l g^{-1} μm^{-1}), assuming that the background absorbance was equivalent to initial cells (Neale et al. 1998b). $S = 0.76$ implies that the dose of UVR needed to cause a certain effect will be about 5 times larger than that needed without increased MAAs (Garcia-Pichel 1994). Recent work by Neale et al. (1998b) provides optical evidence that MAAs are spectrally specific UV sunscreens in *Gymnodinium sanguineum* ($S = 0.80$) and that the measured screening performance exceeds model expectations. The optical pathlength of phytoplankton, being dependent upon size, will also be affected by ultrastructural organization (Dubinsky 1992).

As shown by Lesser et al. (1994), part of the response of phytoplankton to UV radiation is a new equilibrium between damage and repair processes. Therefore, in evaluating the relative importance of the optical protection

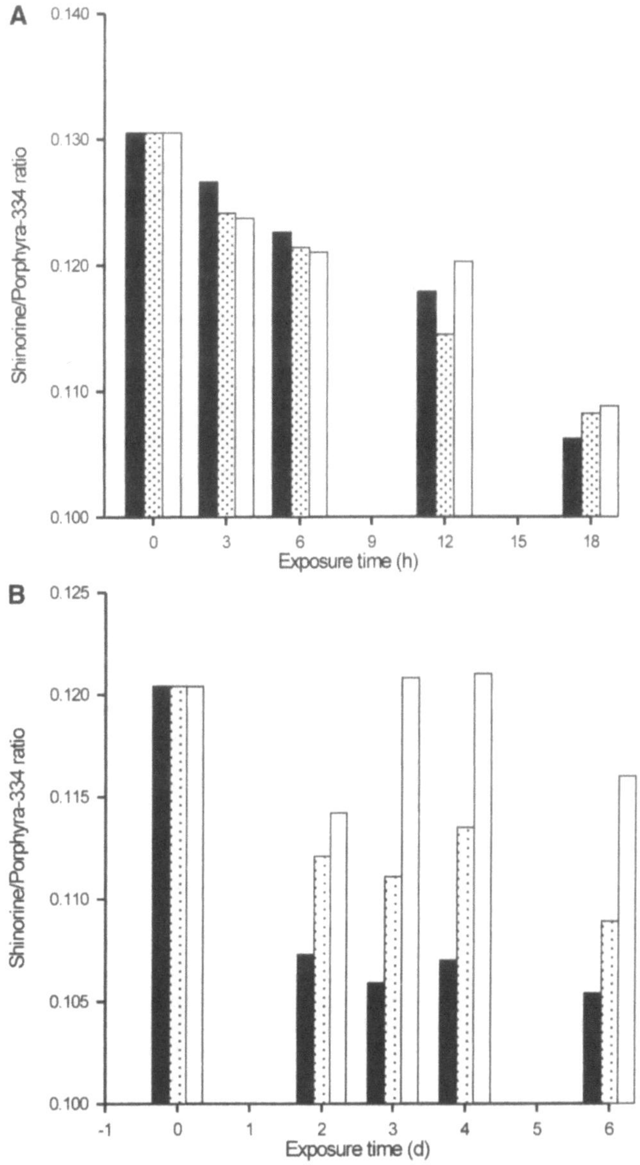

Fig. 8 Ratio for shinorine/porphyra-334 concentration per cell for: **A** short-term experiment and **B** long-term experiment (*black columns* PAR + UVR, *columns with points* PAR + UVA, *empty columns* PAR)

against harmful effects, the kinetics of MAAs accumulation needs to be considered. The terrestrial cyanobacterium *Gleocapsa* sp. showed a lag phase of about 36 h, followed by approximately 48 h during which concentrations continuously rose to a maximum (García-Pichel et al. 1993). In contrast to this cyanobacterium, *Thalassiosira* sp. showed a rapid induction mechanism as well as an efficient enzymatic system for MAAs synthesis. Similar results have been observed in several temperate dinoflagellates (Carreto et al. 1989, and unpublished results), in the marine macroalgae *Chondrus crispus* (Karsten et al. 1998) and in some marine Antarctic diatoms (Helbling et al. 1996). In comparison, our *Thalassiosira* sp. isolate showed a rate of accumulation of 0.088 μg MAAs (μg C)$^{-1}$ day^{-1}, 8 times higher than that reported by Helbling et al. (1996) for other Antarctic isolates of *Thalassiosira* sp. Therefore, we conclude that the efficiency of *Thalassiosira* sp. to synthesize and accumulate high concentrations of MAAs is related to increased resistance to UVB.

An interesting point to be stressed are the different formation kinetics of porphyra-334 and shinorine upon UVB and UVA exposure. Our results suggest that the species has the capacity to react by producing photoprotective compounds in the time scale of hours–days, which fits reasonably with the time scale of changes in irradiance and vertical mixing. Our data support previous field observations made in Potter Cove (Ferreyra et al. 1998), which show a significant positive correlation between UVB irradiance and MAAs concentration present in the natural phytoplankton community within the scale of 1 day.

Acknowledgements We sincerely acknowledge the logistic team at Jubany Station-Dallmann laboratory in Antarctica for their kind cooperation during the 1998–1999 campaign and to Silvia Rodríguez for microscopical examinations and counts. This research was part of a collaboration between the Instituto Antártico Argentino (Argentina) and the Alfred Wegener Institute for Polar and Marine Research (Germany). It was also supported by the Consejo Nacional de Investigaciones Científicas y Técnicas (Argentina) with a fellowship to M.H., and by the Instituto Nacional de Investigación y Desarrollo Pesquero (Argentina) and the Aquatoxsal Project (INCO-DC).

References

Bandaranayake WM (1998) Mycosporines: are they nature's sunscreens? Natural products reports, vol 15. The Royal Society of Chemistry, London, pp 159–173

Carreto JI, De Marco SG, Lutz VA (1989) UV-absorbing pigments in the dinoflagellates *Alexandrium excavatum* and *Prorocentrum micans*. Effects of light intensity. In: Okaichi T, Anderson DM, Nemoto T (eds) Red tides: biology, environmental science and toxicology. Elsevier Science, Oxford, pp 333–336

Carreto JI, Carignan MO, Daleo G, De Marco SG (1990a) Occurrence of mycosporine-like amino acids in the red-tide dinoflagellate *Alexandrium excavatum*: UV-photoprotective compounds? J Plankton Res 12:909–921

Carreto JI, Lutz VA, De Marco SG, Carignan MO (1990b) Fluence and wavelength dependence of mycosporine-like amino acids synthesis in the dinoflagellate *Alexandrium excavatum*: In: Graneli E, Sundströn B, Elder L, Anderson DM (eds) Toxic marine phytoplankton. Elsevier Science, Oxford, pp 275–279

Carreto JI, Carignan MO, Montoya NG (in press) Comparative studies on mycosporine-like amino acids, paralytic shellfish toxins and pigment profiles of the toxic dinoflagellates *Alexandrium tamarense*, *Alexandrium catenella* and *Alexandrium minutum*. Mar Ecol Prog Ser

Davidson AT, Bramich D, Marchant HJ, McMinn A (1994) Effects of UVB irradiation on growth and survival of Antarctic marine diatoms. Mar Biol 119:507–515

Davidson AT, Marchant HJ, Mare WK de la (1996) Natural UVB exposure changes the species composition of Antarctic phytoplankton in mixed cultures. Aqua Microb Ecol 10:299–305

Dubinsky Z (1992) The functional and optical absorption cross-sections of phytoplankton photosynthesis. In: Falkowski PG, Woodhead AD (eds) Primary productivity and biogeochemical cycles in the sea. Plenum Press, New York, pp 31–45

Dunlap WC, Yamamoto Y (1995) Small-molecule antioxidants in marine organisms: antioxidant activity of mycosporine-glycine. Comp Biochem Physiol 112:105–114

Dunlap WC, Chalker BE, Oliver JK (1986) Bathymetric adaptations of reef-building corals at Davies Reef, Great Barrier Reef, Australia. III. UV-B absorbing compounds. J Exp Mar Biol Ecol 104:239–248

Dunlap WC, Chalker BE, Banaszak AT (1989) Biochemical photoadaptation in vision: UV-absorbing pigments in fish eye tissues. Comp Biochem Physiol B 93:601–607

Ferreyra GA, Schloss IR, Abele-Oeschger D (1998) UV-absorbing compounds in surface waters of Potter Cove: preliminary results. In: Wiencke C, Ferreyra GA, Arntz W, Rinaldi C (eds) The Potter Cove coastal ecosystem, Antarctica. (Reports on Polar Research, vol 299) Alfred Wegener Institute for Polar and Marine Research, Bremerhaven, pp 316–321

Franklin LA, Yakovleva I, Karsten U, Lüning K (1999) Synthesis of mycosporine-like amino acids in *Chondrus crispus* (Florideophyceae) and the consequences for sensitivity to ultraviolet B radiation. J Phycol 35:682–693

Garcia-Pichel F (1994) A model for shelf-shading in planktonic organisms and its implications for the usefulness of ultraviolet sunscreens. Limnol Oceanogr 39:1704–1717

García-Pichel F, Wingard CE, Castenholz RW (1993) Evidence regarding the UV sunscreen role of a mycosporine-like compound in the cyanobacterium *Gloeocapsa sp.* Appl Environ Microbiol 59:170–176

Guillard RL, Ryther JH (1962) Studies of marine planktonic diatoms. I. *Cyclotella nana* Hustedt and *Detonula convervacea* Cleve. Can J Microbiol 8:229–239

Helbling EW, Chalker BE, Dunlap WC, Holm-Hansen O, Villafañe VE (1996) Photoacclimation of Antarctic marine diatoms to solar ultraviolet radiation. J Exp Mar Biol Ecol 204:85–101

Holm-Hansen O, Riemann B (1978) Chlorophyll a determination: improvements in methodology. Oikos 30:438–447

Karentz D (1994) Ultraviolet tolerance mechanisms in Antarctic marine organisms. In: Weiler CS, Penhale PA (eds) Ultraviolet radiation in Antarctica: measurements and biological effects. American Geophysical Union, Washington, pp 93–110

Karentz D, Mc Euen FS, Land MC, Dunlap WC (1991) Survey of mycosporine-like amino acid compounds in Antarctic marine organisms: potential protection from ultraviolet exposure. Mar Biol 108:157–166

Karsten U, Franklin LA, Lüning K, Wiencke C (1998) Natural ultraviolet radiation and photosynthetically active radiation induce formation of mycosporine-like amino acids in the marine macroalga *Chondrus crispus* (Rhodofphyta). Planta 205:257–262

Karsten U, Bischof K, Hanelt D, Tug H, Wiencke C (1999) The effect of UV radiation on photosynthesis and UV-absorbing substances in the endemic Arctic macroalga *Develaraea ramentacea* (Rhodophyta). Physiol Plant 105:58–66

Lesser MP (1996) Acclimation of phytoplankton to UVB radiation: oxidative stress and photoinhibition of photosynthesis are not prevented by UV absorbing compounds in the dinoflagellate *Prorocentrum micans*. Mar Ecol Prog Ser 132:287–297

Lesser MP, Cullen JJ, Neale PJ (1994) Photoinhibition of photosynthesis in the marine diatom *Thalassiosira pseudonona* during acute exposure to ultraviolet B radiation: relative importance of damage and repair. J Phycol 30:183–192

Lubin D, Frederick J, Booth C, Lucas T, Neuschuler D (1989) Measurement of enhanced springtime ultraviolet radiation at Palmer Station Antarctica. Geophys Res Lett 16:783–785

Marchant HJ, Davidson AT, Kelly GJ (1991) UVB protecting pigments in the marine alga *Phaeocystis pouchetii* form Antarctica. Mar Biol 9:391–375

Neale PJ, Cullen JJ, Davis RF (1998a) Inhibition of marine photosynthesis by ultraviolet radiation: variable sensitivity of phytoplankton in the Weddell-Scotia Confluence during the austral spring. Limnol Oceanogr 43:433–448

Neale PJ, Banaszak AT, Jarriel CR (1998b) Ultraviolet sunscreens in *Gymnodinium sanguineum* (Dinophyceae). Mycosporine-like amino acids protect against inhibition of photosynthesis. J Phycol 34:928–938

Oren A (1997) Mycosporine-like amino acids as osmotic solutes in a community of halophilic cyanobacteria. Geomicrobiol J 14:231–240

Prézelin BB, Moline MA, Matlick HA (1998) ICECOLORS'93: spectral UV radiation effects on Antarctic fragile ice algae. Antarct Res Ser 73:45–83

Riegger L, Robinson D (1997) Photoinduction of UV absorbing compounds in Antarctic diatoms and *Phaeocystis antarctica*. Mar Ecol Prog Ser 160:13–25

Roy S (2000) Strategies for the minimisation of UV-induced damage. In: De Mora S, Demers S, Vernet M (eds) The effects of UV radiation in the marine environment. (Environmental Chemistry Series) Cambridge University Press, Cambridge, pp 177–205

Shibata K (1969) Pigments and a UV-absorbing substance in corals and a blue-green alga living in the Great Barrier Reef. Plant Cell Physiol 10:325–335

Smith LC, Prézelin BB, Bakev KS, Bidigare RR, et al (1992) Ozone depletion: ultraviolet radiation and phytoplankton biology in Antarctic waters. Science 255:952–959

Takano S, Uemura D, Hirata Y (1979) Isolation and structure of a 334 nm UV absorbing substance, porphyra-334 from red alga *Porphyra tenera* Kjellman. Chem Lett 1979:419–420

Tsujino I, Yabe K, Sekikawa I (1980) Isolation and structure of a new amino acid, Shinorine, from the red alga *Chondrus yendoi* Yamada et Mikami. Bot Mar 23:65–68

Vernet M, Whitehead K (1996) Release of ultraviolet-absorbing compounds by the red-tide dinoflagellate *Linguladinium polyedra*. Mar Biol 127:35–44

Vernet M, Neori A, Haxo FT (1989) Spectral properties and photosynthetic action in red-tide populations of *Prorocentrum micans* and *Gonyaulax polyedra*. Mar Biol 103:365–371

Vernet M, Brody E, Holm-Hamsen O, Mitchell BG (1994) The response of Antarctic phytoplankton to ultraviolet radiation: absorption, photosynthesis and taxonomic composition. Antarct Res Ser 62:143–158

Villafañe VE, Hebling EW, Holm-Hansen O, Chalker BE (1995) Acclimatization of Antarctic natural phytoplankton assemblages when exposed to solar ultraviolet radiation. J Plankton Res 17:2295–2306

Polar Biol (2001) 24: 719–723
DOI 10.1007/s003000100270

Markus Frederich · Franz Josef Sartoris
Hans-O. Pörtner

Distribution patterns of decapod crustaceans in polar areas: a result of magnesium regulation?

Abstract Nearly all decapod crustaceans found in Antarctic waters south of the Antarctic Convergence are caridean shrimps (Natantia) while the group of Reptantia is largely absent in this area. Progress in the development of a physiological hypothesis is reported, which explains this distribution pattern based on differences in the regulation of magnesium levels in the haemolymph ($[Mg^{2+}]_{HL}$) and on the Mg^{2+} dependence of threshold temperatures below which cold-induced failure of cardiac and ventilatory performance occurs. Previous studies had shown that an increase in oxygen consumption and activity levels in the cold can be induced by experimental reduction of $[Mg^{2+}]_{HL}$ in different reptant decapod species. In the present study, we tested the potential of these experimental findings for predicting the effect of low $[Mg^{2+}]_{HL}$ in nature, and investigated temperature-induced changes in oxygen consumption in two species with low but different $[Mg^{2+}]_{HL}$ from southern Chile, *Halicarcinus planatus* and *Acanthocyclus albatrossis* ($[Mg^{2+}]_{HL} = 10.7$ and 21.6 mmol l^{-1}, respectively). In accordance with previous findings, low $[Mg^{2+}]_{HL}$ levels were associated with a reduction of thermal sensitivity and a higher metabolic rate in the cold. A model is developed which describes how $[Mg^{2+}]_{HL}$ reduction caused a threshold temperature (pejus temperature, Tp) to fall, which characterises the onset of cold-induced failure in oxygen supply to tissues. This threshold temperature is interpreted, not only to indicate the limits of cold tolerance, but also of geographical distribution. Tp is shifted towards lower temperatures in Natantia, which are efficient $[Mg^{2+}]_{HL}$ regulators. In contrast, Reptantia, which are poor $[Mg^{2+}]_{HL}$ regulators, appear unable to colonise the permanently cold water of the Antarctic due to insufficient capacity of cardiac performance and, therefore, largely reduced scope for activity at high $[Mg^{2+}]_{HL}$.

Please note: When citing this article, please use the reference to the original source in the journal Polar Biology as given on the top of this page.

M. Frederich · F.J. Sartoris · Hans-O. Pörtner (✉)
Alfred-Wegener-Institute for Polar and Marine Research,
Columbusstrasse, 27568 Bremerhaven,
Germany
E-mail: hpoertner@awi-bremerhaven.de
Tel.: +49-471-48311307
Fax: +49-471-48311149

Present address: M. Frederich
Brigham and Women's Hospital,
Harvard Medical School, Boston, USA

Introduction

Early investigations of the Antarctic shelf fauna seemed to indicate that decapod crustaceans are very rare south of the Antarctic Convergence (Yaldwyn 1965). However, more recent studies, especially in the Weddell Sea, revealed a well-established decapod fauna with eight different species out of seven families (Arntz and Gorny 1991; Arntz et al. 1994), which all belong to the group Caridea (Natantia). The group of Reptantia, which is otherwise very successful in colonising different and even extreme habitats, is almost completely absent in the high Antarctic. Different hypotheses were developed to explain the absence of this whole group of crustaceans, mainly based on larval, ecological or geohistorical patterns. Nevertheless, none of these hypotheses provides a commonly accepted explanation (reviewed by Arntz et al. 1994). This paper reports recent progress in the development of a physiological hypothesis, which explains the absence of Reptantia in the Antarctic based on their inability to regulate magnesium in the haemolymph ($[Mg^{2+}]_{HL}$) at low levels, thereby exacerbating problems of oxygen supply to tissues due to limitations of cardiac and ventilatory performance at low temperatures (Frederich et al. 2000a).

Results and discussion

Crustacean haemolymph is usually more or less isoionic to seawater except for Mg^{2+} (Robertson 1960; Mantel and Farmer 1983). Mean $[Mg^{2+}]_{HL}$ of caridean shrimps is found at 5–12 mmol l^{-1}, far below the magnesium concentration of seawater (53 mmol l^{-1}). In contrast, most Reptantia regulate $[Mg^{2+}]_{HL}$ at levels between 30 and 50 mmol l^{-1}, close to seawater concentrations (Walters and Uglow 1981; Tentori and Lockwood 1990). This difference is of special interest since there is a close relationship between $[Mg^{2+}]_{HL}$ and activity. In crustaceans, low $[Mg^{2+}]_{HL}$ values are clearly correlated with higher activity levels, while high $[Mg^{2+}]_{HL}$ values are usually combined with lower activity levels. This holds true for comparisons of activity levels between different species, as well as for comparisons between long-term periods with different levels of activity, like hibernation, within the same species (Robertson 1953, 1960; Kayser 1961; Walters and Uglow 1981; Morritt and Spicer 1993; Spicer et al. 1994). This correlation between $[Mg^{2+}]_{HL}$ and activity has been known for a long time. Pantin (1946) suggested a formula with high $[Mg^{2+}]$ to anaesthetise invertebrates, and Robertson (1953) described species such as *Maja squinado* with their high $[Mg^{2+}]_{HL}$ as "living in a semi-narcotized state".

Elevated activity levels require an adequate oxygen supply to tissues by ventilation and circulation. Several studies have shown that oxygen may become a limiting factor at extreme temperatures and, consequently, both low and high threshold temperatures were defined by the onset of anaerobic metabolism in both invertebrates and fish (Zielinski and Pörtner 1996; Sommer et al. 1997; Pörtner et al. 1998, 2000; Sommer and Pörtner 1999; van Dijk et al. 1999 ; Frederich and Pörtner, 2000). Cold adaptation elicits a shift of both threshold temperatures to lower values to enable the animals to remain fully active at low temperatures.

In previous experimental studies of the effects of temperature and various $[Mg^{2+}]_{HL}$ levels on the physiology of reptant decapod crustaceans, $[Mg^{2+}]_{HL}$ was reduced from 45 to 8 mmol l^{-1} by exposing specimens of *M. squinado*, *Hyas araneus* and *Eurypodius latreillei* to seawater with artificially reduced $[Mg^{2+}]$ of approximately 6 mmol l^{-1} (compared to 53 mmol l^{-1} in natural seawater). Animals were subsequently incubated at different temperatures and motor activity, heart rate, oxygen consumption or oxygen tensions in the haemolymph were investigated. As described in detail elsewhere (Frederich et al. 2000a), heart rate of control animals in *H. araneus* ($[Mg^{2+}]_{HL}$ 46 mmol l^{-1}) decreased linearly between 7.2 and –1.2°C by 72.7 ± 13.0%. Heart rate in $[Mg^{2+}]_{HL}$-reduced animals ($[Mg^{2+}]_{HL}$ 8 mmol l^{-1}) decreased significantly less – by only 37.2 ± 17.5%. A different pattern was visible in *E. latreillei*. Heart rate and oxygen consumption decreased linearly in control animals ($[Mg^{2+}]_{HL}$ 45 mmol l^{-1}) and in $[Mg^{2+}]_{HL}$-reduced animals ($[Mg^{2+}]_{HL}$ 8 mmol l^{-1}). Nevertheless, both parameters were found at significantly elevated levels at all temperatures in $[Mg^{2+}]_{HL}$-reduced animals.

The availability of oxygen to tissues depends upon oxygen uptake by ventilation and on oxygen distribution via haemolymph circulation (cardiac output). These two parameters showed a strong temperature dependence in *M. squinado* (as described in detail in Frederich et al., 2000b). Cardiac output decreased linearly during progressive cooling between 12 and 8°C. An overproportional drop led to very low values between 7 and 0°C. This overproportional drop was also visible in ventilation and occurred between 7 and 6°C, followed by a transient increase around 4°C. Arterial PO_2 remained almost constant between 12 and 8°C and decreased slightly between 8 and 6°C in accordance with reduced cardiac and ventilatory activities. The lowest PO_2 values were reached below 1°C when ventilation had stopped. The overproportional drop in ventilation and circulation could be shifted to lower temperatures when reducing the $[Mg^{2+}]_{HL}$ from 39.9 ± 4.6 to 6.1 ± 0.7 mmol l^{-1}. These previous findings are summarised schematically in Fig. 1, which demonstrates the two patterns of shifting performance of different parameters due to the reduction in $[Mg^{2+}]_{HL}$.

After these initial findings, the question remained open as to whether an experimental reduction of $[Mg^{2+}]_{HL}$ has any relevance for being able to predict the ability of reptant decapods to survive in cold ocean environments. We therefore chose to compare two decapod crustacean species with different, but relatively low, $[Mg^{2+}]_{HL}$, the two South American intertidal crabs *Halicarcinus planatus* and *Acanthocyclus albatrossis*. Both species display lower $[Mg^{2+}]_{HL}$ than most other brachyurans: 21.6 mmol l^{-1} in *A. albatrossis* and only 10.7 mmol l^{-1} in *Halicarcinus planatus*. Both species live

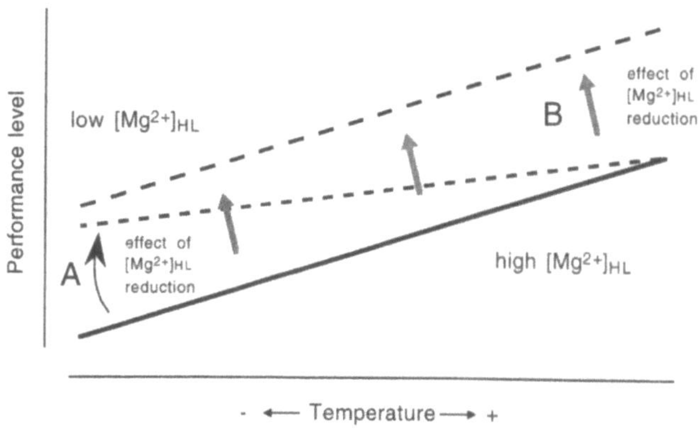

Fig. 1 Schematic presentation of a shift in performance parameters crucial for survival (motor activity, heart rate, oxygen consumption) which is induced by lowering $[Mg^{2+}]_{HL}$ experimentally in various reptant decapod species, evidenced by analyses in *Hyas araneus, Maja squinado* and *Eurypodius latreillei.* Lowering $[Mg^{2+}]_{HL}$ in *E. latreillei* resulted in a parallel upward shift (*B*) while in other investigated species the effect of reduced $[Mg^{2+}]_{HL}$ was visible only at lower temperatures (*A*). The reason for this difference remains unclear (based on data by Frederich et al. 2000, 2000a)

in the same environment and inhabit a comparable ecological niche. The reasons for the different capacities of $[Mg^{2+}]_{HL}$ regulation remain obscure since no data are available on the autecology of the two species. In accordance with findings obtained in *M. squinado* at experimentally reduced $[Mg^{2+}]_{HL}$, thermal sensitivity of oxygen consumption in the cold differed between the two species, reflected by different Q_{10} values (Fig. 2). In *A. albatrossis*, the species with higher $[Mg^{2+}]_{HL}$, oxygen consumption decreased, with a Q_{10} of 6.8 between 8.8 and –1°C, while the Q_{10} of 1.3 in *Halicarcinus planatus* was significantly lower in the same temperature range ($n=5$ per species and group, ANCOVA, $P<0.05$). In support of the patterns depicted in Fig. 1, oxygen consumption of the species with the lower $[Mg^{2+}]_{HL}$, *Halicarcinus planatus*, was threefold higher at subzero temperatures than in *A. albatrossis*.

As summarised in Fig. 1, an elevation of activity can be elicited by a reduction of $[Mg^{2+}]_{HL}$ especially at cold temperatures. Owing to the reduction of thermal sensitivity, $[Mg^{2+}]_{HL}$-reduced crabs displayed elevated heart rates and oxygen consumption values and reacted faster to experimental stimulation, regardless of whether $[Mg^{2+}]_{HL}$ was reduced experimentally or in nature (Fig. 2). Animals were able to survive experimental reductions of $[Mg^{2+}]_{HL}$ for several months, which indicates that elevated $[Mg^{2+}]_{HL}$ levels are not essential for survival, at least under laboratory conditions.

The first report of an effect of experimental $[Mg^{2+}]_{HL}$ reduction on crustaceans is given by Holliday (1980), who described *Cancer magister* held in Mg^{2+}-free seawater as being "easily excited and aggressive". This corresponds to the results reported in this paper. Elevated excitability and activity might be explained by the inhibitory effect of Mg^{2+} on synaptic transmitter release at the nerve-muscular junction, and by the competitive

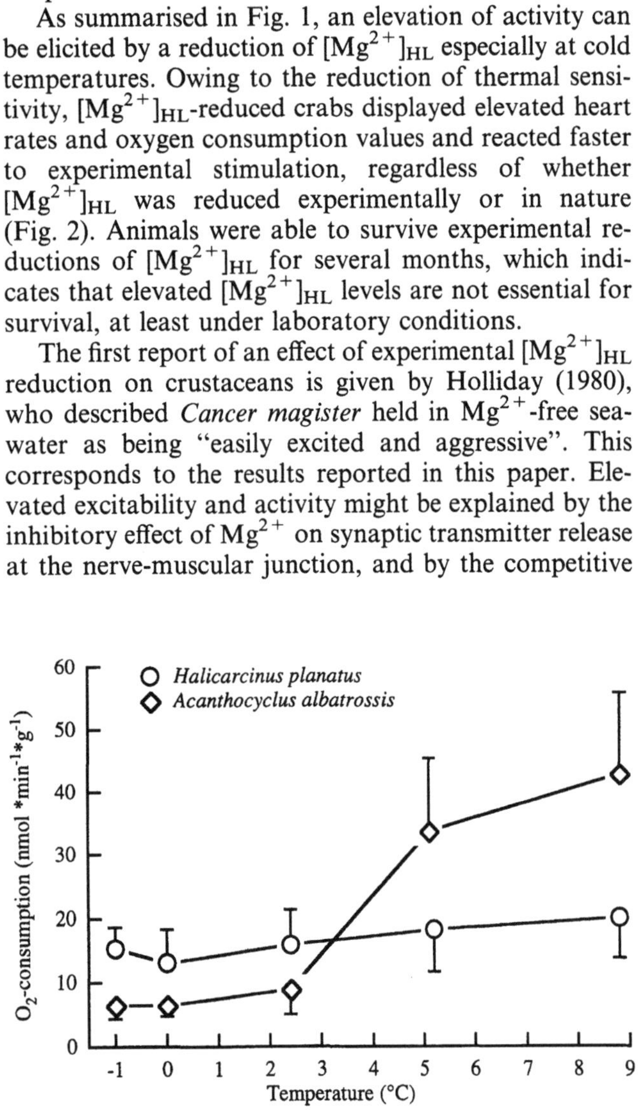

Fig. 2 Oxygen consumption of *Halicarcinus planatus* and *Acanthocyclus albatrossis* at different temperatures. Thermal sensitivity was found to be significantly lower in *H. planatus* at $[Mg^{2+}]_{HL}=10.7$ mmol l^{-1} compared to *A. albatrossis* at $[Mg^{2+}]_{HL}=21.6$ mmol l^{-1} ($n=5$ per species and group, ANCOVA, $P<0.05$)

inhibition of calcium binding (Katz 1936; Del Castillo and Katz 1954; Wernig 1972; Dudel et al. 1982; Howarth and Levi 1998). In accordance with the findings already outlined, elevated levels of Mg^{2+} seem to act more as a relaxant than as an anaesthetic (Sartoris and Pörtner 1997; Frederich et al. 2000a).

The question of why Reptantia regulate $[Mg^{2+}]_{HL}$ to higher levels compared to Natantia is still open. Few studies indicate that there might be differences in the mechanisms of $[Mg^{2+}]_{HL}$ regulation between Caridea and Brachyura (Holliday 1980; Franklin et al. 1987) but further investigations are necessary to elucidate the details.

The anaesthetising effect of Mg^{2+} in the cold becomes important in the context of the concept that oxygen is an important factor for defining threshold values at extreme temperatures. At both low and high critical temperatures, animal metabolism switches from aerobiosis to anaerobiosis, and anaerobic endproducts accumulate (reviewed by Pörtner et al. 1998, 2000). In these studies the ambient oxygen did not become limiting, but oxygen supply to tissues did. Therefore, ventilatory and cardiac performance may play an important role in defining thermal tolerance. As shown previously, arterial haemolymph PO_2, which represents the pressure head for oxygen diffusion into tissues, is correlated with heart and ventilation rates (Frederich and Pörtner, 2000). The oxygen-binding properties of haemocyanin may change with temperature also; however, since the haemocyanin acts as an oxygen buffer to support maintenance of body fluid PO_2, analysis of PO_2 gives the most relevant information on the quality of tissue oxygenation.

The findings of decreasing levels of haemolymph PO_2 and thus progressive oxygen limitation towards high and low temperature extremes led us (Frederich and Pörtner, 2000) to develop a model in accordance with the "law of tolerance" (Shelford 1913, 1931, Fig. 3). The limiting parameter, arterial PO_2, and the progressive failure of cardiac and ventilatory performance towards extreme temperatures characterise various sections of the tolerance range (Fig. 3A). Within the optimum range, oxygen availability and, in consequence, scope for aerobic activity is maximal. The onset of a decrease in arterial haemolymph PO_2 at pejus temperatures (Tp; Frederich and Pörtner, 2000; pejus = getting worse) reflects reduced aerobic scope for activity and is accompanied by progressively insufficient cardiac and ventilatory performance. Transitions from pejus to pessimum ranges at critical temperatures (Tc, Pörtner et al. 1998, 2000) are characterised by the onset of anaerobic metabolism and, in consequence, the accumulation of anaerobic endproducts. Within pessimum ranges, survival becomes a question of time due to insufficient oxygen availability. The upper and lower Tc encompass the range of physiological tolerance.

For *M. squinado*, upper and lower Tp values agree well with the annual minimum and maximum mean temperatures of the environment in the investigated

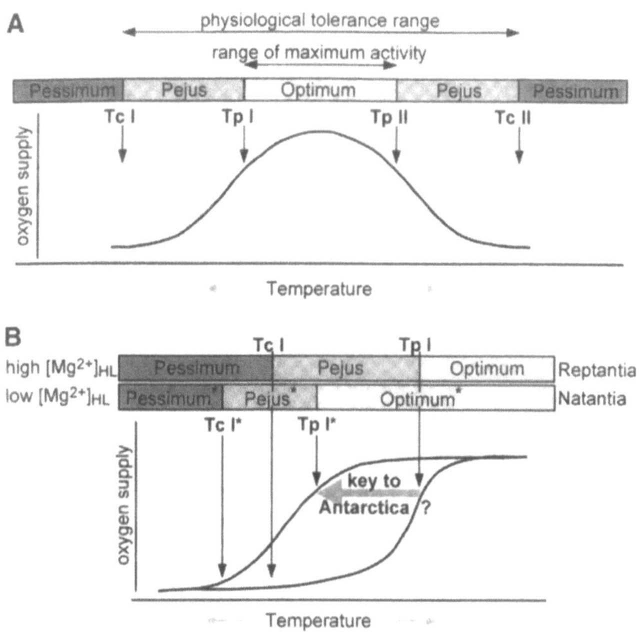

Fig. 3A, B Diagram of physiological threshold temperatures inserted into the law of tolerance. **A** Pejus temperatures (*Tp*) and critical temperatures (*Tc*) are set by the availability of oxygen to tissues and, therefore, depend on ventilatory and circulatory performance (modified after Frederich and Pörtner, 2000). **B** The higher efficiency of $[Mg^{2+}]_{HL}$ regulation in Natantia shifts both pejus and critical threshold values to lower temperatures compared to Reptantia with less effective $[Mg^{2+}]_{HL}$ regulation. This shift is most likely a crucial preadaptation for surviving colder temperatures and, therefore, the physiological key for the colonisation of Antarctic shelves by crustaceans

population (Dauvin et al. 1991; Sournia and Birrien 1995) and, therefore, it can be inferred that the Tp values may limit the ecological tolerance range. In this ecological tolerance range, full scope for activity is available to the animals. In consequence, cold adaptation requires an adaptational shift of both Tp and Tc values towards lower temperatures to maintain sufficient aerobic activity. This likely involves an adjustment of mitochondrial density and capacity (Pörtner et al. 1998, 2000). As a further step forward in the development of the model depicted in Fig. 3A, it appears that high $[Mg^{2+}]_{HL}$ prevents the shift of Tp and Tc to subzero temperatures and thereby limits the capacity of cold adaptation (Fig. 3B).

In summary, the present study shows that slow-moving and sluggish species of Reptantia can be triggered to become more active by decreasing $[Mg^{2+}]_{HL}$ to levels comparable to those found in caridean shrimps. Survival in the natural environment requires sufficient aerobic scope for escape, predation or reproduction. While cold temperatures minimise activity, low $[Mg^{2+}]_{HL}$ levels shift the pejus temperature and, accordingly, the optimum range towards lower values (Fig. 3B). A concomitant shift of the critical temperature is likely. Small differences in cold tolerance by just 1 or 2°C result (Fig. 3) and lead to an exclusion of Reptantia from areas with extremely low temperatures in both the Arctic and the Antarctic, where they are only found at temperatures close to 0°C and above (Frederich et al. 2000).

It appears likely that efficient $[Mg^{2+}]_{HL}$ regulation in Natantia was an important preadaptation for surviving the cooling of Antarctica during the Tertiary (Clarke and Crame 1989; Clarke 1990) or allowed successful recolonisation after glaciation cycles (Brey et al. 1996). In contrast, Reptantia with less efficient $[Mg^{2+}]_{HL}$ regulation are narcotised by the combination of low temperatures and high $[Mg^{2+}]_{HL}$ and, in consequence, may have been unable to remain in the Antarctic or to recolonise the shelves after glaciation. Nonetheless, the ability to maintain sufficient levels of motor activity in the cold is obviously only one precondition for survival in the Antarctic environment, which needs to be complemented by adequate lifestyle adaptations.

Acknowledgements Financial support for this work was provided to M. Frederich by a travel grant for his stay in Chile from the DAAD (D/97/10464).

References

Arntz WE, Gorny M (1991) Shrimp (Decapoda, Natantia) occurrence and distribution in the eastern Weddell Sea, Antarctica. Polar Biol 11:169–177

Arntz WE, Brey T, Gallardo VA (1994) Antarctic zoobenthos. Oceanogr Mar Biol Annu Rev 32:241–304

Brey T, Dahm C, Gorny M, Klages M, Stiller M, Arntz WE (1996) Do Antarctic benthic invertebrates show an extended level of eurybathy? Antarct Sci 8:3–6

Clarke A (1990) Temperature and evolution: Southern Ocean cooling and the Antarctic marine fauna. In: Kerry KR, Hempel G (eds) Ecological change and conservation. Springer, Berlin Heidelberg New York, pp 9–22

Clarke A, Crame JA (1989) The origin of the Southern Ocean marine fauna. In: Crame JA (ed) Origins and evolution of the Antarctic biota. Geol Soc Spec Publ no. 47:253–268

Dauvin JC, Joncourt M, Birrien JL (1991) Température et salinité de leau de mer au large de Roscoff de 1988 à 1990. Cah Biol Mar 32:545–550

Del Castillo J, Katz B (1954) Quantal components of the endplate potential. J Physiol 124:560–573

Dijk PLM van, Tesch C, Hardewig I, Pörtner HO (1999) Physiological disturbances at critically high temperatures. A comparison between stenothermal Antarctic, and eurythermal temperate eelpouts (Zoarcidae). J Exp Biol 202:3611–3621

Dudel J, Parnas I, Parnas H (1982) Neurotransmitter release and its facilitation in crayfish. III. Amplitude of facilitation and inhibition of entry of calcium into the terminal by magnesium. Pflügers Arch 393:237–242

Franklin SE, Teinsongrusme B, Lockwood APM (1987) Inhibition of magnesium secretion in the prawn *Palaemon serratus* by ethacrynic acid and by ligature of the eyestalks. In: Schmidt-Nielsen K, Bolis L, Maddrell SHP (eds) Comparative physiology, water, ions and fluid mechanics. Cambridge University Press, Cambridge, pp 173–193

Frederich M, Pörtner HO (2000) Oxygen limitation of thermal tolerance defined by cardiac and ventilatory performance in the spider crab, *Maja squinado* (Decapoda). Am J Physiol 279:R1531–R1538

Frederich M, Sartoris FJ, Arntz WE, Pörtner HO (2000a) Haemolymph Mg^{2+} regulation in decapod crustaceans: physiological correlates and ecological consequences in polar areas. J Exp Biol 203:1383–1393

Frederich M, DeWachter B, Sartoris FJ, Pörtner HO (2000b) Cold tolerance and the regulation of cardiac performance and hemolymph distribution in *Maja squinado* (Crustacea: Decapoda). Physiol Zool 73:406–415

Holliday CW (1980) Magnesium transport by the urinary bladder of the crab, *Cancer magister*. J Exp Biol 85:187–201

Howarth FC, Levi AJ (1998) Internal free magnesium modulates the voltage dependence of contraction and Ca transient in rabbit ventricular myocytes. Pflügers Arch 435:687–698

Katz B (1936) Neuro-muscular transmission in crabs. J Physiol 87:199–221

Kayser C (1961) The physiology of natural hibernation. Pergamon Press, Oxford

Mantel LH, Farmer LL (1983) Osmotic and ionic regulation. In: Mantel LH (ed) The biology of Crustacea, vol 5. Academic Press, New York, pp 53–161

Morritt D, Spicer JI (1993) A brief re-examination of the function and regulation of extracellular magnesium and its relationship to activity in crustacean arthropods. Comp Biochem Physiol 106A:19–23

Pantin CFA (1946) Notes on microscopical techniques for zoologists. Cambridge University Press, Cambridge

Pörtner HO, Hardewig I, Sartoris FJ, van Dijk PLM (1998) Acid-base balance, ion regulation and energetics in the cold. In: Pörtner HO, Playle R (eds) Cold ocean physiology. Cambridge University Press, Cambridge, pp 88–120

Pörtner HO, van Dijk PLM, Hardewig I, Sommer A (2000) Levels of metabolic cold adaptation: tradeoffs in eurythermal and stenothermal ectotherms. In: Davison W, Howard Williams C (eds) Antarctic ecosystems: models for wider ecological understanding. Caxton Press, Christchurch

Robertson JD (1953) Further studies on ionic regulation in marine invertebrates. J Exp Biol 30:279–296

Robertson JD (1960) Osmotic and ionic regulation. In: Waterman TH (ed) The physiology of Crustacea, vol I. Academic Press, New York, pp 317–339

Sartoris FJ, Pörtner HO (1997) Elevated haemolymph magnesium protects intracellular pH and ATP levels during environmental stress in the common shrimp *Crangon crangon*. J Exp Biol 200:785–792

Shelford VE (1913) Animal communities in temperate America. University of Chicago Press, Chicago

Shelford VE (1931) Some concepts of bioecology. Ecology 12:455–467

Sommer A, Pörtner HO (1999) Exposure of *Arenicola marina* (L.) to extreme temperatures: adaptive flexibility of a boreal and a subpolar population. Mar Ecol Prog Ser 181:215–226

Sommer A, Klein B, Pörtner HO (1997) Temperature induced anaerobiosis in two populations of the polychaete worm *Arenicola marina* (L.). J Comp Physiol 167B:25–35

Sournia A, Birrien JL (1995) La série océanographique côtière de Roscoff (Manche occidentale) de 1985 à 1992. Cah Biol Mar 36:1–8

Spicer JI, Morritt D, Taylor AC (1994) Effect of low temperature on oxygen uptake and haemolymph ions in the sandhopper *Talitrus saltator* (Crustacea: Amphipoda). J Mar Biol Assoc UK 74:313–321

Tentori E, Lockwood APM (1990) Haemolymph magnesium levels in some oceanic Crustacea. Comp Biochem Physiol 95A:545–548

Walters NJ, Uglow RF (1981) Haemolymph magnesium and relative heart activity of some species of marine decapod crustaceans. J Exp Mar Biol Ecol 55:255–256

Wernig A (1972) The effects of calcium and magnesium on statistical release parameters at the crayfish neuromuscular junction. J Physiol 226:761–768

Yaldwyn JC (1965) Antarctic and subantarctic decapod Crustacea. In: Mieghem J van, Oye P van (eds) Biogeography and ecology in Antarctica. The Hague, pp 323–332

Zielinski S, Pörtner HO (1996) Energy metabolism and ATP free-energy change of the intertidal worm *Sipunculus nudus* below a critical temperature. J Comp Physiol 166B:492–500

Polar Biol (2001) 24: 642–649
DOI 10.1007/s003000100258

Bodil A. Bluhm · Thomas Brey · Michael Klages
Wolf E. Arntz

Occurrence of the autofluorescent pigment, lipofuscin, in polar crustaceans and its potential as an age marker

Abstract In crustaceans, the lack of reliable methods often prevents the determination of individual age. The quantification of the autofluorescent age pigment, lipofuscin, has revealed promising results in boreal and tropical species. We studied the presence of morphological lipofuscin and its possible application as an age marker in five Arctic and five Antarctic species, comprising decapods, amphipods and a euphausiid. Lipofuscin granules were located in the brain, using confocal fluorescence microscopy, and quantified from digital images. The pigment was found in 94 of 100 individuals and in all 10 species, and granules occurred in easily detectable amounts in 5 species. Two scavenging amphipod species, the Antarctic *Waldeckia obesa* and the Arctic *Eurythenes gryllus*, revealed the most conspicuous and numerous granules. There was a broad, though weak, correlation of lipofuscin concentration with individual body size within a species, but not with absolute body size of one species compared to another. In larvae of the decapod *Chorismus antarcticus*, lipofuscin accumulation was quantified over the 1st 4 months after larval release. Morphological lipofuscin is a potential index of age in those investigated species with a sufficient accumulation rate of the pigment.

Introduction

Individual age is a basic parameter in population dynamics, required for calculating growth, mortality, productivity, maturation etc. The few attempts to determine age in polar crustaceans (e.g. Bregazzi 1972; Rakusa-Suszczewski 1982; Gorny et al. 1993; Klages 1993) have predominantly used size-based data, as crustaceans lack permanent hard structures carrying potential age markers. The separation of size-frequency distributions into cohorts is hampered by interference between age classes and moulting stages (Klages 1993) and insufficient resolution in long-lived slow-growing species owing to variation in individual growth (France et al. 1991). Currently, a promising alternative approach is to quantify the autofluorescent age pigment, lipofuscin (Belchier et al. 1994, 1998; Sheehy et al. 1994, 1995, 1996; O'Donovan and Tully 1996).

The existence of morphological lipofuscin (as opposed to soluble or extractable age pigments) has been demonstrated in post-mitotic tissue such as nerve tissue, heart and muscles of various invertebrates and vertebrates (e.g. Donato and Sohal 1978; Bassin et al. 1982; Hunter and Vetter 1988; Clarke et al. 1990; Girven et al. 1993). Free radical-induced lipid peroxidation processes are thought to lead to the formation of lipofuscin (Sohal 1981; Zs-Nagy 1988). Accumulation of the pigment over the lifespan of an individual appears to be a universal correlate of animal senescence. For crustaceans, a solvent extraction method (Ettershank 1983, 1984) was used to quantify lipofuscin in several species (Hirche and Anger 1987; Sheehy and Ettershank 1988) including the Antarctic krill, *Euphausia superba* (Ettershank 1983, 1984; Berman et al. 1989, Nicol et al. 1991). Sheehy (1996), however, demonstrated that soluble autofluorescence bears no quantitative relationship to lipofuscin. The author concluded that the histologically based quantitative method with concurrent visual microscopic verification of in situ morphological lipofuscin's fluorescence and histochemical attributes is the only reliable method for quantification. Several recent studies have been attempted to assess the applicability of morphological lipofuscin as an age marker (O'Donovan and Tully 1996; Belchier et al. 1998; Sheehy et al. 1998). Besides

Please note: When citing this article, please use the reference to the original source in the journal Polar Biology as given on the top of this page.

B.A. Bluhm (✉) · T. Brey
M. Klages · W.E. Arntz
Alfred-Wegener-Institut für
Polar- und Meeresforschung,
Columbusstrasse,
27515 Bremerhaven, Germany
E-mail: fsbab@uaf.edu

E. superba (Sheehy 1990a), no other species from high latitudes has been studied until now.

The two major aims of this study are: (1) to investigate the occurrence of lipofuscin in selected polar crustacean species, and (2) to assess the potential of image analysis of morphological lipofuscin as a quantification procedure for the age-marker pigment. Five species each from the Arctic (two amphipod and three decapod species) and the Antarctic (one amphipod, one euphausiid and three decapod species), covering different taxa, feeding types and bathymetric distribution ranges, were examined for the occurrence of lipofuscin and its age-dependent accumulation.

Materials and methods

Sampling

Specimens from the Antarctic (Table 1) were caught between 170 m and 2,100 m depth during the expedition ANT XV/3 (EASIZ II, 1998) of R/V Polarstern to the eastern Weddell Sea. Average annual temperature close to the sea bottom ranges from 0.4°C (Circumpolar Deep Water) to −1.8°C (Antarctic Surface Water), with seasonal variability generally < 0.8°C (Hellmer and Bersch 1985; Arntz et al. 1992). Specimens from the Arctic (Table 1) were caught during the expeditions ARK XIII/1 + 2 (1997) to the northern Barents and Greenland Seas, and ARK XV/1 (1999) to the Greenland Sea at depths between 140 and 5,550 m. Average annual temperatures close to the sea bottom in these areas are + 0.5°C to −1.5°C (Coachman and Aagard 1979; Loeng 1989) with minimal seasonal variation. All investigated specimens were fixed in 4% buffered formaldehyde-seawater solution immediately after sorting the catches, and were stored in the solution until analysis in summer 1999.

Life maintenance

Several females of *Chorismus antarcticus* were caught with fertilised eggs attached to the pleopods in January to March 1998. In September/October 1998, 11 females released zoea I-stage larvae, which were transferred to 100-ml plastic bottles. The larvae were fed with live *Artemia* spp. nauplii and water was replaced every other day. They were reared at 0 ± 0.2°C. At intervals, some larvae were fixed in 4% buffered formaldehyde-seawater solution for lipofuscin analysis. After 4 months all larvae were fixed, or had died.

Sample preparation

Sex, wet weight and body size were recorded (carapace length in decapods to the nearest 0.1 mm, and body length from rostrum to the end of the telson to the nearest millimetre in

Table 1 Species investigated in the present study. Depth ranges and feeding types are added from the literature if available

Species	Taxonomy	Sampling area	Depth range (m) Min.	Max.	Feeding type
Arctic					
Eurythenes gryllus (Lichtenstein, 1822)	Lysianassidae s.l., Amphipoda	Greenland Sea	0[c]	6500[c]	Scavenger[h]
Uristes sp. Dana, 1849	Lysianassidae s.l., Amphipoda	Greenland Sea	?	5550[a]	Scavenger[a]
Sabinea septemcarinata Sabine, 1821	Crangonidae, Decapoda	Barents Sea	30[g]	300[g]	Predator[a]
Sclerocrangon ferox Sars, 1877	Crangonidae, Decapoda	Barents Sea	100[g]	1000[g]	Predator[a]
Pandalus borealis Krøyer, 1844	Pandalidae, Decapoda	Greenland Sea	150[g]	700[g]	Predator, omnivorous[l]
Antarctic					
Waldeckia obesa (Chevreux, 1905)	Lysianassidae s.l., Amphipoda	Weddell Sea	0[d]	900[a]	Scavenger[a]
Euphausia superba Dana, 1850	Euphausiidae, Euphausiacea	Weddell Sea		Pelagic	Herbivorous filterfeeder/ icescraper[j]
Nematocarcinus lanceopes Bate, 1888	Nematocarcinidae, Decapoda	Weddell Sea	600[b]	2505[e]	Predator, omnivorous[k]
Notocrangon antarcticus (Pfeffer, 1887)	Crangonidae, Decapoda	Weddell Sea	5[i]	1320[e]	Predator, omnivorous[f]
Chorismus antarcticus (Pfeffer, 1887)	Hippolytidae, Decapoda	Weddell Sea	5[i]	915[e]	Predator, omnivorous[f]

[a]Personal observations
[b]Arntz and Gorny (1991)
[c]Barnard and Karaman (1991)
[d]DeBroyer and Klages (1990)
[e]Gorny (1999)
[f]Gorny et al. (1992)
[g]Heegard (1941)
[h]Ingram and Hessler (1983)
[i]Kirkwood and Burton (1988)
[j]Reid et al. (1999)
[k]Wenner (1979)
[l]Wienberg (1981)

amphipods and euphausiids). Brains of the investigated species were prepared for lipofuscin measurement essentially as described for various decapod species by Sheehy (1989, 1990a) and Sheehy and Wickins (1994). The brains were dissected and dehydrated in ascending ethanol concentrations from 70% to 100%. For the larvae of *C. antarcticus*, complete heads were embedded due to the small size. Unstained, serial, 6-µm resin (Technovit, Kultzer) sections were prepared following standard embedding procedures excluding, however, mounting medium. Figure 1 shows how brains were orientated and serially sectioned.

Fluorescence microscopy

Sections were analysed with an inverted Leica TCS NT confocal microscope, using 488 nm excitation wavelength of a KrAr laser (Omnichrome) and a ×40 oil immersion lens with a high numerical aperture of 1.25. The sections were additionally viewed using an HBO 50 mercury lamp and 365 nm, as well as 450-nm excitation filters fitted to the same microscope to assess the colour of the fluorescence. In decapods, the (posterior) lateral somacluster of the olfactory lobe was localised (nomenclature after Sandeman et al. 1992); this is composed of cell bodies of ascending olfactory projection neurons (Schmidt and Harzsch 1999). Lipofuscin has proven to be especially dense in this area in decapod crustaceans (Sheehy 1989, 1990b; Sheehy et al. 1995, 1998). In amphipods, we identified regions of high lipofuscin density, so far not described in the literature, and afterwards used them for analysis. In *E. superba*, the entire brains were scanned for lipofuscin. Six, approximately equidistant, sections of the investigated brain areas were selected for lipofuscin analysis. Digital images of 1,024×1,024 pixels resolution (250×250 µm frame area) were recorded, applying Kalman averaging of 4 images each to reduce noise. Photomultiplier intensity, laser power and offset were adjusted whenever necessary.

Lipofuscin quantification

Image analysis was carried out using "Image" software (National Institute of Health). Autofluorescent lipofuscin granules were discriminated using manual greyscale thresholding. The total area

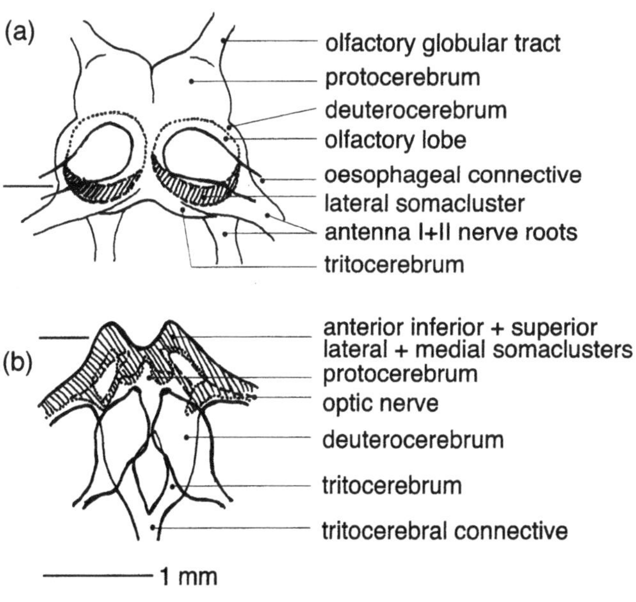

Fig. 1 Schematic views of the brains of a medium-sized **a** caridean decapod, and **b** lysianassoid amphipod. *Hatchings* mark the tissue areas used for lipofuscin quantification. *Horizontal lines* indicate the planes of sectioning

fraction of lipofuscin granules in the binarised selected area of the images was calculated by dividing the area of lipofuscin granules by the total area of analysed tissue, multiplied by 100. Finally, the geometric average area fraction over all analysed sections per specimen was computed (which – in line with stereological convention – corresponds to volume fraction in % lipofuscin).

Histochemistry

The characteristic lipophilia of lipofuscin was confirmed in selected sections of each species by staining with Sudan Black. The resin sections were rehydrated in graded ethanols (95%, 90%, 80%, 70%) for 2 min each. They were then stained by immersion in a saturated solution of filtered Sudan B in 70% ethanol for at least 30 min (modified after Romeis 1968 and Sheehy and Wickins 1994). After rinsing excess stain with ethanol and water, the sections were dried on a heating table for 5 min. The same sections that had been recorded for lipofuscin autofluorescence were recorded digitally for Sudan B staining. Again, no cover glass was applied.

Results

Morphological lipofuscin was found in 94 of 100 individuals and in all species (Table 2). Granules were detected in considerable amounts in five of the ten investigated species, and lipofuscin concentrations ranged from <0.01 to 1% lipofuscin area fraction (excluding *Waldeckia obesa* which cannot be directly compared, see below). The two scavenging amphipod species, *Eurythenes gryllus* and *W. obesa*, revealed the most conspicuous and largest granules with the most intense fluorescence. Among the decapod species, comparatively high concentrations were found in *Notocrangon antarcticus*, *C. antarcticus* and *Sclerocrangon ferox* (max. 0.2% area fraction). Only single small granules were discriminated in the amphipod *Uristes* sp., in the decapods *Pandalus borealis*, *Sabinea septemcarinata* and *Nematocarcinus lanceopes* and in the euphausiid *Euphausia superba*.

In the investigated decapod species, the autofluorescent, roundish to irregularly shaped granules of predominantly 1–5 µm in size were observed in the posterior lateral somacluster of the olfactory lobe (Fig. 1), as known from other decapod species. In amphipods, lipofuscin granules of comparable shape and average size but of higher maximum size, probably due to aggregation, were associated with the anterior inferior lateral and medial somaclusters (and associated bridge), and the anterior superior lateral and medial somaclusters (nomenclature after MacPherson and Steele 1980; Fig. 1). In *Eurythenes gryllus*, lipofuscin granules concentrated in dense aggregations in which single granules seemed to merge. These aggregations were more or less evenly distributed over the somaclusters. Granules were also found in neuropils and nervecords, though in less dense aggregates. In *W. obesa*, however, lipofuscin concentrated in the transition zone between the somaclusters and the associated neuropils, and was not found in the clusters themselves. For analysis, a different portion of the tissue was therefore selected in this spe-

Table 2 Lipofuscin concentrations in individuals of ten polar crustacean species: summary of the survey results. Lipofuscin concentration means and standard deviations of the image analysis of six brain sections per specimen are given. Maximum sizes and estimates for maximum ages of the species are added from the literature if available [*BL* body length (Amphipoda, Euphausiacea), *CL* carapace length (Decapoda), *fe* female, *ma* male, *N* sample size; *bold type* conspicuous concentrations of lipofuscin]. Note that most estimates of maximum age of species are based on size-data and should for reasons discussed in the text be interpreted with caution

Species	Depth of sampling (m)	*N*	Size range analysed (mm CL/BL)		Max. size in study area (mm CL/BL)	Wet weight range analysed (g)		Lipofuscin concentration range (% area fraction) ± SD		Estimated max. age of species (years)
								Min.	Max.	
Arctic										
Eurythenes gryllus	2500	3	33	50	140[g]	0.9	3.2	0.17 ± 0.04	**1.04** ± 0.21	9(fe)/4(ma)[h]
Uristes sp.	5550	2	5	8	14[a]	0.005	0.013	< 0.01	0.01 ± < 0.01	?
Sabinea septemcarinata	160	6	8.8	16.4	BL80[k] CL18.0[a]	0.8	4.0	0	0.02 ± < 0.01	4–6[k]
Sclerocrangon ferox	160–300	4	12.8	21.7	29.9[a]	3.3	11.4	< 0.01	**0.07** ± 0.03	4–6[k]
Pandalus borealis	300	6	8.1	22.4	25[j]	0.5	8.2	0	0.01 ± < 0.01	6[i]
Antarctic										
Waldeckia obesa	400–800	32	19	29	33[d]	0.21	0.45	0.44 ± 0.07	**2.79** ± 0.61	6(–9)[e]
Euphausia superba	400–1400	6	39	41	70[i]	0.7	1.0	0	< 0.01	6[c]
Nematocarcinus lanceopes	1680	7	22.1	29.7	35.0[b]	4.5	11.1	< 0.01	0.01 ± < 0.01	?
Notocrangon antarcticus	245–440	20	9.5	24.4	27.1[b]	0.8	12.9	< 0.01	**0.21** ± 0.04	?
Chorismus antarcticus	245–440	14	1.8	19.2	21.6[b]	< 0.1	5.8	< 0.01	**0.21** ± 0.05	7–10[f]

[a]Personal observations
[b]Arntz and Gorny (1991)
[c]Berman et al. (1989)
[d]Chapelle (1995)
[e]G. Chapelle, personal communication
[f]Gorny et al. (1993)
[g]Ingram and Hessler (1983)
[h]Ingram and Hessler (1987) (age at maturity)
[i]Marschall (1988)
[j]Teigmark (1983)
[k]Weslawski (1987)

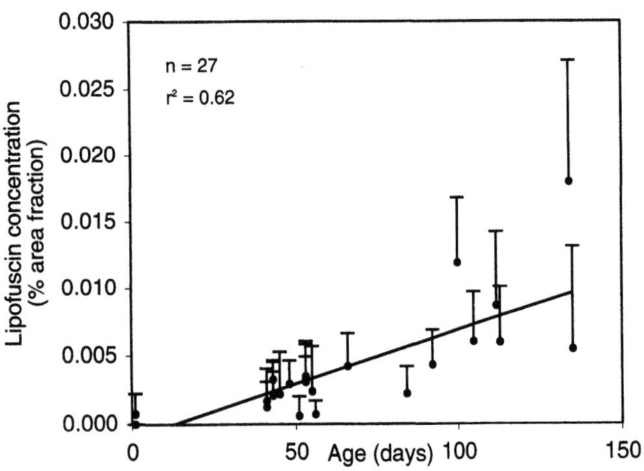

Fig. 2 Relationship between age and lipofuscin in the lateral somacluster of the olfactory lobe of *Chorismus antarcticus* larvae. Means and standard deviations of the image analysis of six brain sections per specimen are given

cies. This increased absolute values of concentrations and, thereby, made them incomparable to other species, but reduced the standard deviation between sections of one individual. In the euphausiid, lipofuscin neither occurred in somaclusters nor in nerve cords and neuropils in considerable amounts.

In laboratory-reared larvae of *C. antarcticus*, age-dependent accumulation of lipofuscin was quantified over the 1st 4 months after larval release (Fig. 2). Granules were, however, inconspicuous and comparatively small (1–2 μm in diameter, max. 4 μm). No lipofuscin was found in freshly released larvae, whereas a concentration of up to 0.02% area fraction (SD = 0.008% area fraction), an order of magnitude lower than maximum concentrations found in adults, was measured in 4-month-old individuals. Within the analysed age range, a linear regression fits the data best.

Among the species with considerable amounts of lipofuscin, small and supposedly young individuals within a species tended to contain less lipofuscin than much larger, heavier and supposedly older ones, as shown for *W. obesa*, *Notocrangon antarcticus* and the protandrous hermaphrodite, *C. antarcticus*, in Fig. 3. Within similar sizes, however, different lipofuscin concentrations were observed. Among the species with negligible amounts of lipofuscin, no such trend was measured. In amphipods, the two larger species contained considerably more pigment than the small spe-

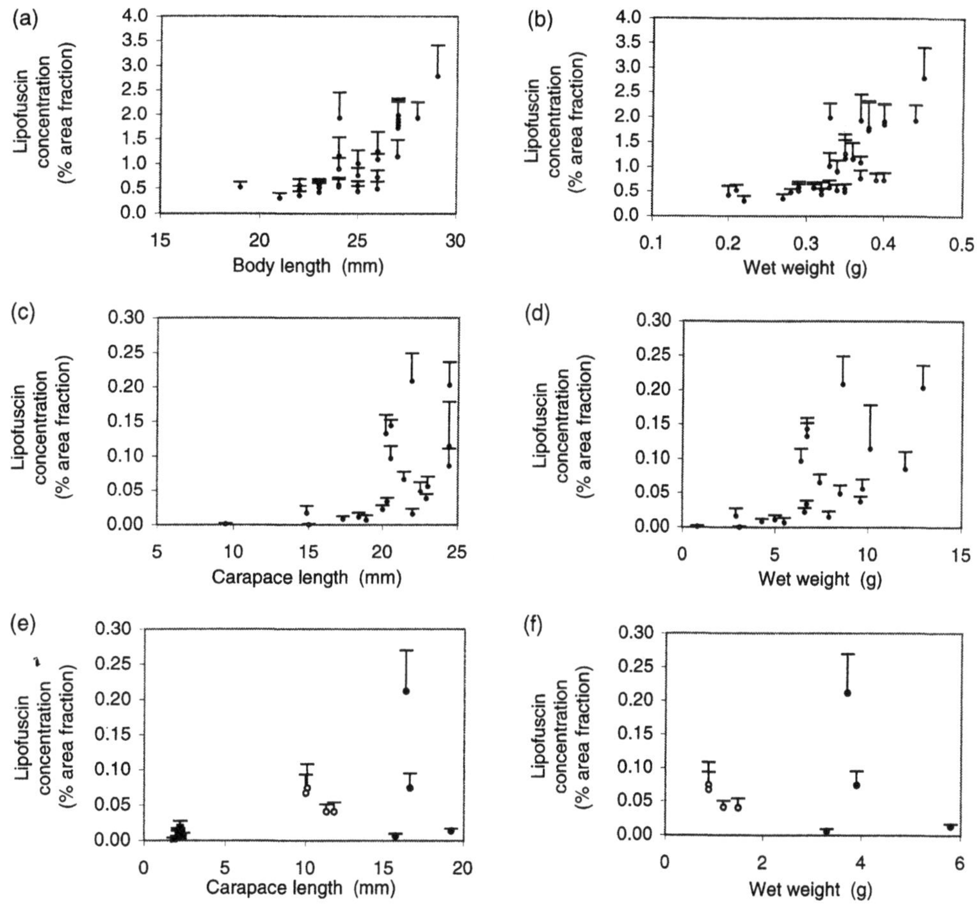

Fig. 3 Relationship between body size and body weight, respectively, and lipofuscin concentration in **a**, **b** the amphipod *Waldeckia obesa* ($n = 32$), **c**, **d** the decapod *Notocrangon antarcticus* ($n = 20$), and **e**, **f** the decapod *Chorismus antarcticus* ($n = 14$) (*dots* females, *circles* males, *squares* larvae). Means and standard deviations of the image analysis of six brain sections per specimen are given. Larvae of *C. antarcticus* were not weighed to avoid damage

cies, although it must be noted that the maximum size of *Uristes* sp. is not definitely known as yet. *Nematocarcinus lanceopes*, however, with the largest absolute body size of the investigated decapod species, contained considerably lower lipofuscin concentrations than the smaller species *Notocrangon antarcticus* and *C. antarcticus*. Highest maximum concentrations within a species were found in female specimens compared to males (Fig. 3).

Discussion

This study confirms the occurrence of morphological lipofuscin in polar crustaceans for the first time. All investigated Arctic and Antarctic species displayed at least some lipofuscin granules. The granules were conspicuous in five of the ten investigated species, with considerable variability in density and fluorescence intensity between species. Unlike the lipofuscin distribution in all decapods investigated so far, lipofuscin in the amphipod *W. obesa* did not occur within the somaclusters but between somaclusters and neuropils. Up to this point we have no explanation for this phenomenon. Those specimens without obvious pigment granules do not necessarily lack the pigment, but granules may not have been detectable with the applied methodology. Although we tried to include specimens close to the known maximum size of the species, age-pigment concentrations ranged an order of magnitude lower than measured in species from lower latitudes (Sheehy et al. 1994, 1998; Belchier et al. 1998).

Lipofuscin has been called a physiological age marker, associated with "subjective" (life)time passing at an individual's or species' pace, as opposed to objective time passing at the same pace universally. "Subjective" time is a function of the metabolic "clock", which in turn depends on genetic as well as environmental parameters. Formation and accumulation of lipofuscin may, therefore, be affected by spatial and temporal environmental variability (Sheehy et al. 1996). Temperature has been identified as one of the primary factors, as metabolic rate increases with temperature. O'Donovan and Tully (1996), for example, found a significant difference in lipofuscin accumulation of European lobsters kept at 8°C and 13°C. Hence, the rate of formation and accumulation of lipofuscin in polar crustaceans can be expected to be low compared to lower latitudes.

Food availability and, therefore, caloric intake are the second important parameter affecting metabolism and lipofuscin formation processes. After starvation of individuals of *W. obesa* for 64 days, Chapelle et al. (1994) measured a decrease of oxygen consumption to

60% which, after feeding, went straight up again. These findings indicate that short-term lipofuscin accumulation rates may vary considerably. On longer time scales, however, short-term metabolic variability will not affect overall lipofuscin accumulation rates.

Generally, food availability over the year is lower in the study areas of this investigation than in shelf areas of lower latitudes (summarised in Arntz et al. 1992). This, in combination with low temperatures and, consequently, comparatively low standard metabolism (Chapelle and Peck 1995), is assumed to lead to the slow growth and high longevity observed in many polar invertebrates. Slow ageing processes cause low lipofuscin accumulation rates (Nakano et al. 1995), and hence the comparatively low observed amounts of lipofuscin found in this study are not surprising.

Beside environmental variability, genetically determined factors inherent to a species and/or individual, such as the activity level or feeding type, can affect lipofuscin accumulation. Our data, however, are inconclusive in this respect. *Eurythenes gryllus*, for example, known for its strong swimming ability (Ingram and Hessler 1987; Hargrave et al. 1995), displays high lipofuscin concentrations whereas *Euphausia superba*, also permanently active, lacks conspicuous granules. Looking at feeding types, the scavenging amphipods contain more lipofuscin than the predatory decapods, as well as the predominantly herbivorous euphausiid. Obviously, an insufficient number of species has been investigated so far to discern a clear pattern. Moreover, we know little about differences between taxa when regarding metabolic processes leading to lipofuscin generation. Hence, the crucial factor causing the pronounced differences between species remains unknown.

There are two possibilities for calibrating lipofuscin concentration against age: (1) analysis of a group of individuals of known age, covering as much of a species' lifespan as possible; (2) statistical separation of a lipofuscin-frequency distribution, analogous to the commonly used size-frequency distributions, into distinct groups representing age classes (Sheehy et al. 1998).

For obvious logistic reasons, calibration by an age-lipofuscin relation is rarely possible in polar species. We measured age-dependent lipofuscin accumulation directly in laboratory-reared *C. antarcticus* larvae over the 1st 4 months after larval release (11 months after the eggs have become attached to the pleopods; Gorny et al. 1993). Lipofuscin concentrations were very low and granules were inconspicuous, as also observed by Sheehy (1990a) and Sheehy et al. (1994) in *Cherax quadrinatus* during the initial months. As in this study, Sheehy (1990b), O'Donovan and Tully (1996) and Sheehy et al. (1998) observed linear accumulation in juveniles but cautioned against extrapolation of this pattern to total life span. Sheehy (1992) and Sheehy et al. (1994), however, observed apparently decelerating accumulation with advancing age.

Slow growth in mature individuals combined with individual variability in growth lead to a merging and pile-up of age classes towards the larger size classes in size-frequency distributions (e.g. Arntz and Gorny 1991; Sheehy et al. 1998), which hamper statistical separation of age classes. Present data indicate that, in those species, the relation between age and lipofuscin is much less variable than the relation between age and body size, thus making lipofuscin-frequency distributions more suitable for age-class separation (Sheehy et al. 1994, 1998).

Our findings are encouraging for the future use of lipofuscin for age determination in those of the investigated species displaying a sufficiently high increase in lipofuscin concentration over their life spans. None of the factors potentially influencing lipofuscin formation are presently considered sufficient to reject the applicability of the method. Stable environmental temperatures and long life spans integrating short-term variations could possibly be helpful in lipofuscin analysis. A study is currently underway to assess population-age distribution in the amphipod *W. obesa* and the decapod *Notocrangon antarcticus*, by applying a lipofuscin age index.

Acknowledgements This project was financed by the German Research Foundation DFG (Br 1220/4-1). The help and support of the "Polarstern" crew and colleagues on board and in the A.W.I. is acknowledged. Special thanks are due to Kerstin Beyer, Katrin Premke and Evelin Siebert for technical assistance. We are grateful to M.R.J. Sheehy and an anonymous referee for the improvement of an earlier draft of the manuscript. This is A.W.I. publication no. 1779.

References

Arntz WE, Gorny M (1991) Shrimp (Decapoda, Natantia) occurrence and distribution in the eastern Weddell Sea, Antarctica. Polar Biol 11:169–177

Arntz WE, Brey T, Gerdes D, Gorny M, Gutt J, Hain S, Klages M (1992) Patterns of life history and population dynamics of benthic invertebrates under high Antarctic conditions of the Weddell Sea. In: Colombo G, Ferrari I, Ceccherelli VU, Rossi R (eds) Marine eutrophication and population dynamics. Proceedings of the 25th European Marine Biology Symposium. Olsen & Olsen, pp 221–230

Barnard JL, Karaman GS (1991) The families and genera of marine gammaridean Amphipoda (except marine Gammaroidea). Rec Aust Mus 13 [Suppl]:886

Bassin ABK, Terblanche SE, Oelofsen W (1982) A comparative study on the effects of ageing and training on the levels of lipofuscin in various tissues of the rat. Comp Biochem Physiol 71A:369–374

Belchier M, Shelton PMJ, Chapman CJ (1994) The identification and measurement of fluorescent age pigment abundance in the brain of a crustacean (*Nephrops norvegicus*) by confocal microscopy. Comp Biochem Physiol 108B:157–164

Belchier M, Edsman L, Sheehy MRJ, Shelton PMJ (1998) Estimating age and growth in long-lived temperate freshwater crayfish using lipofuscin. Freshwater Biol 39:439–446

Berman MS, McVey AL, Ettershank G (1989) Age determination of Antarctic krill using fluorescence and image analysis of size. Polar Biol 9:267–271

Bregazzi PK (1972) Life cycles and seasonal movements of *Cheirimedon femoratus* (Pfeffer) and *Tryphosella kergueleni* (Miers) (Crustacea: Amphipoda). Br Antarct Surv Bull 30:1–34

Chapelle G (1995) Estimating size of amphipods in life cycle studies: what to measure and what for? Pol Arch Hydrobiol 42:295–302
Chapelle G, Peck LS (1995) The influence of acclimation and substratum on the metabolism of the Antarctic amphipods *Waldeckia obesa* (Chevreux 1905) and *Bovallia gigantea* (Pfeffer 1888). Polar Biol 15:225–232
Chapelle G, Peck LS, Clarke A (1994) Effects of feeding and starvation on the metabolic rate of the necrophagous Antarctic amphipod *Waldeckia obesa* (Chevreux, 1905). J Exp Mar Biol Ecol 183:63–76
Clarke A, Kendall MA, Gore DJ (1990) The accumulation of fluorescent age pigments in the trochid gastropod *Monodonta lineata*. J Exp Mar Biol Ecol 144:185–204
Coachman LK, Aagard K (1979) Physical oceanography of Arctic and Subarctic seas. In: Hermann Y (ed) Marine geology and oceanography of the Arctic seas. pp 1–72
DeBroyer C, Klages M (1990) Studies on amphipod biology. Polarforschung 68:113–115
Donato H, Sohal RS (1978) Age related changes in lipofuscin-associated fluorescent substances in the adult male housefly, *Musca domestica*. Exp Gerontol 13:171–179
Ettershank G (1983) Age structure and cyclical annual size change in the Antarctic krill, *Euphausia superba* Dana. Polar Biol 2:189–193
Ettershank G (1984) A new approach to the assessment of longevity in the Antarctic krill *Euphausia superba*. J Crust Biol 4:295–305
France R, Holmes J, Lynch A (1991) Use of size-frequency data to estimate the age composition of crayfish populations. Can J Fish Aquat Sci 48:2324–2332
Girven RJ, Gauldie RW, Czochanska Z, Woolhouse AD (1993) A test of the lipofuscin technique of age estimation in fish. J Appl Ichthyol 9:82–88
Gorny M (1999) On the biogeography and ecology of the Southern Ocean decapod fauna. In: Arntz WE, Ríos C (eds) Magellan-Antarctic: ecosystems that drifted apart. Sci Mar 63 [Suppl 1]:367–382
Gorny M, Arntz WE, Clarke A, Gore DJ (1992) Reproductive biology of caridean decapods from the Weddell Sea. Polar Biol 12:111–120
Gorny M, Brey T, Arntz WE, Bruns T (1993) Growth, development and productivity of *Chorismus antarcticus* (Pfeffer) (Crustacea: Decapoda: Natantia) in the eastern Weddell Sea, Antarctica. J Exp Mar Biol Ecol 174:261–275
Hargrave BT, Phillips GA, Prouse NJ, Cranford PJ (1995) Rapid digestion and assimilation of bait by the deep-sea amphipod *Eurythenes gryllus*. Deep Sea Res I42:1905–1921
Heegard PE (1941) The zoology of East Greenland. Decapod crustaceans. Meddr Grønland 121:1–72
Hellmer H, Bersch M (1985) The Southern Ocean. A survey of oceanographic and marine metereological research work. Ber Polarforsch 26:115
Hirche H-J, Anger K (1987) The accumulation of age pigments during larval development of the spider crab, *Hyas araneus* (Decapoda, Majidae). Comp Biochem Physiol 88B:777–782
Hunter JR, Vetter RD (1988) Accumulation of age pigments (lipofuscin) in two cold-water fishes. Fish Bull 86:401–407
Ingram CL, Hessler RR (1983) Distribution and behavior of scavenging amphipods from the central North Pacific. Deep Sea Res 30:683–706
Ingram CL, Hessler RR (1987) Population biology of the deep-sea amphipod *Eurythenes gryllus*: inferences from instar analyses. Deep Sea Res 34:1889–1910
Kirkwood JM, Burton HR (1988) Macrobenthos species assemblages in Ellis Fjord, Vestfold Hills, Antarctica. Mar Biol 97:445–457
Klages M (1993) Distribution, reproduction and population dynamics of the Antarctic gammaridean amphipod *Eusirus perdentatus* Chevreux, 1912 (Crustacea). Antarct Sci 5:349–359
Loeng H (1989) Ecological features of the Barents Sea. In: Rey L, Alexander V (eds) Proceedings of the sixth conference of the Comité Arctique International 13–15 May 1985. Brill, Leiden, pp 327–365
MacPherson BR, Steele VJ (1980) Microanatomy of the central nervous system of *Gammarus setosus* Dementieva (Amphipoda). The supraoesophageal ganglion (brain). Crustaceana 38:113–120
Marschall H-P (1988) The overwintering strategy of Antarctic krill under the pack-ice of the Weddell Sea. Polar Biol 9:129–135
Nakano M, Oenzil F, Mizuno T, Gotoh S (1995) Age-related changes in the lipofuscin accumulation of brain and heart. Geronology 41:69–79
Nicol S, Stolp M, Hosie GW (1991) Accumulation of fluorescent age pigments in a laboratory population of Antarctic krill *Euphausia superba* Dana. J Exp Mar Biol Ecol 146:153–161
O'Donovan V, Tully O (1996) Lipofuscin (age pigment) as an index of crustacean age: correlation with age, temperature and body size in cultured juvenile *Homarus gammarus* L. J Exp Mar Biol Ecol 207:1–14
Rakusa-Suszczewski S (1982) The biology and metabolism of *Orchomene plebs* (Hurley, 1965) (Amphipoda: Gammaridea) from McMurdo Sound, Ross Sea, Antarctic. Mar Biol 1:47–54
Reid K, Watkins JL, Croxall JP, Murphy EJ (1999) Krill population dynamics at South Georgia 1991–1997, based on data from predators and nets. Mar Ecol Prog Ser 177:103–114
Romeis B (1968) Mikroskopische Techniken. Oldenbourg, München
Sandeman D, Sandeman R, Derby C, Schmidt M (1992) Morphology of the brain of crayfish, crabs, and spiny lobsters: a common nomenclature for homologous structures. Biol Bull 183:304–326
Schmidt M, Harzsch S (1999) Comparative analysis of neurogenesis in the central olfactory pathway of adult decapod crustaceans by *in vivo* BrdU labeling. Biol Bull 196:127–136
Sheehy MRJ (1989) Crustacean brain lipofuscin: an examination of the morphological pigment in the fresh-water crayfish *Cherax cuspidatus* (Parastacidae). J Crust Biol 9:387–391
Sheehy MRJ (1990a) Widespread occurrence of fluorescent morphological lipofuscin in the crustacean brain. J Crust Biol 10:613–622
Sheehy MRJ (1990b) Individual variation in, and the effect of rearing temperature and body size on the concentration of fluorescent morphological lipofuscin on the brains of freshwater crayfish, *Cherax cuspidatus* (Crustacea: Parastacidae). Comp Biochem Physiol 96A:281–286
Sheehy MRJ (1992) Lipofuscin age pigment accumulation in the brains of ageing field- and laboratory-reared crayfish *Cherax quadricarinatus* (von Martens) (Decapoda: Parastacidae). J Exp Mar Biol Ecol 161:79–89
Sheehy MRJ (1996) Quantitative comparison of *in situ* lipofuscin concentration with soluble autofluorescence intensity in the crustacean brain. Exp Geront 31:421–432
Sheehy MRJ, Ettershank G (1988) Extractable age pigment-like autofluorescence and its relationship to growth and age in the water-flea *Daphnia carinata* King (Crustacea: Cladocera). Aust J Zool 36:611–625
Sheehy MRJ, Wickins JF (1994) Lipofuscin age pigment in the brain of the European lobster *Homarus gammarus* (L.). Microsc Anal 40:23–25
Sheehy MRJ, Greenwood JG, Fielder DR (1994) More accurate chronological age determination of crustaceans from field situations using the physiological age marker, lipofuscin. Mar Biol 121:237–245
Sheehy MRJ, Cameron E, Marsden G, McGrath J (1995) Age structure of female giant tiger prawns *Penaeus monodon* as indicated by neuronal lipofuscin concentration. Mar Ecol Prog Ser 117:59–63
Sheehy MRJ, Shelton PMJ, Wickins JF, Belchier M, Gaten E (1996) Ageing the European lobster *Homarus gammarus* by the lipofuscin in its eyestalk ganglia. Mar Ecol Prog Ser 143:99–111
Sheehy MRJ, Caputi N, Chubb C, Belchier M (1998) Use of lipofuscin for resolving cohorts of western rock lobster (*Panulirus cygnus*). Can J Fish Aquat Sci 55:925–936
Sohal RS (ed) (1981) Age pigments. Elsevier, Amsterdam

Teigmark G (1983) Populations of the deep-sea shrimp (*Pandalus borealis* Krøyer) in the Barents Sea. Fiskeridir Skr Ser Havunders 17:377–430

Wenner EL (1979) Distribution and reproduction of nematocarcinid shrimp (Decapoda: Caridea) from the northwestern North Atlantic. Bull Mar Sci 29:380–393

Weslawski JM (1987) Distribution of Decapoda (Crustacea) in south Spitsbergen coastal waters with remarks on their ecology and breeding biology. Pol Polar Res 8:121–134

Wienberg R (1981) On the food and feeding habits of *Pandalus borealis* Krøyer. Arch Fischereiwiss 31:123–137

Zs-Nagy I (ed) (1988) Lipofuscin 1987: state of the art. Proceedings of an International Symposium held in Debrecen, Hungary, 26–30 August

Polar Biol (2001) 24: 663–669
DOI 10.1007/s003000100266

**Jens Kowalke · Marcos Tatián · Ricardo Sahade
Wolf Arntz**

Production and respiration of Antarctic ascidians

Abstract Solitary ascidians are the most abundant group of mega-epibenthic animals below 20 m in Potter Cove, King George Island. The present work deals with aspects of growth and respiration to explain this dominance. High growth rates and longevity (maximum ages between 3.1 and 10.6 years) make them effective colonisers after destructive events. Low basal metabolism (between 0.023 and 0.057 ml O_2 h^{-1} per g ash-free dry mass, $T = 1°C$) elevates the scope for growth. These properties, together with other factors, allow these animals to tolerate disturbances better than other filter-feeding groups.

Introduction

Antarctic benthic communities are dominated by suspension-feeding animals. The Weddell Sea shows a dominance of sponges and bryozoans, whereas the Bellingshausen Sea is mainly dominated by bryozoans (Starmans 1997). McMurdo Sound is dominated by sponges (Dayton et al. 1974), the shallow area around Signy Island by bryozoans, whereas locations investigated in the South Shetland archipelago revealed a high abundance of ascidians (Sáiz-Salinas et al. 1997, off Livingston Island; Sahade et al. 1998a, Potter Cove; Jazdzewski et al. 1986, Admiralty Bay).

Please note: When citing this article, please use the reference to the original source in the journal Polar Biology as given on the top of this page.

J. Kowalke (✉) · W. Arntz
Alfred Wegener Institute for Polar and Marine Research, Columbusstrasse, 27515 Bremerhaven, Germany
E-mail: jkowalke@t-online.de

M. Tatián · R. Sahade
Universidad Nacional de Córdoba,
Cátedra de Anatomía Comparada,
Avenida Vélez Sársfield 299, 5000 Córdoba, Argentina

Ascidians are an exclusively marine group of animals, whose main distribution area is the upper sublittoral. They play an important role in colonising both hard and soft substrates, where they sometimes are among the dominant epibenthic animals (Turon 1988).

At first sight, Potter Cove seems to be an unfavourable habitat for filter-feeding benthos. Water column primary production is very low (Schloss 1997) and the benthos has to cope with sinking inorganic sediments of glacial origin that dilute an already scarce planktonic food supply (Varela 1998; Schloss et al. 1999). Resuspended benthic material is, due to benthic diatoms and detritus, of greater nutritional value (Schloss et al. 1999), but still contains large amounts of inorganic particles (Kowalke 1998) deposited from the meltwater. Filter feeders in Potter Cove are thus forced to handle large quantities of particles of low average nutritional value, but have to prevent the blockage of the filtering apparatus due to an overload of particles.

Despite these environmental properties, the most abundant group of mega-epibenthic animals below 20 m depth are the filter-feeding solitary ascidian species, *Ascidia challengeri* (Herdman 1882), *Cnemidocarpa verrucosa* (Lesson 1830), *Corella eumyota* (Traustedt 1882), *Pyura setosa* (Sluiter 1905) and *Molgula pedunculata* (Herdman 1881). Only the infaunal bivalves, *Laternula elliptica* and *Yoldia eightsii* reach similar values in number and biomass at shallower depths around 15 m (Kowalke and Abele 1998; Sahade et al. 1998a).

In this work, we present new data on production and respiration, which are discussed in the light of the dominance of this filter-feeding group.

Materials and methods

Investigation area

Potter Cove is an inlet of the larger Maxwell Bay system at King George Island, South Shetlands. The inner cove, the main distribution area of ascidians, stretches over 2 km^2 and has a maximum depth of 50 m. The majority of the bottoms are made up of fine sediments, whereas the glacial front consists of moraine deposits (Fig. 1). These fine sediments stem from glacial meltwater, which enters the cove via two main creeks on the southern, and one on the northern shore. Maximum values measured for the biggest creek alone range up to 60 kg s^{-1}; the mean lies around 5 kg s^{-1} (Varela 1998).

Growth rates

Growth rates of the four dominant solitary ascidian species, *A. challengeri*, *Cnemidocarpa verrucosa*, *Corella eumyota* and *M. pedunculata*, were calculated by using the specimens collected by M. Rauschert, who exposed artificial hard substrata in Maxwell Bay in February 1982. The substrata, two plastic plates of 600 cm^2 size and an aluminium buoy of 1,250 cm^2 surface, were fixed close to the bottom at 30 m depth. In April 1985, the substrata were recovered and the animals collected, preserved and stored in 70% ethanol.

To obtain fresh weight, the preserved animals were weighed and a conversion factor of 1.33 was used (Mills et al. 1982). The ash-free dry weight was calculated using the conversion factors given by Kühne (1997) for those ascidian species.

A size-frequency distribution in ten different groups (see Kühne 1997) was established to discriminate between age classes.

The growth rates were computed, establishing a linear growth curve on logarithmically transformed values for age and body mass of the pooled mean of the size classes.

To compare the non-linear growth of different species, the growth performance

$$Y = \log(M_{max}/A_{max})$$

was calculated (Brey 1995). M_{max} is the maximum mass in kilojoules and A_{max} the maximum age in years. The maximum mass (in grammes and kilojoules) was taken from Kühne (1997) and Sahade (1999) and the maximum age was calculated using maximum mass and the established linear growth curve.

Respiration

The individuals of the ascidians, *M. pedunculata* and *Cnemidocarpa verrucosa*, were collected by divers in Potter Cove between January and February 1998. The animals were kept individually in flow-through aquaria at a temperature of 1°C. Ten days were allowed for acclimatisation before starting the experiments.

A one-electrode open-flow system with filtered seawater (0.2 μm, 60 ml/h) was used to determine the oxygen consumption (eight individuals of *Cnemidocarpa verrucosa* and seven individuals of *M. pedunculata*). A permanently aerated reservoir contained the water to which a membrane electrode was calibrated to 100% oxygen saturation. This water passed through a 175-ml experimental chamber. The O_2 depletion in the outflowing water was continuously recorded by an oxygen electrode (Eschweiler, Kiel), and plotted on a Linseis printer in % against the 100% saturation of the inflowing water. The consumption in millilitres/hour was calculated using an oxygen solubility table (Grasshoff 1983), for a salinity of 37‰ and a temperature of 1°C, respectively.

The difference in the O_2 concentration is dependent on the metabolic rate of the animal and the flow speed of the water, but not on the experimental chamber size. Animals reduce the oxygen content in the chamber until a steady-state concentration is reached. As bacterial-free seawater was used, a correction for background respiration was not necessary. Here the curves showed an acclimatisation of the animals to the experimental conditions after 3 h and analysis readings were done after this time.

For the experiments, the animals were inserted into the cooled experimental chambers and recleaned with a continous stream of 0.2-μm filtered seawater. The measurements were done in darkness

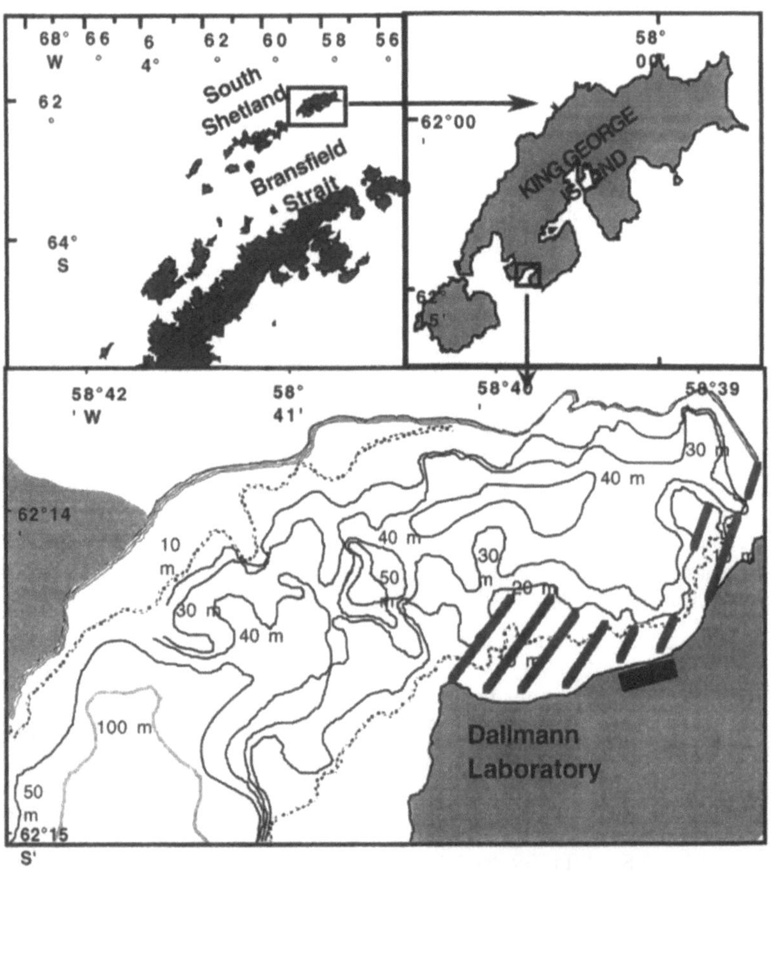

Fig. 1 Map of King George Island with Potter Cove and Maxwell Bay; *shaded lines* in Potter Cove show the location of frequent ice groundings

at a flow speed of 108 ml/h at 1.0°C for 8–10 h. After the experiments, dry mass and ash-free dry mass of each animal was determined by drying at 60°C for 48 h and burning at 500°C for 4 h, respectively. The respiration rate (RR) is expressed as a function of body mass using the allometric equation, which was fitted using a least-squares regression with ln-transformed values:

$$\ln \mathrm{RR} = a + b \cdot M$$

where M is the body mass and b the linear regression coefficient.

Results

In total, 47 individuals of *A. challengeri*, 19 individuals of *Corella eumyota*, 27 individuals of *Cnemidocarpa verrucosa* and 37 individuals of *M. pedunculata* had colonised the substrata after 3 years of exposure. Table 1 shows the dominant species of the soft bottoms of Potter Cove at 30 m depth.

The size-frequency distribution of the four species is shown in Fig. 2. Arrows mark possible annual classes. In the two species *A. challengeri* and *Corella eumyota*, three classes can be discriminated, and in *Cnemidocarpa verrucosa* and *M. pedunculata* two classes, which lead to the regression equations in Table 2. Interestingly, the flat species, *A. challengeri* and *Corella eumyota* live longer than the erect species *Cnemidocarpa verrucosa* and *M. pedunculata*, but the latter grow faster (Fig. 3 and Table 2). The calculated maximum ages range between 3.1 and 10.6 years (Table 2).

Figure 4 shows the respiration rates in a double-log plot. *Cnemidocarpa verrucosa* consumes 0.023 ml O_2 h^{-1} and *M. pedunculata* 0.057 ml O_2 h^{-1} at 1°C (standardised to an animal of 1 g ash-free dry mass), which is significantly less than the rates of species living in boreal waters (Table 3).

Discussion

Growth

It is difficult to obtain growth rates for slow-growing species without hard parts; such parts can be used as markers for age determination. Ascidians belong to these problematical taxa and exact results could only be obtained documenting the growth individually in situ in short intervals over various years. Long-term experiments in Antarctica have only been carried out in McMurdo Sound, where sponges form the bulk of macro-epibenthos (Dayton 1989). The period of this study covers 3 years only and the maximal size was not reached (Table 2). It is somewhat difficult to classify the ascidians into age classes, but as reproduction of the species is strongly seasonal (Kühne 1997; Sahade et al. 1998b) and the pelagic life of ascidian larvae is short (Svane 1983), the existence of cohorts can be assumed. Stanwell-Smith et al. (1987) reported Antarctic ascidian

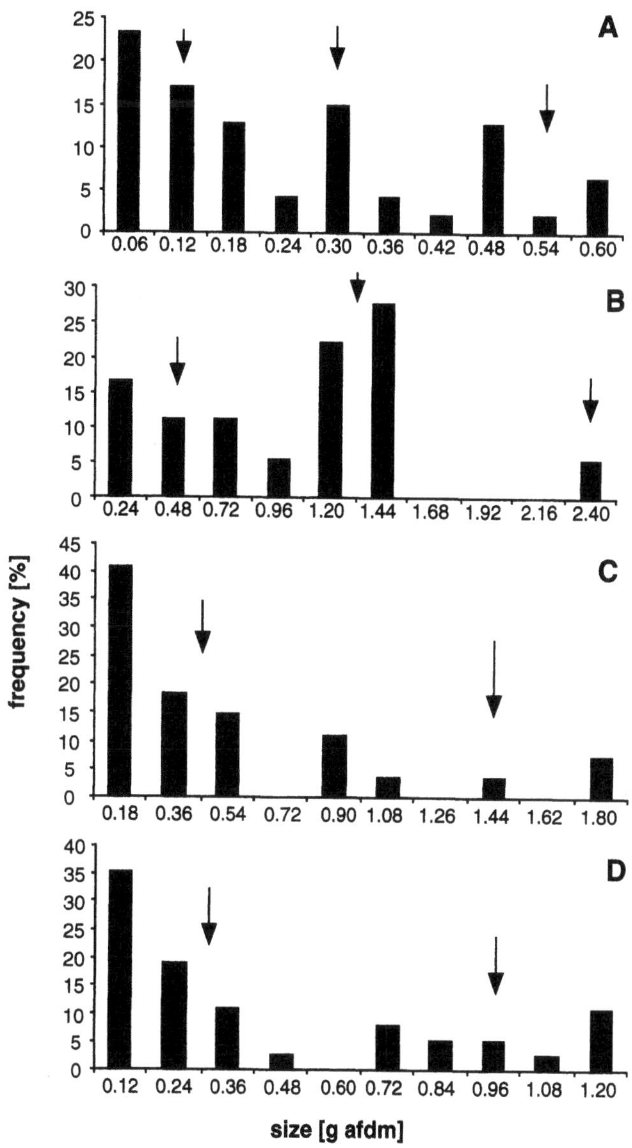

Fig. 2A–D Size-frequency distribution of ascidians recovered in April 1995 (about 30 m depth). Species *Ascidia challengeri* (**A**; $n = 47$), *Corella eumyota* (**B**; $n = 19$), *Cnemidocarpa verrucosa* (**C**; $n = 27$) and *Molgula pedunculata* (**D**; $n = 37$); *arrows* mark possible annual classes (three for *A. challengeri* and *Corella eumyota* and two for *Cnemidocarpa verrucosa* and *M. pedunculata*). Separation of cohorts was made on a time basis and is of preliminary character (see text)

Table 1 Dominant species of the soft bottoms of Potter Cove in 30 m depth; from Sahade et al. (1998a)

Species	Individuals/m^2	Dry mass/m^2
Ascidia challengeri	5.5	8.8
Cnemidocarpa verrucosa	1.8	10.3
Corella eumyota	3.4	17.7
Molgula pedunculata	16.2	14.6
Pyura setosa	0.9	
Laternula elliptica	0.4	4.8
Mycale acerata	0.02	1.5
Isodictya kerguelensis	0.10	1.7
Pennatulacea sp.	0.9	
Others	< 1	

Table 2 Maximal mass in kJ (data from Kühne 1997), calculated maximal age (A_{ma}) and growth performance (Ψ) of the Antarctic species in comparison to boreal ascidians and different Antarctic invertebrates

Species	M_{max}	A_{max}	Ψ	Source
Ascidia challengeri	41.4	10.6	0.59	This work; Kühne (1997)
Cnemidocarpa verrucosa	227.3	3.4	1.83	This work; Kühne (1997)
Corella eumyota	108.7	5.5	1.30	This work; Kühne (1997)
Molgula pedunculata	110.7	3.1	1.55	This work; Kühne (1997)
Boreal species				
Styela coriacea	2.45	2	0.09	Breuer (1995)
Ciona intestinalis	1.03	1	0.01	Petersen et al. (1995)
Pyura tesselata		>11		Svane and Lundälv (1983)
Ascidiella scabra		1.5		Svane and Lundälv (1983)
Ascidia mentula	30.41	7	0.64	Svane and Lundälv (1983)
Molgula manhattensis	0.22	1	–0.65	Breuer (1984)
Antarctic invertebrates				
Adamussium colbecki	523	20	1.42	Brey and Clarke (1993)
Lissarca miliaris	0.1	7	–1.94	Brey and Clarke (1993)
Yoldia eightsii	9.6	49	–0.7	Brey and Clarke (1993)
Bovallia gigantea	72.6	5	1.16	Brey and Clarke (1993)
Ophionotus victoriae	14.1	22	–0.19	Dahm (1995)
Magellania fragilis	2.1	49	–1.36	Brey et al. (1995)

tadpole larvae occurring occasionally; this might support the assumption of short larval stages. The authors are aware of the preliminary character of the results obtained.

The hard substrata were exposed in February 1982 and recovered in April 1985. As the panels were still free of macroscopic epifauna in November 1982 (M. Rauschert, personal communication), it is hypothesised that colonisation began by the two ascidian species, *A. challengeri* and *Corella eumyota*, as both species spawn in austral summer (Kühne 1997), thus presumably leading to a total of three colonisation periods in 1982, 1983 and 1984. The maximum age of the oldest individual ascidian would be 29 months.

As *M. pedunculata* and *Cnemidocarpa verrucosa* reproduce at the beginning of the austral winter (Sahade et al. 1998b), two colonisation events are assumed for these species – the winters of 1983 and 1984; the corresponding maximal age on the panels would be 24 months. These colonisation periods coincide with the 1982/1983 El Niño-Southern Oscillation (ENSO) event, during which Dayton (1989) found heavy recruitment in McMurdo Sound. He speculated that due to the "susceptibility of the larvae to resuspension and transport by benthic and strong wind driven currents, the presence or absence of these events might be the main cause". This might be the explanation of why Rauschert did not find any fouling on the panels in November 1982, and colonisation started not before summer 1982/1983. Growth performances of the Antarctic species are much higher than those of their boreal counterparts, with the exception of *A. mentula*. It is normally assumed that polar benthic species have low annual growth, large final size and prolonged longevity as was shown, e.g., for bivalves (Ralph and Maxwell 1977), nudibranchs (Wägele 1988), echinoids (Brey 1991) and most sponges (Dayton et al. 1974). However, these ascidians are, as are the bivalve *Adamussium colbecki* and the isopod *Bovallia gigantea* (Brey and Clarke 1993), exceptions for the low annual growth of Antarctic species. This might be related to long periods during the year of sufficient food availability in nearshore environments. Interestingly, the flat species, filtering with lower efficiencies (Kowalke 1999), live longer than the erect forms. Boreal species generally live up to 2 years, occasionally 3–4 years (Goodbody and Gibson 1974), so *M. pedunculata* and *Cnemidocarpa verrucosa* seem to be similar to boreal species in their life histories, whereas *Ascidia challengeri* and *Corella eumyota* exhibit a more "Antarctic" pattern with prolonged longevity and slower growth.

Respiration

Effects of low temperatures on metabolism are still under discussion. Metabolism, compared to tropical species, should be lowered to a factor of 9 according to van't Hoff's law. Even if this simple relationship cannot easily be used to compare morphologically similar species living under different historical and ambient conditions, it holds true as a trend (Thiel et al. 1996). Clarke (1983) and Clarke and Prothero-Thomas (1997) stated that the effects of low resting metabolism of polar species might outweigh increased costs of protein synthesis. Hence, species investing less energy in maintaining their basic metabolic needs might grow faster than related ones from warmer waters, and increase their scope for growth (Ahn and Shim 1998).

However, Brey and Clarke (1993), in a comprehensive compilation of productivity/biomass relations between invertebrate species from polar to tropical latitudes, could not find any cold adaptations with respect to population dynamics. This is related to food availability – short pulses of high primary production during austral spring are followed by prolonged periods of low phytoplankton concentrations during the rest of the year. However, recent investigations have shown that minute particles are, at least in near-shore waters, available throughout the year (Clarke and Leakey 1996; Schloss et al. 1999). These may be in sufficient quantities for some groups of suspension feeders such as ascidians to fully utilise their potential growth capacity, and show high growth performances.

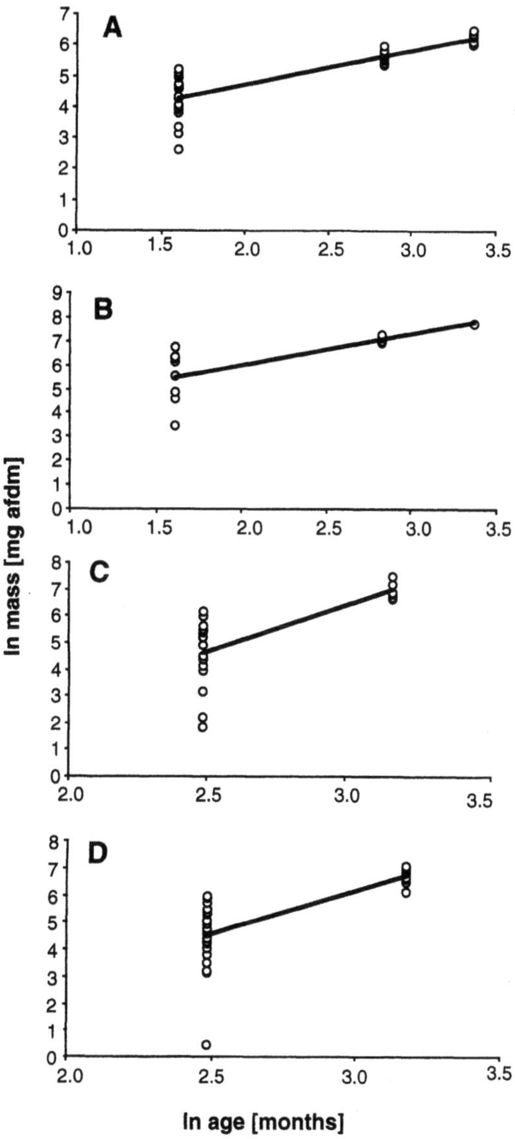

Fig. 3A–D Growth rates of the species in a double-log plot; *x-axis* ln age in months; *y-axis* ln mass in mg ash-free dry mass. **A** *Ascidia challengeri*, $Y = 1.097X + 2.506$ ($r = 0.86$, $F = 132.78$, $P = 0.0001$); **B** *Corella eumyota*, $Y = 1.315X + 3.357$ ($r = 0.77$, $F = 23.04$, $P = 0.0002$); **C** *Cnemidocarpa verrucosa*, $Y = 3.518X - 4.176$ ($r = 0.70$, $F = 24.08$, $P = 0.0001$); **D** *Molgula pedunculata*, $Y = 3.202X - 3.489$ ($r = 0.74$, $F = 40.02$, $P = 0.0001$)

The filtration efficiency (amount of water pumped per ml O_2 consumed) of both species lies within the range of ascidians from warmer waters (Table 3). The high value of *Phallusia mamillata* (19.4) is due to the high pumping rate, which is related to organ dry mass (without tunica).

This confirms the findings of Thomassen and Riisgård (1995), who stressed the importance of the habitat, which dominates biological factors. As 0.46 mg C corresponds to 1 ml O_2, a minimum of 0.04 mg C l^{-1} must be available for *Cnemidocarpa verrucosa* and 0.16 mg C l^{-1} for *M. pedunculata*, to fulfil the maintenance requirements. These values are easily reached in

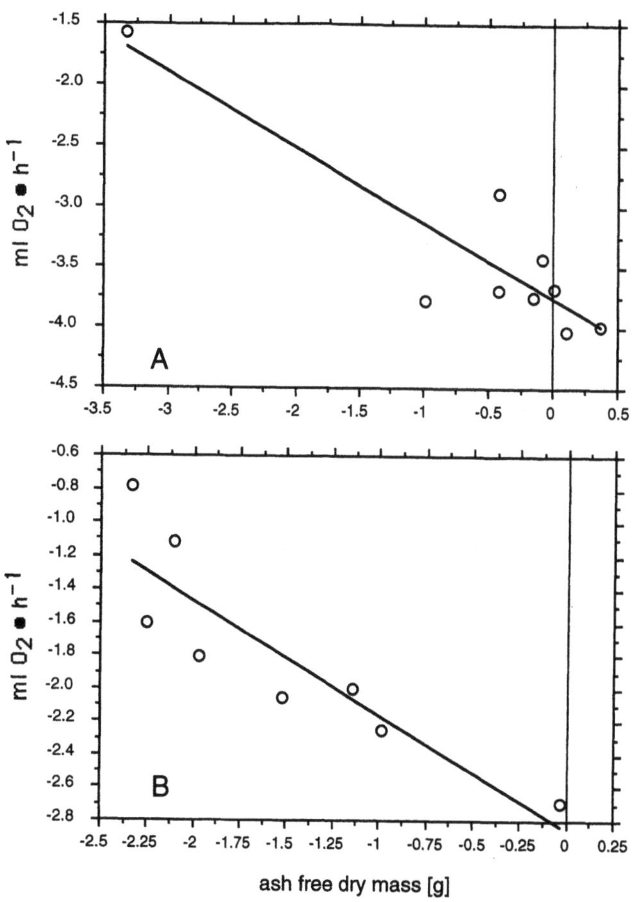

Fig. 4A, B Regression equations (ln-transformed) for respiration in relation to ash-free dry mass in ml/h. **A** *Cnemidocarpa verrucosa* $Y = -0.621 \cdot Ln_M - 3.760$ ($n = 9$; $r = 0.89$; $F = 27.88$; $P = 0.0011$); **B** *Molgula pedunculata* $Y = -0.689 \cdot Ln_M - 2.851$ ($n = 8$; $r = 0.89$; $F = 21.69$; $P = 0.0035$)

Table 3 Respiration rates (*RR* in ml/h dry mass), pumping rates (*PR* in ml/h dry mass) and filtration effectiveness (*FE*) for the Antarctic species in comparison to boreal ascidians

Species	RR	PR	FE	Source
Cnemidocarpa verrucosa	0.016	203	12.7	This work; Kowalke (1999)
Molgula pedunculata	0.039	108	2.8	This work; Kowalke (1999)
Boreal species				
Pyura stolonifera	0.098	480–1260	4.9–12.9	Klumpp (1984); Stuart and Klumpp (1984))
Ciona intestinalis	0.361	2760	7.6	Shumway (1978); Randløv and Riisgård (1979)
Phallusia mamillata	0.324	6300	19.4	Fiala-Médioni (1979)
Styela plicata	0.196	10680	5.4	Fisher (1976); Fiala-Médioni (1978)

Potter Cove (Schloss et al. 1999). Thus, food availability does not hinder the ascidians in exploiting their potential growth capacity effectively.

Conclusions

Ascidians form the majority of macro-epibenthic species on the muddy bottoms of the inner, and the gravel-dominated bottoms of the outer Potter Cove. They seem to be positively selected due to their ability to withstand the physical disturbances better than other groups such as sponges. High growth performances make them effective colonisers after destructive events. The southern shore of Potter Cove is influenced by the regular occurrence of drifting ice. Bottoms down to 3 m depth are regularly abraded by grounding ice blocks from the Forcade Glacier surrounding two-thirds of Potter Cove, whereas big growlers, reaching down to 20 m, occur irregularly and devastate the bottom of the southern shore. This impact reaches the habitat of the ascidians, which are able to recolonise the bottoms more effectively than other macro-epibenthic species. Only the infaunal bivalves *Laternula elliptica* and *Yoldia eighsii* reach similar values in biomass at shallower depth, where growler groundings are too frequent for a lasting ascidian colonisation. Low filtration performance combined with squirting prevents overloads of inorganic sediment in the peribranchial cavity and, thus, suffocation.

Furthermore, ascidians do not seem to be attractive prey organisms. They were only found in the stomach contents of the fish *Notothenia neglecta* (Barrera-Oro and Casaux 1990) and the ophiuroid *Ophionotus victoriae* (Dahm 1995). The energy contents of five ascidian species from Potter Cove were measured by Kühne (1997) and yielded values between 5 and 9 kJ/g dry mass. This is considerably lower than those of two Antarctic anthozoan species (between 15 and 17 kJ/g dry mass; Slattery and McClintock 1995) and the bivalve *L. elliptica* (19 kJ/g dry mass; Kowalke 1998).

Acknowledgements We are grateful to Martin Rauschert for the ascidian samples, the Alfred Wegener Institute and the Instituto Antárctico Argentino for logistical support, and Tom Brey, Rainer Knust and two anonymous reviewers for critical comments on the manuscript.

References

Ahn I-Y, Shim JH (1998) Summer metabolism of the Antarctic clam, *Laternula elliptica* (King and Broderip) in Maxwell Bay, King George Island and its implications. J Exp Mar Biol Ecol 224:253–264

Barnes DKA, Clarke A (1995) Seasonality of feeding ecology in polar suspension feeders. Pol Biol 15:335–340

Barrera-Oro E, Casaux R (1990) Feeding selectivity in *Notothenia neglecta*, Nybelin, from Potter Cove, South Shetland Islands, Antarctica. Antarct Sci 2:207–213

Breuer G (1984) Ökologische Untersuchung an Ascidien in einem eutrophierten Meeresabschnitt (Kieler Förde). Thesis Universität Kiel, 110 pp

Brey T (1991) Population dynamics of *Sterechinus antarcticus* (Echinodermata: Echinoidea) on the Weddell Sea shelf and slope, Antarctica. Antarct Sci 3:251–256

Brey T, Pearse J, Basch L, McClintock J, Slattery M (1995) Growth and production of *Sterechinus neumayeri* (Echinoidea: Echinodermata) in McMurdo Sound, Antarctica. Mar Biol 124 (2):279–292

Brey T (1995) Temperature and reproductive metabolism in macrobenthic invertebrates. Mar Ecol Prog Ser 125:87–93

Brey T, Clarke A (1993) Population dynamics of marine benthic invertebrates in Antarctic and Subantarctic environments: are there unique adaptations? Antarct Sci 5:253–266

Clarke A (1983) Life in cold water: the physiological ecology of polar marine ectotherms. Oceanogr Mar Biol Annu Rev 21:341–453

Clarke A, Leakey RJG (1996) The seasonal cycle of phytoplankton, macronutrients, and the microbial community in a nearshore Antarctic marine ecosystem. Limnol Oceanogr 41:1281–1294

Clarke A, Prothero-Thomas E (1997) The influence of feeding on oxygen consumption and nitrogen excretion in the Antarctic nemertean *Parbolasia corrugatus*. Phys Zool 70:639–649

Dahm C (1995) Ökologie antarktischer Ophiuriden. Rep Pol Res 194:1–289

Dayton PK (1989) Interdecadal variation in an Antarctic sponge and its predators from oceanographic climate shifts. Science 243:1484–1486

Dayton PK, Robillard GA, Paine RT, Dayton LB (1974) Biological accommodation in the benthic community at McMurdo Sound, Antarctica. Ecol Monogr 44:105–128

Fiala-Médioni A (1978) Filter Feeding ecology of benthic invertebrates (Ascidians). IV. Pumping rate, filtration rate, filtration efficiency. Mar Biol 48:243–249

Fiala-Médioni A (1979) Effects of oxygen tension on pumping, filtration and oxygen uptake in the ascidian *Phallusia mammillata*. Mar Ecol Prog Ser 1:49–53

Fisher TR (1976) Oxygen uptake of the solitary tunicate *Styela plicata*. Biol Bull 151:297–305

Goodbody I, Gibson J (1974) The biology of *Ascidia nigra* (Savigny). V. Survival in populations settled at different times of the year. Biol Bull 146:217–237

Grasshoff E (1983) Methods of seawater analysis. Verlag Chemie, Weizenheim

Jazdzewski K, Jurasz W, Kittel W, Presler E, Presler P, Sicinski J (1986) Abundance and biomass estimates of the benthic fauna in Admiralty Bay, King George Island, South Shetland Islands. Polar Biol 6:5–16

Klumpp DW (1984) Nutritional ecology of the ascidian *Pyura stonolifera*: influence of body size, food quantity and quality on filter-feeding, respiration, assimilation efficiency and energy balance. Mar Ecol Prog Ser 19:269–284

Kowalke J (1998) Energieumsätze benthischer Filtrierer der Potter Cove, Antarktis. Ber Polarforsch 286:1–147

Kowalke J (1999) Filtration in Antarctic ascidians – striking a balance. J Exp Mar Biol Ecol 242:233–244

Kowalke J, Abele D (1998) A first record of the soft bottom infauna community of Potter Cove. Ber Polarforsch 299:106–112

Kühne S (1997) Solitäre Ascidien in der Potter Cove (King George Island, Antarktis). Ihre ökologische Bedeutung und Populationsdynamik. Rep Pol Res 252, 153 pp

Mills EL, Pittmann K, Munroe B (1982) Effect of preservation on the weight of benthic invertebrates. Can J Fish Aquat Sci 39:221–224

Peterson JK, Schou O, Thor P (1995) Growth and energetics in the ascidian *Ciona intestinalis*. Mar Ecol Prog Ser 120:175–184

Ralph R, Maxwell JGH (1977) Growth of two Antarctic lamellibranchs: *Adamussium colbecki* and *Laternula elliptica*. Mar Biol 42:171–175

Randløv A, Riisgård HU (1979) Efficiency of particle retention and filtration rate in four species of ascidians. Mar Ecol Prog Ser 1:55–59

Sahade R (1999) Patrones y procesos en una comunidad epi-bentónica antártica: el caso de caleta potter. PhD Thesis, Universidad Nacional de Córdoba

Sahade R, Tatián M, Kowalke J, Kühne S, Esnal G (1998a) Benthic faunal associations on soft substrates at Potter Cove, King George Island, Antarctica. Polar Biol 19:85–91
Sahade R, Tatián M, Kowalke J, Kühne S, Esnal G (1998b) Epifaunal communities in Potter Cove, King George Island, Antarctica. In: Wiencke C, Ferreyra G, Arntz W, Rinaldi C (eds) The Potter Cove coastal ecosystem, Antarctica. Ber Polarforsch 299:123–131
Sáiz-Salinas JI, Ramos A, García FJ, Troncoso JS, San Martin G, Sanz C, Palacín C (1997) Quantitative analysis of macrobenthic soft bottom assemblages in South Shetland waters (Antarctica). Polar Biol 17:393–400
Schloss I (1997) Escalas temporales-espaciales de variabilidad del fitoplankton costero antártico. PhD Thesis, Universidad de Buenos Aires
Schloss I, Ferreyra G, Mercuri G, Kowalke J (1999) Potential food availability for benthic filter feeders in an Antarctic coastal shallow environment: a sediment trap study. In: Arntz W, Rios C (eds) Magellan-Antarctic. Ecosytems that drifted appart. Sci Mar 63 [Suppl 1]:99–111
Shumway SE (1978) Respiratory, pumping activity and heart rate in *Ciona intestinalis* exposed to fluctuating salinities. Mar Biol 48:235–242
Slattery M, McClintock JB (1995) Population structure and feeding deterrence in three shallow water antarctic soft corals. Mar Biol 122:461–470
Stanwell-Smith D, Hood A, Peck L, (1987) A field guide to the pelagic invertebrate larvae of the maritime Antarctic. British Antarctic Survey, Cambridge
Starmans A (1997) Vergleichende Untersuchungen zur Ökologie und Biodiversität des Megaepibenthos der Arktis und Antarktis. Ber Polarforsch 250:1–150
Stuart V, Klumpp DW (1984) Evidence for food-resource partitioning by kelp-bed filters feeders. Mar Prog Ser 16:27–37
Svane I (1983) Ascidian reproductive patterns to long-term population dynamics. Sarsia 68:249–255
Svane I, Lundälv T (1983) Persistence in ascidian populations: Long-term population dynamics and reproductive pattern of *Pyura tessellata* (Forbes) in Gullmarfjorden on the Swedish coast. Sarsia 67 (4):249–257
Thiel H, Pörtner HO, Arntz WE (1996) Marine life at low temperatures – a comparison of polar and deep sea characteristics. In: Uiblein F, Ott J, Stachowitsch M (eds) Deep-sea and extreme shallow water habitats: affinities and adaptations. Biosystematics and Ecology series. Book 11, Österreichische Akademie der Wissenschaften, Vienna:183–219
Thomassen S, Riisgård HU (1995) Growth and energetics of the sponge *Halichondria panicea*. Mar Ecol Prog Ser 128: 239–246
Turon X (1988) Distribución ecológica de las ascidias en las costas de Cataluña e Islas Baleares (Mediterráneo Occidental). Misc Zool 12:219–236
Varela L (1998) Hydrology of Matias and Potter Creeks. In: Wiencke C, Ferreyra G, Arntz W, Rinaldi C (eds) The Potter Cove coastal ecosystem, Antarctica. Ber Polarforsch 299:33–39
Wägele H (1988) Riesenwuchs kontra spektakuläre Farbenpracht – Nudibranchia der Antarktis. Nat Mus 118:46–53

Polar Biol (2001) 24: 869–874
DOI 10.1007/s003000100298

Christian Bock · Franz J. Sartoris · Rolf-M. Wittig
Hans-O. Pörtner

Temperature-dependent pH regulation in stenothermal Antarctic and eurythermal temperate eelpout (Zoarcidae): an in-vivo NMR study

Abstract Temperature-dependent adjustments of intracellular pH are thought to play a major role in the maintenance of protein function. Comparative studies were carried out in two species from the same fish family (Zoarcidae), the stenothermal Antarctic eelpout (*Pachycara brachycephalum*) and the eurythermal eelpout (*Zoarces viviparus*), to find out whether pH regulation is modified by temperature in the closely related species and to what extent the respective pattern differs between eurytherms and stenotherms. Previous invasive studies had compared individual animals sampled at various temperatures and suggested that a decrease in intracellular pH (pHi) values occurs at rising temperatures, as predicted by the alpha-stat hypothesis of acid-base regulation. The present study used non-invasive in vivo ^{31}P-NMR spectroscopy in non-anaesthetized, unrestrained fish for long-term online recordings in individual specimens. Control spectra obtained at $T = 0°C$ for *P. brachycephalum* and at 12°C for *Z. viviparus* indicated low stress conditions, as well as a high stability of energy and acid-base status over time periods longer than 1 week. Temperature changes had no influence on the concentration of high-energy phosphates like phosphocreatine or ATP. Temperature-induced pH changes were monitored continuously in a range between 0 and 6°C for polar, and 12 and 18°C for temperate eelpout. A pHi change of around –0.015 pH units/°C was observed within both species, in accordance with the alpha-stat hypothesis; however, extrapolation to the same temperature revealed different set points of pH regulation in the two species. These findings confirm that an alpha-stat pattern of pH regulation can be found in stenothermal Antarctic animals, at set points deviating from an alpha-stat pattern, however, in a between-species comparison.

Please note: When citing this article, please use the reference to the original source in the journal Polar Biology as given on the top of this page.

C. Bock (✉) · F.J. Sartoris · Rolf-M. Wittig · Hans-O. Pörtner
Alfred Wegener Institute for Polar and Marine Research,
Columbusstrasse, 27568 Bremerhaven, Germany
E-mail: cbock@awi-bremerhaven.de
Tel.: +49-471-48311380
Fax: +49-471-48311149

Introduction

The alpha-stat hypothesis of acid-base regulation implies that pH regulation in poikilotherms should maintain a constant degree of protonation (α) of imidazole groups in proteins despite changes in body temperature (Reeves 1985). A pH_i change of around –0.018 pH units/°C is expected to support the alpha-stat pattern and ensure protein function at fluctuating temperatures. With some exceptions, an alpha-stat pattern of intracellular pH regulation has been found in several marine ectotherms (invertebrates and fish) from temperate latitudes (for a review see Pörtner et al. 1998) where animals are exposed to wide diurnal and seasonal temperature fluctuations. The question arises as to whether the same holds true for Antarctic species which live within a very narrow temperature window throughout the year. Answering this question will lead to a more profound insight into the overall homeostatic importance of temperature-dependent pH regulation and the relevance of this process in polar species.

In vivo ^{31}P-NMR (nuclear magnetic resonance) spectroscopy has become a helpful tool for physiological studies on fish, and has been applied especially to the determination of changes in energy metabolism and intracellular pH during hypoxia in freshwater species (e.g. Van den Thillart et al. 1989a, b; Van Ginneken et al. 1995; Borger et al. 1998). A few in-vivo NMR studies have been carried out on marine fish – the goby *Pomatoschistus* sp. (Grøttum et al. 1998) and *Harpagifer antarcticus* (Moerland and Eggington 1998). The latter study focused on intracellular pH and temperature, although online pH measurements at different temperatures were not performed. These investigations used small animals (≤ 5 cm body length) that were anaesthetized and oriented in a vertical position in narrow-bore standard high-resolution NMR spectrometers. These measurements may not reflect the normal physiological status of the animals. Therefore, we developed an experimental set-up for NMR studies of marine fish using a horizontal magnet which allows for controlled and well-defined physiological conditions and continuous measurements during long-term experiments.

Invasive studies compared pH values measured in different individuals of the eurythermal North Sea eelpout (*Zoarces viviparus*) sampled at various temperatures, and supported the conclusion that a decrease in intracellular pH values occurs, as predicted by the alpha-stat hypothesis (Van Dijk et al. 1997). The aim of the present study was to carry out long-term online recordings during changing temperatures in the same individuals and to investigate the α-stat pattern of pH regulation, not only in eurythermal eelpout (*Z. viviparus*) but also in the stenothermal Antarctic eelpout (*Pachycara brachycephalum*). Species of the fish family Zoarcidae can be found around the world in all latitudes, including the deep sea and polar regions. This family is therefore most suitable for the comparison of closely related species from different latitudes. The investigation of pH_i regulation is suitable for answering the question as to whether and to what extent adaptive flexibility with respect to temperature change found in eurythermal temperate zone fish is still present in stenothermal Antarctic fish.

Materials and methods

Antarctic eelpout (*P. brachycephalum*) were caught in Admiralty Bay, Antarctica during the cruise ANT XV/III (1998) and kept in seawater aquaria at 0°C. Three to six animals (average length 30 cm, average weight 60 g) were used for the in vivo studies. Temperate eelpout were collected in the Baltic Sea by commercial fishermen in 1998 and kept in aquaria at 12°C. Animals were fed once a week and starved for 1 week before experiments. The animals were placed in a Plexiglas flow-through chamber with variable slide barriers on each side to position the animals in the centre of the chamber (Fig. 1a). Animals were conscious (not anaesthetized), unrestrained and free to move inside the chamber during the whole experimental time. The chamber was inserted into the spectrometer and a constant flow of seawater (1.5 l/min) was supplied by hydrostatic pressure from a seawater reservoir (ca. 50 l). After passing the chamber, the water was caught in a second tank and pumped back to the reservoir (Fig. 1b). Temperature was measured directly inside the chamber and in the water tanks using a fluoroptic thermometer (Luxtron 504, Polytec, Waldheim, Germany). Temperature data were sampled continuously (sampling rate 10/min, MacLab system, AD Instruments, Australia) and stored on a computer. Temperature control (stability ± 0.3°C in the range between 0° and 30°C) was achieved using a cryostat connected to the upper water reservoir and by feeding the tubings through jackets perfused with coolant.

All experiments were carried out using a 4.7 T magnet with actively shielded gradient coils (Bruker Biospec 47/40 DBX System), equipped with a 20-cm actively decoupled ^{1}H-resonator for signal excitation during imaging and a 5-cm triple-tuned (^{1}H-^{31}P-^{13}C) surface coil (passively decoupled) for signal reception. The coils were oriented orthogonal to each other to prevent cross talking. The 5-cm surface coil was placed directly onto the chamber wall and positioned close to the tail of the animal for ^{31}P-NMR spectroscopy (Fig. 1a).

Pilot scans were collected just before temperature variation and directly after reaching the desired temperature using a multi-slice gradient echo technique in all three directions to control the position of the animal (matrix 256x192, FOV = 12 cm, slice thickness 2 mm, TR = 100 ms, TE = 10 ms, scan time 25 s). For anatomical studies, multi-slice RARE images were performed in coronal and transversal directions (matrix 256x192, FOV = 12 cm, 4 slices with a slice thickness of 1 mm, TR = 2000 ms, TE = 10 ms, rare factor of 8, scan time 1.5 min).

In vivo ^{31}P-NMR spectra were acquired continuously over 1,200 scans, consisting of a 100 µs bp pulse, a repetition delay of 0.52 s resulting in a measurement time of 10 min. All spectra were processed automatically applying a user program with size = 16 K, line broadening = 5 Hz and an automatic baseline correction. The spectra were calibrated to phosphocreatine (PCr) used as an internal standard. Typical line width of PCr was 15 Hz. Signal integration and chemical shifts were calculated using an automatic fit routine (mdcon, Bruker Analytical, Rheinstetten) and compared with results determined by automatic peak picking. A calibration curve for the calculation of pH values from the chemical shift of the inorganic phosphate (Pi) signal was obtained from standard solutions at different pH, and temperatures at 0 and 20°C, with a simulated intracellular ion content. The temperature-dependent pKs for inorganic phosphate and temperature correction factors were obtained from Kost (1990).

The animals were allowed to recover for at least 24 h under control conditions inside the magnet before starting temperature incubations. Steady-state conditions were confirmed by constant in vivo ^{31}P-NMR spectra. Temperature was increased in two steps, from 0 to 3°C and from 3 to 6°C in steps of 1°C over 3 h (12 to 18°C for *Z. viviparus*). When a new steady state was reached (12°C) the temperature was kept constant for at least 12 h. Temperature was decreased to 0°C again at the end of the experiment and the animal was brought back into the aquarium.

Results

The experimental set-up allowed the investigation of Antarctic eelpout under control conditions at 0°C for more than 1 week (data not shown). MR images were obtained from unrestricted and unanaesthetized fish without any movement artefacts, and they revealed that the fish were usually resting on the ground oriented towards the incoming water (Fig. 2). Almost no surrounding water was detected due to the high flow rate in comparison with the detection time of the surface coil. Using a cross coil set-up reduced the surrounding water signal and therefore increased the signal to noise ratio compared with measurements with a single resonator. Different organs are visible inside the fish (assigned as liver, gills, stomach and gall bladder), indicating the good anatomical resolution of the MR images.

Recovery from handling stress occurred within 2 h after setting up the experiment, indicated by constant in vivo ^{31}P-NMR spectra (data not shown). Figure 3 shows a typical in vivo ^{31}P-NMR spectrum of Antarctic eelpout (*P. brachycephalum*) at 0°C compared to temperate eelpout (*Z. viviparus*) at 12°C obtained in the same set-up. Low stress is indicated by the high phosphocreatine over inorganic phosphate ratio (mean value around 11) and the high β-ATP signal compared to the α- and

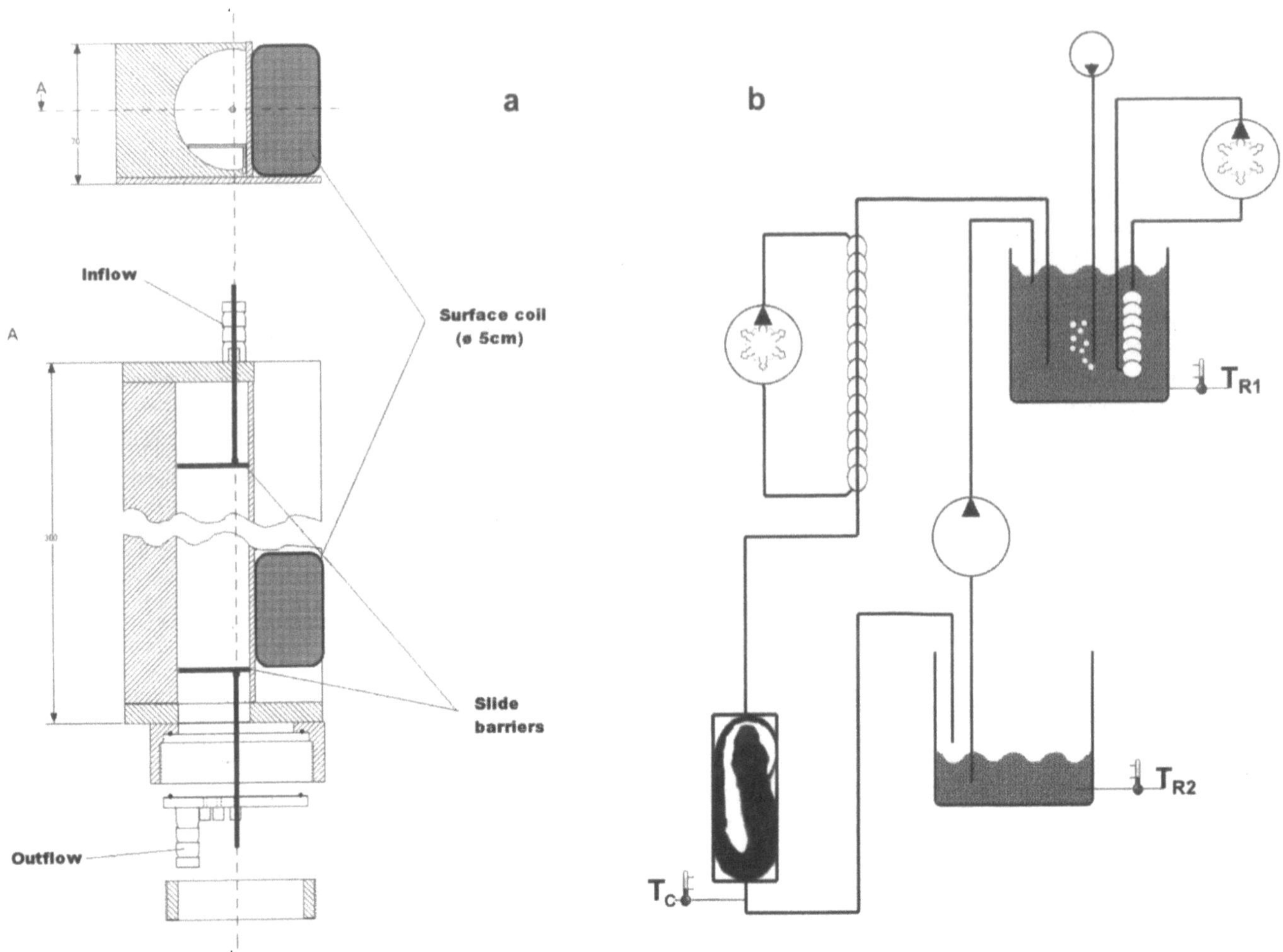

Fig. 1 Flow-through chamber (**a**) for in-vivo NMR measurements of unanaesthetized and unimmobilized fish (max. length 40 cm). The *graph* shows the direction of flow, the position of the surface coil for NMR spectroscopy and the slide barriers to optimize the position of the animals in the sensitive volume of the surface coil. **b** Experimental set-up. The chamber was placed into the spectrometer and a constant flow of seawater (up to 1.5 l/min) was supplied by hydrostatic pressure. Seawater (approx. 50 l) was bubbled with air in the header tank; temperature was measured in the chamber (T_C) and in the water reservoirs (T_{R1} and T_{R2}) using a fluoroptic thermometer connected to a MacLab system. Temperature control was achieved by a cryostat connected to the upper water reservoir and the cooling jackets of the tubings

γ-ATP signals, which also include ADP. In addition to PCr, inorganic phosphate and the three ATP resonances, glycerophosphoethanolamine (GPE), glycerophosphocholine (GPC), phosphoethanolamine (PE) and phosphocholine (PC) signals could be identified in eelpout spectra. Furthermore, a high level of an unknown phosphate compound (X) could be detected in the phosphodiester region (PD). In contrast, only a small amount of GPC and GPE, as well as PC and PE, were visible in *Z. viviparus*, and signal X was not found at all in the in vivo ^{31}P-NMR spectrum of the eurythermal eelpout.

A typical time course of intracellular pH variations at rising temperatures is presented in Fig. 4. An intracellular pH of about 7.4 was measured at 0°C. The scatter of individual values resulted from the small inorganic phosphate signal under control conditions and the insensitivity of the automatic fit routine for small signals in combination with small changes in chemical shifts. An acidification could be observed immediately after changing the temperature from 0°C to 3°C, as well as from 3°C to 6°C. At both temperatures a new steady state value was reached within 3 h. Figure 5 summarizes the mean values of intracellular pH of *P. brachycephalum* at 0, 3 and 6°C compared to pH changes observed in *Z. viviparus* at 12, 15 and 18°C. The decrease in pH with temperature calculated from these data was about ΔpH/°C = –0.015/°C in *P. brachycephalum* and ΔpH/°C = –0.014/°C in *Z. viviparus*.

Discussion

The aim of this study was the online monitoring and comparison of changes in intracellular pH and energy status with temperature in the stenothermal Antarctic eelpout *P. brachycephalum* and the eurythermal temperate eelpout *Z. viviparus*. All animals survived heat exposure and no mortalities occurred during the subsequent recovery period. 6°C is well below the critical

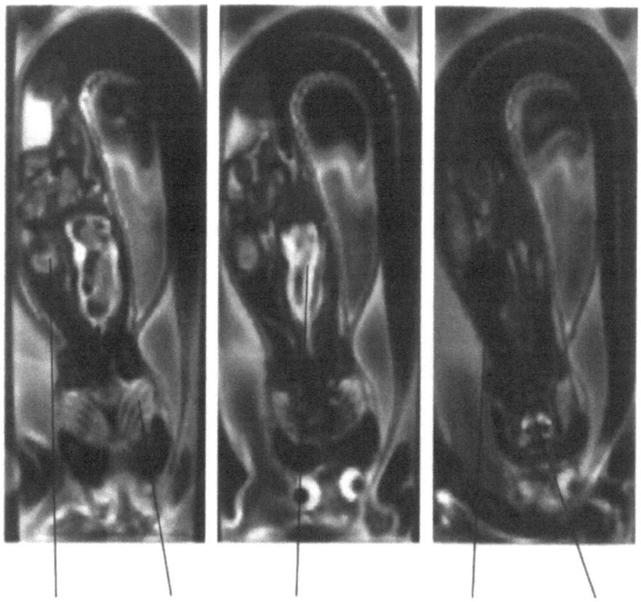

Fig. 2 In-vivo MR images from the Antarctic eelpout *Pachycara brachycephalum* under control conditions (obtained with RARE; parameters: matrix 256×192, FOV = 12 cm, 4 slices with a slice thickness of 1 mm, TR = 2,000 ms, TE = 10 ms, rare factor of 8, scan time 1.5 min). Different organs are visible inside the fish (assigned as gall bladder, gills, stomach, liver and brain), indicating the good quality of the MR images, obtained in seawater as a highly conductive medium

temperature of *P. brachycephalum* found at about 9°C, according to changes in oxygen consumption and anaerobic metabolism (Van Dijk et al. 1999). Since no anaesthetics were used, their potential influence on blood and tissue acid-base parameters as reported for rainbow trout (Iwama et al. 1989) and in the Antarctic fish *Pagothenia borchgrevinki* (Ryan 1992) could be excluded.

Spectra obtained from *Pachycara brachycephalum*, as well as from *Z. viviparus*, indicated low stress conditions, reflected by low Pi signals and stable resting pH_i. In contrast to *Z. viviparus*, high amounts of phosphomono- and diesters could be detected in *Pachycara brachycephalum*. Van den Thillart et al. (1989a) found phosphodiester signals in in vivo ^{31}P-NMR spectra of carp and goldfish. They also recorded ^{31}P-NMR spectra from different tissue extracts and concluded that these signals mainly accrued from the lateral red muscle. The lateral red muscle of fish displays a higher free fatty acid content, a much greater lipase activity and a faster lipid turnover than the white muscle (George 1962); the authors therefore interpreted these phosphodiesters to be intermediate products of phospholipid formation (Van den Thillart et al. 1989a). This may suggest that Antarctic eelpout have a higher free fatty acid content, reflecting a faster lipid turnover or greater lipase activity than in temperate species. A higher free fatty acid content, as well as a faster lipid turnover, might be associated with metabolism used to preserve membrane fluidity (Storelli et al. 1998). Alternatively, the high phosphodiester content may reflect the higher mitochondrial density in Antarctic eelpout (Johnston et al. 1988).

pH plays an important role in the maintenance of protein function during temperature change. The constant dissociation state of histidine residues within proteins, especially in active sites of enzymes, is seen as a key factor in this process (α-stat hypothesis, Reeves 1985). A shift in intracellular pH with changing body temperature may compensate for the temperature-dependent change of pK values of histidine imidazole groups ($\Delta pK/\Delta T \sim -0.018/°C$). The temperature-dependent flexibility of acid-base regulation is interpreted as reflecting metabolic adjustments and thereby adaptive flexibility with respect to temperature; however, this pattern is only found within the range of thermal tolerance (e.g. Sommer et al. 1997). In our study, we did not observe any significant changes in the levels of high-energy phosphates like PCr and ATP during temperature incubations between 0 and 6°C (12 and 18°C in temperate eelpout), thus confirming the good physiological conditions inside the magnet and also reflecting the insensitivity of white muscle energy status to thermal stress, possibly supported by an α-stat pattern of pH_i regulation.

A mean pH_i change of –0.015 pH/°C, as observed below the critical temperature in *Pachycara brachycephalum*, is close to the value expected according to an α-stat pattern of pH regulation. Furthermore, our data agree with those pH values and temperature-induced pH changes found by use of invasive methods (at –0.015 pH/°C in *Pachycara brachycephalum* white muscle; Van Dijk et al. 1999). In contrast to *Z. viviparus* where the contribution of passive mechanisms to the pH change is small (Van Dijk et al. 1997), a large fraction of the temperature-induced pH change in *Pachycara brachycephalum* is caused by passive physico-chemical buffering (–0.011 pH/°C; Pörtner and Sartoris 1999). Such a large contribution of passive buffering is typically found in polar stenotherms (Pörtner and Sartoris 1999).

Moerland and Eggington measured the pH_i of the stenothermal Antarctic teleost *H. antarcticus* using ^{31}P-NMR, and concluded that their data were not consistent with the α-stat hypothesis (Moerland and Eggington 1998). This conclusion arose from a comparison of pH_i values between unrelated species, since in their own study pH_i was only determined at one temperature (1°C) and compared with published data from different organisms obtained at different temperatures. pH_i values compared between species were relatively independent of temperature below 17°C. In fact, the comparison of pH_i values from Antarctic eelpout (0–6°C) with those found in the closely related Baltic eelpout (12–18°C) also reveals a lower slope (dotted line, –0.009 pH/°C, Fig. 5). At first sight, this is in line with the conclusion of Moerland and Eggington. Nevertheless, both eelpout species showed a pH_i change in accordance with α-stat pH regulation (see Fig. 5, –0.015 and –0.014 pH/°C, respectively), disclosing a misleading generalisation

Fig. 3 In vivo ^{31}P-NMR spectra of *Pachycara brachycephalum* (*a*) and *Zoarces viviparus* (*b*) muscle tissue under control conditions. The spectrum of Antarctic eelpout showed additional signals in the phosphodiester region, assigned as *GPE*, *GPC*, and their monomers *PE* and *PC*. An unknown signal *X* was also observed

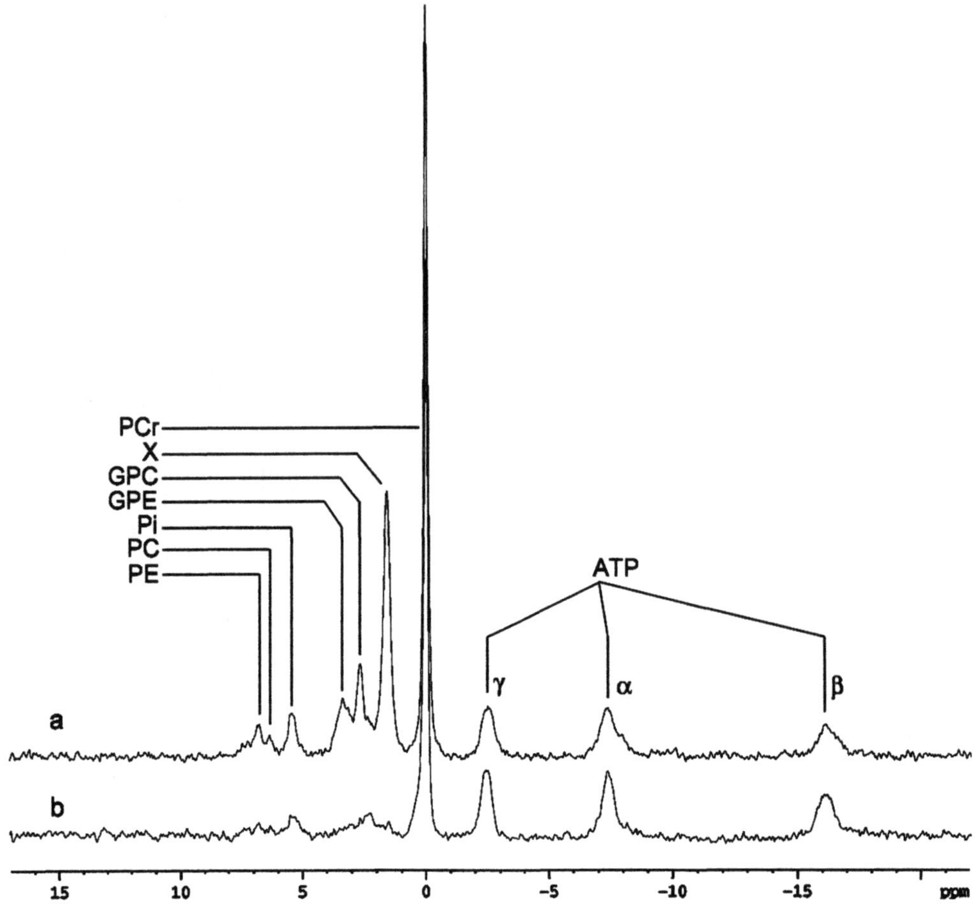

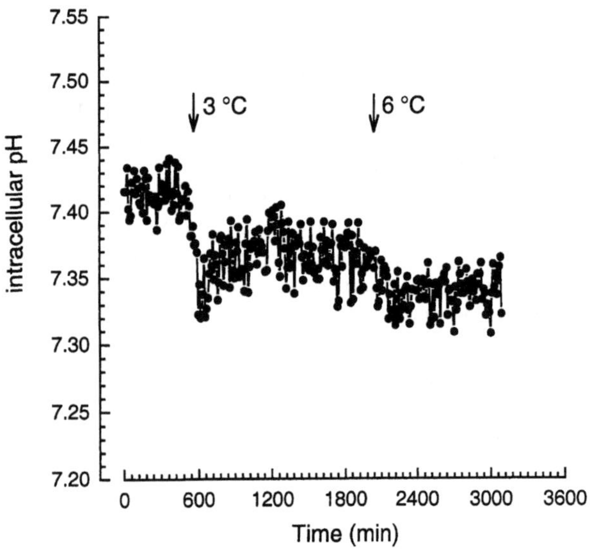

Fig. 4 Typical time course of pH changes at rising temperatures obtained from in vivo ^{31}P-NMR spectra of *Pachycara brachycephalum*. A slight acidification could be observed developing with increasing temperatures (0 to 3°C and 3 to 6°C)

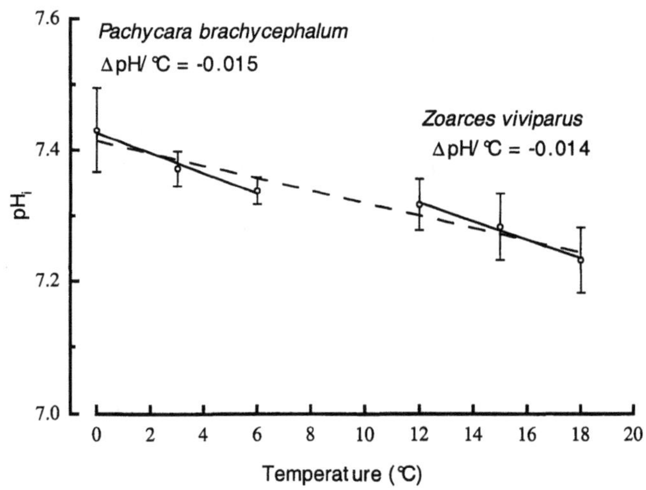

Fig. 5 Intracellular pH values derived from in-vivo ^{31}P-NMR spectra of Antarctic eelpout at 0°C ($n=6$), 3°C ($n=5$) and 6°C ($n=3$) in comparison to pH values from Baltic Sea eelpout at 12°C ($n=3$), 15°C ($n=3$) and 18°C ($n=2$, mean ± range). Spectra obtained 3 h after the change of temperature were used for the calculation of mean values. The decrease in pH with increasing temperature occurred by $\Delta pH/°C = -0.015$ for *Pachycara brachycephalum* and $\Delta pH/°C = -0.014$ for *Zoarces viviparus* and reflects α-stat pH regulation, whereas the comparison between species shows a slope (*broken line*, $\Delta pH/°C = -0.009$) lower than expected from the α-stat pattern (see text)

based on comparisons between species. It appears that cold-adapted species operate at a lower set point of intracellular pH than expected from an extrapolation of pH values from temperate or warm-adapted species. Alternatively, Pörtner and Sartoris (1999) suggested that a higher mitochondrial density causes lower mean values of intracellular pH. Further study must elaborate on whether this may explain the apparent non α-stat pattern seen in the between-species comparisons.

A recent in vivo ^{31}P-NMR study on common carp at different temperatures also found an α-stat pattern of pH regulation (Borger et al. 1998). These authors reported an immediate change in pH upon exposure to the new temperature and a pH significantly different from control conditions after a long-term temperature change by 8°C (25°C to 33°C) over 4 weeks (Borger et al. 1998). This confirms and reflects exactly our finding of a rapid temperature-dependent pH_i change in Antarctic eelpout. The immediate pH_i change with temperature in *Pachycara brachycephalum* is in line with a large contribution of passive mechanisms to temperature-induced pH_i changes.

In conclusion, and in contrast to previous reports in the literature, our study demonstrates that pH_i regulation in at least some Antarctic fish occurs as predicted by the α-stat hypothesis. This observation emphasizes the importance of a finely tuned adjustment of intracellular pH, even in polar animals, which experience minor temperature changes compared to species living in temperate zones. Future research should compare the role of intracellular pH in the regulation of metabolic pathways and energy turnover in closely related species from various, including polar, latitudes.

Acknowledgements Support by a BMBF grant (project 03PL02A, NMR laboratory) is gratefully acknowledged. A contribution to the ELOISE project, Effects of climate-induced change on marine coastal fishes (CLICOFI, ENV-CT97-0596), funded by the European Union. ELOISE publication no. 141.

References

Borger R, Boeck G de, Van Audekerke J, Dommisse R, Blust R, Van der Linden A (1998) Recovery of the energy metabolism after a hypoxic challenge at different temperature conditions: a ^{31}P-nuclear magnetic resonance spectroscopy study with common carp. Comp Biochem Physiol A 120:143–150

George JC (1962) A histochemical study of red and white muscle of mackerel. Am Midl Nat 68:487–494

Grøttum JA, Erikson U, Grasdalen H, Staurnes M (1998) In vivo ^{31}P-NMR spectroscopy and respiration measurements of anaesthetized goby (*Pomatoschistus* sp.) pre-exposed to ammonia. Comp Biochem Physiol A 120:469–475

Iwama GK, McGeer JC, Pawluk MP (1989) The effects of five fish anaesthetics on acid-base balance, hematocrit, blood gases, cortisol, and adrenaline in rainbow trout. Can J Zool 67:2065–2073

Johnston IA, Camm JP, White MG (1988) Specialisation of swimming muscles in the pelagic Antarctic fish, *Pleurogramma antarcticum*. Mar Biol 100:3–12

Kost GJ (1990) pH standardization for phosphorus-31 magnetic resonance heart spectroscopy at different temperatures. Magn Res Med 14:496–506

Moerland T, Eggington S (1998) Intracellular pH of muscle and temperature: insight from in vivo ^{31}P NMR measurements in a stenothermal Antarctic teleost (*Harpagifer antarcticus*). J Therm Biol 23:275–282

Pörtner HO, Sartoris FJ (1999) Invasive studies of intracellular acid-base parameters: quantitative analyses during environmental and functional stress. In: Eggington S, Taylor EW, Raven JA (eds) Regulation of acid-base status in animals and plants. Society for Experimental Biology Seminar Series. Cambridge University Press, Cambridge, pp 69–98

Pörtner HO, Hardewig I, Sartoris FJ, Van Dijk PLM (1998) Energetic aspects of cold adaptation; critical temperatures in metabolic, ionic and acid base regulation? In: Pörtner HO, Playle R (eds) Cold ocean physiology. Society for Experimental Biology Seminar Series. Cambridge University Press, Cambridge, pp 88–120

Reeves RB (1985) Alphastat regulation of intracellular acid-base state? In: Gilles R (ed) Circulation, respiration, and metabolism. Springer, Berlin Heidelberg New York, pp 414–423

Ryan SN (1992) Susceptibility of the Antarctic fish *Pagothenia borchgrevinki* to MS-222 anaesthesia. Polar Biol 11:583–589

Sommer A, Klein B, Pörtner HO (1997) Temperature induced anaerobiosis in two populations of the polychaete worm *Arenicola marina*. J Comp Physiol 167B:25–35

Storelli C, Acierno R, Maffia M (1998) Membrane lipid and protein adaptations in Antarctic fish. In: Pörtner HO, Playle R (eds) Cold ocean physiology. Society for Experimental Biology Seminar Series. Cambridge University Press, Cambridge, pp 166–189

Van Dijk PLM, Hardewig I, Pörtner HO (1997) Temperature dependent shift of pH_i in fish white muscle: contribution of passive and active processes. Am J Physiol 272:R84–R89

Van Dijk PLM, Tesch C, Hardewig I, Pörtner HO (1999) Physiological disturbances at high temperatures: a comparison between stenothermal Antarctic and eurythermal temperate eelpouts (Zoarcidae). J Exp Biol 202:3611–3621

Van Ginneken V, Van den Thillart G, Addink A, Erkelens C (1995) Fish muscle energy metabolism measured during hypoxia and recovery: an in vivo ^{31}P-NMR study. Am J Physiol 268:R1178–1187

Van den Thillart G, Van Waarde A, Muller HJ, Erkelens C, Addink A, Lugtenburg J (1989a) Fish muscle energy metabolism measured by in vivo ^{31}P-NMR during anoxia and recovery. Am J Physiol 256:R922–R929

Van den Thillart G, Körner F, Van Waarde A, Erkelens C, Lutgenburg J (1989b) A flow through probe for in vivo ^{31}P NMR spectroscopy of unanaesthetized aquatic vertebrates at 9.4 Tesla. J Magn Res 84:573–579

List of Referees

Abele	D.	University of Bremen	Bremen	Germany
Anger	K.	Alfred Wegener Institute	Helgoland	Germany
Ballesteros	M.	Centro de Estudios Avanzados	Blanes	Spain
Bargelloni	L.	University of Padova	Padova	Italy
Barnes	D. K. A.	University of Cork	Cork	Ireland
Boero	F.	University of Lecce	Lecce	Italy
Bouillon	J.	Université Libre de Bruxelles	Bruxelles	Belgium
Brandt	A.	University of Hamburg	Hamburg	Germany
Breeman	A. M.	University of Groningen	Groningen	Netherlands
Brey	T.	Alfred Wegener Institute	Bremerhaven	Germany
Buck	K.	Marine Laboratories	Moss Landing	USA
Bückmann	D.	University of Ulm	Ulm	Germany
Cadée	G.	Nederlands Instituut voor Onderzoek der Zee	Texel	Netherlands
Cattaneo-Vietti	R.	University of Genova	Genova	Italy
Clarke	A.	British Antarctic Survey	Cambridge	UK
Coleman	C. O.	Humboldt University	Berlin	Germany
Conlan	K.	Canadian Museum of Nature	Ottawa	Canada
Davenport	J.	University College	Cork	Ireland
Dayton	P. K.	Scripps Institution of Oceanography	La Jolla	USA
De Beer	D.	MPI for Microbiology	Bremen	Germany
De Broyer	C.	Institut Royal de Sciences Naturelles	Bruxelles	Belgium
De Wachter	B.	University of Antwerpen	Antwerpen	Belgium
Di Prisco	G.	Inst. of Protein Biochemistry and Enzymology	Napoli	Italy
Dieckmann	G.	Alfred Wegener Institute	Bremerhaven	Germany
Eastman	J.	Ohio University	Athens	USA
Ellington	R.	Tallahassee University	Florida	USA
Ferguson	W.	University of Pretoria	Pretoria	South Africa
Fiege	D.	Forschungsinstitut Senckenberg	Frankfurt	Germany
Fischer	A. E.	University of Mainz	Mainz	Germany
Gambi	M. C.	Stazione Zoologica A. Dohrn	Napoli, Ischia	Italy
George	K.	University of Oldenburg	Oldenburg	Germany
Gili	J.-M.	Instituto de Ciencias del Mar (CSIC)	Barcelona	Spain
González	O. E.	University of Ensenada	Ensenada	Mexico
Gosselin	M.	Université de Québec	Montréal	Canada
Gulliksen	B.	University of Tromsö	Tromsö	Norway
Gutt	J.	Alfred Wegener Institute	Bremerhaven	Germany
Hanelt	D.	Alfred Wegener Institute	Helgoland	Germany
Hochberg	E.	Museum of Natural History	Sta. Barbara	USA
Hopkins	T. S.	North Carolina State University	Releigh	USA
Hopkins	T. L.	University of South Florida	Tampa	USA
Hughes	R.	University of Wales	Bangor	UK
Jazdzewski	K.	University	Lodz	Poland
Kirst	G. O.	University	Bremen	Germany
Klages	M.	Alfred Wegener Institute	Bremerhaven	Germany
Kock	K. H.	BFA für Fischerei	Hamburg	Germany

Krapp	F.	Museum König	Bonn	Germany
Marquardt	J.	University of Oldenburg	Oldenburg	Germany
Mills	C. E.	University of Washington	Washington	USA
Montiel	P.	British Antarctic Survey	Cambridge	UK
Noest Hegseth	E.	University of Tromsö	Tromsö	Norway
Nöthig	E.-M.	Alfred Wegener Institute	Bremerhaven	Germany
O'Donovan	V.	Marine Institute	Newport	Ireland
Patarnello	T.	University of Padova	Padova	Italy
Paul	R. J.	University of Münster	Münster	Germany
Pedrós	C.	Instituto de Ciencias del Mar (CSIC)	Barcelona	Spain
Piatkowski	U.	Institut für Meereskunde at the University	Kiel	Germany
Piepenburg	D.	University of Kiel	Kiel	Germany
Poore	G.	Museum Victoria	Melbourne	Australia
Priddle	J.	British Antarctic Survey	Cambridge	UK
Pusceddu	A.	University of Ancona	Ancona	Italy
Rakusa-Suszczewski	S.	Polish Academy of Sciences	Warszawa	Poland
Rozbaczylo	N.	Pontificia Universidad Católica	Santiago	Chile
Rumohr	H.	Institut für Meereskunde at the University	Kiel	Germany
San Martín	G.	University of Madrid	Madrid	Spain
Schuchert	P.	Muséum D'Histoire Naturelle	Genf	Switzerland
Schulenburg	H.	University of Münster	Münster	Germany
Serra	J.	University of Barcelona	Barcelona	Spain
Sheehy	N. R. J.	University of Leicester	Leicester	UK
Shirley	T. C.	School of Fisheries and Ocean Sciences	Juneau	USA
Slattery	M.	University of Mississippi	Jackson	USA
Stanyon	R.	National Cancer Institute	Frederic, MD	USA
Storch	V.	University of Heidelberg	Heidelberg	Germany
Svavarsson	J.	University of Rejkjavik	Rejkjavik	Iceland
Taylor	A. C.	University of Glasgow	Glasgow	UK
Tilzer	M.	University of Konstanz	Konstanz	Germany
Türkay	M.	Forschungsinst. Senckenberg	Frankfurt	Germany
Vader	W.	University of Tromsö	Tromsö	Norway
van den Thillart	G.	University of Leiden	Leiden	Netherlands
van der Linden	A.	University of Antwerpen	Antwerpen	Belgium
Vecchione	M.	National Museum of Natural History	Washington	USA
Weslawski	J. M.	Institute of Oceanology	Sopot	Poland
Whiteley	N.	University of Wales	Bangor	UK
Wiencke	C.	Alfred Wegener Institute	Bremerhaven	Germany
Wilson	R.	Institut für Meereskunde at the University	Kiel	Germany

Subject Index

MIX
Papier aus verantwortungsvollen Quellen
Paper from responsible sources
FSC® C105338

If you have any concerns about our products,
you can contact us on
ProductSafety@springernature.com

In case Publisher is established outside the EU,
the EU authorized representative is:
Springer Nature Customer Service Center GmbH
Europaplatz 3, 69115 Heidelberg, Germany

Printed by Libri Plureos GmbH
in Hamburg, Germany